# Sustainability Matters

## Environmental and Climate Changes in the Asia-Pacific

# Sustainability Matters

## Environmental and Climate Changes in the Asia-Pacific

Editors

**Lye Lin-Heng**

**Victor R Savage**

**Kua Harn-Wei**

**Chou Loke-Ming**

**Tan Puay-Yok**

National University of Singapore

NUS National University of Singapore

World Scientific

*Published by*

World Scientific Publishing Co. Pte. Ltd.
5 Toh Tuck Link, Singapore 596224
*USA office:* 27 Warren Street, Suite 401-402, Hackensack, NJ 07601
*UK office:* 57 Shelton Street, Covent Garden, London WC2H 9HE

**British Library Cataloguing-in-Publication Data**
A catalogue record for this book is available from the British Library.

**SUSTAINABILITY MATTERS**
**Environmental and Climate Changes in the Asia-Pacific**



ISBN 978-981-4719-13-1

In-house Editor: Amanda Yun

Typeset by Stallion Press
Email: enquiries@stallionpress.com

## PREFACE AND ACKNOWLEDGEMENTS

### Introduction to the MEM Programme

The National University of Singapore (NUS)'s Master of Science (Environmental Management) "MEM" Programme was launched in July 2001. This is a multi-disciplinary and inter-disciplinary graduate programme to provide education in environmental management for senior and mid-level managers in corporations, institutions, and government and non-government organisations. It aims to equip graduates with the necessary knowledge and skills to properly manage the environment and to deal with the challenges of an environmentally conscious society and international market. It enables graduates to assume responsible and influential roles in the public and private sectors, and to make environmentally sound decisions that support sustainable development and livelihoods.

The MEM Programme has the collaboration of **nine** Faculties and Schools in the NUS, and is hosted by the School of Design and Environment (SDE). The other partners in this program are: the Faculties of Arts and Social Sciences, Engineering, Law, and Science; the NUS Business School, the Lee Kuan Yew School of Public Policy, the Saw Swee Hock School of Public Health, and the Yong Loo Lin School of Medicine.

The Programme is administered by two Co Directors from the staff of the host Faculty/School, the SDE, and led by a Programme Management Committee comprising a representative from each collaborating Faculty/School. The Programme has formal links with the Yale School of Forestry and Environmental Studies (FES) and Duke University's Nicholas School of the Environment and Earth Sciences, providing for collaboration in teaching and research, as well as academic staff (faculty) and student exchanges. It has a distinguished Advisory Committee, chaired by Prof Tommy Koh, Ambassador-at-Large and former Dean of the Law School, comprising national and international leaders in the environment, from academia, industry, government and non-government organisations.

The MEM programme can be undertaken either full-time (one academic year) or part-time (two academic years). It welcomes applicants from any discipline with a good honours degree from a reputable university, who meet the required standards for proficiency in the English language. Candidates are required to complete a 40 modular credit programme, of which the main components are a group of seven Core Modules, taught by staff from the different faculties/schools in NUS. These modules are:

- Business and the Environment
- Environmental Economics and Public Policy
- Environmental Law
- Environmental Management and Assessment
- Environmental Planning
- Environmental Science
- Environmental Technology

Students must also pursue substantive research, completing either a Study Report of 10,000 words plus one elective module, or write a Dissertation of 20,000 words, under guided supervision. In addition, students attend a series of seminars on current issues relating to the environment, across a wide spectrum of disciplines, conducted by experts in the relevant fields, to broaden their perspectives and widen their horizons. These are held every fortnight, and are also open to the public, as part of the University's efforts towards public education on the environment.

Through the years, the programme has been enriched by students from many disciplines and many nations — from Asia to Europe and North America, including the Middle East and Africa. Our graduates now hold positions in many different organisations, public and private, in various parts of the world. They bring the benefit of their multi-disciplinary experience in the MEM programme, to build a better and more sustainable world. The MEM's Alumni Association plays an active role in the programme's activities, continuing to maintain close ties with alumni members as well as the teaching staff. In the past few years, it has organised public seminars in partnership with non-government organisations, as well as eco-trips abroad for its members.

This book is the fifth volume of some of the best research papers done by students in the programme for the academic years from 2012/2013 and 2013/2014. They have been edited for publication. Where two names appear as authors, the first name will be that of the student, with the supervisor's name thereafter, as co-author. The programme thanks these students and their supervisors for their valuable contributions to research in the many facets of environmental management, and also acknowledges those students whose works do not appear in this volume; emphasizing that every student in this programme has contributed to its success. Each paper has served to enhance our understanding of the environment and compelled us to think of how we can manage it better.

**Acknowledgements**

Many people have contributed to the success of this programme. We wish to thank them most sincerely. They are:

1. NUS President Prof Tan Chorh Chuan and Provost Prof Tan Eng Chye for continuing to support and fund the teaching of this programme.
2. The Dean of the School of Design and Environment, Prof Heng Chye Kiang, and his team, for hosting the programme since its inception, and continuing to support it administratively and financially.
3. The Deans of the other eight Faculties and Schools in NUS who are partners and collaborators in this programme. They have supported us by mounting relevant courses in their faculties/schools and recognising the contributions of their staff in teaching, supervision of research, and in examining our students.
4. The representatives of each Faculty and School, who serve on the Programme Management Committee (PMC). Their enthusiasm and passion for the environment and for this programme have helped in immeasurable ways to ensure its sustainability and its constant improvement.
5. Our teaching staff from the different schools/faculties, adjunct teaching staff, as well as other staff members who served as supervisors, examiners, and contributed in many different ways to the programme.
6. Members of our Advisory Committee, led by Prof Tommy Koh as Chair.
7. Shell Companies in Singapore for being the programme's stalwart supporter since its early years, providing bursaries for needy students and prizes for the best students. Shell also bears the costs of this publication and the earlier volumes in this series. We wish to especially thank Shell's Chairperson, Ms Goh Swee Chen for her support and encouragement. Ms Goh is a member of the MEM's Advisory Committee, taking over in November 2014 from former CEO Mr Lee Tzu Yang upon his retirement in October 2014. Mr Lee has been our stalwart supporter since June 2002 and we wish to record our special thanks to him, and wish him well on his retirement. Thanks are also due to Shell's Mr Jason Leow, General Manager, Communications, and Ms Caroline Loke, Manager, Social Investment, for their continued support and encouragement, and also

Ms Mavis Kuek, former General Manager, External Affairs and Communication, who has now moved to Shell's corporate office in the UK.
8. The Tan Chay Bing Education Foundation and the National University of Singapore Society for providing scholarships and other financial assistance for our students.
9. Our alumni, who are spread across the globe in different capacities and disciplines, who are working to improve the environment, and whom we call upon to assist in various ways from time to time.
10. The many individuals we have called upon to speak to our students at fortnightly seminars and who willingly shared their expertise and time on Friday evenings from 6–8 pm.

As has earlier been mentioned, this is the fifth volume of the best student papers from the NUS MEM programme. The sixth volume will soon follow. We wish to congratulate the writers and their supervisors for their hard work and to thank them for sharing their research findings and proposed solutions. These papers are a fine contribution to the vast field of sustainable environmental management.

My colleagues and I in the MEM programme are heartened that NUS has strengthened its focus on environmental research and teaching. It now has a multi-disciplinary Bachelor in Environmental Studies (BES) program, jointly hosted by the Faculties of Arts & Social Sciences (FASS) and the Faculty of Science. Environmental Studies is also one of the areas of focus in the new Yale–NUS programme. Other initiatives include the establishment of the NUS Environmental Research Institute (NERI), the Centre for Sustainable Asian Cities, the Tropical Marine Science Institute (TMSI), the NUS Energy Studies Institute (ESI), and the Asia-Pacific Centre for Environmental Law (APCEL). It is hoped that these initiatives can be further strengthened so that NUS will be a centre of excellence in environmental studies for Asia and the world.

**Associate Professor Lye Lin-Heng**
Chair, M.Sc. (Env. Mgt.) Programme Management Committee
School of Design and Environment,
&
Director, Asia-Pacific Centre for Environmental Law,
Faculty of Law
National University of Singapore

# MESSAGE FROM GOH SWEE-CHEN, CHAIRPERSON, SHELL COMPANIES IN SINGAPORE

I extend heartfelt congratulations to the faculty and students from the Master of Science (Environmental Management) [MEM] programme on their fifth volume of exemplary student research papers. This collection addresses broad-ranging topics on environmental and climate changes in Asia, offering thought-provoking research into what matters for the sustainability of the world we live in. This rich content also reflects the diversity of the student mix in the MEM programme; an advantage that allows the exploration of sustainability challenges in rural and urban environments. I believe that learning begins with curiosity — this is the spirit that the students of MEM have demonstrated.

Fields such as environmental science and engineering offer tangible solutions to fulfil our commitment to continue human progress in a responsible and sustainable manner. People engaged in these fields need to be unafraid to challenge known norms to uncover potential breakthroughs to persistent challenges in the balance between development and sustainability. Research and analysis, such as the ones that have been consolidated here, pave the way for a better future by generating viable applications to help improve societies of today.

At Shell, we see ourselves responsible for providing the world with the energy that keeps it progressing. We understand that energy is vital to modern living. Without it, we could lose access to other essential aspects of our daily lives, like clean water, food, and healthcare. In planning our strategies for tomorrow, we too depend on sound science to solve real world challenges. This latest volume of studies adds on to the progress being made daily, and Shell is happy to continue its support for the MEM programme.

**Goh Swee-Chen**
Chairperson, Shell Companies in Singapore
April 2015

## MESSAGE FROM TOMMY KOH, CHAIRMAN, MEM ADVISORY COMMITTEE

It gives me great pleasure to once again, welcome another publication from the students of the NUS Master's in Environmental Management (MEM) programme. The MEM programme is a unique and thoughtfully crafted programme that draws on the talents of staff members from nine faculties and schools in NUS. It is multi-disciplinary and inter-disciplinary, exposing students to the many facets and perspectives of environmental studies, including environmental technology, science, law, economics, planning, business, and ethics. It focuses on effective management of the environment — this can only be achieved with a good understanding of its many dimensions, not just in theory, but in practice.

Through the years I have been impressed by the quality of the students, who come from different disciplines and from many countries. I know they have learnt much from the programme, and from each other. Our graduates pursue successful careers in government, academia, non-government agencies, and in the corporate sector, in countries across the world.

A strong research component is built into the MEM programme. Apart from inculcating the skills of research as a discipline, the results of these efforts are useful contributions to knowledge. This volume contains 18 papers, focussing on challenges relating to Biodiversity, Environmental Management, Green Business, Urban Studies, and Climate Change. It is hoped that they will be studied by policy makers, and that the recommendations proposed can be implemented in relevant countries.

I congratulate the students and their supervisors, for the chapters that appear in this volume, and commend the teachers in the MEM programme for their dedication, not just in teaching and supervision, but also in collating and editing yet another fine contribution to environmental studies. I wish the programme continual success and look forward to the next volume in *Sustainability Matters*!

**Tommy Koh**
Ambassador-at-Large
&
Professor, Faculty of Law
Chair, M.Sc. (Env. Mgt.) Programme Advisory Committee
National University of Singapore

## MESSAGE FROM HENG CHYE-KIANG, DEAN, SCHOOL OF DESIGN AND ENVIRONMENT, NUS

I have great pleasure in writing this message for the fifth compilation of the best Dissertations and Study Reports of students in the M.Sc. (Environmental Management) (MEM) programme at the National University of Singapore (NUS), hosted by my school. The multi-disciplinary and inter-disciplinary nature of the programme is clearly evident from the variety of subjects covered by the 18 chapters in this book, which are arranged under the headings of: Biodiversity; Environmental Management; Green Business, Urban Studies; and Climate Change. Each chapter is based on a research project undertaken by an MEM student under the guidance of a supervisor. The successful production of this book bears another testimony to the strength and effectiveness of the multi-faculty effort which is behind the running of the MEM programme. The series of compilations of MEM Dissertations and Study Reports has now become well-established among the key writings on the broad field of environmental management.

I congratulate the students whose Dissertations and Study Reports have been included in this book. They come from many disciplines and from many countries. For most of them, it would be their first time doing substantive research, and this exercise must have been an invaluable experience. It is also admirable that they were able to take up the task of summarizing their Dissertations and Study Reports into the succinct chapters in this book despite their holding full-time positions in government or industry organisations in Singapore and other countries around the world. I also congratulate their supervisors for the success of their guidance and support. I wish to express my appreciation to the Editorial team — they have worked hard to produce a volume each year now. The fact that this volume now contains 18 chapters also indicates that more students are producing excellent research in this programme.

The MEM programme continues to do well in all aspects; it attracts good students from Asia and beyond — students from all academic backgrounds. The graduates hold responsible positions in government, industry, and academia in many countries. We continue to hear of the achievements they have made. The enthusiasm of the team of professors and practitioners who teach on the programme remains high. We also continue to see a high level of interest from the alumni in the wellbeing of the programme and the welfare of the students. I am grateful to the Shell Group of Companies in Singapore for providing the grant which made the publication of this book possible, and for supporting its publication since the series began.

I recommend this book to all who want to find information on contemporary issues of environment and development in a broad range of academic and policy fields, covered in a multi-disciplinary manner, drawing on the experience in various countries worldwide.

**Professor Heng Chye-Kiang**
Dean,
School of Design and Environment
National University of Singapore

# ABOUT THE AUTHORS

**ANG Guorong Albert** holds a B.Sc. in Education (Hons) from Nanyang Technological University, Singapore, and did his post-graduate degree at the National University of Singapore, earning an M.Sc. in Environmental Management. His passion in education, environment and psychology found intersections when he taught and headed schools in Singapore and Indonesia where he is able to apply their principles in curriculum design and organisational management. Continually engaged in education, his recent endeavour in the practice and teaching of mindfulness meditation hopes to help individuals and groups achieve cognitive clarity and emotional well-being.

**Denise CHEONG** holds an LL.B. (Hons) and M.Sc. in Environmental Management (Shell Medal and Prize, and Shell Best Dissertation Award winner) from the National University of Singapore. She is a Solicitor of England & Wales, and an Advocate & Solicitor of the Supreme Court of Singapore. She is currently a research fellow at the Centre for International Law, National University of Singapore (CIL), where she focuses on ocean law and policy. She has a particular interest in international environmental law issues as they impact the marine environment. Prior to joining CIL, she worked with Linklaters (Singapore and Tokyo) for over 10 years where she was a member of Linklaters' award-winning global capital markets practice. Prior to joining Linklaters, she worked as a solicitor with Allens in Sydney, Australia, and M/s Lee & Lee, Singapore.

**Julia EMERSON** earned a B.Sc. in Marketing from the University of Dayton, Ohio, in 1993; an M.B.A. from the University of Tulsa, Oklahoma, in 1998 and an M.Sc. in Environmental Management in 2014 from the National University of Singapore. Julia has been working for twenty over years, with the last fifteen years being in the multifamily affordable housing industry, which is what drove her to pursue the Environmental Management degree. Julia had been working on a variety of projects involving property owners and residents, and as a result, could see the potential benefit of introducing an environmental management perspective into the company. Julia is still working in property management, applying her new education to improving the efficiency and cost effectiveness of the properties, and researching on the behavioural aspect of implementation for both owners and residents.

**Efstathios GIANNOUSTAS** holds a B.Eng. (Hons) and an M.Sc. (Hons) in Environmental and Geo-environmental Engineering, both from the University of Wales, Cardiff, UK. He graduated in 2013 from the National University of Singapore (School of Design and Environment) with an M.Sc. in Environmental Management. He is currently working as a Senior Principal Environmental Engineer with CPG Consultants in Singapore in the fields of water resources, drainage, flood risk, environmental impact assessment and water quality studies. Overall he has approximately 10 years of environmental engineering consultancy experience in the United Kingdom and in Singapore, participating in major infrastructure projects such as the Stamford Diversion Canal and studies such as the Drainage Masterplan for Singapore. He is passionate about environmental issues around the world especially in the field of water resources availability.

**KOH Kai-Jie** holds a B.Sc. in Environmental Engineering from Nanyang Technological University, Singapore, and graduated with an M.Sc. Environmental Management from the National University of Singapore in 2013. He has been with Singapore's Housing and Development Board (HDB) for over 7 years, and is currently involved in the upgrading programmes of HDB flats. Previously, he was managing estate matters of HDB Towns. He is dedicated to working on improving the living environment of HDB residents and hopes to contribute towards the research and development of environmental, building, and construction technology in the future.

**LOO Hui-Min** is a geographer and educator who is a passionate driver in environmental education. She has an avid interest in environment sustainability issues, and with clarity of vision, she has woven environmental literacy in her teachings. After spending a few years heading the Geography department in a school, she took on the Ministry of Education Professional Development Leave to pursue her M.Sc. (Environmental Management) at her alma mater. She holds a Postgraduate Diploma in Education from the National Institute of Education, Singapore and a B.A. in Statistics and Geography from the National University of Singapore. Hui-Min has a special interest in the relationship between corporations and the environment, and believes the road to success is growing without damaging the prospects of future generations.

**Himadri MAHAJAN** is currently working with Bharti Airtel Limited, India, as Manager (Sustainability). Her role involves strategizing and implementing the Corporate Sustainability Framework across different functions of the organisation. She drives the process of data collection, stakeholder engagement (internal & external), materiality assessment, and sustainability awareness, and develops the Sustainability Report and Business Responsibility Report. Prior to this, she worked for over 3.5 years with India's apex industry association, the Confederation of Indian Industry's (CII) Centre of Excellence for Sustainable Development. She was primarily associated with the Sustainability Awareness & Outreach team and was involved in planning, conceptualising and executing events/seminars on triple bottom line issues. She is a graduate of the M.Sc. (Environmental Management) programme at the National University of Singapore, and holds a post graduate diploma in Environmental Law from the National Law School of India University (NLSIU), Bangalore. She is also an environmental engineer.

**Nidhi MEHRA** holds a B.Eng. in Civil Engineering from Government Engineering College Aurangabad, Maharashtra, India, and graduated from National University of Singapore's (NUS) M.Sc. (Environmental Management) Programme in 2013. She worked as a Structural Designer in India and the United States before taking a break from work to raise her children. Her keen interest for doing 'her bit' for the environment led her to join this Master's program at NUS. She is currently working as an Environment Engineer at Shell Eastern Petroleum Pte. Ltd.

**Maitreyee MUKHERJEE** is a Research Assistant at the Institute of Water Policy (IWP), National University of Singapore (NUS). She holds an M.Sc. in Zoology from Calcutta University, India (2005) and an M.Sc. in Environmental Management from NUS (2014), where she was the recipient of the "Shell Best Dissertation Award, 2013–2014". She also has a Bachelor of Education (Teachers' Training programme) from North Bengal University, India (2007). Prior to joining the IWP she worked as research assistant at various laboratories, as well as taught in a middle school in India. Her current research interests include urban water management, water pricing and environmental sustainability in water resource development projects. She has published in peer-reviewed journals like *Water Policy* and *Global Environmental Change.*

**Mallika d/o NAGURAN** graduated from the National University of Singapore with a B.A. in 1988, specialising in English Language and

English Literature. She worked her way up in communications, eventually heading up teams in IT and telecoms sectors, and in doing so developed strong credentials in public relations, publishing, branding and crisis management. Compelled to do her bit to address climate change, Mallika made a mid-career switch in pursuit of the environment. She began to research and write about social and environmental issues, primarily for her own online publication *Gaia Discovery*, which she founded in 2008. In 2013, Mallika graduated with an M.Sc. in Environmental Management from the National University of Singapore (NUS). She then worked as a researcher with the Energy Studies Institute and the Centre for Sustainable Asian Cities at NUS. She is currently employed by a global hospitality group as the Asia Pacific manager of sustainable development and communications. On weekends, Mallika can be found engaged in a number of environmental causes related to biodiversity conservation and restoration including growing new coral gardens in the sea.

**Rosa Celia POQUITA-DU** holds a B.Sc. in Marine Biology from the University of San Carlos, Cebu, Philippines. She received her M.Sc. in Environmental Management from the National University of Singapore in 2013. She has been part of NUS Marine Biology laboratory since 2010 as a Research Staff and was involved with various projects such as environmental controls for coral growth, effects of turbidity on seagrass, enhancing biodiversity through artificial substrates such as seawalls, and effects of vessel-generated waves on corals. She is now attached to the Tropical Marine Science Institute to work on a research project that aims to enhance Singapore's coral reef ecosystem.

**Tayef QUADER** holds a B.Sc. in Environmental Management from North South University, Bangladesh, India, and obtained his M.Sc. in Environmental Management from the National University of Singapore in 2013. He began his career in a textile manufacturing conglomerate in Bangladesh as the Head of Sustainability. His work experience covers a range of areas such as workplace safety and wellness, social health, corporate responsibility and environmental stewardship. He is presently employed at the NUS Saw Swee Hock School of Public Health as a Research Associate and continues to pursue his interest in social health and workplace safety issues.

**Ellen May Zanoria REYNES** holds a B.Sc. in Chemical Engineering, majoring in Sugar Technology from the University of the Philippines in Los Banos (2005), and an M.Sc. in Environmental Management from the National University of Singapore (2013). She has been involved in the energy/environment field for almost 10 years now, with a focus on socio-economic safeguards and sustainable development benefits. She has worked on the ground as a project developer of climate change mitigation projects, implementing waste-to-energy projects in Southeast Asia, and looking into several methodologies and tools to ensure that 100+ projects were on par with UNFCCC standards and that proper monitoring protocols were implemented. She is now with The Gold Standard Foundation where she started as the Regional Manager for Southeast Asia in 2009, and has since moved on to take care of global key accounts and business development, focusing on strategic partnerships and fundraising for new initiatives and programmes.

**Marra Lin TEASDALE-HENSBY** holds a B.A. (Hons) in Business and Psychology from Trent University, Ontario, Canada. She spent her career in Canada working for Fortune 500 companies as an Environmental Stewardship and Sustainability professional specialising in Extended Producer Responsibility. In 2012, she moved to Singapore and enrolled in the National University of Singapore's Master of Science (Environmental Management) programme graduating in 2013. Since then, she has been volunteering as a Project Specialist with the Global Product Stewardship Council, a not-for-profit global forum for product stewardship development. She is currently working for Williams Sonoma Inc. as their Social Responsibility Manager located in Singapore.

**Thida TUN** grew up in Myanmar and studied Business Management via distance-learning, while she was working as a full-time staff of Happy Land Co. Ltd., in the local Fast Moving Consumer Goods sector. She went on to attend Ritsumeikan Asia Pacific University, Japan, in 2006, majoring in Sustainable Development. Upon graduation, she moved to Singapore and worked as a consultant in the travel industry. She returned to school and attended the National University of Singapore's M.Sc. (Environmental Management) programme, from which she graduated in 2013. During her student life in Japan, she was awarded the Oita Prefecture Scholarship (2007–2008), JASSO Honour Scholarship (2009–2010) and Tuition Reduction Scholarship (2006–2010). She received the Tan Chay Bing Bursary for her postgraduate study in Singapore. She is currently working for Toyota Tsusho Asia Pacific in Singapore, focusing on the strategic sustainable business development of Toyota's automobile business in the Mekong region.

**ZHANG Qiang** was born in the Sichuan Province of China which is famous for pandas. He obtained an LL.B. from Sichuan University, China, and an M.Sc. in Environmental Management (MEM) from the National University of Singapore (NUS) in 2008 and 2013, respectively. He has keen interests in the environment, sustainable development and the aviation industry (civil aviation in particular). Between 2008 and 2012, he worked as an Immigration Officer at an international airport in China. After his graduation from NUS, he joined a leading waste management company in Singapore where he is primarily responsible for sustainable business development.

# ABOUT THE SUPERVISORS AND EDITORS

**CHOU Loke-Ming** retired in October 2014 and currently holds joint appointments as Adjunct Research Professor at both the Tropical Marine Science Institute, and the Department of Biological Sciences, National University of Singapore. His research interests include coral reef ecology and restoration, and integrated coastal management. He studied coral reefs in Okinawa and the ASEAN region and currently serves as a member of the Joint Group of Experts on the Scientific Aspects of Marine Environmental Protection. He has provided consultancy services in the field of marine environment management to international agencies like the United Nations Environment Programme (UNEP), Food and Agriculture Organisation (FAO), World Bank, WorldFish Center, as well as many national agencies. Presently a Fellow of the Singapore National Academy of Sciences and an Honorary Fellow of the Singapore Institute of Biology, Prof Chou has over 100 publications in international journals and over 50 book chapters focusing on the marine environment.

**Audrey CHIA** is an Associate Professor of Management and Organisation Department at the NUS Business School. She also holds a Joint Associate Professor position at the NUS Saw Swee Hock School of Public Health. She received her Ph.D. from the University of Texas at Austin, USA. Her first degree was in Philosophy and English, from the National University of Singapore. Her research and teaching interests are multi-disciplinary. For the UCLA–NUS Executive MBA and the NUS MBA, she offers courses in leadership, change management and international management strategies. She has published in various journals such as the *Academy of Management Executive*, *Science Communication*, *Jurimetrics* and the *Journal of Business Ethics*. She has also co-authored two books: *The Three Paradoxes: Working Women in Singapore*, and *Culture and Management: A Casebook*. Her research has been presented at international conferences on management, applied psychology and socio-economics. Her current interests are in workplace diversity, and the repair of reputation and relationships.

**Namrata CHINDARKAR** is an Assistant Professor at the Lee Kuan Yew School of Public Policy, NUS. She received her Ph.D. from the School of Public Policy, University of Maryland, concentrating in international development policy. Her field of interest is international development and policy areas such as poverty and inequality, water and energy, policy impact evaluation, public health, and gender and development. She holds an M.A. in Development Studies from the University of Manchester, U.K., and an M.Phil. in Social Sciences from the Tata Institute of Social Sciences, Mumbai.

**KUA Harn-Wei** is an Associate Professor in the Department of Building, NUS, for which he is currently the Deputy Head (Research). He obtained his Ph.D. from the Massachusetts Institute of Technology (MIT), USA, in sustainable building technology and policy. His current research interests include technology and policy issues related to sustainable building materials. Specifically, he devises sustainability assessment methods for existing and emergent materials, and uses the results of these assessments to design integrated sustainability policies to improve and promote these materials. At the department, he is co-leading a research cluster on building materials, in which he focuses on creating bio-based materials that sequester and utilize carbon dioxide. He is an MIT Martin Family Society Fellow for Sustainability, an MIT Carroll L. Wilson Fellow and a World Cities Summit Young Leader. He is also a Council Member of the Southwest Community Development Council of Singapore.

**LIM Lei-Theng** LLB (Hons) (NUS), LLM (Harvard, USA) is an Associate Professor and Deputy Director of the Legal Skills Programme at NUS Law and has been a member of the Law Faculty's Asia-Pacific Centre for Environmental Law (APCEL) since its inception in 1996. She is a coach of several moot teams and her interests in environmental law have had to take a back seat to her teaching focus on skills and clinical legal education, but she retains an interest in climate change law and the dispute resolution aspects of environmental law.

**LYE Lin-Heng** LLB (Sing), LLM (King's College, London, UK), LLM (Harvard, USA) is an Associate Professor at the Faculty of Law, NUS, and Director of its Asia-Pacific Centre for Environmental Law (APCEL). Her areas of teaching and research are in Environmental Law and Property Law. She chaired the Steering Committee for the MEM Programme and now chairs its Programme Management Committee. She is also a member of the Programme Management Committee for the NUS Bachelor in Environmental Studies (BES). She is former Vice-Dean of the Law School at NUS, former Vice-Chair of the IUCN Academy of Environmental Law and was a member of its Board of Governors, representing Southeast Asia for many years. She is a Visiting Associate Professor at the Yale School of Forestry & Environmental Studies, and also teaches at the Faculty of Law, University of Sydney. She was Honorary Legal Advisor to the Nature Society Singapore, for many years. She is a member of the Ministry of Environment and Water Resources Committee on Clean Drinking Water Standards for Singapore. She is also a member of the Strata Titles Board, and a former Board Member of the Housing and Development Board (HDB).

**Harvey NEO** is a social-economic geographer who has interests in the political economy of the livestock industry, food geographies and green urban development. He is also interested in animal geographies and has authored papers along these research themes. He teaches courses relating to nature, society and development at the Department of Geography, National University of Singapore. He is also Editor of *Geoforum* and Associate Editor of *Regional Studies, Regional Science*. He holds a Ph.D. from the Graduate School of Geography, Clark University, Massachusetts, USA.

**George OFORI** received both his doctoral degree and a higher doctorate degree from the University of London, UK. He is also a Fellow of the Ghana Academy of Arts and Sciences. He is currently a Professor at the National University of Singapore, and Director of the multi-disciplinary M.Sc. (Environmental Management) programme. He is a Fellow of the Chartered Institute of Building, UK; Fellow of the Royal Institution of Chartered Surveyors, UK; and Fellow of the Society of Project Managers, Singapore. He is the Deputy Chairman of the Construction Sector Transparency Initiative (CoST). He is also a Chair Professor at Tsinghua University, China. He has authored several papers which have been published in international refereed journal papers, conference papers, books, chapters in books, and published reports. He has been invited to deliver keynote papers at many international conferences. His main research area is construction industry development, focusing on the improvement of the construction industries of developing countries. He was the Founding Co-ordinator of the CIB Working Commission 107 (W107) on Construction in Developing Countries of the International Council for Research and Innovation in Building and Construction (CIB) (in 1997 to 2007). He has been a consultant to many governments and international agencies.

**Rick REIDINGER** is an Adjunct Senior Lecturer at the School of Design and Environment, NUS. He teaches the core courses, "Business and Environment" and "Environmental Management and Assessment" in the MEM programme. He has an M.Sc. from the London School of Economics and Political Science, UK, and more than 20 years of experience in the environmental industry, primarily based in Asia. He is currently the CEO of Eco Special Waste Management, Singapore's largest hazardous waste management company, with 270 staff and annual revenues of US$45 million. Prior to joining Eco, Rick was Regional Director of AECOM's environmental consulting business in Asia, worked at SembCorp, Singapore's waste management business, and spent 10 years with Environmental Resources Management (ERM), Singapore, the global environmental consulting company.

**T. S. Gopi RETHINARAJ** is an Associate Professor in the School of Natural Sciences and Engineering, at the National Institute of Advanced Studies (NIAS), Bangalore. His research and teaching interests include energy security, climate policy, civilian and military uses of nuclear energy, and international security. Before joining NIAS, he was a faculty member at the Lee Kuan Yew School of Public Policy at the National University of Singapore from 2005 to 2014. He received his Ph.D. in nuclear engineering from the University of Illinois at Urbana-Champaign (Illinois), USA, in 2005, and a M. Sc. in Physics from Bharathidasan University (Tamil Nadu), India, in 1995. Dr Rethinaraj was a research associate at the Illinois Program in Arms Control, Disarmament and International Security (ACDIS), USA from 1999 to 2005, and also worked as a science correspondent for the Indian Express in Mumbai, Insia from 1995 to 1999. He has written on science, technology, and security issues for *Jane's Intelligence Review* and Reuters, and was a visiting fellow at the Bulletin of the Atomic Scientists, Chicago, USA, in 1999.

**Victor R. SAVAGE** is an Associate Professor and Director of the Alumni Relations Office, at the National University of Singapore (NUS). He is the Deputy Director of the Masters of Science in Environmental Management Programme's (MEM) Management Committee. A/P Savage's research interest focuses mainly on Singapore and the Southeast Asian region — on historical and cultural landscapes, sustainable environments, environmental education, sustainable urban development and cross-cultural issues. Among his books are: *Western Impressions of Nature and Landscape in Southeast Asia* (1984); *The Naga Awakens: Growth and Change in Southeast Asia* (co-edited with Lily Kong and Warwick Neville, 1998); *Environmental and*

*Climate Change in Asia: Ecological Footprints and Green Prospects* (jointly edited with Lye Lin-Heng, 2012); *Singapore Street Names: A Study of Toponymics* (jointly authored with Brenda Yeoh, 2013). A/P Savage is a board member of the *Singapore Journal of Tropical Geography* and an International Editorial Board member of *Geografiska Annaler* (Sweden) and Editor-in-Chief of the *Open Journal of Ecology.*

**TAN Puay-Yok** is an Associate Professor in the Department of Architecture at the School of Design and Environment of the National University of Singapore. He is the Programme Director for the Masters of Landscape Architecture and Co-Director for the Masters of Science (Environmental Management) programmes. Prior to joining academia, Puay-Yok held senior positions in public service in the areas of urban greenery management, policies and research. His research, teaching, and professional activities, focus on the policies, science, and practices of urban greening and the ecology of the built environment. He currently serves in the Editorial Advisory Board of *Landscape and Urban Planning*, *Urban Ecology*, and *CITYGREEN*, a periodical of the Centre for Urban Greenery and Ecology for which he was co-founding Editor. His other professional activities include serving as reviewer for a number of national and international grants, advisory committee member for land use developments, jury member for design competitions, and as a Planning Appeals Inspector for the Ministry of National Development, Singapore. He also consults on landscape design and planning projects in the region as a means of transferring knowledge from academia to practice.

**Jesuthason (Dodo) THAMPAPILLAI** holds a Personal Chair in Environmental Economics at Maccquarie University, Australia, and an Adjunct Professorship in the same field at the Swedish University of Agricultural Sciences in Uppsala, Sweden. Presently, he is a Professor at the Lee Kuan Yew School of Public Policy, National University of Singapore. In 2005, he was included in a list of Eminent Environmental Economists by UNESCAP and has been a member of the UNEP Expert Group in Environmental Economics. He has over 100 publications, including seven books and nine refereed monographs. Prof Thampapillai's current research focus is on macroeconomics and the environment.

**Peter TODD** received his B.Sc. (1st Class Hons) in Ecological Sciences from the University of North London, UK, in 1997 and completed his Ph.D. in Marine Ecology at NUS in 2002. After a two and half year postdoctoral period at Napier University, UK, he returned to NUS as an Instructor and then Lecturer. He is now Assistant Professor in the Department of Biological Sciences. He is an experimental marine ecologist who specialises in designing, building, implementing, and analysing novel approaches to investigations. His basic research focuses on three model taxa groups: corals, crabs and (giant) clams. In his applied research he tries to better understand, and ultimately improve, Singapore's impacted coastal environments. Peter has published over 100 peer-reviewed scientific papers. He is also an Academic Editor for *PLOS ONE*, Subject Editor for *Oikos*, and Associate Editor for the *Raffles Bulletin of Zoology*, as well as being Review Editor for *Aquatic Biology* and an Editorial Board Member for the *Journal of Experimental Marine Biology and Ecology.*

**Dennis WICHELNS** is an Agricultural and Natural Resource Economist, with experience in academia, research institutes, production agriculture, and international consulting. He has served on the faculty of several colleges and universities, and has conducted research in several countries in Asia and Africa. Dr Wichelns has directed two research centres and has served as Principal Economist with the International Water Management Institute. He is co-Editor-in-Chief of *Agricultural Water Management* and the Founding Editor-in-Chief of *Water Resources & Rural Development.* His areas of research interest include crop production, water resources, food security, and livelihoods, with a particular focus on agricultural and natural resource issues in lower income countries.

**WONG Khei-Mie, Grace** is a Senior Lecturer at the School of Design and Environment, NUS. Her academic qualifications include B.Sc. (Estate Management) (Hons) and M.Sc. (Property & Maintenance Management) from NUS, as well as M.Phil. (Housing Economics) and Ph.D. (Housing Economics) from the University of Cambridge, UK. Since she first joined NUS in 1989, Dr Wong has taught a wide range of core, GEM (General Elective Modules), and elective modules, and has won numerous teaching awards including the NUS Annual Teaching Excellence Awards; the School of Design & Environment Teaching Excellence Awards; as well as GEM and SS Incentive Scheme Awards. Besides teaching, as the Programme Director of the B.Sc. (Real Estate) programme as well as the B.Sc. (Real Estate) Minor and B.Sc. (Urban Studies) Minor programmes, Dr Wong further contributes to curriculum design, development and administration. Dr Wong is also a Fellow of the NUS Teaching Academy with effect from 2014. Her teaching research publications, which focus on her pedagogical initiatives, are featured in *Ideas on Teaching, CDTL Brief, CDTLink*, as well as in conferences for teaching and learning in higher education.

## CONTENTS

# ENVIRONMENTAL AND CLIMATE CHANGES IN ASIA: LESSONS IN HISTORY AND GAME CHANGERS IN ECONOMICS, POLITICS AND SCIENTIFIC RESEARCH

VICTOR R. SAVAGE, LYE LIN-HENG, KUA HARN-WEI,
CHOU LOKE-MING and TAN PUAY-YOK

## 1. Introduction

In its November 2014 review, the Intergovernmental Panel on Climate Change (IPCC) gave the dire warning that global warming under current trends will lead to global disaster. Its prognosis is that temperatures are likely to rise to 4°C over pre-industrial times, key greenhouse gases are at their highest in 800,000 years, and current GHG emissions has to drop to 70 percent worldwide by 2050 — in just 35 years. If nothing is done to change this, the outcome of these global warming trends will lead to a spate of droughts, floods, rising sea levels, and species extinction with consequences on food security, migration patterns, economic issues, housing, human health and other socio-cultural issues.

Clearly, the world community seems too preoccupied with domestic economic issues, nationalistic political concerns, and urban social challenges that have blinded political leaders and government administrators from seeing current climate and environmental changes from a global and longer term perspective. Nothing brings home more vividly the problems cities and countries face with regard to environmental challenges than climate change. Environmental changes are slow and dictated by long-term geologic time frames. But human-induced climate change seems to be picking up in speed and their impacts are now quotidian reality around the world. Yet many sceptics think that climate change issues are ephemeral natural phenomenon — and those in political and economic power have control and sway over common good resources and public opinion. The fracas in November 2014 by Australian politicians on President Obama's reference to the Great Barrier Reef being damaged by Climate Change is one clear example that the Australian government and officials are arch sceptics about the realities of Climate Change impacts. It is no wonder the current Australian government under Tony Abbott repealed the Australian carbon tax set by an earlier administration; demonstrating that the country is a clear climate change sceptic. In the corporate world, ExxonMobil has spent US$29.9 million since 1997 to promote the denial of climate change (Hertsgaard, 2014). Despite all the scientific prognostications of rising $CO_2$ emissions and real world climate change induced natural hazards, there still remains an uneven global representation of the causes and impacts of climate change.

Time is clearly at the global community's disadvantage. We live out the mantra of being a globalised community undergoing globalisation; where borders are porous and geography is dead. But in reality, every country is acting in narrow nationalistic interest and competing on the assumption of a fixed global pie and trying to maximise selfish benefits for their own national good. The idea of the "common good" and global "common pool" resources seems like a pipe dream in which every state pays lip service to recognising environmental problems at global forums but acts selfishly in national terms. This is the global 'tragedy of the commons'. Without enlightened mindset changes of global leaders, corporate titans and government bureaucrats, we will continue to see a recycling of old ideas which are not representative of, nor pertinent to, the bewildering speed of environmental changes taking place around us.

Given this negative and sceptical view on climate change in certain quarters of the global community, there needs to be focused, concerted, and sustained efforts to inform the misinformed and alter the mindsets of decision makers to think globally and act locally on environmental challenges. This fifth volume of students' theses from the NUS *Masters in Science (Environmental Management)* (MEM) Programme, once again gives readers a glimpse of the academic curiosity that undergirds our students with regard to nature, the environment and climate change. The studies provide a wide range of environmental themes and subjects and cover a gamut of disciplinary vantage points — legal issues, engineering, social science perspectives, economic intersections, scientific deliberations and public policy interventions. In reading through the articles we hope that readers will discern the MEM's multidisciplinary and interdisciplinary vantage points, the academic appropriateness and rigour in employed academic methods, the grounded nature of concepts and theories, and the real world outcomes of public policies. One might ask: what are the values of MEM's environmental studies? Will it change the global or national situation? Perhaps not in the immediate future — but no exposure to environmental education can be seen as wasted and irrelevant. Education has long term individual and public ramifications on knowledge, the way one thinks, passionate activism and concerns undergirded by secular ethics and religious morals.

We believe there are three outcomes of these studies. Firstly, environmental education is a powerful long term vehicle of social and cultural change — and as educators, one hopes that the dissemination of knowledge, techniques and ideas will make inroads in pertinent ways in the national and global political and social fabric of cities, regions and states. Secondly, the MEM education touches on many diverse aspects of environmental management. It allows students to see varied perspectives of the complex environmental issues. Their essays are an embodiment of this multi-disciplinary lens. Environmental problems are multi-dimensional, multi-tiered and multi-lateral. We need to move students out of the silos of tunnel-view disciplinary knowledge and expose them to wider and holistic interventions. Thirdly, given that many of our students will work in government, private corporations and non-government organisations, their exposure to the many perspectives of 'managing the environment' provides applied and practical lessons and not just conceptual and theoretical knowledge. When the students embark on careers, they bring with them grounded expertise and, hopefully, valuable insights which have been distilled through their many interactions with varied speakers from governments and corporations. Most of all, our students of varied nationalities in Asia and further afield provide fervent nationalistic and local perspectives to global environmental challenges.

## 2. The Historical Lessons

Over the last decade, several academic tomes have been instrumental in alerting state leaders, governments, and the public about the serious problems that climate change can cause. The best examples are found in history — and unfortunately because they are remote in time, their impacts have not resonated with current politicians, administrators, entrepreneurs and the public about their disastrous impacts on societies and civilisations. Several books give interesting glimpses of what has taken place in the past. William Rosen's (2014) book, *The Third Horseman* discusses the impact of climate change on the great famine in Europe in the $14^{th}$ century. In his masterly work, *Global Crisis*, Geoffrey Parker (2013), details the massive catastrophes arising from climate change and "The General Crisis" of the $17^{th}$ century. Unlike Rosen's work, the $17^{th}$ century climate change crisis was not about global warming but about the "little ice age" that plagued the whole world. The 17$^{th}$ century climate change crisis had devastating impacts around the world — from America to Japan, Chile to Europe, and Russia to Australia. Closer to home, Victor Lieberman (2007, 2009) in his thought provoking two-volume work, ***Strange Parallels: Southeast Asia in Global Contexts*** discussed the impacts of both climate change and Eurasian (Europe and Asian states) political changes in relation to Southeast Asia's expanding and collapsing civilisations. This was the first major historical study on Southeast Asia where climate change was used to interpret historical changes apart from wars, political intrigue, strategic jostling and power plays.

The take home lessons from these highly detailed studies have lessons for all of us. Firstly, these historical studies demonstrate that the rise and fall of human civilisations and states of the past had less to do with merely political and social issues, and a lot more to do with massive climate and environmental changes taking place at particular periods of time. Using what historians call a "cross sectional" approach in history (events at a particular point in time), Rosen and Parker's works underscore the devastating impacts of climate change in history around the world in particular centuries. Rosen's (2014) underscores this cross-sectional view of historical catastrophes when he states: "The great conceit of history is that humanity's worst disasters occur within some identifiable and discrete time frame".

In the case of Lieberman's (2009) monumental study of mainland Southeast Asian states over a period of 900 years (from 800 to 1830), he was able to show how expanding civilisations correlated with good climate and harvests and collapsing civilisations fell under periods of global warming, drought and agricultural disasters. He notes how "wrenching crisis" gripped Southeast Asia between 1240–1450 due to a "climatic downturn" that led to "land degradation and slowing reclamation" while in the previous 300–400 years, there was rapid climatically assisted agrarian expansion (Lieberman, 2009). Without food, famine became the norm, political strains grew, institutional weaknesses developed with land shortages, cities crumbled, and populations were decimated in the region.

Secondly, one might see that climatic change disasters spare no societies and civilisations, no religious landscapes, nor political systems. Whether there is a 2°C increase or decrease in temperature, the impacts are devastating for harvests — Parker (2013) showed that the decrease of 2°C in temperatures in the 17$^{th}$ century resulted in an 80% reduction in grain harvests. Parker's work also showed how the carrying capacities of land could not support the populations in China (150 million), India (116 million) and Europe (100 million) in the $17^{th}$ century (Parker, 2013:23). In Asian countries where rice is the grain and where generally one acre of rice field yields 6 tons of food; an 80% fall in harvests spells massive disaster for rural Asian populations. If temperatures rise to 4°C, famines and starvation for many societies will become rampant and human populations will be reduced. In his study of civilizations, Linden (2007) convincingly argues that of all the natural catastrophes, drought arising from climate change was the nail in the coffin of many ancient civilisations.

The ideas underscoring the inaction of leaders and politicians today is the false sense of confidence arising from technology and science in dealing with climate change. Unfortunately, many global and national leaders feel that time is on our side because we can 'fix' climate change before it becomes a global disaster. The counter argument to the pessimistic lessons of history is that the climate change disasters cited in works like Parker's and Linden's are remote in time, when technology and science were still in their infancy compared to today. We live in a technological time in history where organic ecological issues of Gaia are being supplanted by machines, robots and man-made technical systems.

Thirdly, in a growing and expanding urban scenario, globally and in Asia, escalating issues of food insecurity are likely to hit urban populations first, due to both food shortages and spiralling food prices. In many ways, political commentators see the Arab spring in 2012–2013 as a climate change issue due to flood and drought induced increases in food prices in cities, which led to riots and urban unrest, and finally, the toppling of many governments. In almost completely urbanised countries like Singapore where 90% of food is imported, any global food crisis will have disastrous consequences for the city-state. One must remember that in times of crisis, food producing countries are likely to ban food exports. Russia, for example, banned its wheat exports in 2012. Populous Asian countries like China, India, Indonesia, Pakistan, the Philippines and Bangladesh will need to be more wary of the impacts of climate change on their national food security.

## 3. The Science of Climate Change

> "The Greeks' discovery of nature made possible the invention of science. China's failure to develop science can be attributed in part to lack of curiosity, but the absence of a concept of nature would have blocked the development of science in any case."
>
> ~Richard Nisbett (2003)

> "There is an invisible man — the possibility of whose existence haunts science, just as there is an invisible Earth that haunts our scientific knowledge of nature and our planet."
>
> ~Jacob Needleman (2012)

One of the few subjects which bring together the sciences and other disciplines in the social sciences, law, business studies, architecture, estate management, engineering, medicine, and policy issues, is the environment. Though its entry as an academic discipline is fairly new, environmental studies can be said to be the mother of all academic disciplines. That is why the educationalist, David Orr (2004) notes that "all education is environmental education".

The problem with applying science to environmental issues is the danger in expecting science to give us definitive answers to problems. There is no doubt that science can answer the causes and processes of climate changes — for one, it has given us good predictability of El Niño and La Niña, and their global impacts. We are slowly piecing together the various natural components (seas, soil, gases, species, etc.) of how James Lovelock's Gaia organism operates and functions. Clearly the predictability of science in explaining natural processes will be of great help for humankind in avoiding natural catastrophes and hazards. Even the predictability of landslides has been made clearer due to scientific applications and has helped to reduce human and property costs. The many scientific articles in *Science* and *Nature* demonstrate our increasing common understanding behind climate and environmental changes. They have helped to provide a better understanding that climate change is neither an ephemeral occurrence; nor is it, ironically, a purely physical induced process.

Given that climate is often seen as a product of nature and physical geographical processes, it is now difficult for the scientific community to come to terms with physical processes that are partly or wholly generated by anthropocentric causes. Current global changes are now increasingly seen as human-induced. Thus, it is not surprising that the scientific community has grudgingly accepted a new geological phase called the "*Anthropocene*" (human induced phase) that has superseded the *Holocene* (see Corlett, in press). Hence if one accepts that global physical processes are human related, the traditional scientific enquiry from the perspective of the 'pure' or 'hard' sciences (physics, chemistry, biology) will be inadequate; social sciences, humanities, and the soft sciences must also come to the fore for us to adequately deal with the complexity of the challenges faced.

After all, if human beings are generators of physical geographical processes, it seems logical that one cannot apply the laws of science to human activities and human motivations. It is this change in interdisciplinary enquiry, and the inclusion of the social sciences (including law, urban planning, architecture, and business), that seems to be the greatest current hurdle in understanding environmental and climate changes. It would mean that there is a need for more cooperation between the hard and soft sciences, and a greater need for dialogue between scientists, engineers and social scientists. As Richard Corlett (in press) states: "biophysical scientists are going to need allies from the social sciences if they are going to influence policy and practice, and these alliances are likely to be most effective if they have deep roots in shared research and teaching".

The nature of environmental challenges in the future demand that environmental studies be geared more towards interdisciplinary perspectives. We can divide the future of environmental students into three areas. Firstly, we need to continue basic research in science to understand the principles of varied environmental processes. The nature of both macro and micro scale research will give us insights into how nature functions under natural conditions or under simulated changing conditions. Secondly, we need interdisciplinary research across different disciplines to better comprehend the grounded nature and complexities of environmental processes. The recent edited book by Jared Diamond and James Robinson (2010), *Natural Experiments of History*, is a step in the right direction. This was a book trying to marry quantitative methods (statistics) with comparative methods in the humanities such as history. They note that historians make a lot of subjective and vague statements about time, processes and historical events (more, less, growing, declining, human-nature issues and changes) which can be better verified with statistical data, quantitative methods and detailed analysis on causes and effects. Jared Diamond (2010), for example, studied deforestation in the Pacific Islands, especially Easter Island, and concluded that climatic issues and geographic island variations were not the only reasons for deforestation. He found that islands with strong kings opposed deforestation, hence preserving environmental sustainability, while islands with weak or federal chiefs were prone to deforestation.

Thirdly, we need applied research in many arenas as environmental and climate changes are destroying environments, dwindling biodiversity, impacting human landscapes and killing human beings. Even in historical research, there is a need to see things in a realistic and larger perspective.

Historians need to devise methods of "doing science" which are "observing, describing and explaining the real world" according to Diamond and Robinson (2010). Historical perspectives can no more be seen as mere documentations of what went on in the past, but also as explanations of lessons one needs to learn. While one may not be a believer in the view that "history repeats itself", there are still lessons in history that can be learnt when dealing with contemporary challenges.

Science to a large extent becomes invigorated and stimulated by major technological and game changers in history. According to Alvin Toffler, human history has witnessed three major waves: the First Wave was the agricultural revolution; the Second Wave was the industrial revolution; and now the global community is involved in the Third Wave, the information-digital revolution. Each revolution has had both pluses and minuses in terms of positive and negative impacts on society and the environment. Benjamin Cohen (2009) in his engaging book, *Notes from the Ground*, demonstrates how agriculture in America set a train of scientific understanding in investigating the scientific knowledge of soils, managing agriculture (best practices), and bringing together nature and culture with science and agriculture which undergirds "the relationships between ideas and practices in the environment" (Cohen, 2009).

We like to add a Fourth Wave, the environmental and climate change revolution, which had its initiation in the latter half of the $20^{th}$ century with Rachel Carson's book, *Silent Spring*. Ironically because of increasing climate change and environmental challenges, industrialists, entrepreneurs, scientists and academics have been busy finding new mitigation and adaptive mechanisms, devices, ideas and methods to deal with global climate and environmental changes. Our attempts at finding energy reduction systems in household appliances, industrial plants, building systems, urban planning, transport systems, and alternative energy have created a boom in human creativity and innovation. Ironically, out of necessity, the global community has become environmentally sensitive, nature friendly and energy concerned. Nowhere in human history has environmental awareness been so fore-fronted in human thinking. Yet, two centuries of ambivalence and nonchalance to $CO_2$ rises has left a global community with multiple perceptions — scientists have expressed growing alarm, the poor are resigned to fatalism, the rich thrive on selfish greed, and government leaders cling to personal power politics.

## 4. Pragmatic Political Policies

One of the pertinent issues of climate and environmental change is finding leaders who have percipient views, practical programmes and the political will to address challenging issues. Three relatively new leaders in Asia, China's President Xi Jinping, India's Prime Minister Narendra Modi and Indonesia's President Joko Widodo, provide hope that they will act decisively on issues pertaining to development programmes within environmental-friendly perspectives. China's joint pledge with the US to curb carbon emissions took the world by pleasant surprise even though environmental sceptics wonder about the vague Chinese commitment of maintaining peak carbon emissions by "around 2030". But China needs to take drastic steps because air pollution is affecting her economy. Over 10 percent of the gross domestic product in China is attributed to health and mortality burdens resulting from air pollution (Geall, 2014). India on the other hand is looking to solar power to reduce its $CO_2$ emissions by increasing solar power fivefold by 2022 (Nirmala, 2014). Indonesia has increased its subsidised fuel price by 30 percent in November 2014 (the subsidised fuel accounts for one fifth of total government spending) to help pay for government subsidies to the poor, build infrastructure, and hopefully reduce unnecessary private transportation and conserve energy. Ironically all three countries are taking bold steps for reducing carbon emissions because it affects their economies, the quality of life, and the mortality rates of their citizens.

Finally, leaders are beginning to realise that climate change impacts, besides killing biodiversity, also kill people. The issue of both Indonesia and Malaysia's subsidised fuel has greatly impeded a realistic market assessment of their economic development over several decades. Both countries are moving in the right direction not only for their own national development, but more importantly, for more responsible environmental reasons. Finally both countries are lessening their fuel subsidies and using their financial savings to undergird expenditures in other social services for the poor and financing infrastructure development. Fuel subsidies in both countries have not only provided an unrealistic assessment of their economic development, it has also allowed for fuel wastage, $CO_2$ increases, and unrealistic benchmarks of productivity.

The unfortunate political credo at all levels of society (village, provincial, urban state) is to recall Norwegian playwright Henrik Ibsen's phase "vital lie" which is the way people try to hide painful truths. One area in environmental issues where this is perpetually practiced is the acknowledgement of the full costs of ecological impacts due to our ignorance of the marketplace transactions: this is where people often endorse the 'vital lie' based on "what we don't know or can't see does not matter". As Daniel Goleman (2009) asserts, "our indifference to the consequences of the sum total of what we buy and do, and our unexamined habits as consumers, drive a vast number of threats to the environment and to health". These personal and societal mindsets need urgent changing if we are going to accept that our personal consumer habits can have profound impacts on ecosystems in other places and countries. While one might acknowledge recently the economic and cultural translations of globalisation, the environment and ecological aspects of globalisation have always been with Gaia since organisms developed on Earth.

## 5. The Persistence of Capitalism

The recent blockbuster book, *Capital in the Twenty-First Century*, by the French economist Thomas Picketty (2014) sheds light on the endemic nature of the capitalistic system. In his 200-year analysis of French and German economic history, Picketty (2014) noted that the capitalistic system generates inequality which transcends political, cultural and social influences. Specifically, the market economy has created inequality because "wealth accumulated in the past grows more rapidly than output and wages. This inequality expresses a fundamental logical contradiction. The entrepreneur inevitably tends to become a rentier, more and more dominant over those who own nothing but their labour. Once constituted, capital reproduces itself faster than output increases. The past devours the future" (Piketty, 2014). In short, once the market economy takes root in a society "the evolution of prices and wages, or incomes and wealth, obeys an autonomous economic logic having little or nothing to do with the logic of politics or culture" (Piketty, 2014). Ironically, only the two World Wars brought the national inequality in Europe to some sort of social and economic equalisation — seemingly a dismal scenario for the French economist when he states "the decline in the capital/income ratio between 1913 and 1950 is the history of Europe's suicide, and in particular of the euthanasia of European capitalists" (Piketty, 2014).

Despite his critics, Piketty's disturbing conclusion on capitalism has had wide ramifications of attempts to develop social equity amongst peoples and countries around the world — and this inability to close the income gap and economic inequality in capitalistic systems which dominate the global economy has serious impacts on the way natural resources, energy, and food are used, bought, sold and distributed around the world, and its outcomes on ecology, landscapes and environments. Indeed, the World Bank notes that 22 percent of global emissions is already covered by some form of carbon pricing (Hertsgaard, 2014). This only delays the reduction of carbon emissions through economic means but does not mitigate global $CO_2$ emissions.

If we accept John Piketty's (2014) dismal economic conclusions on capitalism it seems difficult to see how, without dramatic or catastrophic changes, the global community can have equitable and sustainable systems. Furthermore, how can such persistent economic inequalities lead to the management of environmental and climate changes? Also, widespread consumerism can be seen to be fuelling and strengthening capitalism — and consumer capitalism may in turn increase the scale and pace of consumerism. Yet, if consumerism and capitalism are impervious to political, cultural and social influences, it means the ecological footprint of rich and developed communities will continue to be enlarged at the expense of poorer communities. That is, while the rich can afford to consume more, the negative environmental impact caused by their consumptions may end up hurting the poor more, because the poor is less able to protect themselves against the effects of such impacts. Furthermore, the rich get richer and the poor become poorer and never the twain will meet. Hence, Sir Nicholas Stern is right in stating that climate change and the pricing of carbon is "the greatest market failure in history".

## 6. Changing Socio-Cultural Mindsets

In his interesting book, *The Geography of Thought*, the sociologist Richard Nisbett (2003) compared the thought and ideas of the West and East and concluded they were generally very different. Unlike Edward Said's (1978) *Orientalism*, this book was not meant to be an ideological denunciation of western imperial thought on Asia, nor the suppression of Asian thinking through western colonialism. *The Geography of Thought* was a seemingly objective exposition that westerners and easterners think differently due to their culture and societal conditioning.

While some readers might view Nisbett's (2003) thesis as cultural stereotyping, there are clearly some conclusions which need to be confronted frankly. There is no doubt that with globalisation, urbanisation, information technology and popular culture, the cultural differences between East and West might be closing. The wealth differences and widening consumer habits are now between cities and the hinterlands/rural areas. The city, as we are reminded by Aihwa Ong (2011), is also "the exclusive site of capitalism or postcolonial activism" that is subject to national and global forces. The irony is that cities only contain 1 percent of the world's land area but are consumers of 70 percent of the world's food, energy, natural resources, manufactured goods — besides being centres of pollution, sewerage and garbage. The new environmental challenge is trying to make cities sustainable given that these nodes of living are the new norm of human life.

The biggest challenge of our world is how environmental economists, corporations and governments can develop an "ecological phase of capital" by internalising social and environmental costs within the "structure of the private market" which John Foster (2002) argues is contradictory and unthinkable. Following the developed country's pathways to development, developing economies in Asia are using the same economic logic of "pricing the planet" which is like "economic imperialism vis-à-vis nature" (Foster, 2002). Indeed, China and its corporations are viewed as the new imperialists because their trade packs and infrastructure investments around the world (Africa, Australia, Middle East, and Latin America) are all geared towards obtaining more natural resources, energy sources, and raw materials for its own economic development. Not only does it have a large domestic $CO_2$ emission, it also is developing an equally large global ecological footprint and so-called embodied emissions. The continuing attempts to deconstruct nature into economic commodities and materials (i.e. natural capital) for purely human gain, greed and selfish needs are likely to go against organic nature, the ecological principles of biodiversity, the natural laws of Gaia, and the broader holistic interrelationships of a Green planet.

Clearly in response to the Washington consensus, China's entry into the global market has created what Stefan Halper (2010) calls the "Beijing Consensus" which, among other issues, creates new domestic environmental impacts and global ecological ramifications. Like several other countries, China has modified capitalism to suit its authoritarian political system in what is seen as "command-and-control capitalism", viewed variously as 'illiberal capitalism', 'market authoritarianism' and 'state capitalism', that has created a new centre of economic autonomy away from western based capitalism. Unlike democratic systems in the west, the Chinese government can act quicker and more effectively in all it does. Its total command of domestic government also means there are little checks and balances on any problems with regard to climate change, pollution and biodiversity. The Chinese factor is changing political relationships around the world and creating new geopolitical realignments. Yet, ironically, if Chinese philosophy and traditional religion in *Daoism* (its human–nature harmony) could filter through in national and global thinking, the benefits for creating a more sustainable world could become a closer reality.

## 7. Reflections

> "Many ancient writings speak of the Earth as an angel. But perhaps the Earth is the embryo of an angel, an angel on the way to being born."
>
> ~Jacob Needleman (2012)

Never in the history of the mass media, does one see advertisements in leading popular Journals (*Time, Bloomberg Businessweek, The Economists, Fortune, Forbes*) carrying all sorts of environmental messages, warnings and admonitions that are addressed at consumers and a somewhat indifferent public. The World Wide Fund for Nature (WWF) appeals to consumers subtly through their "Help Save the Lipstick" campaign in order to ensure tropical forests are not being replaced by palm oil estates, the source of lipstick manufacture. In another campaign, WWF requests its readers to "dare the world to save the planet". The Effie Awards for consumerism notes that "72% of consumers want to buy green products" but "only 17% do". The Wildlife Conservation Society states that "96 Elephants are killed every day in Africa" and asks the global community to "Take Action Now" or else we "have to shamefully admit to our children that we stood by as elephants were driven to extinction". The Rainforest Alliance Certified reminds readers to take small actions today under the theme "Small Actions, Big Changes" in protecting wildlife habitats and curbing climate change. The environmental NGO, *The Nature Conservancy* best sums up nature's contributions when it states "Nature Saves Lives". Other advertisements demonstrate some current acute environmental issues. Hitachi requests social innovation for water (desalination) technologies the company offers for the human future with this quote "If we could drink the ocean wouldn't we have enough water?"; while the energy management company, Delta, addresses climate change challenges by enticing readers with an advertisement stating "Greener buildings start with smarter solutions".

Given the focus on consumerism in these advertising Green campaigns, we need to look more closely at economic issues and our capitalist system. It is difficult to shut down the capitalist system given that four-fifths of the world's states thrive on it for economic growth and development. Therein lies the global contradiction. How can the global community solve an intractable issue? The biggest economic dynamos (China, Germany, and the US) are also the biggest consumers of coal power. In China alone, coal power killed 670,000 people in 2012 due to pollution and smog. The Chinese government has to be worried because by 2019, 31 percent of global $CO_2$ will come from China as opposed to 15 percent from the US. The newly emerging economies, the BRIC states (Brazil, Russia, India and China) are trying to close the economic gap with the developed countries by only one means — capitalist-driven economic development and subsidised energy.

In Hegelian terms, unfortunately, global capitalism is the 'tested' economic thesis which will prevail given that there is no alternative developmental antithesis. The economic antithesis in 'planned economies' and communist systems, which dominated 70 years of global state histories, have failed dismally with the fall of Eastern Europe communist states (collapse of the Berlin Wall in 1989) and the former Russian Federation in 1991. Yet, while its political manifestations have been defeated, the ideological message of socialism in equality, non-materialism and community development needs to be sustained in an altered and modified capitalistic system. An individual-centred, greed-orientated, materially-driven capitalistic economic system, will undermine any national gains in adaptation and mitigation programmes in climate change. The clock is ticking quicker now with $CO_2$ rates reaching 400 ppm in 2013, which is dangerously close to the scientific view that 550 ppm is likely to be reached in 2050, a critical benchmark and — according to Tim Flannery — a point of no return.

The silver lining in changing global political and economic attitudes in climate change must come from the US–China (with 40% of global emissions) announcement in November 2014 to tackle climate change. Despite its sceptics and detractors, one has to applaud both these global powers for their 'historic', 'turning point', and the 'positive signal' of their joint agreement. Barack Obama's 2014 comments on Australia's climate change impacts on the Great Barrier Reef might not have gone down well with the Australian government or public, but these moves and opinions are contrary to global opinion on the need for every country to curb severe carbon emissions nationally. The unfortunate moral of the story is that countries are still adopting self-interested nationalistic perspectives rather than all inclusive global 'common-pool' views.

Besides the historical lessons of environmental and climate change impacts and disasters, there are real life examples of supposed 'primitive' indigenous societies that have lessons one can learn from. In his thought provoking book, *The World Until Yesterday*, Jared Diamond (2012) provides much food for thought about how traditional societies operate, organize and manage themselves in varied environments — many of these ecosystems are 'hostile' and difficult for human living — from the tribal groups in the jungles of New Guinea to the deserts of the Kung Bushmen of the Kalahari desert. There are many lessons of how traditional societies adopt many methods (environmental and cultural adaptation) for survival. Despite hazardous environments, environmental dangers ranks only as the third major cause of death, after disease and human violence, for traditional societies (Diamond, 2011).

Comparing ourselves to traditional communities, urban dwellers are often smug about their superior intelligence. However, in our modern states and cities, one finds many monolingual communities (Britain, USA). Yet, amongst the primitive societies of New Guinea, multilingualism is the norm. Small groups usually speak several languages as a means of survival for trading purposes, developing alliances, providing access to resources and even for getting a spouse (Diamond, 2011). If language was used as a barometer of intelligence, then such developed modernized communities might rank lowly. The point here is that the skills and adaptive intelligence for survival cannot be reduced to only our modern barometers of intelligence. We need to look beyond our silos of urban living to realize that even in the deserts and jungles, human communities have established a wide variety of survival intelligence, cultural adaptations and practical knowledge. And, most important of all, a genuine respect for Nature. When you see how many tribal groups have lived for centuries in small ecosystems (see Diamond, 2011), one can realise that they must have developed the knowhow and management skills for sustainable living without damaging their ecosystems and, more importantly, without living on the largesse of natural resources obtained from an extended ecological footprint. Having said that, the questions remain: how able are we in learning these lessons on sustainable living from these tribal groups, and how willing are we to change our lifestyle or consumer habit to ensure that humanity will never reach the point of no return?

## References

Carson, Rachel (1962) *Silent Spring*, Boston, Massachusetts: Houghton Mifflin Harcourt.

Cohen Benjamin J. (2009) *Notes from the Ground: Science, Soil & Society in the American countryside*, New Haven: Yale University Press.

Corlett, Richard (in press) "The Anthropocene concept in ecology and conservation" in *Trends in Ecology & Evolution*, pp. 1–6.

Diamond, Jared (2011) *The World Until Yesterday*, London: Allen Lane.

Diamond, Jared and Robinson, James (2010) (eds.) *Natural Experiments of History*, Cambridge, Harvard University Press.

Foster, John Bellamy (2002) *Ecology against Capitalism*, New York: Monthly Review Press.

Geall, Sam (2014) "What next for China after historic climate deal?", *The Straits Times*, 15 November, p. A45.

Goleman, Daniel (2009) *Ecological Intelligence*, London: Allen Lane.

Halper, Stefan (2010) *The Beijing Consensus*, New York: Basic Books.

Hertsgaard, Mark (2014) "If its good enough for big oil..." in *Bloomberg Businessweek,* November 17–23, pp. 10–11.

Lieberman, Victor (2009) *Strange Parallels Southeast Asia in Global Context,* c. 800–1830, Vol. 2, New York: Cambridge University Press.

Linden, E. (2007) *The Winds of Change: Climate, Weather, and the Destruction of Civilizations*, New York: Simon and Schuster.

Needleman, Jacob (2012) *The Unknown World: notes on the Meaning of the Earth*, New York: Penguin.

Nirmala, Ganapathy (2014) "Sun's finally shining on India's climate change fight", *The Straits Times*, 24 November, p. A9.

Nisbett, Richard E. (2003) *The Geography of Thought*, New York: Free Press.

Ong, Aihwa (2011) "Introduction Worlding Cities, or the Art of being Global" in Ananya Roy and Aihwa Ong (eds) *Worlding Cities: Asian Experiments and the Art of being Global*, Oxford: Wiley-Blackwell, pp. 1–26.

Orr, David (2004) *Earth in Mind,* Washington: Island Press.

Parker, Geoffrey (2013) *Global Crisis*, New Haven: Yale University Press.

Picketty, Thomas (2014) *Capital in the Twenty-First Century*, Cambridge, Massachusetts: The Belknap Press.

Rosen, William (2014) *The Third Horseman*, New York: Viking.

Said, Edward W. (1978) *Orientalism*, New York: Pantheon Books.

# Part 1

# Biodiversity

# 1

# UNDERSTANDING HARMFUL ALGAL BLOOM (HAB) OCCURRENCES IN MANILA BAY, PHILIPPINES

ROSA CELIA POQUITA-DU and PETER ALAN TODD

Harmful algal blooms (HAB) are a yearly event in the waters of Manila Bay, Philippines. The very first major bloom was caused by a toxic dinoflagellate species, *Pyrodinium bahamense var. compressum*, which is known to cause Paralytic Shellfish Poisoning (PSP). This chapter reviews HAB episodes in the bay recorded from the 1980s to 1990s, and compares them with the more recent events in the 21$^{st}$ century. *Pyrodinium* blooms occurred for almost a decade from 1988 until 1998, before being succeeded by a non-toxic species, *Noctiluca scintillans*, that was first recorded in the year 2001. No PSP cases were recorded between then and the year 2010. More recent bloom episodes, however, were caused by another toxic dinoflagellate species, *Alexandrium minutum*. The numbers of recorded PSP cases were higher in the 1980s and 1990s bloom events compared to the more recent toxic blooms. The decrease in incidences of PSP do not necessarily mean that the ecological health of the bay has improved, but instead could be due to changes in the physico-chemical properties of the water favouring one phytoplankton species over another. Even though sea surface temperatures and rainfall are known to be associated to HABs, the present study found no significant correlations among the variables tested. Due to the complexity of the environmental issues identified in Manila Bay, it is essential to conduct more comprehensive studies on the biological, physical, and chemical components of the bay to facilitate a clearer understanding of HAB episodes.

## 1. Introduction

Almost every coastal country in the world is affected by Harmful Algal Blooms (HABs), commonly known as *red tides* (Anderson, 2007), and they represent one of the major marine environmental problems in Southeast Asia (SEA). The phenomenon is caused by the proliferation of phytoplankton species which results from a combination of physical, chemical and biological factors including natural mechanisms of species dispersal, pollution-related nutrient enrichment of water bodies, climatic shifts, and transport of algal species via ship ballast water (Anderson, 1998 & 2007; Smayda, 1989; Hallegraeff, 1993). Coastal waters in the SEA region are increasingly contaminated by pollutants from various sources, sometimes reaching levels that cause the rejection of shellfish for human consumption and export (Chua *et al.*, 1989). The Philippines is one of the countries greatly affected by HABs, with almost yearly recurrences at several coastal areas. Since 1999, the country has recorded the highest number of Paralytic Shellfish Poisoning (PSP) incidences caused by *Pyrodinium bahamense var. compressum* blooms in the region (Azanza and Taylor, 2001). PSP results from the ingestion of a potent endotoxin called *saxitoxin* (STX), which is produced by some species of dinoflagellates. These toxins are accumulated in the tissues of bivalve organisms (e.g. mussel, oysters) and are harmful when consumed by humans. Saxitoxin is classified as a neurotoxin which is capable of causing paralysis to the victim within 24 hours after ingestion (Popkiss *et al.*, 1979).

The Philippines has experienced a considerable number of red tide outbreaks in the coastal waters in all three major islands: Luzon, Visayas and Mindanao. This phenomenon has been continuously monitored over the last three decades starting from the 1980s to the present. Up to the year 2001, the country has experienced a total of 42 toxic algal blooms which resulted in a total of 2,107 PSP cases with 117 fatalities (Furio & Gonzales, 2004). Reported economic loss in the Philippines during the 1983 bloom (the very first HAB occurrence in the country) was approximately USD 5 million. The first bloom that occurred in Manila Bay in 1988 resulted in losses of USD 300,000 per day during the height of the episode (Corrales and Maclean, 1995; Azanza and Taylor, 2001).

HABs have brought tremendous negative impacts, not just to the marine ecosystem, but also to the economy and human health; therefore, understanding bloom dynamics can potentially improve existing mitigation measures and further minimize economic losses. This chapter (1) examines the recorded toxic and non-toxic algal blooms in Manila Bay from 1988 to 2011, and discusses the mechanisms of each causative species; (2) evaluates the roles of two environmental factors, i.e. Sea Surface Temperature (SST) and rainfall, in initiating bloom episodes in the bay and; (3) recommends steps towards improving existing management strategies.

## *Manila Bay*

Manila bay is a semi-closed body of water situated in the western part of Luzon between 14.23° and 14.87°N and 120.53° and 121.03°E (Figure 1.1). Its coastline extends to approximately 190 km coastline with a surface area of 1,700 km$^2$, estimated volume of $2.89 \times 10^{10}$ m$^3$ and an average depth of

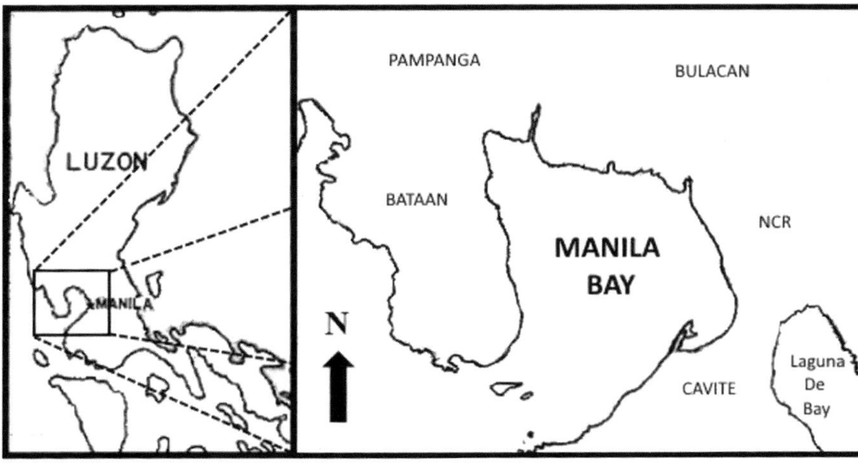


Fig. 1.1. Map of Manila Bay showing the neighboring provinces.

17 m (Jacinto, *et al.*, 1998; Velasquez, *et al.*, 2002). The climate of the bay is characterized by two pronounced seasons: (1) dry from November to May, and (2) wet from June to September. Generally, August is the rainiest and February the driest. Three major wind regimes are observed in the bay: (1) Northeast monsoon from October to January, (2) south-easterly from February to May, and (3) south-westerly trade winds from June to September. The circulation pattern of the surface water is influenced by the interplay of the surface wind stress, tidal forces, and freshwater discharges (De las Alas and Sodusta, 1985). It receives drainage directly from almost 17,000 $km^2$ of watershed composed of 26 catchment areas. A total of 131 rivers and creeks drain into the Bay and high organic loading from these rivers led to the eutrophication of the waters (Bajarias and Relox, 1996). It is bounded by coastal cities and municipalities of Bulacan and Pampanga in the North, Cavite and National Capital region or NCR (including Manila, Pasay, Paranaque, Las Pinas and Navotas) in the East, Bataan in the West and Northwest. Also, within the watershed are the non-coastal cities and municipalities of the NCR (Quezon City, Caloocan, Makati, Pasig, Marikina, Mandaluyong, Muntinlupa, Valenzuela, Malabon, San Juan, Pateros and Taguig), provinces of Nueva Ecija and Tarlac in Region 3, and Rizal and Laguna in Region 4. There are several residential areas near the Bay region that stretch from the low lying coastal zone to upland areas. A large number of migrants have moved to the urban areas around Manila Bay, which has contributed to poverty. Many live in shanty

houses in waterways (by the rivers and coastal zones), under bridges and abandoned railways (Ashby, 2008).

Manila Bay plays an important role to the economy of the country due to its numerous contributions (Jacinto *et al.*, 2006). The bay supports several industries from shipping ports to major fishing grounds and numerous aquaculture farms (Prudente *et al.*, 1994). Not only does it have high economic value, but it also had been declared as a critical habitat in Metro Manila as it supports numerous endemic wildlife species, migratory birds, as well as mudflats and mangrove resources (Su *et al.*, 2009). The bay used to be one of the major sources of fishes, shellfishes and other aquaculture products for exports, but the volume of desirable species has diminished considerably over time. Massive economic and biological overfishing in the bay occurred as early as 1948 (Muñoz, 1991). It is considered to be eutrophic due to the accumulated large amounts of nutrients, mostly coming from domestic discharges, particularly near Pasig River watershed (Jacinto *et al.*, 1998), and some from indiscriminate discharge of effluents from industries around the area (Jacinto, *et al.*, 2006). Results from the study conducted by Chang *et al.* (2009) showed high concentrations of nitrogen species, particularly ammonium, in Pasig River which indicated the river to be the main source of nutrient input to the bay. It receives all outflow from Manila's sewerage system and only about a quarter of the entire population of Manila (approximately 1.14 million) is served by the communal sewerage system (Gunnerson and Cuellar, 1988).

## 2. Methods

Most of the data were obtained from government reports prepared by Bureau of Fisheries and Aquatic Resources (BFAR) — Marine Biotoxins Monitoring Unit (MBMU) and the Philippine Atmospheric, Geophysical and Astronomical Services Administration (PAG-ASA) for the records of HAB occurrences in Manila Bay, and the Average Monthly Rainfall respectively. Other data sources included the International Research Institute (IRI) Climate Data Library[1] for the Average Monthly SST data and also published quantitative studies on HAB occurrences in Manila Bay.

[1]http://iridl.ldeo.columbia.edu/

### *Acquisition of Secondary Data*

(i) *Record of PSP cases in Manila Bay*

Data of HABs in Manila Bay were obtained from BFAR Central Office in Quezon City, Philippines. The monitoring unit mainly focused on harmful algal blooms associated with shellfish poisoning. Acquired data include: (a) number of illnesses & deaths associated with PSP cases in Manila Bay from 1988 when it first occurred in the area until the recent date (mid 2012); (b) causative phytoplankton species for each bloom; (c) implicated shellfish and levels of PSP toxins; (d) specific locations where the bloom started and; (e) dates/duration of shellfish ban. The sampling protocol applied by MBMU in detecting PSP events employs measurements of two key parameters which include: (a) toxicity level in shellfish meat samples and; (b) presence of PSP causing organisms. Collection of samples is on a regular basis wherein the frequency of sampling depends on the site. Field sites which are representative of the whole area of concern and specifically chosen based on the logistic reports of red tide occurrence and availability of shellfish resources are already established. One of the primary sampling stations in the Philippines is Manila Bay, which covers Cavite, Navotas (Metro Manila), Bulacan, and Bataan. Sampling for Manila Bay sites is conducted twice a week. Plankton samples are collected using $20\,\mu$m plankton net vertically hauled from varying depths. Hauling depths are dependent on the total depth of the sampling station and the length of the plankton net used. Vertical hauling is made starting at 2 m above the bottom. Samples of shellfish meat are collected from the shellfish farms; however wild shellfish are collected if farms are not present. The samples are stored for transportation to the MBMU laboratory in Quezon City. PSP causing dinoflagellates are identified using a compound light microscope and subsequently quantified to be reported as number of cells per litre (L). Levels of toxin from the shellfish samples are determined by performing the mouse assay technique and values are reported as micrograms ($\mu$g) STX equivalent (eq) per 100 grams (g) of shellfish meat. Results obtained from the monitoring are used to determine whether or not to issue a shellfish ban in the sampled areas. BFAR established a Shellfish Bulletin in their official website and post shellfish ban notices twice a month. The document specifies which areas of the Philippines are free from toxic red tides and which areas are under the shellfish ban. A shellfish ban entails banning activities which involve harvesting, selling and consuming of bivalve shellfishes. The regulatory limit set for the Philippines is $60\,\mu$g

STX eq per 100 g shellfish meat. Whenever laboratory results show toxin levels greater than the threshold set, a shellfish ban will be automatically issued. Areas under the shellfish ban are constantly monitored on a weekly basis until toxin levels fall below the limit for three consecutive weeks and an advisory declaring the area free from toxic red tide is issued; the area is then subsequently opened for shellfish harvesting. Details of the sampling protocol were obtained from Bajarias *et al* (2006) and through personal communication with the head of BFAR-MBMU, Juan Relox Jr.

(ii) *Annual rainfall*

Data time series of the monthly average rainfall (expressed in millimetres (mm) per month) of Manila Bay (Manila Port Area Sampling Station) from 1988 to 2011 were obtained from Philippine Atmospheric, Geophysical and Astronomical Services Administration (PAG-ASA) which is a government agency mandated to monitor atmospheric, geophysical, and astronomic information in different locations within the country. The agency issues on a routine basis, through the Climatology and Agrometeorology Branch, meteorological and climatological publications and other processed information for interested parties to obtain via requests, provided these data will not be used for commercial purposes. Compiled climate data were transferred into a separate excel sheet and plotted for each corresponding year of occurrence.

### *Extraction of Sea Surface Temperature (SST) Data*

Time series of SST (in degree Celsius (°C)) data was retrieved from an online data library provided by the International Research Institute (IRI) Climate Data Library.[2] There are several ways of finding datasets in the data library, one of which is retrieval "*By Source*" which opens a list of datasets organized by different original sources. Integrated Global Ocean Services System Products Bulletin (IGOSS) was selected as the source which provides different datasets and variables, including Global SST data, based on Reynolds and Smith. Version 2 of the optimum interpolation (OI) analysis of SST was chosen over version 1 as the former reduces past errors, and improvements of bias correction were made for the version used (Reynolds *et al.*, 2002). The OI version 2 SST analysis is produced monthly on a 1° lat/long spatial grid. Monthly SST data were retrieved by selecting the desired grids: (X) Longitude 120.5 East, (Y) Latitude 14.5

[2]http://iridl.ldeo.columbia.edu/

North (Location of Port Area Station, Manila Bay), and Time from January 1988 to December 2011.

## 3. Results & Discussion

### *HABs from 1988 to 2011 and Implications of Bloom Episodes*

The very first documented algal bloom in Manila Bay waters was caused by the toxic dinoflagellate species, *P. bahamense var. compressum* in 1988. Similar HABs reoccured almost every year until 1998 (Figure 1.2). This species is known to cause shellfish poisoning which can be fatal to other organisms including humans. During the toxic blooms in 1997 and 1998, both *Gymnodinium catenatum* and *P. bahamense* were listed as the causative species. Shellfish bans lasted up to 8 or 9 months in 1992 and 1993, but only about 5 to 7 months in other HAB events. No recorded PSP cases happened between 1999 and 2009 (personal communication, J. Relox, May 22, 2011).); however, persistent blooms of the non-toxic species, *Noctiluca scintillans*, were observed in 2001 (Borja & Furio, unpublished; Furuya *et al.*, 2006a). As monitoring protocols conducted by MBMU are only limited to monitoring toxic algal species, detailed information about the blooms of *N. scintillans* is limited. Blooms of *N. scintillans* took

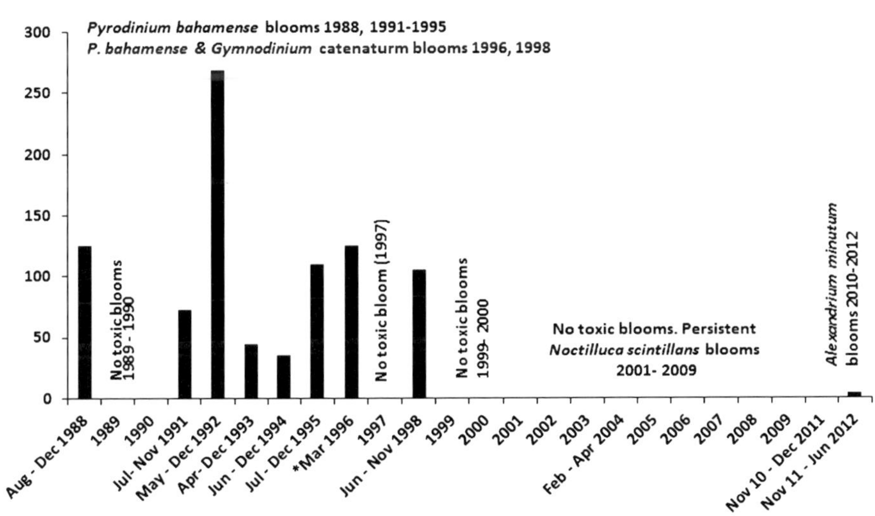


Fig. 1.2. Reported PSP cases (illnesses and deaths) in Manila Bay from 1988 to 2011.

over from *P. bahamense* starting from 1998 and dominated from June to September (Jacinto *et al.*, 2006), almost the same time as *P. bahamense* were expected to bloom. A previous study conducted by Furuya *et al.* (2006a) highlighted that blooms of *N. scintillans* were observed in 2004 from late February until late April, and in March 2002. Another shift of causative toxin-producing organism was observed in Manila Bay during the episodes recorded in 2010 and 2011. Blooms of a toxic dinoflagellate species, *Alexandrium minutum* were observed during the months of November in 2010 and 2011, lasting for 2 months and 7 months, respectively. There were no reported PSP cases in 2010; however, four PSP cases were recorded in the 2011 *Alexandrium* bloom.

The greatest number of PSP cases happened during the 1980s and 1990s bloom events (i.e. compared to 2010 and 2011). The highest number of PSP cases was in 1992, where 269 local residents where reported ill, with 11 casualties. Toxicity levels recorded in that year reached 1270 $\mu$g STX eq per 100 grams of shellfish (Figure 1.3), which is beyond the regulatory limit of 60 $\mu$g STX eq per 100g of shellfish meat. The highest recorded toxicity level (6,351 $\mu$g STX eq per 100g) was recorded in 1994. There were 36 PSP cases reported during the said year. Toxicity levels recorded for the 2010 and 2011 *Alexandrium* blooms were lower compared to *Pyrodinium* blooms. There were only a few cases of shellfish poisoning caused by *A. minutum*. A detailed list of the PSP cases and some other relevant information on

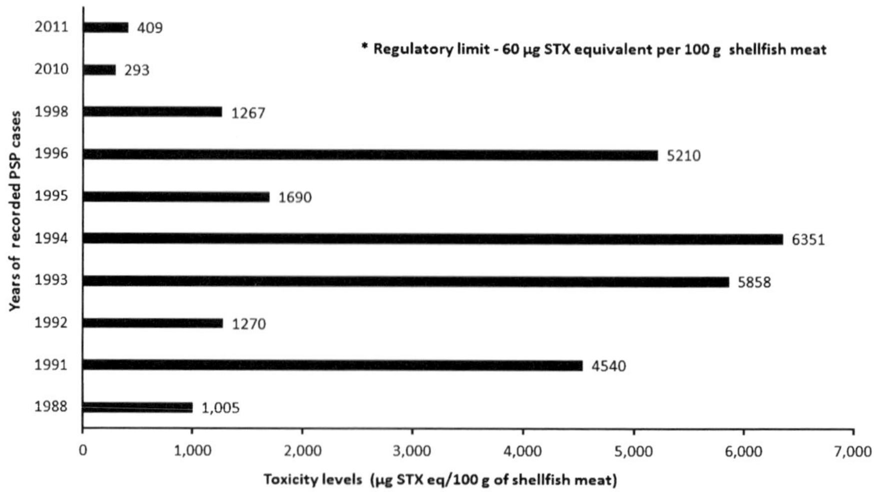


Fig. 1.3. Toxicity levels of the recorded PSP cases in Manila Bay.

Table 1.1. Record of PSP cases with the corresponding values of annual SST and rainfall.

| Year | PSP cases | Annual SST | Annual rainfall |
|---|---|---|---|
| 1988 | 125 | 29.27 | 196.73 |
| 1989 | 0 | 28.65 | 162.06 |
| 1990 | 0 | 28.62 | 207.03 |
| 1991 | 73 | 28.63 | 152.89 |
| 1992 | 269 | 28.78 | 152.93 |
| 1993 | 45 | 28.57 | 147.30 |
| 1994 | 36 | 28.77 | 176.28 |
| 1995 | 110 | 29.09 | 207.55 |
| 1996 | 125 | 28.91 | 145.09 |
| 1997 | 0 | 28.79 | 191.38 |
| 1998 | 105 | 29.50 | 174.63 |
| 1999 | 0 | 29.15 | 236.08 |
| 2000 | 0 | 28.96 | 287.34 |
| 2001 | 0 | 29.17 | 152.68 |
| 2002 | 0 | 28.87 | 220.18 |
| 2003 | 0 | 28.77 | 170.15 |
| 2004 | 0 | 28.64 | 128.84 |
| 2005 | 0 | 28.88 | 132.48 |
| 2006 | 0 | 28.82 | 174.60 |
| 2007 | 0 | 28.94 | 164.79 |
| 2008 | 0 | 28.71 | 183.31 |
| 2009 | 0 | 28.46 | 245.52 |
| 2010 | 0 | 29.28 | 163.43 |
| 2011 | 4 | 28.30 | 270.69 |

each HAB year is presented in Table 1.1. Examining the values of toxicity levels shown in Figure 1.3, it can be observed that *A. minutum* produced relatively less toxins — despite which, it still caused shellfish poisoning to humans who happened to eat contaminated shellfish as the levels were already beyond the regulatory limit. The levels of toxicity measured during the *P. bahamense* blooms from 1988 to 1998 were much higher compared to the recent blooms caused by *A. minutum* in 2010 and 2011. One potential reason for the lower toxicity levels of the recent episodes by *A. minutum* is a change of the nutrient flux ratio in Manila Bay waters. Since PSP toxins are nitrogenous compounds, a reduced amount of toxin per cell in a PSP-producing dinoflagellate is observed under N limitation (Boyer *et al.*, 1987; Anderson *et al.*, 1990). Conversely, toxin production is higher under P limitation (Boyer *et al.*, 1987, Anderson *et al.*, 1990; and Guisande *et al.*, 2002). The shift of land-based activities surrounding Manila Bay could have influenced a change in the chemical property of the waters into an N-limited environment. Most agricultural countries such as Philippines, have now been industrialized with a corresponding decrease in the number of farms. In Central Luzon, there was a decrease of 2.7% of the number of agricultural farms in 2002, starting from 1991.[3] The year 2002 was the latest available data published on the National Statistics Office's (NSO) website. Generally, the economies of the Asian countries have developed significantly for the past decades, with growth rates of industry and exports far exceeding the growth rates of agricultural sectors (Bautista and De Rosa, 1996).

Species competition happens naturally between living organisms and plays a major role in their survivability, especially under harsh conditions. Different organisms exhibit various types of mechanisms in order to compete with co-existing species. Based on Shellford's Law of Tolerance, the range of tolerance of a species to local environmental conditions limits its distribution; so that the species will be abundant in areas within its optimum range or, rare if the species experiences physiological stress when certain environmental factors are not within its tolerance limits.[4] In the case of *P. bahamense*, it was *able* to dominate over other phytoplankton species for so many years by producing harmful toxins which act as feeding deterrents to higher predators and subsequently increase grazing pressure on other co-existing species. However, an organism's success is based on a set of environmental conditions, may it be biotic factors, abiotic factors or a combination of factors, which determines its success.[5] Organisms respond naturally to the changes of its surrounding environment so that each organism responds differently to various factors according to their degree of tolerance to withstand them.

### *Affected Coastal Areas and Their Characteristics*

HAB episodes are generally seen in almost the same area in every year of occurrence (Figure 1.4). These coastal provinces include Bataan, Cavite and a few cities within NCR (particularly Navotas, Las Piñas & Parañaque). Most of the affected areas are engaged with aquaculture activities and close to urban centers. The events in 1988, 1991, 1994, 1996 and 1998 occurred only in Bataan. The *Pyrodinium* bloom events in the years 1992 and 1993 hit three areas; namely Bataan in the west, the coastal cities of Navotas in Metro Manila, as well as in Cavite, which lies along the eastern shores of

---

[3]http://www.census.gov.ph/content/review-agriculture-sector-central-luzon
[4]http://www.biog1105-1106.org/demos/106/unit08/media/3a.lawofminimum.pdf
[5]http://en.wikipedia.org/wiki/Shelford's_Law_of_Tolerance

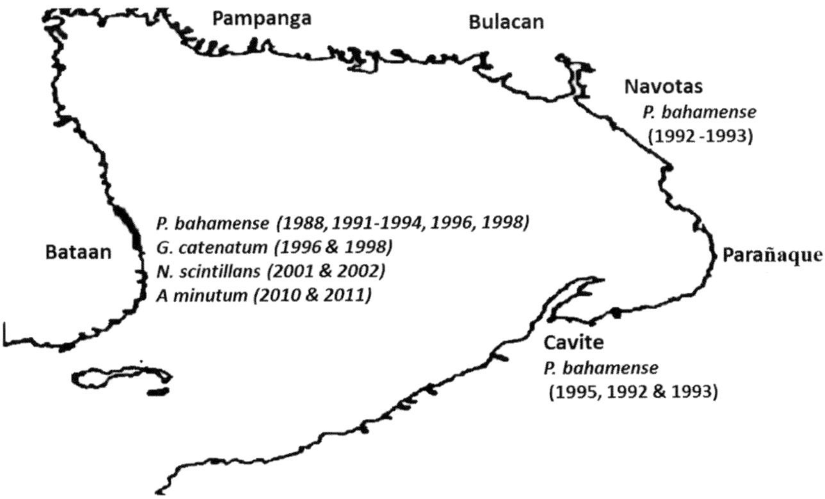


Fig. 1.4. Affected coastal areas with the corresponding years of PSP cases.

Manila Bay. For the year 1995, it was only seen in Cavite coastal waters. On the other hand, the blooms of the green *Noctiluca*, which has been prevalent in the bay since 2001, was observed particularly in the western side (Borja and Furio, unpublished). As indicated by Furuya *et al.* (2006a), there was a large scale bloom of *N. scintillans* off Limay, Bataan in March 2002.

Bloom occurrences at Bataan are more frequent than at other coastal areas. Bataan is subject to multiple environmental issues from improper domestic and major pollutive industries' waste disposal; degradation of the upland forests (due to slash and burn and unsustainable lumber production) and lowland forests (due to quarrying activities in the coastal zone), which subsequently cause soil erosion; and waste deposited in the coastal waters from aquaculture or mariculture activities (GEF/UNDP/IMO/PEMSEA, 2007). The municipalities of Limay and Orion in Bataan experienced the most number of recurring algal blooms in the province. The source of bloom formations in the entire Manila Bay came generally from Bataan. Moreover, shellfish products such as mussels and oysters, which are the usual vectors of PSP toxins, are cultured in the coastal waters of Bataan. Siltation and shallowing of the near-shore are also common problems in Bataan which are consequences of unsound resource-use practices in both inland and coastal areas. The problem of siltation has been around even before the area started receiving *lahar* (volcanic material) deposits brought about by the Mount Pinatubo eruption in 1991, which aggravated the problem. Vast quantities of minerals were brought to the surface environment, including copper, zinc, chromium, nickel, as well as massive amounts of toxic heavy metals such as lead, arsenic and cadmium.[6] A year after the Mt. Pinatubo eruption, an accidental oil spill from a tanker going to the refinery in Limay occurred. This incident immediately resulted in pollutive materials entering the Bataan coastal waters. So far, the highest total volume of oil spill in the bay was from the two major oil spill events in 1999 which occurred in both Metro Manila and Limay, Bataan (Jacinto *et al.*, 2006). Records of oil spill events in Manila Bay from 1990 to 2005 identified Bataan and Metro Manila as having the greatest number of oil spill occurrences, with 8 and 19 incidents, respectively (Jacinto *et al.*, 2006).

The National Capital Region (comprising of Navotas, Pasig and Parañaque River areas) on the other hand, is characterized as highly urbanized and industrialized areas surrounding Manila Bay. A scene of urban centers with proliferating shanty houses can be found in the metropolis. There is a large amount of informal settlements which are believed to contribute substantially to the deterioration of the water quality. NCR houses 39 special economic zones and technological parks which make up about 30% of the country's Gross Domestic Product (GDP). Various sectors are concentrated in the area: such as agriculture, forestry and aquaculture sectors; industrial sectors which include manufacturing and mining industries; and the largest portion, the services sector, that comprises of trade and financial centres.[7] Approximately 69% of Metro Manila's land area is situated on high elevation while the remaining 31% rest in flood-prone zone (i.e. Manila, Navotas, Malabon, and other parts of Caloocan). Particularly, Navotas and Malabon are the coastal towns that are most prone to flooding during high tides; in the east, the vulnerable areas are Pasig, Marikina, Pateros and Taguig. These areas experience frequent flooding caused by the overflow of the Pasig and Marikina rivers which then contribute nutrients to the bay. Nutrient concentrations (ie. nitrate, nitrite, ammonia and phosphate) were reported to be increasing at the eastern part of the bay especially near Pasig and Parañaque Rivers (Velasquez and Jacinto, 1995; PRRP, 1999). The Pasig River alone is estimated to receive about 168 metric tons per day of domestic sewage from 11 areas within Metro Manila (PRRP, 1998).

Lastly, in 1992 and 1995, the coastal waters of Cavite City also experienced toxic algal blooms, caused mainly by *P. bahamense*. Like the

[6]http://en.wikipedia.org/wiki/Mount_Pinatubo
[7]http://www.emb.gov.ph/mbemp/dloads/mbcs%2002mvw.pdf

other affected areas, the province of Cavite also engages in aquaculture and agricultural activities. Agriculture is the main land use, with 49.38% of the total provincial land area engaged in some form of farming.[8] The economy of the province relies mostly on agriculture despite the urbanization and industrialization initiated in the 1990s. There are two zones identified within Cavite, (1) extensive open-water aquaculture and urbanized environment (Northern Cavite) and, (2) limited aquaculture, extensive agriculture, tourism and natural environment (Southern Cavite). Cavite is the leading industrial zone located outside Metro Manila, with more than 31 industrial estates available for light and heavy manufacturing activities including the Cavite Export Processing Zone, which is one of the biggest contributors to the country's GDP particularly exports receipts. As a consequence of the complete transformation of the province into an industrial hub, a rapid increase of population occurred with its majority living in the urban areas. Due to the increasing industrial estates in the province, housing subdivisions have also increased which encourage people to migrate and seek job opportunities.[9]

## 4. Mechanisms of the Causative Organisms

This section provides more detailed discussions of each causative organisms' morphology and mechanisms to explain their dominance over a particular period leading to a bloom in the bay.

### *P. Bahamense var. Compressum Bloom Episodes (1988–1998)*

Algal blooms in Manila Bay during the 1980s were mostly caused by *P. bahamense* Plate *var compressum* Böhm (Steidinger *et al.*, 1980) which is a thecate, chain-forming and toxin-producing species. Except for the bloom events that occurred in 1996 & 1998, *G. catenatum* dominated at the same time with *P. bahamense*. A study on the mechanism of *P. bahamense* by Villanoy *et al* (1996) postulated that thermal stratification and vertical stability of the water initiates bloom formation. Azanza and Taylor (2001) observed that El Niño events coincided with the *Pyrodinium* blooms in the Indo-West Pacific such as in Samar–Leyte in 1982–1983, and

[8]http://www.cavite.gov.ph/home/multimedia%20files/SEPP/2009/Economic%20Sector-Agriculture.pdf
[9]https://indangbeauty.wordpress.com/category/about-cavite-province/

Manila Bay in 1997–1998. However, this hypothesis was negated as the El Niño Phenomenon in 1997 did not trigger any bloom events in the entire Manila Bay (Azanza, 1997) despite the favourable conditions of high water temperature. The growth of *P. bahamense* is known to respond to changes of temperature. In a study by Gedaria *et al.* (2007), *P. bahamense* was observed to grow within a temperature range of 23 to 36°C, with optimum growth at 25°C. Previous researchers had observed a slower growth rate as well as a constant decrease in toxin concentration with higher temperature conditions. This observation could potentially explain why there were no cases of PSP during the El Niño event in 1997 as an increase of temperature might have caused *P. bahamense* to produce less toxins as a stress response. However, another study observed a significant effect of temperature on toxin content such that toxicity increased by three-fold as the temperature decreased to 22°C (from the optimum of 28°C Usup *et al.*, 1994). Hence, the varying response of this species to change of temperature only proves that SST is not solely the initiating factor of *Pyrodinium* blooms.

### *Gymnodinium catenatum* Bloom Episodes (1996 and 1998)

Among the dinoflagellate species which are known to contain PSP toxins, *G. catenatum* is the only species which is not armoured. Extensive blooms of this species is thought to be triggered by high rainfall and influx of dissolved organic matter (DOM), i.e. humic substances (Doblin *et al.*, 1998). In culture experiments of *G. catenatum*, Hallegraeff *et al.* (1995) revealed doubling growth of this species with addition of soil extract to the culture medium. It is known that soil extract produces chelating effects, which could decrease toxic concentrations of metal ions (e.g. copper) or increase availability of other metal ions required as nutrients, such as iron and manganese (Anderson and Morel, 1978; Sunda, 1989). According to Hallegraeff *et al.* (1995), the key environmental variables that initiate *G. catenatum* blooms are water temperature, windstress and most of all, rainfall events which contribute to the input of nutrients from terrestrial runoff.

These previous observations suggest that *G. catenatum* can co-exist with *P. bahamense* as they seem to be initiated and maintained by both high water temperature and high amounts of rainfall. In Tasmanian waters, *G. catenatum* appear to develop in a favourable water temperature window particularly between January and June (January to March is characterised by drought) and the first significant rainfall event aggravated the bloom formation (Hallegraeff *et al.*, 1995). These are the same conditions observed by Villanoy *et al.* (1996) and Bajarias and Relox (1996) where bloom events of *P. bahamense* were usually preceded by high water temperature and further triggered by rainfall events, particularly during the southwest monsoon.

## *Noctiluca Scintillans* Bloom Episodes (2001–2009)

*Noctiluca scintillans* has dominated *P. bahamense* since the last toxic bloom occurrence in 1998. Blooms of *N. scintillans* are a common phenomenon associated with eutrophic waters such as in Jakarta Bay (Adnan, 1989) and the upper Gulf of Thailand (Suvapepun, 1989). *Noctiluca scintillans* produces pinkish red discoloration in various temperate and subtropical waters but green discoloration in tropical waters on the other hand (Fukuyo *et al.*, 2011). There are two classifications of this species which are mainly due to either the presence of a tiny green flagellate endosymbiont (green *Noctiluca*) or its absence (red *Noctiluca*), which cause the difference in color during bloom formation. The species of *N. scintillans* in Manila Bay contained a green flagellate as an endosymbiont; hence, it is classified as green *Noctiluca*. Furuya *et al.* (2006a) used laboratory cultures of *N. scintillans* to show that the symbiosis gave an advantage to the organism in times of food limitation; hence, it dominated co-existing diatom and dinoflagellate species. They also found that although the symbiosis secured survival of the green *Noctiluca*, it did not stimulate proliferation even with sufficient light and nutrient salts. Another mechanism of the green *Noctiluca* is phagotrophy, which favors active growth and allows them to gain dominance over co-existing algal species. However, this particular mechanism is facultative according to the results of an experiment by Saito *et al.* (2006) where isolated cultures of non-feeding strains from the upper Gulf of Thailand grew photoautotropically for years, but they also fed on *Dunaliella tertiolecta*, which indicated that phagotrophy is an optional mechanism of green *Noctiluca* to survive. This observation revealed the role of phagotrophy exhibited by the green *Noctiluca* in controlling other phytoplankton population by acting as grazers (Fukuyo *et al.*, 2011). Hansen *et al.* (2004) showed that phagotrophy of green *Noctiluca* on the *P. bahamense* cells contributed to approximately 30% of its direct growth. This ability to graze on co-existing species could have allowed the green *Noctiluca* to dominate during the months when blooms of *Pyrodinium* were anticipated to occur but never happened and continued to persist in the bay since 2001.

*Noctiluca* species are less harmful compared to other HAB species; however, they can still cause massive fish kills due to oxygen depletion which can significantly impact fisheries (Huang and Qi, 1997; and Harrison *et al.*, 2011). This consequence is very well recognised and has been responsible for major caged-fish kills (Smayda, 1997; Navqi *et al.*, 1998). *Noctiluca scintillans* is considered harmful to fish because it is capable of massive slime production during the bloom decay process, which can clog fish gills (Subrahmanian, 1985). This species is known to produce a simple ichthyotoxin of high ammonia (NH3) concentration which is contained in their vacuoles (Elbratcher and Qi, 1998) and are detrimental to the fish population (Smayda, 1997). Most fish kill events happen in confined areas or embayments with less water circulation and are further caused by the following situations: (1) unconsumed food particles deposited in the substrate of a fish cage — this promotes aerobic decomposition, which leads to competition for available oxygen for consumption — and (2) accumulation of excreted materials from fishes in fish cages, especially when there is lack of water circulation, which further cause oxygen depletion. The massive fish kill in Bolinao Pangasinan (a coastal community at the northern tip of Luzon) in 2002 seemed to coincide with the increase proliferation of fish cages and pens in the area (Azanza *et al.*, 2006). Increased nutrient concentration from the years 1995 to 2001 was attributed to a more than 300% increase in fish pens and cages (San Diego-McGlone and Ranches, 2003) in Bolinao. These observations demonstate the major effects of the unregulated utilization of resources to the coastal environment.

Blooms of *N. scintillans* are formed by assemblages of bouyant cells on the water surface which are generally transported by the prevailing winds, tides and oceanic currents (Le Fevre and Grall, 1970; Schaumann *et al.*, 1988; Huang and Qi, 1997). There have been reports of massive fish kills in 2002 in different coastal communities in Luzon Island including the eastern side of Manila Bay where some fishermen observed water discoloration in the area between May to June; however the causative organism was never confirmed. It was in 2001 that a large scale bloom of *N. scintillans* was seen in the bay for almost the entire year (Furuya *et al.*, 2006a). The cells of *Noctiluca* could have been advected to the eastern side through the prevailing southeasterly wind from February to May (Furuya *et al.*, 2006a). Also, there was a separate report of fish kills in Bataan between the last week of June to the first week of July 2002 (BFAR-MBMU) but water samples collected and analyzed during that event showed a monospecific bloom of *Alexandrium sp.* instead. As there was a spike of the monthly average rainfall in July 2002 (from 37.4 mm in June to approximately 1468.0 mm in July), the heavy downpour could have caused the excess nutrient inputs to the bay from terrestrial runoff which resulted in oxygen depletion and subsequently, massive fish kills.

Bloom duration and intensity of the green Noctiluca have been observed to be increasing in Manila Bay and in other parts of Southeast Asia such as in the Gulf of Thailand and Jakarta Bay (Furuya *et al.*., 2006a). However, quantitative descriptions of the blooms of green *Noctiluca* in Manila Bay are not known (Hansen, *et al.*, 2004; Furuya, *et al.*, 2006b; Harrison *et al.*, 2011). Bloom dynamics of the green *Noctiluca* included phagotrophy (Sriwoon *et al.*, 2008) as well as photoautotrophy under food-limited water environments (Harrison *et al.*, 2011). These observations imply a progressive euthrophication of Southeast Asian waters which provide a competitive advantage to the green *Noctiluca* over co-existing phytoplankton species through increased food availability (Harrison *et al.*, 2011).

## *Alexandrium minutum* Bloom Episodes (2010–2011)

The bloom dynamics of *A. minutum* in Manila that previously caused shellfish poisoning in the area is not very well understood. Like *P. bahamense*, *A. minutum* also undergoes a dormancy period; but it is a mandatory stage so it does not proceed with germination even when environmental conditions are favourable (Anderson, 1998). According to Anderson (1980), the duration of the dormancy period among *Alexandrium* species is generally considered a time for physiological "maturation" which varies from species to species of the same genera. The duration of this interval is only known for *A. tamarense*, which undergoes dormancy for 4 to 6 months at 4°C. However, when temperatures increase, the period will be shortened to 1–3 months (Anderson, 1980). Hence, it is critical to understand the dormancy stage of the causative *Alexandrium* species to better forecast the timing of recurrent blooms (Anderson, 1998).

In the Philippines, incidents of *A. minutum* bloom have so far occurred only in the coastal waters of Bataan and Pangasinan (Bolinao) in 2010 and 2011. Several studies have confirmed that *A. minutum* usually blooms in shallow salt ponds and coastal bays (Anderson *et al.*, 1983; Su *et al.*, 1993; Ho and Hodgkins, 1993; Takeuchi *et al.*, 1995; Giacobbe *et al.*, 1996). The past two bloom episodes of *A. minutum* happened in Bataan waters where shallowing of the nearshore is widespread. Only a few patches of lowland vegetation is left to protect the coastal areas from soil erosion, and sediments carried via runoff are accumulating, possibly causing the shallowing of the coastal area. Siltation is also a huge problem in Bataan because the coastal area is being utilised for aquaculture and industrial developments. According to Lim *et al.* (2006), *A. minutum* exhibits a shade-adaptation strategy which allows it to adapt to a very low light intensity and still grow optimally. Bolinao coastal waters have experienced a bloom of *A. minutum* for the first time in 2003, but it did not bloom for several years until 2011, together with Bataan. The conditions of the water quality in both areas were comparable, with net fish cages being very common. Similar to Bataan, Bolinao also suffers high levels of siltation (Bach *et al.*, 1998; Terrados *et al.*, 1998). Furthermore, *A. minutum* is classified as euryhaline due to its ability to withstand a wide range of salinity. According to Lim *et al.* (2004; 2007), *A. minutum* blooms have been found in low salinity waters. The Philippines has been experiencing high rainfall, which has caused flooding incidents, especially in Metro Manila areas. With heavy rainfall, large amounts of freshwater input cause salinity fluctuations in Manila Bay's waters and possibly favour *A. minutum*. While nutrients play a great role in the survivability of a particular organism, Hutchinson (1944) pointed out that some events could not be explained by simply looking at the associated abiotic factors but also the possibility of the organisms to produce other active compounds known as allelochemicals. *Alexandrium* is known to exhibit allelopathy which may be more important to the success of this species than the PSP toxin, saxitoxin (Fistarol *et al.*, 2004). Allelopathy refers to any processes involving secondary metabolites released by plants, microorganisms, viruses and fungi that affect competing organisms.[10] Allelochemicals have been suggested to have great influence in species competition, succession and bloom formation or even maintenance of the species in an area (Pratt, 1966; Keating, 1977; Rice, 1984; Lewis, 1986; Wolfe, 2000; Rengefors and Legrand, 2001; Vardi *et al.*, 2002; Legrand *et al.*, 2003).

## 5. Environmental Parameters Associated with Bloom Episodes

Blooms of *P. bahamense* were mostly observed at the onset of the wet season, i.e. June to September, as previously discussed by Bajarias and Relox (1996). A few bloom events however, started a little earlier before

[10] www-ias.uca.es/bylaws.htm#CONSTI

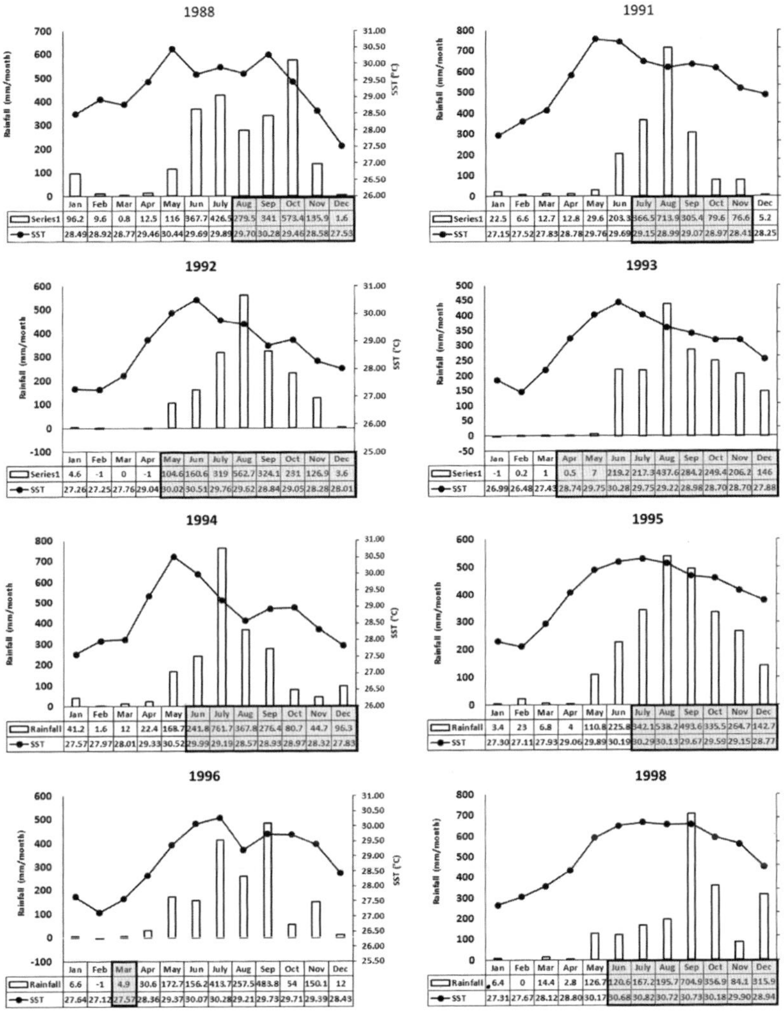


Fig. 1.5. Monthly average SST and Rainfall during the years of toxic blooms caused by *P. bahamense var. compressum.*

the onset of the wet season, such as the events in 1992, 1993 and 1996. The periods of blooms (highlighted in Figure 1.5 below) were generally preceded by relatively high values of sea surface temperature (between 28 and 31°C) and blooms continued to persist until the colder months, which are typically from October to January during the Northeast monsoon.

According to Villanoy *et al.* (1996) the formation and maintenance of *Pyrodinium* blooms were due to the increase in thermal stratification and vertical stability of the water column Vertical stratification caused by anomalous sea surface temperatures is said to be one of the major factors that influence HABs according to Moore *et al.* (2008). Increased stratification caused by warming reduces vertical mixing and favours the maintenance of phytoplankton assemblages in the upper water column (Doney, 2006). In addition, utilization of nutrient supply at the water surface in a thermally stratified ocean is rapid (Behrenfield *et al.*, 2006). This condition favours growth of certain phytoplankton species which allows them to dominate and bloom at the water surface. Species with lower nutrient requirements or those with the ability to drift to nutrient-rich regions when the availability of food is getting scarce will survive (Schofield and Falkowski, 2004). However, the bloom pattern which coincides with the Southwest monsoon described by Bajarias and Relox (1996) was not observed during the years of 1989, 1990 and, more importantly, in 1997, when an El Niño event occurred. Environmental parameters during the mentioned years are shown in Figure 1.6. It was previously thought that red tides caused by toxic dinoflagellates, including *P. bahamense*, could increase in intensity and duration during or after an El Niño phenomenon (MacLean, 1989). The El Niño event in 1997 to 1998 brought severe drought to the country. This observation suggests an important role of heavy rainfall in the initiation or maintenance of bloom by providing the limiting nutrient that enhances growth of phytoplankton species.

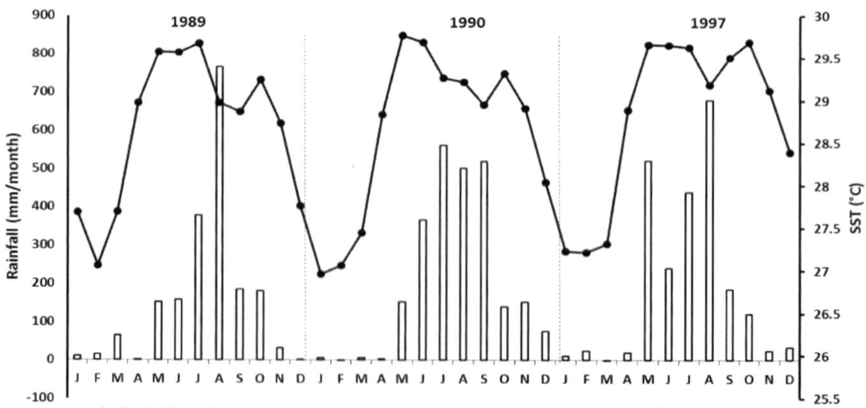


| | J | F | M | A | M | J | J | A | S | O | N | D | J | F | M | A | M | J | J | A | S | O | N | D | J | F | M | A | M | J | J | A | S | O | N | D |
|---|---|---|---|---|---|---|---|---|---|---|---|---|---|---|---|---|---|---|---|---|---|---|---|---|---|---|---|---|---|---|---|---|---|---|---|---|
| Rainfall | 11 | 16 | 64 | 0.7 | 153 | 159 | 378 | 765 | 185 | 181 | 31 | 1.2 | 5.6 | -1 | 5.6 | 3.2 | 152 | 367 | 561 | 501 | 520 | 141 | 152 | 77 | 10 | 24 | -1 | 20 | 521 | 240 | 439 | 680 | 185 | 121 | 24 | 33 |
| SST | 28 | 27 | 28 | 29 | 30 | 30 | 30 | 29 | 29 | 29 | 29 | 28 | 27 | 27 | 27 | 29 | 30 | 30 | 29 | 29 | 29 | 29 | 29 | 28 | 27 | 27 | 27 | 29 | 30 | 30 | 30 | 29 | 30 | 30 | 29 | 28 |

Fig. 1.6. Monthly average SST and Rainfall during the years when no PSP cases were reported (ie. 1989, 1990, and 1997).

It is worth noting that the two major oil spills that occurred in Manila Bay in 1999 (particularly in Metro Manila and Bataan) could potentially be one source of some major changes in the water quality which then affected the phytoplankton community assemblages. A study on the indirect effects of contaminants (including petroleum hydrocarbons) on aquatic environment done by Fleegera *et al.* (2003) revealed both top-down and bottom-up effects. In their experiment, they observed an increase of primary producer assemblages following a contaminant-induced impact on grazers. Many studies support the hypothesis that there is a very strong top-down effect caused by water contaminants. Predators or grazers seem to be more sensitive to contaminants than their prey as they suffer significant mortality, which allows the prey to increase in abundance. On the other hand, these contaminants influence the base of the food chain which may be affected by the contaminant-induced modifications of the rate of nutrient flux for primary producers. Petroleum hydrocarbons contain a significant amount of organic matter (Peterson *et al.*, 1996) which contribute to the enrichment of nutrient availability in a water system. These conditions could have favoured an optimum environment for another phytoplankton species (in this case, *N. scintillans)* to dominate *P. bahamense. Noctiluca scintillans* is a non-toxic dinoflagellate species which is often associated with excess food supply or a eutrophic condition. This species is also a heterotrophic (non-photosynthetic) species, which engulfs its prey (also known as phagotrophy), and is primarily composed of plankton including diatoms and other dinoflagellates.[11] Due to the imbalanced proportions of nutrient ratios, other organisms are able to use organic compounds to supplement their uptake of inorganic substances. Hence, heterotrophy is one of the major strategies used in a high proportion by algae in eutrophic environments (Burkholder *et al.*, 2008). Based on the field observations by Furuya *et al.* (2006) during the bloom event in 2004, proliferation of *Noctiluca* was observed during the months of February until April of that year. Average SST on these months ranged from 27°C to 29°C and rainfall values of 40.50 mm, 0 mm and −1.0 mm from February, March and April, respectively (highlighted in Figure 1.7). However, climate parameters during the earlier *Noctiluca* bloom in March 2002 showed a relatively low SST (27.49°C) and less amount of rainfall (7.8 mm).

Low rainfall during the bloom events of *N. scintillans* suggests that the species did not solely rely on their nutrient requirements from inputs

[11] http://en.wikipedia.org/wiki/Noctiluca_scintillans

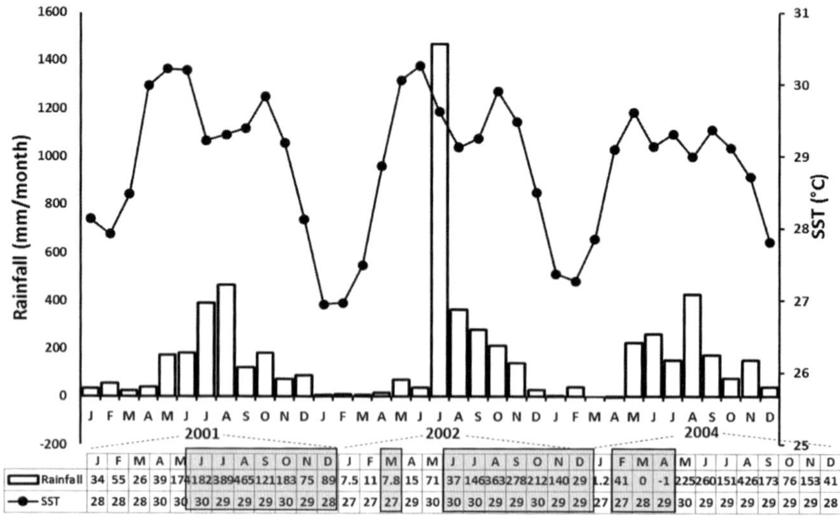


Fig. 1.7. Monthly rainfall and SST during the *N. scintillans* bloom occurrences.

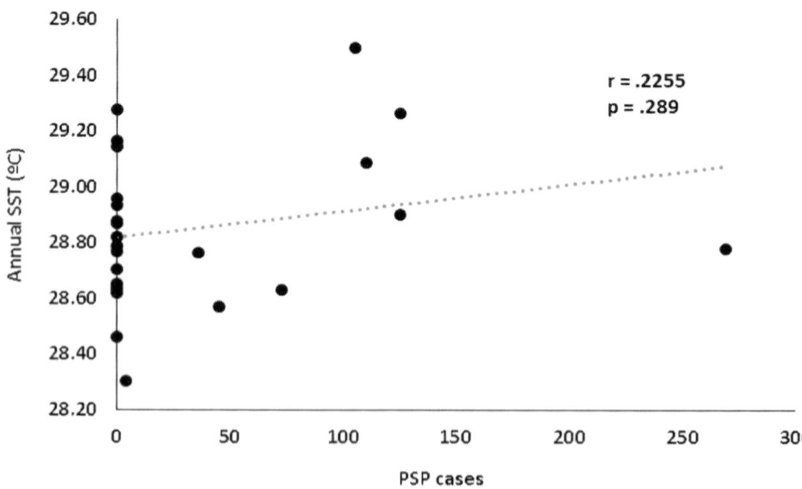


Fig. 1.8. Correlation between annual SST and PSP cases.

through runoff but on other strategies too; such as phagotrophy in order to survive in a competitive environment. Due to a very limited availability of data, the present study was not able to add to knowledge regarding the specific conditions during the bloom events of *N. scintillans* in Manila Bbay from 2001–2009. Lastly, the more recent HAB events in the bay were

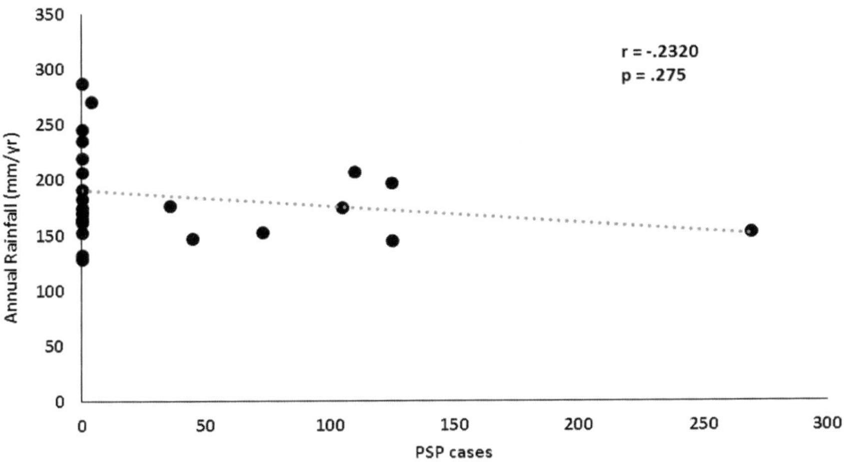


Fig. 1.9. Correlation between annual rainfall and PSP cases.

Table 1.2. Summary of correlation analyses between annual SST and rainfall to PSP cases.

| | | Correlation of PSP cases with annual SST and rainfall marked correlations are significant at $p < .05000$ | |
|---|---|---|---|
| Variable | | Annual SST | Annual rainfall |
| PSP cases | N | 24 | 24 |
| | r | .2255 | −.2320 |
| | p | .289 | .275 |

caused by *A. minutum*, which is another PSP-causing dinoflagellate species. *Alexandrium* blooms in 2010 and 2011 typically started towards the end of the year, having moderate amounts of rainfall and SST ranging from 28°C to 29.5°C. Blooms of *A. minutum* were observed occurring for only two years recently, hence there are only a few data to describe it.

A summary of the PSP cases in each year of occurrence with the corresponding values of the average annual SST and rainfall is shown in Table 1.1. Standard correlation of the number of PSP cases yearly with annual SST and rainfall (mm per year (yr)) was performed but established no significant relationship between the variables (Figures 1.8–1.9 and Table 1.2). However, these findings are not conclusive as the data for SST and rainfall were only correlated to the number of PSP cases and not HAB occurrences. There are some years when there were no PSP cases recorded but non-toxic HABs were observed in the bay. Regardless of whether it is toxic-related or not, regular monitoring of the density of algal cells (i.e. target plankton species) with corresponding physico-chemical parameters should be done to further elucidate the relationship between variables.

## 6. Conclusions and Recommendations

A more holistic approach, incorporating all the relevant components (i.e. biological, chemical and physical aspects) should be employed in order to achieve a clearer understanding of HAB trends in Manila Bay. The decreasing number of PSP cases over time does not necessarily suggest improved ecological health of the bay but, rather, could indicate a potential change in the physico-chemical properties of the water that favour one phytoplankton species over another. Although no significant correlation between rainfall and PSP cases was detected in this study, it is critical to investigate the quality of rainfall that the bay is receiving, considering the amount of atmospheric pollution from the different industries, ships and other automobiles which are emitted within the vicinity of the bay. Rainfall is known to react with atmospheric pollutants to form acidic compounds such as nitric and sulphuric acids that are detrimental to aquatic organisms. Ultimately, these compounds could potentially intensify the nutrient concentrations of Manila Bay which then favour algal bloom formations. Hence, constant monitoring of the physico-chemical parameters are essential for explaining bloom formation. In addition, biological aspects should also be given additional attention so that the overall ecological health of the bay can be assessed. Explaining the shift in the causative organisms can be very challenging, but, it will be possible with long term and a more comprehensive environmental monitoring data. The monitoring protocol should not only focus just on toxin analysis of shellfish products, but also integrate a thorough water quality analysis of the identified monitoring sites. More importantly, good governance that promotes transparency, accountability and participation among various stakeholders is an essential step in solving the complex environmental issues related to harmful algal blooms.

## References

Anderson, D.M. (1980). "Effects of temperature conditioning on development and germination of *Gonyaulax tamarensis* (Dinophyceae) hypozygotes", *Journal of Phycology*, **16**, 166–172.

Anderson, D.M. (1998). "Physiology and Bloom Dynamics of Toxic *Alexandrium* species, with Emphasis on Life Cycle Transitions" in: Anderson, D. M., Cembella, A. D. & Hallegraeff, G. M. (Eds.), *Physiological Ecology of Harmful Algal Blooms*, Springer-Verlag Berlin Heidelberg, Germany, pp. 29–48.

Anderson, D.M. (2007). "The ecology and oceanography of harmful algal blooms: multidisciplinary approaches to research and management". UNESCO, Paris. *IOC Technical Series 74*, IOC/2007/TS/74.

Anderson, D.M., Chisholm, S.W. & Watras, C.J. (1983). "The importance of life cycle events in the population dynamics of *Gonyaulax tamarensis*", *Marine Biology*, **76**, 179–190.

Anderson, D.M., Kullis, D.M., Sullivan, J.J., Hall, S. & Lee, C. (1990). "Dynamics and physiology of saxitoxin production by the dinoflagellates Alexandrium spp", *Marine Biology*, **1004**, 511–524.

Ashby, M. (2008). "Megacities, migration and Manila", *Geography Review*, 2–5.

Azanza, R.V. (1997). "Contributions to the Understanding of the Bloom Dynamics of Pyrodinum bahamense var. compressum: A Toxic Red Tide Causative Organism", *Science Diliman*, **9**(1–2), 1–6.

Azanza, R., Baula, I. & Fukuyo, Y. (2006). "Seasonal changes in phytoplankton composition in an extensive fish culture area in Bolinao, Pangasinan, northern Philippines", *Coastal Marine Science*, **30** (1), 85–87.

Azanza, R.V. & Taylor, F.J.R.M. (2001). "Are *Pyrodinium* Blooms in the Southeast Asian Region Recurring and Spreading? A View at the End of the Millennium", *Ambio*, **30**(6), 356–364.

Bach, S.S., Borum, J., Fortes, M.D. & Duarte, C.M. (1998). "Species composition and plant performance of mixed seagrass beds along a siltation gradient at Cape Bolinao, the Philippines", *Marine Ecology Progress Series*, **174**, 247–256.

Bajarias, F.F.A., Juan R. & Relox, J. (1996). "Hydrological and Climatological Parameters Associated With The *Pyrodinidum Blooms* In Manila Bay, Philippines" in: Yasumoto, T., Oshima, Y. & Fukuyo, Y. (Eds.), *Seventh International Conference on Toxic Phytoplankton.* United Nations Educational, Scientific and Cultural Organization, Sendai, Japan, p. 586.

Bautista, R.M. & De Rosa, D.A. (1996). "Agriculture And The New Industrial Revolution in Asia". Paper prepared for the *Seventh Biennial Conference on U. S. — Asia Economic Relations*, May 14–17, 1996.

Behrenfield, M.J., O'Malley, R.T. & Siegel, D.A., McClain, C.R., Sarmiento, J.L., Fieldman, G.C., Milligan, A.J., Falkowski, P.G., Letelier, R.M. & Boss, E.S. (2006). "Climate-driven trends in contemporary ocean productivity. *Nature"*, **444**, 752–755.

Boyer, G.L., Sullivan, J.J., Anderson, R.J., Harrison, P.J. & Taylor, F.J.R. (1987). "Effects of nutrient limitation on toxic production and composition in the marin dinoflagellate Protogonyaulax tamarensis", *Marine Biology*, **96**, 123–128.

Burkholder, J.M., Glibert, P.M. & Skelton, H.M. (2008). "Mixotrophy, a major mode of nutrition of harmful algal species in eutrophic waters", *Harmful Algae*, **8**, 77–93.

Chang, K.-H., Amano, A., Miller, T.W., Isobe, T., Maneja, R., Siringan, F., Imai, H. & Nakano, S.-I. (2009). "Pollution Study in Manila Bay: Eutrophication and Its Impact on Plankton Community" in: Obayashi, T., Isobe, T., Subramanian, A., Suzuki, S., Tanabe, S. (Eds.), *Interdisciplinary Studies on Environmental Chemistry- Environmental Research in Asia*, TERRAPUB, pp. 261–267.

Chua, T.E., Paw, J.N. & Guarin, F.Y. (1989). "The Environmental Impact of Aquaculture and the Effects of Pollution on Coastal Aquaculture Development in Southeast Asia", *Marine Pollution Bulletin*, **20**(7), 335–343.

Corrales, R.A. & MacLean, J.L. (1995). "Impacts of harmful algae on seafarming in the Asia-Pacific areas", *Journal of Applied Phycology*, **7**, 151–162.

De las Alas, J.G. & Sodusta, J.A. (1985). "A model for the wind driven circulation in Manila Bay", *Natural and Applied Science Bulletin*, **37**, 159–170.

Doblin, M.A., Blackburnb, S.I. & Hallegraeffa, G.M. (1998). "Growth and biomass stimulation of the toxic dinoflagellate *Gymnodinium catenatum* (Graham) by dissolved organic substances", *Journal of Experimental Marine Biology and Ecology*, **236**, 33–47.

Doney, S.C. (2006). "Plankton in a warmer world", *Nature*, **444**, 695–696.

Elbratcher, M. & Qi, Y.Z. (1998). "Aspects of Noctiluca (Dinophyceae) population dynamics" in: Anderson, D.M., Cembella, A.D., Hallegraeff, M.G. (Eds). *Physiological ecology of harmful algal blooms*, Springer-Verlag, Berlin, pp. 315–335.

Fistarol, G.O., Legrand, C., Selander, E., Hummert, C., Stolte, W. & Granéli, E. (2004). "Allelopathy in Alexandrium spp.: effect on a natural plankton community and on algal monocultures", *Aquatic Microbial Ecology*, **35**, 45–56.

Fleegera, J.W., Carmana, K.R. & Nisbetb, R.M. (2003). "Indirect effects of contaminants in aquatic ecosystems", *The Science of the Total Environment*, **317**, 207–233.

Fukuyo, Y., Kodama, M., Omura, T., Furuya, K., Furio, E.F., Cayme, M., Teen, L.P., Ha, D.V., Kotaki, Y., Matsuoka, K., Iwataki, M., Sriwoon, R. & Lirdwitayaprasit, T. (2011). "Ecology and oceanography of harmful marine microalgae (Project-2)" in: Nishida, S., Fortes, M.D. & Miyazaki, N. (Eds.), *Coastal Marine Science in Southeast Asia — Synthesis Report of the Core University Program of the Japan Society for the Promotion of Science (2001–2010)*, TERRAPUB, pp. 23–48.

Furio, E.F. & Gonzales, C.L. (2004). "Toxic Red Tide and Paralytic Shellfish Poisoning Profiles in the Philippines", *Practical Guide on Paralytic Shellfish Poisoning Monitoring in the Philippines*, 3–18.

Furuya, K., Saito, H., Sriwoon, R., Omura, T., Furio, E.F., Borja, V.M. & Lirdwitayaprasit, T. (2006a). "Vegetative growth of Noctiluca scintillans with green flagellate endosymbiont *Pedinomonas noctilucae*", *African Journal of Marine Science*, **28**(2), 305–308.

Furuya, K., Saito, H., Sriwoon, R., Vijayan, A.K., Omura, T., Elsa E. Furio, Borja, V.M., Boonyapiwat, S. & Lirdwitayaprasit, T. (2006b). "Persistent whole-bay red tide of Noctiluca scintillans in Manila Bay, Philippines", *Coastal Marine Science*, **30**(1), 74–79.

Gedaria, A.I., Luckas, B., Reinhardt, K. & Azanza, R.V. (2007). "Growth response and toxin concentration of cultured *Pyrodinium bahamense var. compressum* to varying salinity and temperature conditions", *Toxicon*, **50**, 518–529.

Giacobbe, M.G., Oliva F.D. & Maimone, G. (1996). "Environmental factors and seasonal occurrence of the dinoflagellate *Alexandrium minutum*, a PSP potential producer in a Mediterranean Lagoon", *Estuarine, Coastal and Shelf Science*, **42**(5), 539–549.

Guisande, C., Frangópulos, M., Maneiro, I., Vergara, A.R. & Riveiro, I. (2002). "Ecological advantages of toxin production by the dinoflagellate Alexandrium minutum under phosphorus limitation", *Marine Ecology Progress Series*, **225**, 169–176.

Gunnerson, C.G. & Cuellar, M. (1988). "Case studies. Manila Bay In Wastewater Management for Coastal Cities, The Ocean Disposal Option" in Gunnerson, C. G. Gunnerson (Ed.), *World Bank Technical Paper No. 77.* World Bank, Washington, D. C., pp. 307–317.

Hallegraeff, G.M. (1993). "A review of harmful algal blooms and their apparent global increase", *Phycologia*, **32**, 9–79.

Hallegraeff, G.M., Mccausland, M. & Brown, R. (1995). Prediction of toxic dinoflagellate blooms of *Gymnodinium catenatum* in Southern Tasmanian waters. *Journal of Plankton Research* **17**, 1163–1176.

Hansen, P.J., Miranda, L. & Azanza, R. (2004). "Green *Noctiluca scintillans*: a dinoflagellate with its own greenhouse", *Marine Ecology Progress Series*, **275**, 79–87.

Harrison, P.J. , Furuya, K., Glibert, P.M., Xu, J., Liu, H.B., Yin, K., Lee, J.H.W., Anderson, D.M., Gown, R., Al-azri, A.R. & Ho, A.Y.T. (2011). "Geographical distribution of red and green *Noctiluca scintillans*", *Chinese Journal of Oceanology and Limnology*, **29**(4), 807–231.

Ho, K.C. & Hodgkiss, I.J. (1993). "Characteristics of red tides caused by Alexandrium catenella (Whedon & Kofoid) Balech in Hong Kong" in: *Toxic Phytoplankton Blooms in the Sea.* Smayda, T. & Shimizu, Y. (Eds). Elsevier, Amsterdam, pp. 263–268.

Huang, C. & Qi, Y. (1997). "The abundance cycle and influence factors on red tide phenomena of *Noctiluca scintillans* (Dinophyceae) in Dapeng Bay, the South China Sea", *Journal of Plankton Research*, **19**(3), 303–318.

Hutchinson, G.E. (1944). "Limnological studies in Connecticut. VI. A critical examination of the supposed relationship between phytoplankton periodicity and chemical changes in lake water", *Ecology*, **25**, 3–26.

Jacinto, G.S., Azanza, R.V., Velasquez, I.B. & Siringan, F.P. (2006). "Manila Bay: Environmental Challenges and Opportunities" in: Wolanski, E. (Ed.), *The Environment in Asia Pacific Harbours*, Springer, The Netherlands, pp. 309–328.

Jacinto, G.S., San Diego-McGlone, M.L., Velasquez, I.B. & Smith, V. (1998). "N and P Budget in Manila Bay, Philippines" in: Watson, I., Vigers, G.,

Ong, K.S., McPherson, C., Millson, N. & Gass, D. (Eds.), *ASEAN Marine Environmental: Towards Sustainable Development and Integrated Management in ASEAN, Proceedings of the Fourth ASEAN-Canada Technical Conference on Marine Science*, EVS Environment Consultants, North Vancouver Department of Fisheries, Malaysia, Langkawi Malaysia, pp. 341–345.

Keating, K.I. (1977). "Alleolepathic influence on blue-green bloom sequence in a eutrophic lake", *Science*, **196**, 885–886.

Landsberg, J.H. (2002). "The effects of harmful algal blooms on aquatic organisms", *Review in Fisheries Science*, **10**, 113–390.

Le Fevre, J. & Grall, J.R. (1970). "On the relationships of *Noctiluca* swarming off the western coast of Brittany with hydrological features and plankton characteristics of the environment", *Journal of Experimental Marine Biology and Ecology*, **4**, 287–306.

Legrand, C., Rengefors, K., Fistarol, G.O. & Granéli, E. (2003). "Allelopathy in phytoplankton — biochemical, ecological and evolutionary aspects", *Phycologia*, **42**, 406–419.

Lewis, W.M. Jr (1986). "Evolutionary interpretation of allelochemical interactions in phytoplankton algae", *American Naturalists*, **127**, 184–194.

Lim, P.T., Leaw, C.P. & Ogata, T. (2007). "Morphological variation of two *Alexandrium* species responsible for paralytic shellfish poisoning in Southeast Asia", *Botanica Marina*, **50**, 14–21.

Lim, P.T., Leaw, C.P., Usup, G., Kobiyama, A., Koike, K. & Ogata, T. (2006). "Effects of light and temperature on growth, nitrate uptake, and toxin production of two tropical dinoflagellates: *Alexandrium tamiyavanichii* and *Alexandrium minutum* (Dinophyceae)", *Journal of Phycology*, **42**, 786–799.

Lim, P.T., Leaw, C.P. & Usup, G. (2004). "First incidence of paralytic shellfish poisoning on the east coast of Peninsular Malaysia" in: *Marine Science into the New Millennium: New Perspectives and Challenges.* Phang, S.M., Chong V.C., Ho, S.S. Mokhtar, N. & Ooi, J.L.S. (Eds.)., University of Malaya Maritime Research Center, Kuala Lumpur, Malaysia, pp. 661–667.

Maclean, J.L. (1989). "Indo-Pacific red tides, 1985–1988", *Marine Pollution Bulletin*, **20**, 304–310.

Moore, S.K., Trainer, V.L., Mantua, N.J., Parker, M.S., Laws, E.A., Backer, L.C. & Fleming, L.E. (2008). "Impacts of Climate Variability and Future Climate Change On Harmful Algal Blooms and Human Health. Environmental Health 7 (Suppl 2): S4", *Proceedings of the Centers for Ocean and Health Investigators Meeting*, Woods Hole, Massachussets, United States of America, 24–27 April 2007, pp. 9–12.

Muñoz, J.C. (1991). "Manila Bay: Status of its fisheries and management", *Marine Pollution Bulletin*, **23**, 311–314.

Pasig River Rehabilitation Program (PRRP) (1998–1999). "Annual Reports", *Pasig River Rehabilitation Program.* Department of Environment and Natural Resources, Quezon City, Philippines and Danish International Development Assistance (Danida), Denmark.

Peterson, C.H., Kennicutt, M.C., Green, R.H., Montagna, P., Harper, D.E., Powell, E.N. & Roscigno, P.F. (1996). "Ecological consequences of environmental perturbations associated with offshore hydrocarbon production: a perspective on long-term exposures in the Gulf of Mexico", *Canadian Journal of Fisheries and Aquatic Sciences*, **53**, 2637–2654.

Popkiss, M.E.E., Horstman, D.A. & Harpur, D. (1979). "Paralytic Shellfish Poisoning. A Report of 17 Cases in Cape town", *SA Medical Journal*, 1017–1023.

Pratt, D.M. (1966). "Competition between *Skeletonema costatum* and *Olithodscus luteus* in Narragansett Bay and in culture", *Limnology and Oceanography*, **11**, 447–455.

Prudente, M.S., Ichihashi, H. & Tatsukawa, R. (1994). "Heavey Metal Concentrations in Sediments from Manila Bay, Philippines and Inflowing Rivers", *Environmental Pollution*, **86**, 83–88.

Rice, E.L. (1984). *Allelopathy*, 2nd edn. Academic Press, Orlando, pp. 189–205.

Rengefors, K. & Legrand, C. (2001). "Toxicity in Peridinium aciculiferum — an adaptative strategy to outcompete other winter phytoplankton?", *Limnology and Oceanography*, **46**, 1990–1997.

Saito, H., Furuya, K. & Lirdwitayaprasit, T. (2006). "Photoautotrophic growth of *Noctiluca scintillans* with the endosymbiont *Pedinomonas noctilucae*", *Plankton Benthos Research*, **1**, 97–101.

San Diego-McGlone, M.L.S. & Ranches, G. (2003). "Water quality assessment of Lingayen Gulf", *Philippine Scientist*, **40**, 41–56.

Schaumann, K. Gerdes, D. & Hesse K.J. (1988). "Hydrographic and biological characteristics of a Noctiluca scintillans red tide in the German Bight, 1984", *Meeresforschung*, **32**, 77–91.

Schofield, T.S. & Falkowski, P. (2004). "Historical climate change and ocean turbulence as selective agents of phytoplankton functional groups", *Marine Ecology Progress Series*, **274**, 123–132.

Smayda, T.J. (1989). "Primary production and the global epidemic of phytoplankton bloms in the sea: a linkage?" in: Cosper, E.M., Bricelj, V.M., Carpenter, E.J. (Eds.), *Novel phytoplankton blooms.* Springer, New York, pp. 449–483.

Smayda, T.J. (1997). "What is a bloom? A commentary", *Limnology and Oceanography*, **42**(5), 1132–1136.

Sriwoon, R., Pholpunthin, P. & Lirdwitayaprasit, T. (2008). "Population dynamics of green Noctiluca scintillans (Dinophyceae) associated with the monsoon cycle in the Upper Gulf of Thailand", *Journal of Phycology*, **44**, 605–615.

Subrahmanian, A. (1985). "Noxious dinoflagellates in Indian waters" in: *Toxic dinoflagellates.* Anderson, D.M., White, A.W. and Baden, D.G. (Eds.). Elseveir, New York, pp. 525–528.

Su, H.M., Liao, I.C. & Chiang, Y.M. (1993). "Mass mortality of prawn caused by Alexandrium tamarense in a culture pond in Southern Taiwan" in: *Toxic Phytoplankton Blooms in the Sea.* Smayda, T. & Shimizu, Y. (Eds.). Elsevier, Amsterdam. pp. 329–333.

Su, G.S., Martillano, K.J., Alcantara, T.P., Ragragio, E., Jesus, J.D., Hallare, A. & Ramos, G. (2009). "Assessing Heavy Metals In The Waters, Fish And

Macroinvertebrates In Manila Bay, Philippines", *Journal of Applied Sciences in Environmental Sanitation*, **4** (3), 187–195.

Sunda, W.G. (1989). "Trace metal interactions with marine phytoplankton", *Biological Oceanography*, **6** (5–6), 411–442.

Suvapepun, S. (1989). "Occurrences of red tide in the Gulf of Thailand" in: *Red Tides. Biology, Environmental Science and Toxicology.* Okaichi, T. Anderson, D.M. & Nemoto, T. (Eds.), *Proceedings of the First International Symposium Red Tides*, November 10–14, Takamatsu, Japan. Elsevier, New York, pp. 41–44.

Takeuchi, T., Kokubo, T., Fukuyo, Y. & Matsuoka, K. (1995). "Quantitative relationship among vegetative cells, planozygotes, and hypnozygotes of *Alexandrium catenella* (Dinophyceae) in its blooming season at Tanabe Bay, Central Japan. Abstract", *7th International Conference on Toxic Phytoplankton.* Sendai, Japan.

Terrados, J., Duarte, C.M., Fortes, M.D., Borum, J., Agawin, N.S.R., Bach, S., Thampanya, U., Kamp-Nielsen, L., Kenworthy, W.J., Geertz-Hansen, O. & Vermaat, J. (1998). "Changes in community structure and biomass of seagrass communities along gradients of siltation of SE Asia", *Estuarine, Coastal and Shelf Science*, **46**, 757–768.

Usup, G., Kulis, D.M. & Anderson, D.M. (1994), "Growth and Toxin Production of the Toxic Dinoflagellate *Pyrodinium bahamense var. compressum* in Laboratory Cultures", *Natural Toxins*, **2**, 254–262.

Vardi, A., Schatz, D., Beeri, K., Motro, U., Sukenik, A., Levine, A. & Kaplan, A. (2002). "Dinoflagellate-cyanobacteria communication may determine the composition of phytoplankton assemblage in a mesotrophic lake", *Current Biology*, **12**, 1767–1772.

Velasquez, I.B., Jacinto, G.S. & Valera, F.S. (2002). "The speciation of dissolved copper, cadmium and zinc in Manila Bay, Philippines", *Marine Pollution Bulletin*, **45**, 210–217.

Velasquez, I.B. & Jacinto, G.S. (1995). "ASEAN criteria and monitoring: advances in marine environmental management and human health protection", *Proceedings in ASEAN-Canada Midterm Review.* Conference in Marine Science, Singapore, pp. 278–284.

Villanoy, C.L., Corrales, R.A., Jacinto, G.S., Cuaresma, N.C.T. Jr. & Crisostomo, R. (1996). "Towards the development of a cyst-based model for *Pyrodinium* red tides in Manila Bay" in: *Harmful and Toxic Algal Blooms*, Yasumoto, T., Oshima, Y., Fukuyo, Y. (Eds.), IOC of UNESCO, Paris, pp. 189–192.

Wolfe, G.V. (2000). "The chemical defence ecology of marine unicellular plankton: constraints, mechanisms, and impacts", *The Biological Bulletin*, **198**, 225–244.

# 2

# CITES LEGISLATIVE IMPLEMENTATION: LESSONS FROM AND FOR ASEAN MEMBER STATES

CHEONG WEI-MIN, DENISE and LYE LIN-HENG

The Convention on International Trade in Endangered Species (CITES) places a clear obligation on Parties to enact appropriate national legislation, without which its enforcement would be impossible. In the context of the ten ASEAN Member States, the CITES National Legislation Project, established in 1992 to review and evaluate national legislation based on the CITES legislative criteria, has categorised the national legislation of Brunei, Cambodia, Indonesia, Malaysia, Singapore, Thailand and Vietnam as Category 1 legislation i.e. legislation that is believed to generally meet the requirements of the CITES legislative criteria. Notwithstanding this classification, there are concerns that ASEAN Member States may not be fully complying with the CITES legislative criteria due to the lack of clarity as to its precise scope, resulting in incoherence in the implementation of the applicable legislative requirements. In view of the above, this paper analyses the scope of the CITES legislative criteria and argues that it comprises legislative requirements that stem from both hard and soft law. The paper seeks to draw lessons for ASEAN based on a comparative analysis of the laws of Malaysia and Singapore as well as "model" provisions found in the national legislation of benchmark Parties such as New Zealand, Australia, Canada and Hong Kong.

## 1. Introduction

### *The Illegal Wildlife Trade is a Serious Problem in Southeast Asia*

Southeast Asia is one of the world's biodiversity hotspots as well as one of its hotspots for the illegal trade in wildlife. Felbab-Brown estimated that the illegal trade in wildlife in Southeast Asia is worth US $8 to $10 billion per year.[1] According to Association of South East Asian Nations Wildlife Enforcement Network (ASEAN-WEN), poaching, transit and consumption occurs in all countries across Southeast Asia to varying degrees.[2] Trade

[1] Vanda Felbab-Brown, *The Disappearing Act: The Illicit Wildlife Trade in Asia* (Washington, DC: Brookings Institution, 2011).

[2] ASEAN Wildlife Enforcement Network, *Illegal Wildlife Trade in Southeast Asia* (Bangkok, ASEAN-WEN, 2009).

Records Analysis of Flora and Fauna in Commerce (TRAFFIC) cautioned that the unsustainable and often illegal exploitation of wild plants and animals is having devastating effects on the region's biodiversity, particularly on wildlife species with high commercial value, many of which are now rare, endangered or locally extinct.[3]

### *Role of CITES in Regulating the International Wildlife Trade*

CITES is an international agreement that aims to ensure that the international trade in specimens of wild animals and plants does not threaten their survival. All States in ASEAN are party to CITES. All import, export, re-export and introduction from the sea of species covered by the Convention have to be authorised through a permit system. Parties must designate at least one Management Authority ("MA") to administer the permit system and at least one Scientific Authority ("SA") to advise on the effects of trade on the status of the species. The level of protection given to a species depends on which of the three CITES Appendices it is listed in, with the highest level of protection given to species listed in Appendix I followed by those listed in Appendix II. Appendix III lists species that have been individually listed by Parties that need the cooperation of other Parties in the control of trade.[4]

### *The Need for CITES-Implementing Legislation*

CITES-implementing legislation is essential as CITES is a non-self-executing treaty. CITES places a clear and pressing obligation on Parties to enact appropriate CITES-implementing legislation, without which enforcement of CITES would be impossible. CITES requires that Parties take measures to prohibit trade in specimens of species included in CITES Appendices I, II and III that does not satisfy the conditions laid down by the Convention;[5] to penalise trade in, and possession of, such specimens;

---

[3]TRAFFIC Southeast Asia, *Wildlife Trade in South-East Asia* (Malaysia: TRAFFIC Southeast Asia, 2008).
[4]CITES Secretariat, *What is CITES?*, online: CITES <http://www.cites.org>.
[5]These conditions would encompass those laid down in the main trade-control provisions of CITES i.e. Articles III, IV and V.

and provide for the confiscation or return to the State of export of such specimens.[6]

## *CITES Legislative Criteria and the National Legislation Project*

Given the importance of CITES-implementing legislation, the National Legislation Project ("NLP") was established in 1992 and introduced an evaluation system where legislation would be categorised into Category 1, 2 or 3 based on the CITES legislative criteria.[7] The CITES legislative criteria requires that CITES-implementing legislation must authorise Parties to:

(a) designate one MA and one SA;
(b) prohibit trade in specimens in violation of the Convention;
(c) penalise trade in specimens in violation of the Convention; and
(d) confiscate specimens illegally traded or possessed.

Category 1 legislation is legislation believed to generally meet all the requirements for the implementation of CITES; Category 2 legislation is legislation believed not to meet all but only some of the requirements; and Category 3 legislation is legislation believed not to meet any of the requirements for the implementation of CITES.[8] The CITES compliance regime with its trade sanctions-based approach would then be relied on to induce recalcitrant Parties with Category 2 or Category 3 legislation to enact new laws if necessary.[9]

## *Status of CITES Legislative Implementation in ASEAN Member States*

Seven ASEAN Member States have CITES-implementing legislation that is believed to generally meet the requirements of the CITES legislative criteria ("Category 1 legislation"). The remaining three Member States have

[6]Cyrille de Klemm, *Guidelines for Legislation to Implement CITES* (Cambridge, UK: IUCN, 1993) at 7.
[7]*National Laws for Implementation of the Convention*, Resolution Conf. 8.4 (Rev. CoP15), CITES, eighth meeting of the CoP, (1992) at 1.
[8]*National Laws for the Implementation of the Convention*, CITES, Notification to the Parties, No. 2012/036 (2012) at 1.
[9]*National Laws for Implementation of Convention*, CITES, 16th meeting of the CoP, CoP16 Doc.28 (2013) at 4.

CITES-implementing legislation which is believed generally not to meet some or all of the requirements ("Category 2 legislation" and "Category 3 legislation", respectively).[10]

Notwithstanding that most of the ASEAN Member States have Category 1 legislation, there are concerns that: (i) Member States may not be fully complying with the CITES legislative criteria due to the lack of clarity as to its precise scope; and (ii) there may be a growing degree of incoherence in the way Member States have implemented the legislative requirements of CITES.

This situation arises as apart from the four criteria that constitute the CITES legislative criteria, there is a growing body of soft laws that elaborate on these criteria which Member States may not be complying with.[11] The Secretariat has validated this observation by acknowledging that these hard and soft law requirements have yet to be consolidated but that such a consolidation may be beneficial to Parties handling the legislative requirements more coherently.[12] The usual criterion for distinguishing between hard law and soft law is whether a particular rule or principle is legally binding i.e., results in legally enforceable commitments for States (in the context of international law), in which case it would be hard law. However, in reality, the distinction between the two may not be as clear-cut and definitive.

### *CITES and Soft Law*

This paper takes the view that soft law represents an agreed understanding of the Convention's terms and how they should be implemented; where appropriate requirements set out in soft law should be reflected in national legislation. Soft law is also an important means for CITES to adapt itself to changes in circumstances, many of which could not have been foreseen by the original drafters of the Convention; this happens when recommendations of the Conference of the Parties (CoP) (such as those in the form of Resolutions) which would otherwise not be enforceable, are incorporated into legislation.[13]

---

[10] *National Laws for Implementation of Convention*, CITES, 16th meeting of the CoP, CoP16 Doc.28 (2013), Annex 2 (Rev. 1) at 1, 3, 5.

[11] Rosalind Reeve, "Wildlife Trade, Sanctions and Compliance: Lessons from the CITES Regime" (2006) 82 International Affairs 881 at 895.

[12] *National Laws for Implementation of Convention*, CITES, 16th meeting of the CoP, CoP16 Doc.28 (2013) at 6.

[13] Cyrille de Klemm, *Guidelines for Legislation to Implement CITES* (Cambridge, UK: IUCN, 1993) at 7.

## 2. Analysis of CITES Legislative Requirements for Implementing Legislation

In the absence of a consolidation by the CITES Secretariat of the hard and soft law requirements, this paper attempts to analyse, identify and consolidate the hard and soft law requirements that comprise the CITES legislative criteria, through a review of all valid Resolutions and Decisions of a general application. Through this review, it became clear that many of the Resolutions imposed further legislative requirements which were relevant to the CITES legislative criteria and that each of the four main criteria of the CITES legislative criteria comprises components and sub-components; collectively they form a complex set of criteria against which to evaluate CITES-implementing legislation. This section will deal with each criterion of the CITES legislative criteria in turn.

### *Designate at least one Management Authority and one Scientific Authority*

Article IX underlies this first criterion and the NLP analysis will seek to determine if the Management Authority (MA) and Scientific Authority (SA) have been designated by legislation.

Legislation should stipulate that the SA be "independent" from the MA in accordance with soft law. Unfortunately, soft law does not define this term but seems to introduce the idea of each authority not being subject to the other's authority or jurisdiction so that SAs enjoy at a minimum, a degree of independence that allows them to carry out their duties under CITES effectively.[14]

The NLP analysis will look to see if legislation separates the functions of the MA and SA and provides them the necessary powers to carry out their responsibilities. The functions and powers of the SA and MA, both pursuant to hard and soft law, have not been consolidated. Given the sheer volume of soft laws that have evolved over the years, it would be difficult and infeasible for Parties to implement this requirement in practice.

The NLP analysis will also seek to determine if legislation provides for mechanisms to facilitate coordination and communication between the MA and SA as well as with other government agencies with relevant

[14] *Designation and Role of the Scientific Authorities*, Resolution Conf. 10.3, CITES, 10th meeting of the CoP, (1997) at 1.

competence.[15] There appears to be some soft-law support for the legislative inclusion of such mechanisms.[16]

### *Prohibition of Trade in Specimens in Violation of the Convention*

The second criterion requires legislation to prohibit trade in violation of the Convention. There are several aspects to this: (i) the scope of application; (ii) trade controls; (iii) e-commerce involving wildlife; (iv) permits; (v) marking; and (vi) exemptions and special provisions.

### *Scope of Application*

The NLP analysis will seek to establish if legislation prohibits the illegal trade in specimens of all species included in the CITES Appendices (as amended from time to time).[17] Specimens as defined in the Convention would include animals and plants, live and dead, as well as readily "recognisable parts and derivatives".[18]

#### *Trade controls*

It is critical that legislation prohibits trade in violation of CITES Articles III, IV and V; these being the main provisions to control trade in specimens of CITES-listed species. The NLP analysis will seek to determine if legislation regulates the range of transactions contemplated by the Convention. This includes exports, imports, re-exports and introductions from the sea.[19] Although this is not addressed by the NLP, possession of, and domestic trade in, CITES-listed species should also be prohibited for an effective implementation of the Convention.[20]

---

[15] *National Laws for Implementation of the Convention*, CITES, 12th meeting of the CoP, CoP12 Doc.28 (2002) at 2.

[16] *Compliance and Enforcement*, Resolution Conf. 11.3 (Rev. CoP16), CITES, 11th meeting of the CoP, (2000), Annex 3 at 1.

[17] *Convention on International Trade in Endangered Species of Wild Fauna and Flora*, 3 March 1973, 993 UNTS 243, Articles XV and XVI.

[18] *Trade in Readily Recognizable Parts and Derivatives*, Resolution Conf. 9.6 (Rev. CoP16), CITES, ninth meeting of the CoP, (1994) at 1.

[19] *National Laws for Implementation of the Convention*, CITES, 12th meeting of the CoP, CoP12 Doc.28 (2002) at 2.

[20] Cyrille de Klemm, *Guidelines for Legislation to Implement CITES* (Cambridge, UK: IUCN, 1993) at 57.

The NLP analysis will also seek to determine if legislation reflects the CITES conditions that need to be fulfilled, before permits are granted for the different types of transactions; or at a minimum, if it contains an express provision that requires permits to be issued in accordance with the provisions of the Convention.[21]

There is soft law that amplifies on the implementation of these conditions, for example, by setting out guiding principles and scientific concepts for a SA to take into account when considering whether trade in a particular case, would be detrimental to the survival of a species.[22] Such principles and concepts should be incorporated into CITES-implementing legislation as their purpose is to assist with the uniform implementation of CITES.

*E-commerce*

Given that CITES came into force on 1 July 1975, it is not surprising that CITES does not specifically regulate e-commerce involving wildlife. Recognising the rapid growth in e-commerce of specimens of CITES-listed species and the unique challenges associated with such trade, soft law prescribes that trade controls should extend to prohibiting e-commerce in violation of CITES and that a national unit to monitor and investigate internet-related wildlife crime be established.[23]

*Permits*

The NLP analysis also seeks to establish if national implementing legislation contains a general clause prohibiting any transactions without a valid permit.

The NLP analysis will look to see if there are conditions relating to the form and validity of permits. CITES prescribes only the minimum information that permits need to contain.[24] However, soft law prescribes a standard form as well as other essential details such as their maximum time

---

[21] Cyrille de Klemm, *Guidelines for Legislation to Implement CITES* (Cambridge, UK: IUCN, 1993) at 92–93.

[22] *Non-detriment Findings*, Resolution Conf. 16.7, CITES, 16th meeting of the CoP, (2013) at 1, 2.

[23] *Compliance and Enforcement*, Resolution Conf. 11.3 (Rev. CoP16), CITES, 11th meeting of the CoP, (2000) at 6.

[24] *Convention on International Trade in Endangered Species of Wild Fauna and Flora*, 3 March 1973, 993 UNTS 243, Article VI.

validity period.[25] Given that the permit system constitutes the very core of CITES, such critical details furnished by soft law, need to be incorporated into legislation.

In addition, legislation should (i) empower the MA to revoke, modify and suspend permits, (ii) provide that permits must not be transferable, (iii) make clear that a separate permit is required for each consignment of specimens.[26]

Article X provides that Parties may accept "comparable documentation" which substantially conforms to the requirements of the Convention when trading with non-Party States. Soft law elaborates on what amounts to "comparable" and states that where permits do not meet the prescribed requirements, such documentation should not be accepted.[27]

### *Marking*

Marking helps to ensure that specimens entering international trade are indeed the same as those for which permits are issued. As more and more species and even specimens are becoming subject to different rules, marking is increasingly becoming a necessary identification and enforcement tool.[28]

Article VI.7 provides that where appropriate and feasible, an MA may affix a mark upon any specimen to assist in identifying the specimen. This is the only provision in CITES that relates to marking, and the marking of specimens as a legislative requirement has not been addressed by the NLP; whereas detailed rules have developed as soft law — for example, soft law identifies the types of animals that require marking and specifies how marking should be undertaken.

Legislation should reflect these soft law requirements and prohibit transactions that do not comply with such requirements. The removal, alteration, defacing or erasing of a mark should also be prohibited.[29]

---

[25]See *Permits and Certificates*, Resolution Conf. 12.3 (Rev. CoP16), CITES, 12th meeting of the CoP, (2002).
[26]Cyrille de Klemm, *Guidelines for Legislation to Implement CITES* (Cambridge, UK: IUCN, 1993) at 35, 93.
[27]*Trade with States Not Party to the Convention*, Resolution Conf. 9.5 (Rev. CoP16), CITES, ninth meeting of the CoP, (1994).
[28]Willem Wijnstekers, *The Evolution of CITES: A Reference to the Convention on International Trade in Endangered Species of Wild Fauna and Flora*, 9th, ed. (Hungary: CIC – International Council for Game and Wildlife Conservation, 2011) at 195.
[29]Cyrille de Klemm, *Guidelines for Legislation to Implement CITES* (Cambridge, UK: IUCN, 1993) at 100.

*Exemptions and special provisions*

In prohibiting trade in violation of the Convention, legislation will need to specify the instances when exemptions and special provisions apply.[30] Exemptions are set out in Article VII of the Convention; while special provisions to do with ranching and quotas have developed through soft law, and are now a well-established part of the CITES regime. The effect of an exemption is that the CITES trade controls do not apply. Special provisions do not have the same effect but nonetheless, confer special treatment in a given situation.

Legislation will also need to address how these exemptions and special provisions are to be interpreted and applied, and their effect.[31] There are many soft laws that amplify on or govern the interpretation and application of the exemptions and special provisions. For example, subject to certain exceptions, Article VII.3 provides that trade controls do not apply to specimens that are "personal or household effects". Soft law provides guidelines that define "personal and household effects" as well as sets out categories of exempted specimens.[32]

Notwithstanding the above, Parties are free to decide not to allow any or all such exemptions and may impose special provisions to provide for more restrictive conditions, or to limit the "preferential" effect of such exemptions and special provisions.[33] However, such matters must be addressed in legislation.

## *Penalising Trade in Specimens in Violation of the Convention*

The third criterion of the CITES legislative criteria requires legislation to penalise trade in specimens in violation of the Convention. This requires legislation to (i) identify the prohibited activities and transactions that should be made offences; (ii) prescribe penalties for these offences; and (iii) designate enforcement authorities and agents. The legal basis is rooted

---

[30] Cyrille de Klemm, *Guidelines for Legislation to Implement CITES* (Cambridge, UK: IUCN, 1993) at 95.

[31] Cyrille de Klemm, *Guidelines for Legislation to Implement CITES* (Cambridge, UK: IUCN, 1993) at 95.

[32] *Control of Trade in Personal and Household Effects*, Resolution Conf. 13.7 (Rev. CoP16), CITES, 13th meeting of the CoP, (2004) at 2, 4.

[33] *Convention on International Trade in Endangered Species of Wild Fauna and Flora*, 3 March 1973, 993 UNTS 243, Article XIV.1.

in Article VIII.1(a) and covers not just trade in, but also the possession of, CITES-listed specimens acquired in violation of the Convention.

*Prohibited activities and transactions that should be made offences*

The NLP analysis seeks to verify that legislation clearly lists the activities that are prohibited and specifies that the breach of any prohibition constitutes an offence. In determining what should constitute an offence, the discussion in relation to the second criterion of the CITES legislative criteria is relevant; it helps to identify the activities and transactions that legislation should prohibit. Two examples are provided below.

The import, export, re-export and introduction from the sea, as well as the possession of, and domestic trade in CITES-listed specimens without the required permits should be criminalised.[34] Legislation should also make e-commerce involving, or resulting in the possession of, CITES-listed specimens acquired in violation of the Convention, an offence.

The violations of conditions attached to a permit; trade in, or possession of, CITES-listed specimens with invalid or forged permits; or permits obtained fraudulently; must be made an offence. In this regard, what soft law prescribes as amounting to a valid permit or comparable documentation from non-Parties that is acceptable, must be taken into account in framing the offence.

*Penalties for offences*

Adequate penalties are crucial to the effective enforcement of CITES. However, as the specific penalties to be prescribed is a task best left to the judgment of Parties (in consideration of the circumstances in their respective jurisdictions), this section will highlight principles that are relevant to this issue.

Soft law states that sanctions should reflect the nature and gravity of the crime and that in punishing perpetrators, high priority should be given to offer for sale of specimens of species listed in Appendix I.[35] This suggests that terms of imprisonment may be appropriate in cases involving

---

[34] *Convention on International Trade in Endangered Species of Wild Fauna and Flora*, 3 March 1973, 993 UNTS 243, Article VIII.1.

[35] *Compliance and Enforcement*, Resolution Conf. 11.3 (Rev. CoP16), CITES, 11th meeting of the CoP, (2000) at 6.

specimens of Appendix-I species as well as for offences relating to large numbers of commercially valuable specimens or to punish offenders of multiple offences.[36]

In 2000, the CoP has also suggested that in the context of illegal wildlife trade with links to money laundering, assets should be forfeited to ensure that criminals do not benefit from the proceeds of their crimes.[37] This was an important development that highlighted the need for specific legislative powers to combat wildlife trade with links to money laundering.

### *Designating enforcement authorities and agents*

The NLP analysis also verifies that the departments and agents responsible for enforcing CITES are clearly designated by the legislation and that enforcement agents so appointed are given the necessary powers to carry out their tasks. These include powers to search persons, vehicles and premises; powers to seize specimens; and even powers of arrest.[38] Soft law also states that wildlife enforcement personnel should enjoy equal status with other official enforcement agencies such as the police and Customs.

### *Confiscate specimens illegally traded or possessed*

The NLP analysis will seek to verify that legislation provides for the confiscation of specimens illegally traded or possessed. Article VIII.1(b) provides the legal basis for this fourth criterion of the CITES legislative criteria and states that Parties should take measures "[t]o provide for the confiscation or return to the State of export of such specimens". Factors the NLP considers include whether or not legislation specifies which authorities may confiscate, the extent of their confiscation powers, and the procedures that must be followed.

Where import is refused, soft law provides that exporting or re-exporting Parties should take measures to ensure that such specimens are not re-entered into the illegal trade. Soft law provides that the seizure and confiscation of specimens exported or re-exported in violation of the

---

[36]Cyrille de Klemm, *Guidelines for Legislation to Implement CITES* (Cambridge, UK: IUCN, 1993) at 103.

[37]*Compliance and Enforcement*, Resolution Conf. 11.3 (Rev. CoP16), CITES, 11th meeting of the CoP, (2000) at 7.

[38]*National Laws for Implementation of Convention*, CITES, 16th meeting of the CoP, CoP16 Doc.28 (2013) at 3.

Convention, is preferred over the refusal to import such specimens.[39] Legislation should likewise reflect the above.

CITES provides that confiscated live specimens shall, after consultation with the State of export, be returned to that State at its expense or to a rescue centre (designated by an MA) or such other place as the MA deems appropriate and consistent with the purposes of the Convention.[40] Soft law provides very detailed guidelines on what constitutes disposal that is "appropriate and consistent" with the purposes of the Convention for both live animals and plants.[41] Legislation should reflect these guidelines.

As for the disposal of confiscated and accumulated dead specimens, soft law sets out the circumstances where disposal of Appendix-I species would be permissible; it provides that the disposal of specimens of Appendix-II and Appendix-III species should be done in the "[b]est manner possible to achieve the purposes of the Convention...", and in such a way that the person responsible for the offence does not make any gain from the disposal.[42] Again, legislation should reflect these guidelines.

## 3. A Comparison of the CITES-implementing Legislation of Singapore and Malaysia with that of Benchmark Parties

A comparison of Singapore's CITES-implementing legislation, the Endangered Species (Import and Export) Act[43] ("ESA") and Malaysia's CITES-implementing legislation, the International Trade in Endangered Species Act 2008[44] ("INTESA"), reveals common areas of strengths and weaknesses. Comparisons were also made to provisions from the CITES-implementing legislation of Benchmark Parties, such as Australia, New Zealand, the United Kingdom ("UK"), Canada and Hong Kong, for areas

[39] *Confiscation of Specimens Exported or Re-exported in Violation of the Convention*, Resolution Conf. 9.9, CITES, ninth meeting of CoP, (1994).

[40] *Convention on International Trade in Endangered Species of Wild Fauna and Flora*, 3 March 1973, 993 UNTS 243, Article VIII.4(b).

[41] *Disposal of Confiscated Live Specimens Included in the Appendices*, Resolution Conf. 10.7 (Rev. CoP15), CITES, 10th meeting of the CoP, (1997) Annexes 1 (for animals), 2 (for plants).

[42] *Disposal of Confiscated and Accumulated Specimens*, Resolution Conf. 9.10 (Rev. CoP15), CITES, ninth meeting of the CoP, (1994) at 2.

[43] *Endangered Species (Import and Export) Act* (Cap 92A, 2008 Rev Ed Sing).

[44] *International Trade in Endangered Species Act 2008*, Malaysia, Act 686.

where there is scope for the ESA and INTESA to further strengthen their legislation.

### *Designate at Least one MA and one SA*

Overall, INTESA satisfies relatively more aspects of this first criterion of the CITES legislative criteria than the ESA. INTESA is an example of legislation that attempts to specify the functions and powers of an MA and an SA. Both regimes do not satisfy the requirement for an independent SA and provide for coordination and communication mechanisms among enforcement agencies.

#### *Legislative designation of at least one MA and one SA*

INTESA specifically designates the Lead MA, SA and MAs, but falls short of designating the SAs. Instead, the SAs that were appointed were listed in Malaysia's Biennial Report to the CITES Secretariat.[45] The ESA gives the Director-General of the Agri-Food and Veterinary Services appointed under the Animals and Birds Act[46] (DG) the responsibility of administering the ESA without specifying that the DG functions as both the MA and SA. Benchmark Parties such as New Zealand, Australia and the USA designate their respective MAs and SAs in their CITES-implementing legislation. These Benchmark Parties adopt the CITES terms "MA" and "SA", making it easy to identify these authorities in the relevant countries.

#### *Legislative prescription of MAs' and SAs' functions and powers*

The ESA does not prescribe the DG's functions and only partially prescribes his powers. On the other hand, INTESA provides an example of legislation that attempts to specify the functions and powers of the Lead MA, MAs and SAs, consistent with those envisaged by CITES.[47] However the list is not exhaustive and does not include functions prescribed by soft law such as the interpretation of listings in the CITES Appendices which may be necessary at times due to the dynamic nature of biological nomenclature and a lack of a standard reference. Benchmark Parties such

[45] CITES, *Malaysia's Biennial Report (2009/2010)*, online: CITES <http://www.cites.org>.
[46] *Animals and Birds Act* (Cap. 7, 2002 Rev Ed Sing).
[47] *International Trade in Endangered Species Act 2008*, Malaysia, Act 686, ss 5, 7, 9.

as the USA, Australia, Canada and New Zealand also do not specify the functions and powers of their MAs and SAs in their legislation.

*Designated SA to be independent of MA*

Both Singapore and Malaysia have SAs that are also MAs and accordingly, not "independent" as envisaged by soft law. The EU does not use the word "independent" in its legislation but requires each Member State to designate at least one SA with "[a]ppropriate qualifications whose duties shall be separate from those of any designated management authority". Pursuant to this regulation, the UK has designated the Joint Nature Conservation Committee[48] and the Royal Botanic Gardens Kew[49] as the SAs for fauna and flora, respectively.[50] Apart from possessing the appropriate qualifications, both SAs have duties that are distinct from the designated MAs.

*Legislative provision for coordination and communication mechanisms*

Both the ESA and INTESA do not have provisions setting out mechanisms that facilitate coordination and communication between the MA, SA and relevant government agencies involved in CITES enforcement contrary to softlaw. Such coordination and communication mechanisms are also absent from Benchmark Parties' legislation.

The EU has identified several good practices that promote inter-agency coordination and communication amongst enforcement officials including the use of formal agreements (such as memoranda of understanding) to define roles and procedures and the establishment of an inter-agency committee that meets regularly to discuss ongoing enforcement coordination and strategic matters.[51] Such practices could form the basis for drafting the relevant legislative provisions.

### *Prohibit Trade in Specimens in Violation of the Convention*

As for the second criterion of the CITES legislative criteria, both the ESA and INTESA adopt stricter trade controls than is required by CITES such

[48] *Natural Environment and Rural Communities Act 2006* (UK), c 16, Schedule 4.
[49] Royal Botanic Gardens, Kew, *About Kew*, online: Royal Botanic Gardens, Kew <http://www.kew.org>.
[50] CITES Secretariat, *National contacts & information — United Kingdom*, online: CITES <http://www.cites.org>.
[51] Milieu Ltd and Orbicon Consulting, *Study on the Enforcement of the EU Wildlife Trade Regulations in the EU-25* (Belgium: Milieu Ltd, 2006) at 21–23.

as prohibiting *any* import, export, re-export, introduction from the sea, and transit of scheduled species without the requisite permits. However, it is unclear from a plain reading of the ESA and INTESA which exemptions and special provisions apply (or if they apply at all). Specific provisions to regulate e-commerce involving wildlife are also absent.

### *Scope of application*

Both the ESA and INTESA makes no reference to the CITES Appendices. The ESA's Schedule comprises three Appendices whose coverage is practically identical to the three CITES Appendices while INTESA's Third Schedule reflects many (although not all) of the CITES-listed species. This only became apparent following a comparison of both Schedules with the CITES Appendices.

By contrast, Australia has a provision that deals solely with the listing of CITES species for the purposes of its implementing legislation. It makes clear that such a list "[m]ust include all species from time to time included in any of Appendices I, II and III to CITES. The list must not include any other species".[52] The actual list also contains a statement to the above effect.[53]

### *Trade-control provisions*

Both regimes have implemented stricter trade controls than required by CITES, prohibiting *any* import, export, re-export, introduction from the sea, and transit of scheduled species without the requisite permits. Possession of, and domestic trade in, any scheduled species without a valid permit is also an offence.[54]

However, both regimes lack detailed rules covering the types of permits required for the different transactions and the conditions that need to be satisfied before permits are granted. By comparison, New Zealand's legislation sets out very clearly the permits required for any import, export, re-export and introduction from the sea of endangered, threatened and exploited species as well as the conditions to be satisfied prior to the grant

[52] *Environment Protection and Biodiversity Conservation Act 1999* (Cth), s 303CA.
[53] Austl, Commonwealth, Department of the Environment, *List of CITES Species for the Purposes of the Act* (Canberra: Department of the Environment, 2013).
[54] *Endangered Species (Import and Export) Act* (Cap 92A, 2008 Rev Ed Sing), ss 4, 5; *International Trade in Endangered Species Act 2008*, Malaysia, Act 686, ss 10, 11, 12, 13.

of such permits. The relevant provisions reflect the provisions of Articles III, IV and V of CITES, respectively.[55]

*E-commerce*

Both regimes do not have specific provisions dealing with e-commerce involving wildlife and neither do the Benchmark Parties. The CITES-implementing legislation of the Czech Republic provides that any advertisement offering the sale of a specimen subject to the Act has to be accompanied by a notice. The effect of this notice is that the virtual buyer would be made aware of (i) the CITES documentation proving legal possession of the specimen and (ii) the applicable registration requirements or prohibitions against commercial activity.

The Czech law also requires website owners and/or operators to ensure that such obligatory information is made publicly available for each specimen offered for sale and the Czech Environmental Inspectorate can require the removal of advertisements that do not comply with the relevant regulations. E-permits are issued by the Czech MA to facilitate such online trade and to avoid the fraudulent use of traditional paper permits.[56]

*Permits*

Both regimes lack detailed rules governing the form of permits (from Parties as well as non-Party States) and their effect.

The EU has a specific regulation that is dedicated to prescribing the standard forms to which permits and other documents provided for in its implementing legislation must correspond.[57] This regulation sets out the design and technical specifications for the different permits.

As for the type of documentation required from non-Party States, none of the Benchmark Parties except Hong Kong have provisions in their legislation that set out what is required. The relevant provision

[55] *Trade in Endangered Species Act 1989* (NZ), 1989/18, ss 13–s24.

[56] INTERPOL, *Project Web: An Investigation Into the Ivory Trade Over the Internet Within the European Union* (Lyon: INTERPOL Environmental Crime Programme, 2013) at 15. The actual legislation was not referred to as it is not in English and an English translation of the current legislation is not publicly available.

[57] *Commission Regulation (EU) No 792/2012 of 23 August 2012 laying down rules for the design of permits, certificates and other documents provided for in Council Regulation (EC) No 338/97 on the protection of species of wild fauna and flora by regulating trade therein and amending Commission Regulation (EC) No 865/2006*, [2012] OJ, L 242/13.

reflects the CoP's recommendations on what permits should contain when trading with States not party to the Convention (albeit with some slight modification).[58] These include sufficient identification of the species concerned and certification of the origin of the specimen concerned.

New Zealand makes clear that permits are effective from the date of grant, are non-transferable, valid for not more than six months and are transaction and specimen specific.

### *Marking*

Both regimes also lack rules on when marking is required. In the EU, there are detailed provisions on marking requirements.[59] Focusing on the marking provisions that relate to the international trade of wildlife, the EU has provisions that control such trade by making the issue of import and export permits for certain specimens contingent on them being marked in accordance with marking methods approved or recommended by the CoP.[60]

### *Exemptions and special provisions*

Many Benchmark Parties have detailed exemption regimes in their CITES-implementing legislation. However, it is unclear that from a plain reading of the ESA and INTESA, which CITES exemptions and special provisions apply (or if they apply at all). Given that the Convention allows Parties to take stricter measures than what has been provided in the Convention, it is important for the sake of clarity and certainty that CITES-implementing legislation state clearly the CITES exemptions and special provisions that apply, the conditions that need to be satisfied, their effect and how they are to be applied.

New Zealand's implementation of CITES exemptions takes into account the above factors and serves as a positive example of how CITES exemptions may be implemented. Using the exemption for specimens which

---

[58] *Trade with States Not Party to the Convention*, Resolution Conf. 9.5 (Rev. CoP16), CITES, ninth meeting of the CoP, (1994).

[59] *Commission Regulation (EC) No 865/2006 of 4 May 2006 laying down detailed rules concerning the implementation of Council Regulation (EC) No 338/97 on the protection of species of wild fauna and flora by regulating trade therein*, [2006] OJ, L 166/1, Chapter XVI, Articles 64–68.

[60] *Commission Regulation (EC) No 865/2006 of 4 May 2006 laying down detailed rules concerning the implementation of Council Regulation (EC) No 338/97 on the protection of species of wild fauna and flora by regulating trade therein*, [2006] OJ, L 166/1, Chapter XVI, Articles 64, 65(4), and 66(6).

are the subject of a scientific transfer as an example, New Zealand's CITES-implementing legislation makes clear that the effect of this exemption is that the trade control provisions in the main legislation do not apply to transactions within the scope of the exemption. Legislation then prescribes the four conditions that need to be satisfied, reflecting the conditions in Article VII.6 of CITES.[61] Furthermore, rules set out in subsidiary legislation explain how the exemption is to be applied in practice.[62]

### *Penalising Trade in Specimens in Violation of the Convention*

The ESA and INTESA provide more comprehensively for matters covered by the third criterion of the CITES legislative criteria i.e., offences, penalties and enforcement. In several areas, both regimes have provisions which other ASEAN Member States can benchmark themselves against. That said, both regimes have common areas that can be further strengthened.

#### *Offences*

(i) *Main offences*

The main offences under the ESA and INTESA are similar in scopepenalising any import, export, re-export, introduction from the sea, and transit of scheduled species without the requisite permits. Both regimes also make possession of, and domestic trade in, scheduled species without a valid permit an offence.[63] The ESA, but not INTESA, offers a statutory defence in circumstances where the person charged took all reasonable precautions and exercised due diligence to avoid the commission of the offence.[64]

Apart from the above offences, INTESA also makes it an offence for any commercial operation to produce captive-bred or artificially propagated specimens of any scheduled species, without being registered with an MA.[65] The ESA does not have an equivalent provision nor do the Benchmark Parties.

---

[61] *Trade in Endangered Species Act 1989* (NZ), 1989/18, s 32.
[62] *Trade in Endangered Species Regulations 1991* (NZ), 1991/274, 1, Reg. 3.
[63] *Endangered Species (Import and Export) Act* (Cap 92A, 2008 Rev Ed Sing), ss 4(1), 4(2); *International Trade in Endangered Species Act 2008*, Malaysia, Act 686, ss 10, 11, 12, 13.
[64] *Endangered Species (Import and Export) Act* (Cap 92A, 2008 Rev Ed Sing), s 6.
[65] *International Trade in Endangered Species Act 2008*, Malaysia, Act 686, s 14.

(ii) *Permit offences*

INTESA, as with many Benchmark Parties,[66] makes a failure to comply with any condition attached to a permit or registration an offence, as required by the CITES legislative criteria.[67] The ESA is lacking in this regard; instead, the DG is given the power to cancel a permit if any condition or restriction of the permit is contravened.[68]

Other permit offences covered by both regimes include providing false or misleading information in connection with a permit application or using a permit which is false in a material particular; and altering, forging or defacing a permit or using such a permit.[69]

As discussed earlier, the ESA and INTESA do not prescribe the standard forms for permits nor specify the requirements for comparable documentation from non-Party States. By contrast, the EU and HK have such provisions. To ensure enforceability, the EU requires Member States to impose sanctions for transactions without "appropriate" permits or with "invalid" permits.[70] In Hong Kong, when trading with non-Party States, unless an exemption applies, it is an offence to import a specimen of any scheduled species without a "Convention certifying document" (defined to include documentation from non-Party States that complies with the relevant requirements).[71]

(iii) *Marking offences*

Both regimes make it an offence to contravene the marking requirements of an MA. However, as discussed above, both regimes do not currently specify when marking is required. Australia makes non-compliance with any marking requirements prescribed by the Secretary an offence.[72] The EU makes permit issuance conditional on compliance with its marking requirements (where applicable) but does not require Member States to

---

[66] For example, EU, Australia, the United Kingdom, New Zealand and Hong Kong.
[67] *International Trade in Endangered Species Act 2008*, Malaysia, Act 686, s 16(4).
[68] *Endangered Species (Import and Export) Act* (Cap 92A, 2008 Rev Ed Sing), s 8(1)(a).
[69] *Endangered Species (Import and Export) Act* (Cap 92A, 2008 Rev Ed Sing), s 17; *International Trade in Endangered Species Act 2008*, Malaysia, Act 686, s 44.
[70] *Council Regulation (EC) No 338/97 of 9 December 1996 on the protection of species of wild fauna and flora by regulating trade therein*, [1997] OJ, L 61/1, Article 16(1)(a).
[71] *Protection of Endangered Species of Animals and Plants Ordinance* (Cap 586, 2006, Hong Kong), c 586, ss 5, 11.
[72] *Environment Protection and Biodiversity Conservation Act 1999* (Cth), s 303EU. As of 22 January 2014, no such written determination has been made by the Secretary.

impose sanctions for non-compliance with marking requirements. Both regimes make it an offence to alter, counterfeit, destroy, erase or remove any mark without the MA's consent.[73]

(iv) *Other offences*

The ESA and INTESA have provisions that make it an offence to obstruct enforcement officers in the execution of their duty.[74] Offences committed by bodies corporate and their officers are also punishable. Both regimes also make it an offence to attempt or abet offence.[75] These provisions are not always present in the CITES-implementing legislation of Benchmark Parties.

*Penalties for main offences*

(i) *Fines as a penalty*

The ESA and INTESA impose fines for any person who imports, exports, re-exports or introduces from the sea, any scheduled species without a permit. Both regimes impose fines not exceeding S$50,000 and RM100,000, respectively, for each scheduled species and maximum aggregate fine limits of S$500,000 and RM$1,000,000, respectively.[76]

Similar fines and maximum aggregate fine limits apply for the offence of (i) possession of, and domestic trade in, scheduled species without a valid permit; and (ii) in the case of INTESA, producing captive-bred or artificially propagated specimens of any scheduled species without being registered with an MA.[77] Both countries have over the years also increased the severity of their fines.

Within the context of Singapore's and Malaysia's GDP per capita, the penalties prescribed by the ESA seem significantly less punitive than those prescribed by INTESA. The maximum fine per scheduled species under the ESA (i.e. S$50,000) is approximately 75 per cent of Singapore's GDP per

[73] *Endangered Species (Import and Export) Act* (Cap 92A, 2008 Rev Ed Sing), s 13; *International Trade in Endangered Species Act 2008*, Malaysia, Act 686, s 15.
[74] *Endangered Species (Import and Export) Act* (Cap 92A, 2008 Rev Ed Sing), s 16; *International Trade in Endangered Species Act 2008*, Malaysia, Act 686, s 40.
[75] *Endangered Species (Import and Export) Act* (Cap 92A, 2008 Rev Ed Sing), s 19; *International Trade in Endangered Species Act 2008*, Malaysia, Act 686, s 47.
[76] *Endangered Species (Import and Export) Act* (Cap 92A, 2008 Rev Ed Sing), s 4(1); *International Trade in Endangered Species Act 2008*, Malaysia, Act 686, ss 10, 11.
[77] *Endangered Species (Import and Export) Act* (Cap 92A, 2008 Rev Ed Sing), s 4(2); *International Trade in Endangered Species Act 2008*, Malaysia, Act 686, s 12.

capita. By comparison, under INTESA, the maximum fine per scheduled species (i.e. RM100,000) is approximately three times Malaysia's GDP per capita. Furthermore, Singapore's GDP per capita is almost five times more than that of Malaysia. Yet the ESA's maximum fine per scheduled species and the maximum aggregate fine limit (i.e. S$500,000) are only 30 per cent more than the corresponding maximum fine and maximum aggregate fine limit (i.e. RM1,000,000) under INTESA.[78]

A comparison to the quantum of fines imposed by Benchmark Parties is not helpful as their penalties are imposed on a "per offence" basis rather than "per specimen" basis. This means that the penalties of Benchmark Parties are meted out for a particular offence (e.g., for the import of a CITES-listed species without the requisite permit) but do not take into account the number of specimens involved in that offence.

(ii) *Differentiated penalties*

There is scope within both regimes to reflect the soft-law principle that sanctions must reflect the seriousness and scale of the offence and in punishing perpetrators, high priority should be given to offers or sales of Appendix-I listed species.

The ESA and INTESA impose fines for the main offences on a "per specimen" basis, which means that the penalty for an offence involving many specimens (whether of the same or a different species) will be greater than the penalty for an offence involving fewer specimens. In this way, penalties under both regimes may be said to reflect the scale of an offence. Yet by having maximum aggregate fine limits, these limits place a ceiling on the extent penalties can reflect the scale of an offence. It could also be argued that the fines prescribed do not take into account the seriousness of the offence by distinguishing between the type of species; its commercial value; or whether the offender is a repeat offender.

New Zealand prescribes different penalties for offences committed in respect of endangered, threatened and exploited species, while Hong Kong has specific provisions prescribing a higher penalty for offences relating

[78]The GDP per capita for Singapore and Malaysia are US$51,709 and US$10, 432, respectively, see The World Bank, *Data, GDP per capita (current US$)*, online: World Bank <http://www.data.worldbank.org>. All calculations were done in US dollar terms, using the US exchange rate for Singapore dollar of US$1: S$1.27 and for Malaysian ringgit of US$1: S$3.29 (source: XE Currency Converter, online: XE Currency Converter <http://www.xe.com> as at 12 March 2014).

to specimens of Appendix-I species committed for commercial purposes.[79] These provisions further distinguish between the penalties for offences involving Appendix II and III species.[80] Canada provides for a doubling of the fine where a person is convicted a second or subsequent time.[81]

(iii) *Power to confiscate proceeds of crimes*

Where the commission of an offence results in monetary benefits accruing to the offender, Canada has a provision that allows the court to impose an additional fine equal in amount to the court's estimation of those monetary benefits.

The ESA and INTESA do not provide for forfeiture of financial assets to ensure that criminals do not benefit from the proceeds of their crimes. However, in the context of illegal wildlife trade with links to money laundering, the courts in Singapore and Malaysia have powers under the Corruption, Drug Trafficking and Other Serious Crimes (Confiscation of Benefits) Act and the Anti-Money Laundering and Anti-Terrorism Financing Act 2001, respectively, to forfeit benefits derived from the commission of an offence.[82]

(iv) *Imprisonment as an alternative or additional penalty*

Both regimes prescribe imprisonment as an alternative or additional penalty for all the main offences. The ESA prescribes a maximum term of imprisonment of two years as compared to seven years under INTESA.

INTESA's relatively long maximum term of imprisonment stands out even when compared to penalties prescribed by Benchmark Parties. At the top of the scale, Australia prescribes imprisonment for 10 years for any import or export without a permit.[83] For a similar offence, Canada prescribes imprisonment for a term up to five years,[84] whereas New Zealand

[79] *Protection of Endangered Species of Animals and Plants Ordinance* (Cap 586, 2006, Hong Kong), c 586, ss 5, 6, 7, 8.
[80] *Protection of Endangered Species of Animals and Plants Ordinance* (Cap 586, 2006, Hong Kong), c 586, s 16.
[81] *Wild Animal and Plant Protection and Regulation of International and Interprovincial Trade Act*, SC 1992, c 52, s 22(2).
[82] *Corruption, Drug Trafficking and Other Serious Crimes (Confiscation of Benefits) Act* (Cap 65A, 2000 Rev Ed Sing), s 10(1); *Anti-Money Laundering and Anti-Terrorism Financing Act 2001*, Malaysia, Act 613, ss 55, 59(1).
[83] *Environment Protection and Biodiversity Conservation Act 1999* (Cth), ss 303CC, 303 CD.
[84] *Wild Animal and Plant Protection and Regulation of International and Interprovincial Trade Act*, SC 1992, c 52, s 22(1)(b)(ii).

prescribes imprisonment for a term not exceeding three or five years, depending on whether the offence relates to a specimen of an endangered or threatened species.[85] As with fines and in keeping with the soft-law principle that sanctions must reflect the seriousness and scale of the offence, there is scope to differentiate maximum imprisonment terms for different offences.

*Designating enforcement authorities and agents*

Both the ESA and INTESA confer on enforcement agents, such as the police and customs officers, comprehensive powers of enforcement but do not expressly designate the authorities responsible for enforcing their legislation (e.g., in the case of Singapore, it does not specifically designate the Agri-Food and Veterinary Authority of Singapore ("AVA") as the enforcement authority although it is stated to be so on the CITES website). The US legislation specifically designates the authorities responsible for enforcement and confers powers of enforcement to persons authorised by such authorities.[86]

The powers of enforcement conferred by both regimes on enforcement agents include the powers identified by the NLP such as the power to search persons, vehicle premises; to request information; to inspect documents; to take samples of specimens and even to arrest. In addition, INTESA expressly authorises access to computerised data, and provides the power to require the attendance before an enforcement officer, of persons acquainted with the facts and circumstances of the case. INTESA also expressly provides for parity of enforcement officers with the police (consistent with soft law) and empowers the MA to reward informers.[87]

### *Confiscate Specimens Illegally Traded or Possessed*

Both the ESA and INTESA have provisions that provide for the confiscation or return of specimens exported or re-exported, and disposal of confiscated specimens, as required by the fourth criterion of the CITES legislative criteria. However, both regimes do not reflect the applicable soft laws that amplify on the fourth criterion; such as soft-law guidelines on what constitutes disposal that is "appropriate and consistent" with the purposes of the Convention for both live animals and plants.

[85] *Trade in Endangered Species Act 1989* (NZ), 1989/18, ss 44(2),(3) and 45(2),(3).
[86] *Endangered Species Act*, 35 USC 16 (1973), Sec. 11(e)(1), 11(e)(3).
[87] *International Trade in Endangered Species Act 2008*, Malaysia, Act 686, ss 20(2), 51.

Pursuant to the ESA, where a person is convicted of any of the main offences, the DG will automatically forfeit the scheduled species in respect of which the offence was committed without the need for any further orders.[88] Where a person is prosecuted and acquitted or given a discharge, the scheduled species and related items may only be forfeited by the DG if ordered by the courts.[89] Where a person cannot be prosecuted, the DG has the power to forfeit the scheduled species and related items seized.

By comparison, where a person is prosecuted for an offence under INTESA, forfeiture is permissible regardless of whether the person is convicted for the offence. However, a forfeiture order will only be made if the court is satisfied that an offence under INTESA has been committed and that the scheduled species was the subject matter of the offence.[90]

Both regimes have provisions that permit the DG (in the context of the ESA) or relevant MA (in the context of INTESA) to repatriate any scheduled species so forfeited.[91] However, both regimes do not reflect the position at soft law that (i) the seizure and forfeiture of specimens exported or re-exported in violation of the Convention, is preferred over, the repatriation of such specimens (to avoid re-entry into illegal trade); and (ii) where import is refused, the MAs should take measures to ensure that such specimens are not re-entered into the illegal trade; and. In fact, all Benchmark Parties are also lacking in this regard.

INTESA's regulations providing for the establishment of rescue centres represents a step in the right direction as regards the implementation of the Convention's requirement for the provision of rescue centres for confiscated live specimens.[92] Such provisions are lacking in the ESA.

The ESA's and INTESA's provisions on disposal of confiscated specimens hardly reflect soft-law guidelines on the disposal of live specimens of CITES-listed species and confiscated and accumulated dead specimens. In fact, the CITES-implementing legislation of all Benchmark Parties do not reflect these guidelines.

---

[88] *Endangered Species (Import and Export) Act* (Cap 92A, 2008 Rev Ed Sing), s 15(1).
[89] *Endangered Species (Import and Export) Act* (Cap 92A, 2008 Rev Ed Sing), s 15(2).
[90] *International Trade in Endangered Species Act 2008*, Malaysia, Act 686, s 34(2).
[91] *Endangered Species (Import and Export) Act* (Cap 92A, 2008 Rev Ed Sing), s 15(10); *International Trade in Endangered Species Act 2008*, Malaysia, Act 686, s 34(9).
[92] *International Trade in Endangered Species (Rescue Centre) Regulations 2009*, PU(A) 460/2009, Regs. 4, 5.

## 4. Findings and Recommendations

### *Findings*

*Lessons ASEAN can draw from a comparative analysis of the ESA and INTESA that could benefit Member States with either Category 2 or 3 legislation*

While the ESA and INTESA may not satisfy all the requirements of the CITES legislative criteria notwithstanding its status as Category 1 legislation, there are certainly lessons that can be drawn from both regimes that could benefit not just ASEAN Member States with Category 2 or 3 legislation, but possibly others with Category 1 legislation.

(i) *Functions and powers of CITES authorities are clearly specified*

INTESA provides an example of legislation that attempts to specify the functions and powers of an MA and SA, as well as that of the Lead MA. While the functions and powers listed are clearly not exhaustive, it attempts to distinguish between the functions and powers of an MA and SA and serves as a useful starting point for ASEAN Member States wishing to refine their CITES-implementing legislation in this respect.

(ii) *Adoption of stricter domestic measures*

The ESA and INTESA illustrate how stricter domestic measures (than required by CITES) may be implemented, making the possession of, and domestic trade in, any scheduled species, without a valid permit, an offence. Any scheduled species in transit will also need to be accompanied by the requisite CITES documentation.

(iii) *A range of activities and transactions are prohibited and penalised*

Apart from trade controls, both regimes also provide for a range of other offences that help to ensure an effective implementation of CITES.

INTESA makes non-registration of commercial operations that produce captive-bred or artificially-propagated specimens of any scheduled species one of the main offences. As these operations are sometimes used as fronts for the illegal wildlife trade, INTESA's strict control of such operations can be said to be leading the way for other countries facing similar challenges.

INTESA makes any violation of the conditions attached to a permit an offence. Both INTESA and the ESA make it an offence to apply for a permit using false or misleading information in connection with a permit application or using a permit that is false in a material particular; altering, forging or defacing a permit or using such a permit. It is also an offence under both regimes to contravene the marking requirements of an MA (although both regimes do not specify the marking requirements) or to alter, counterfeit, destroy, erase or remove any mark without the MA's consent.

There are also provisions that make it an offence to obstruct enforcement officers in the execution of their duty and to refuse their requests for information or documents. The Acts also make it an offence to abet the commission or attempt to commit, any offence under the Acts. Offences by body corporates and their officers are also punishable.

(iv) *Enforcement agents have comprehensive powers of enforcement*

Both the ESA and INTESA confer comprehensive powers of enforcement on enforcement officers including the power to search persons, vehicle and premises; to request information; to inspect documents; to take samples of specimens; and even to arrest. In addition, INTESA provides the power to require the attendance before an enforcement officer, of persons acquainted with the facts and circumstances of the case, and empowers the MA to reward informers — increasing the arsenal of enforcement tools available in prosecuting illegal trade in wildlife. INTESA also expressly provides for parity of enforcement officers with the police as required by soft law.

(v) *Penalties have increased in severity and are imposed on a "per specimen" basis*

Both Malaysia and Singapore impose fines on a "per specimen" basis and not on a "per species" or "per offence" basis. This is to be commended as doing so allows penalties to take into account the scale of the offence. Both countries have, over the years, also increased the severity of their penalties. This is in line with the increasing recognition of the serious and far-reaching impacts of the illegal trade in wildlife, and the need for severe penalties to deter wildlife crimes, in particular, those that are committed on a large scale for commercial purposes. Malaysia's current maximum imprisonment term of seven years stands out even when compared to Benchmark Parties.

(vi) *Power to order confiscation of benefits illegally obtained*

Given the links between illicit wildlife trafficking and other forms of illegal trafficking, the financing of rebel groups, corruption and money laundering, both regimes recognise that the issue cannot be solely seen as an environmental issue and courts need to be equipped with the appropriate powers to ensure that criminals do not benefit from the proceeds of their crimes. The courts in Singapore and Malaysia have powers under the Corruption, Drug Trafficking and Other Serious Crimes (Confiscation of Benefits) Act and the Anti-Money Laundering and Anti-Terrorism Financing Act 2001, respectively, to forfeit benefits derived from the commission of an offence involving wildlife trade with links to money laundering.

*Lessons ASEAN can draw from Benchmark Parties that would further strengthen the CITES-implementing legislation of all ASEAN Member States*

It is worth noting that these Benchmark Parties themselves have Category 1 legislation like Malaysia and Singapore, but they offer lessons on how Malaysia and Singapore may enhance compliance with the different aspects of the CITES legislative criteria, particularly in areas where the ESA and INTESA have been found to be lacking. Indeed, these lessons can potentially benefit any other ASEAN Member State wishing to further strengthen their CITES-implementing legislation.

(i) *CITES authorities are clearly designated*

Benchmark Parties such as New Zealand, Australia and the USA designate their respective MAs and SAs in their CITES-implementing legislation. These Benchmark Parties adopt the CITES terms "MA" and "SA", making it easy to identify these authorities in the relevant countries. The USA has language in its legislation that stands out for clearly identifying the authorities responsible for enforcing its CITES-implementing legislation and providing the necessary enforcement powers to persons authorised by such authorities.

(ii) *SA is duly qualified and has duties that are separate from those of MA*

Given the importance of the SA's role as envisaged by the Convention, in particular its role in making non-detriment findings, the aims of the Convention are best served by having an SA that is independent of the MA. Although EU does not use the word "independent" in its legislation, it requires each Member State to designate at least one appropriately qualified SA whose duties shall be separate from those of the MA.

(iii) *Express reference to CITES Appendices*

The Australian approach of expressly referring to the CITES Appendices places beyond doubt that the list of species in the CITES-implementing legislation conforms to the listing in the Convention. However, Australia does not provide for a blanket provision that automatically includes within national legislation, all species included in any of the CITES Appendices from time to time.

(iv) *E-commerce*

None of the Benchmark Parties have specific provisions in their CITES-implementing legislation dealing with e-commerce in wildlife. However, the Czech Republic has specific rules in its CITES-implementing legislation that impose mandatory notification requirements on sellers that seek to protect potential buyers and provide for a system of e-permits.

(v) *Legislation is clear and comprehensive*

Benchmark Parties have legislation which is clear and comprehensive, covering areas that include: (i) the type of permit(s) required for the different transactions and conditions to be satisfied prior to grant; (ii) the form and effect of the permits; (iii) when marking is required; and (iv) applicable exemptions and special provisions — areas where the ESA and INTESA are lacking. New Zealand, the EU and Hong Kong have CITES-implementing legislation that regulate these matters as well as provide for enforcement mechanisms to sanction any non-compliance.

(vi) *Differentiated penalty provisions*

NZ, HK and Canada have adopted a differentiated approach towards their penalty provisions, taking into account factors such as the type of species, its commercial value and whether the offender is a repeat offender.

### *Recommendations*

*Secretariat to provide comprehensive and authoritative interpretation of the CITES legislative criteria*

While this paper attempts to flesh out the detailed legislative requirements underlying the CITES legislative criteria, what is needed is a comprehensive and authoritative interpretation by the Secretariat that sets out the legislative requirements underlying each of the four criteria of the CITES legislative criteria that collectively require CITES-implementing legislation to authorise Parties to (i) designate one MA and one SA; (ii) prohibit trade in specimens in violation of the Convention; (iii) penalise trade in specimens in violation of the Convention; and (iv) confiscate specimens illegally traded or possessed. This would provide the clarity and certainty that is needed to enable Parties to comply with the CITES legislative criteria and avoid incoherence in the way legislation is being implemented. This could take the form of a loose-leaf handbook that will need to be updated regularly to reflect any new legislative requirements as a result of ongoing soft law development.

Furthermore, areas where guidance and assistance from the NLP is required include (i) identifying the basic functions and powers of the MAs and SAs and consolidating their functions and powers as contemplated by soft law; (ii) defining the criteria for an "independent" SA; (iii) setting out the minimum provisions required to effectively regulate e-commerce in wildlife; and (iv) drafting legislative provisions that set out the mechanisms needed to facilitate coordination and communication between the MA, SA and enforcement agencies.

### *Replace the current NLP system of categorising legislation with a more finely calibrated system*

As discussed above, notwithstanding their status as Category 1 legislation, there are components within each criterion of the CITES legislative criteria that both the ESA and INTESA do not comply with. Furthermore, there are areas where even Benchmark Parties were lacking. Accordingly, there needs to be a more finely calibrated system of categorising legislation, with perhaps a scoring system for each criterion of the CITES legislative criteria, where Category 1 legislation will need to satisfy a certain threshold score for each of the four criteria. The basis for the scoring system should be transparent to all Parties and in order to maintain one's status as Category 1 legislation, the onus should be placed on Parties to demonstrate on-going compliance with evolving soft-law requirements.

### *The ESA and INTESA to be amended and strengthened in light of lessons learned from Benchmark Parties*

#### (i) *Designate at least one MA and one SA*

As for the first criterion, the ESA should make clear that the DG of AVA is the MA, and INTESA should also reflect the SAs that have been appointed. Like INTESA, the ESA should, at a minimum, specify the basic functions and powers of the MA and SA, and both regimes should designate "independent" SAs, there the UK legislation can serve as a reference point. Legislation should incorporate mechanisms to facilitate coordination and communication among the MA, SA and enforcement agencies, taking into account best practices highlighted by the EU.

(ii) *Prohibition of trade in specimens in violation of the Convention*

As for the second criteria, INTESA should fully reflect the species listed in the CITES Appendices as has been done in the ESA. Both regimes should make clear that their listings of scheduled species conform to the CITES Appendices and consider the value of having a blanket provision that automatically includes within national legislation, all species included in any of the CITES Appendices from time to time.

Both regimes would benefit from more detailed rules (for example, covering the type of permit(s) required for the different transactions and conditions to be satisfied prior to grant; the form and effect of permits; when marking is required, etc.). The power to enact such rules exists in both regimes. In this connection, care should be taken to ensure that conditions prescribed in CITES permits actually issued should be consistent with, and within the authority of, what is provided in the CITES-implementing legislation.

Given that Malaysia and Singapore have adopted stricter domestic measures in their implementation of CITES exemptions and special provisions, for the sake of clarity and certainty, both regimes should state clearly the exemptions and special provisions that apply, set out the conditions that need to be satisfied and make clear their effect.

In all these areas, there are examples from New Zealand, the EU and Hong Kong that can both inspire and instruct. Both regimes should also specifically regulate e-commerce involving wildlife and in this respect, the Czech Republic's CITES-implementing legislation serves as an example.

(iii) *Penalising trade in specimens in violation of the Convention*

With regard to the third criterion, the penalties prescribed for the main offences in both regimes should distinguish between offences based on the type of species involved, its commercial value, as well as the offender's history of previous convictions for similar offences. Both regimes should also abolish their maximum fine limits that effectively put a ceiling on the amount of fine that may be imposed for offences involving large quantities of specimens of a commercially valuable species.

(iv) *Confiscate specimens illegally traded or possessed*

With regard to the fourth criterion, both regimes should reflect the soft-law position that: (i) where import is refused, the MAs should take measures to ensure that such specimens are not re-entered into the illegal trade; (ii) the seizure and forfeiture of specimens exported or re-exported in violation of the Convention, is preferred over, the repatriation of such specimens (which may result in such specimens being re-entered into illegal trade); and (iii) the disposal of live and confiscated and accumulated dead specimens of scheduled species should reflect soft-law guidelines on the disposal of such specimens.

# Part 2

# Environmental Management

# 3

# ASSESSMENT OF PROTECTION AGAINST SEA LEVEL RISE: A CASE STUDY OF A COASTAL AREA IN SINGAPORE

EFSTATHIOS GIANNOUSTAS*

Changes in the rate of sea level rise will have important impacts on the coastal zone, and will potentially displace and alter ecosystems, and impact infrastructure. As a low-lying densely-populated island, Singapore is bound to face the effects of sea level rise in the years to come, therefore, it needs to prepare its strategy for mitigation and adaptation. As both the land area of Singapore and the infrastructure of the island are extremely valuable, protection measures may be a viable solution for the future. Thus, a cost comparison between the losses in capital stock due to inundation, versus the cost of constructing a seawall to protect a coastal area of Singapore against sea level rise, was undertaken. However, instead of a traditional seawall design, this study investigates the introduction of biodiversity-enhancing measures on the seawall design.

## Introduction

Changes in the rate of sea level rise will have important impacts on the coastal zone, and will potentially displace and alter ecosystems, and impact infrastructure (Pethick, 2001). By the year 2100, 600 million people will be living in coastal areas which lie below the 1 in 1000 year flood level (Ng & Mendelsohn, 2005).

### 1.1 *Global Sea Level Rise*

Estimates by the Intergovernmental Panel on Climate Change (ICCP) for the 20th century show that global average sea level rose at a rate of about $1.7\pm0.5$ mm yr$^{-1}$. From 1961 to 2003, the rate has increased to $1.8\pm0.5$ mm yr$^{-1}$ (IPCC, 2007).

The global rate of sea level rise was calculated at $1.8 \pm 0.3$ mm/yr from 1950 to 2000 (Church *et al.*, 2006). However, other studies have described how the rate of sea level rise increases as temperature increases, and that

*This paper was supervised by Jesuthason Thampapillai.

even if the rate of sea level rise settles at 3.5 mm/yr, the sea level rise from 1990 to 2100 will be 38 cm (Rahmstorf, 2007).

A recent study that incorporated data on sea level change from spacecraft from 1993 to 2010 re-affirmed that oceans globally are rising by just over 3 mm/yr, albeit with some big regional differences (Amos, 2012).

### 1.2 *Global Sea Level Rise Predictions*

Earlier predictions on the mean sea level rise (MSL) varied, and by 2100, MSL rise might range from less than 0.2 m to approximately 0.9 m (Church *et al.*, 2001). Recent predictions have projected that MSL rise for the 21st century will range between 9 and 58 cm (IPCC, 2007). Rahmstorf suggests that the MSL rise can be over 1 m by 2100 and that the very low sea level rise values reported in the IPCC TAR are rather implausible. Therefore, sea level rise predictions may be closer to the "upper range" rather than the lower one.

A recent study found that for Antarctica and Greenland the loss in ice mass increased significantly between 2000 and 2011 compared to 1992–2000 (Shepherd *et al.*, 2012). Only East Antarctic gained in ice mass from 2000–2011. They estimated that the loss in ice-sheet mass from 1992 to 2011 corresponds to a global mean sea level rise of 11.2 mm $\pm$3.8 mm (0.59 mm/yr).

### 1.3 *Regional Sea Level Rise*

Using a reconstructed global sea level between 1870 and 2004, it was found that the rate of sea level rise in Asia accelerated by 0.013 mm/yr during the 20th century (Church *et al.*, 2006). Jevrejeva showed that similar rates of sea level rise had started since the end of the 18th century (Jevrejeva *et al.*, 2008). Trisirisatayawong found that 1940–2004 trends of sea level change in the northern part of the Gulf of Thailand varied between $5.0 \pm 1.3$ and $4.5 \pm 1.3$ mm/yr and are significantly higher than global predictions. Using altimetry data from 1993 to 2009, the values ranged between 3.5 mm/yr and 5.5 mm/yr (Trisirisatayawong *et al.*, 2011).

The estimates for Asia-Pacific predict a MSL rise between 18 and 59 cm (IPCC4) until 2100, with the upper estimate being close to most global predictions, presented by IPCC, but with a significantly increased minimum estimate (18 mm compared to 9 mm). Other data confirm that parts of the Pacific and Indian regions have experienced increases in MSL greater that the global mean. The Andaman Sea and the Pacific Islands have experienced a sea level rise of 3 to 4 mm/yr, compared to 2 and 3 mm/yr for the Gulf of Thailand and Singapore (Church *et al.*, 2006).

## 1.4 *Sea Level Rise in Singapore*

Singapore experienced the effects of rising sea levels in 1974 when a rare astronomical event caused the tides to rise 3.9 meters, more than double the usual level (Wong, 1992). This showed that Singapore can be vulnerable to sea level rise especially if coupled with extreme high tide conditions. Yearly mean temperatures of Singapore have increased significantly in the last 60 years (WEATHERwise, 2012), so a vulnerability study to assess the possible effects of climate change, including sea level rise, was commissioned by the Ministry of Environment and Water Resources (SSNC, 2010). Phase 1 of the study showed that both lower and upper estimates for Singapore are higher than the IPCC AR4 scenario projections by 25% and 10% respectively (Table 3.1) (SSNC, 2010).

A rise of 0.65 m can result in loss of coastal land; however, 70–80% of the island is currently surrounded by seawalls, which can protect against erosion, but not necessarily against overtopping (SSNC, 2010).

### 1.4.1 *Storms & sea level anomalies*

Possible changes in the rate and severity of storms due to climate change is still controversial and no consensus has been reached. Results from a model for the 21st century revealed that storm surge frequency and magnitude are expected to decrease in southern Europe (Marcos *et al.*, 2011). However, storm surges can have an effect on sea level change even in the protected waters of Singapore. Tkalitch reported on extreme sea levels in the Singapore Strait during the northeast monsoon. These events, although not periodic, are generally associated with surges due to monsoon winds, (November to January). When these events overlap with spring tides, the

Table 3.1. Climate change predictions for Singapore based on Singapore's second national communication for climate change.

| Climate change Projections (in 2100 compared to present) | IPCC AR 4 projections | Phase 1 study, local projections |
|---|---|---|
| Change in average Temperature°C | +1.7 to +4.4 (A1B Scenario, SE Asia) | +2.7 to +4.2 (A1B Scenario) |
| Change in Rainfall (%) | −2% to +15% (A1B Scenario, SE Asia) | No discernible trend. More studies needed. |
| Change in mean sea level (m) | +0.18 to +0.59 (All IPCC Scenarios, Global) | 0.24 to +0.65 (# IPCC Scenarios) |

*Source*: Singapore's Second National Communication.

increase in sea level above the normal mean can be up to 80 cm (Tkalitch *et al.*, 2009).

Anomalies were observed in December 1999 and 2001 (+43 and +50 cm) and in December 2006 (+60 cm) (Tkalitch *et al.*, 2009). However, the majority lay between −20 to +20 cm. Therefore, sea level anomalies are closely linked with storm surges and they should be taken into account when examining sea level rise and land inundation in Singapore.

## 1.5 *Rates of Sea Level Rise: Summary*

Table 3.2 summarises some of the key predictions of sea level rates. The upper limits of the estimates from the Singapore study are lower than the rates Rahmstorf predicted and Church estimated, but higher compared to the IPCC. Conversely, the observed past data have smaller rates ranging from 2 to 6 mm/yr approximately.

Some of the rates in Table 3.2 were directly derived by using the total sea level rise for 100 years, and assuming that the rate of change per year will be equal (i.e. 0.24 m/100 yr = 2.4 mm/yr). However, as described in Table 3.2, the rate of sea level rise is unlikely to be constant as it will fluctuate with temperature.

Table 3.2. Summary of various observed and predicted sea level rise rates.

| Predicted $\alpha$ (mm/yr) | Area | Observed $\alpha$ (mm/yr) | Area |
|---|---|---|---|
| 2 | Global Lower Estimate (Church 2001) | $5.0 \pm 1.3$ | Gulf of Thailand |
| 9 | Global Maximum Estimate (Church 2001) | $4.5 \pm 1.3$ | Gulf of Thailand |
| 1.8 (0.18 m in 100 yr) 2.4 (0.24 in 100 yr) | Singapore IPCC Singapore Study | 2 | Singapore (Church 2006) |
| 5.9 (0.59 m in 100 yr) 6.5 (0.65 in 100 yr) | Singapore IPCC Singapore Study | 3 | Singapore (Global altimetry data 1993–2010) |
| 3.5 (0.35 m in 100 yr) | Global (Rahmstorf not conservative) | $1.8 \pm 0.3$ mm/yr | Global, Church |
| 12 (1.2 m in 100 yr) | Global (Rahmstorf conservative) | 1.7 mm/yr | Global, Church (20$^{th}$ century) |

## 1.6 *Coastal Protection*

Coastal development is spreading along the World's coasts. Sea levels are rising, so major future asset losses are expected (Abel *et al.*, 2011) as 10% of the global population live in areas with elevations less than 10 m above sea level (McGranahan *et al.*, 2007).

### 1.6.1 *Protection approaches*

In most cases a mixture of techniques and methods are used to protect or adapt to rising sea levels. The two opposite methods are protection and retreat. In between lies accommodation or adaptation (Tol, 2009). Traditionally, protection of coastal areas from flooding is approached from an engineering perspective (Bas *et al.*, 2011). However, new approaches are emerging that are making use of natural processes to preserve or even enhance biodiversity.

### 1.6.2 *Hard structures*

Hard structures are used extensively to protect inland areas from sea water inundation. They are designed to keep the shoreline in a fixed position or to prevent flooding when sea levels are high. The most commonly used structures are seawalls, retaining structures, or revetments (MACUCCSP, 2009).

Seawalls are vertical impermeable barriers constructed mainly from concrete and are designed to withstand the strongest storm waves and prevent sea water from overtopping. In Singapore, due to the large difference in tidal levels, the seaward side of the seawall can be both under water or part of the beach depending on tide conditions. Modern seawalls usually have sloping faces to reduce reflection (Chadwick & Morfet, 1993).

Retaining structures vary in design and can take the form of sheet piles which are driven downward into the sand, or they can sometimes be constructed as long, cylindrical, sand-filled tubes. They are similar to seawalls, and the aim is to withstand the strongest storm waves and prevent sea water from overtopping (MACUCCSP, 2009).

Revetments are walls whose seaward facing side is not vertical but at a slope (similar to a modern seawall). Their slope matches the natural beach profile better so they are more effective at dissipating energy from waves (MACUCCSP, 2009). Thus, they are less likely to cause beach erosion and are also less prone to failure.

### 1.6.3 *Re-alignment/retreat*

*Managed re-alignment* or *managed retreat* (Rupp-Armstrong & Nicholls, 2007) is most commonly associated with protection of inter-tidal areas such as mud flats and salt marshes in estuaries (Chadwick & Morfet, 1993). The principle is to allow areas that were converted to different use to return to their natural state and coverage, therefore creating natural barriers to seawater intrusion of such salt-marsh and intertidal areas (Zanuttigh, 2011). Planned retreats provide space for coastal ecosystems to invade as sea rises (Abel *et al.*, 2011). An optimal retreat generally requires a longer time than shore protection (Titus *et al.*, 2009) because the economic investments in infrastructure, and human investments in businesses and communities, can have useful lifetimes of many decades (MACUCCSP, 2009).

An example of a managed realignment scheme is in Medmerry (UK), where new defences further inland from the coast were constructed to allow a new intertidal area to form (MMRS, 2012). Studies have shown that realignment can be viable. In fact, Zanuttigh suggests that cost benefit analyses typically show a net advantage of managed realignment over other constructed defence options (Zanuttigh, 2011). Further work showed that managed realignment for a location in Humber (UK) can become more economically efficient if the time horizon over which the measures were considered is longer (Turner *et al.*, 2007).

### 1.6.4 *Ecosystem engineering/modified seawall structures*

There are options for shore protection that may combine hard engineering structures and retreat measures with ecosystem engineering and other modifications that can improve biodiversity.

A new approach for coastal protection is ecosystem engineering which utilises ecosystem engineering species that achieve civil engineering objectives in coastal protection (Bas *et al.*, 2011). This can be achieved either by using species that trap sediment and damp wave activity, or by modifying hard substrates in seawalls to enhance their ecological functions (Bas *et al.*, 2011).

Examples include the use of oyster beds, mussel beds, willow floodplains, marram grass and other grasses such as eelgrass. Transplanted intertidal eelgrass areas can increase sediment accretion by 5 to 7 mm of sediment during the growing season in the Wadden Sea (Boss *et al.*, 2007), whilst sediment accretion was present on perennial seagrasses such as *Posidonia* beds at 2 mm per year in the Mediterranean Sea (Gacia & Duarte, 2001).

Also, there is evidence to show that soil accretion rates in mangrove forests are currently keeping pace with mean sea level rise (Alongi, 2008). Therefore it is possible to strengthen or partly replace engineering measures to protect the coastline with ecological elements (Bas *et al.*, 2011). For Singapore, these ecological elements can be mangroves or seagrasses.

### 1.7 *Approaches in Singapore*

#### 1.7.1 *Current shore protection in Singapore*

Hard structures are used at most of the length of the Singapore coastline. In general, sloping seawalls have been constructed in areas with high wave activity, and vertical walls in sheltered areas. In some cases, mangrove plants have colonised the base of the seawall.

#### 1.7.2 *Modified seawall projects*

Due to the small land area, hard protection methods appear to be appropriate for Singapore. However, since 2007, Singapore has been studying modified seawalls that can prevent land inundation and provide biodiversity opportunities. The "Garden Cities in the Sea" project aims to reduce the negative impacts of urban development on biodiversity as they create habitats for the marine life in man-made infrastructure (GCS, 2012).

Another project was undertaken that aimed to design innovative coastal protection using ecosystem-based approaches. Researchers tested whether complex concrete tiles attached to seawalls can enhance biodiversity (Hong, 2012). Understanding how to improve the value of seawalls as substitutes of natural habitats is important for intertidal biodiversity conservation on modified shorelines (Todd, 2009).

As the value and need for land is so significant in Singapore, it can be concluded that this modified seawall design is an appropriate coastal protection measure to be used in this study. Therefore, the cost of constructing this biodiversity enhanced seawall was compared against the loss of capital stock due to the inundation of the selected site.

## Study Area

### 2.1 *Tide Levels*

Information on the tide levels was taken from the tide tables produced by The Marine Port Authority (MPA) of Singapore for 2012. The Tanjong

Pagar tide station was chosen for this assessment as the reference station, as it is the closest station to the study area approximately 5 km away.

The ML (mean level) for the Tangong Pagar Station is set at 1.652 m whilst the means for high tides during a spring (MHWS) and neap (MHWN) tide are set at 2.9 m and 2.2 m respectively. Similarly, the low tide levels for a spring and a neap tide (MLWS and MLWN) are 1.3 m and 0.6 m respectively. However, according to the National Environment Agency (NEA), maximum high tides can reach up to 3.3 m. This maximum value of 3.3 m was used in the analysis as it represents the highest tide condition. Figure 3.1 explains how the different tide levels are measured and interpreted.

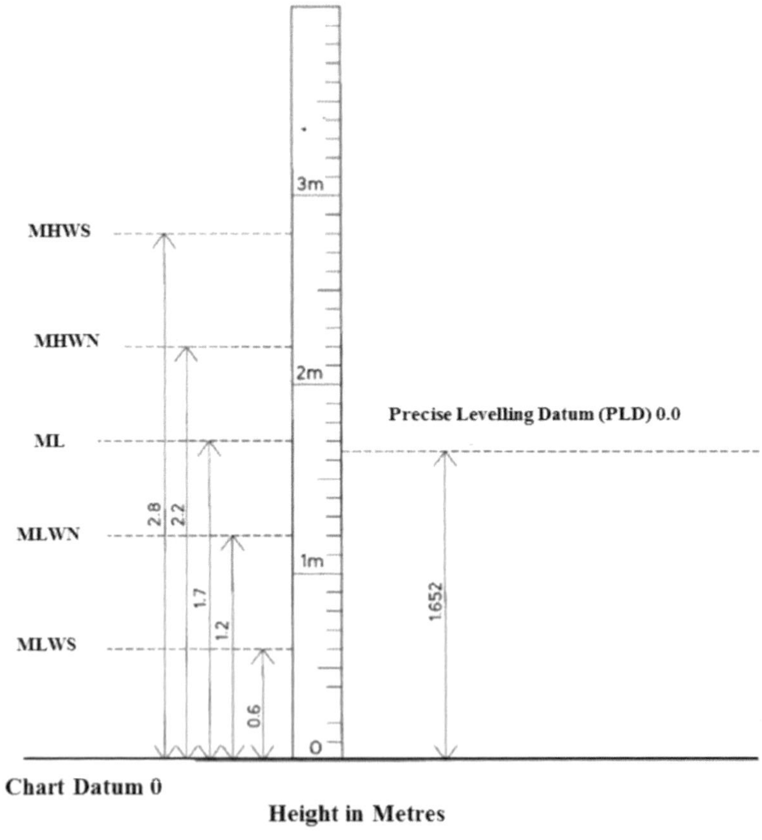


Fig. 3.1. Relationship between various levels at Tanjong Pagar tidal station (*Source*: Singapore Tide Tables, 2012).

Table 3.3. Proposed sea level rise scenarios for Singapore coupled with spring high tides.

| | Scenario 1 | Scenario 2 | Scenario 3 | Scenario 4 | Scenario 5 |
|---|---|---|---|---|---|
| Rise in 100 yr (m) | +0.24 m | +0.65 m | +1.00 m | +0.65 m | +1.00 m |
| Sea level rise (mm/yr equivalent) | 2.4 | 6.5 | 10 | 6.5 | 10 |
| Tide Stage | Max. Spring High Tide | Max. Spring High Tide | Max. Spring High Tide | Max. Spring High Tide | Max. Spring High Tide |
| Sea Level Anomaly | No | No | No | Yes | Yes |

## 2.2 *Sea Level Rise Scenarios*

Five sea level rise scenarios were chosen as they combine the findings of the local study, coupled with extreme conditions such as maximum spring tides, sea level anomalies and sea level rise (Table 3.3). These scenarios represent a range of estimates and were not selected based on their probability of occurring. Scenario 1 assumes that sea levels in Singapore will increase by 0.24 m. Scenario 2 assumes that sea levels in Singapore will increase by 0.65 m. For Scenarios 3, 4 and 5, a sea level rise of 1 m as described in section 1.2 is assumed, with no sea level anomaly in Scenario 3 and sea level anomalies in Scenarios 4 and 5.

## 2.3 *Site Selection and Description*

In section 1.4.1 it was shown that the area most vulnerable to the effects of storm surges due to the north-east monsoon is the Singapore Straits near the south, south-eastern coasts of Singapore, as it is more exposed. In addition, according to the SSNC for Climate Change, land reclamation levels for new coastal reclamation sites have been set at a minimum of 1.25 m above the astronomical tide (Chang, 2006).

These land reclamation levels refer to developments after 1991, so it is possible that pre-1991 reclamation was not undertaken using the 1.25 m clearance. Therefore, an area in Singapore that was reclaimed before 1991 and may be vulnerable to storm surges was selected. In addition, the site selection was based on the availability of ground level information which was used to create the ground model of the site.

The site selected covers an area of 380 hectares (ha) and borders Joo Chiat Road South in the west, the coastline of East Coast Park in the

Fig. 3.2. Calculation of area of the proposed site using One Map (courtesy of One Map Singapore).

south, Changi Road in the north and Frankel Avenue and Marine Vista in the east. The area of the site was calculated using AutoCAD software and the OneMap tool, (Figure 3.2). The East Coast Park and the East Coast Parkway were completed in the 70 s and 80 s. The south side of the site is surrounded mainly by reclaimed sandy beaches and also coastal structures such as artificial headlands.

## 2.4 *Study Site Ground Elevations*

To derive the ground model, ground levels were obtained by:

(1) Sewer interpretation plans (SIP) produced by the Public Utilities Board (PUB). The ground levels from the SIP are based on sewer manhole cover level information and topographical surveys. The plan does not include building levels. The SIP broadly covers the area highlighted in a dotted line hatched pattern in Figure 3.3, which is almost 90% of the total area of the selected site. Due to the sensitivity of the information, the detailed data from the SIP cannot be presented.

Fig. 3.3. Approximate site boundary and coverage of SIP (dotted line hatched pattern) and topographic map (solid line dotted pattern) (courtesy of One Map Singapore).

(2) The remaining ground levels (Figure 3.3, in solid line dotted pattern) were calculated using a hard copy of the 1:50000 scale topographic map of Singapore, produced by the Ministry of Defence (MD) in 2006. The contour interval of the map is 20 metres whilst the vertical datum of the ground levels is based on the Mean Sea Level (ML) at Keppel Harbour. The spot height information available on the topographic map was very limited (Table 3.4), therefore the triangulation process for the location south of Marine Parade was less accurate compared to the rest of the site. The spot heights were referenced against ML, using Figure 3.1.

The survey method used in Singapore is the SVY21 reference system which sets the mean sea level (ML) at 100.000 m. The SIP plan uses the SVY21 whilst the levels in the topographic map were based on ML. Therefore, the tide levels and the topographic map levels were converted into the SVY21 system. Specifically:

The chart datum based on SVY21 is 100.000 m − 1.652 m = 98.348 m, so the ML will be 98.348 + 1.7 = 100.048 and the MHWS = 98.348 +

Table 3.4. Spot heights from 1:50000 scale topographic map of Singapore used in the study.

| location | Ground level from topographic map (m) |
|---|---|
| Marine Parade NPP Flyover | 3.08 |
| East Coast Rd — Siglap Canal | 4.00 |
| East Coast Parkway (close to Island Resort) | 2.64 |

Table 3.5. Conversion of chart datum, ML, MHWS and max spring tide levels to the SVY21 reference system.

| No | | PLD reference | SV21 reference |
|---|---|---|---|
| 1 | Chart Datum | 0 | 98.348 m |
| 2 | ML | 1.652 m | 100.000 m |
| 3 | MHWS | 2.9 m | 101.248 m |
| 4 | Max. Spring High Tide | 3.3 m | 101.648 m |

2.8 = 101.48 m. Using the highest tide (as shown, it can reach up to 3.3 m) the level is then 98.348 + 3.3 = 101.648 m (Table 3.5). This worst case tide condition (No. 4 in Table 3.5) was used to represent sea levels without any sea level rise or sea level anomaly conditions.

The spot height information from the MD topographic map was converted to the SVY21 together with the SIP ground level information (Table 3.6).

The SIP plan was then digitised and all the spot levels were assigned the correct x, y and z coordinates. Subsequently, the spot levels were inserted into ArcGIS software and the ground profile of the area was re-produced in raster format (Figure 3.4). The elevation varies from 101.8 m to 118.0 m, and this figure will be referred to as "baseline". It should be stressed that no building levels were available; therefore the elevations represent the ground levels and roads.

## 2.5 *Detailed Land Use of the Selected Site*

The land uses of the selected site were then identified. The identification was undertaken in order to investigate how representative the site is compared

Table 3.6. Conversion of topographic map spot heights to the SVY21 reference system.

| Location | Ground level from topographic map (m) | Converted to SV21 |
|---|---|---|
| Marine Parade NPP Flyover | 3.08 | 103.128 |
| East Coast Rd — Siglap Canal | 4.00 | 104.048 |
| East Coast Parkway (close to Island Resort) | 2.64 | 102.688 |

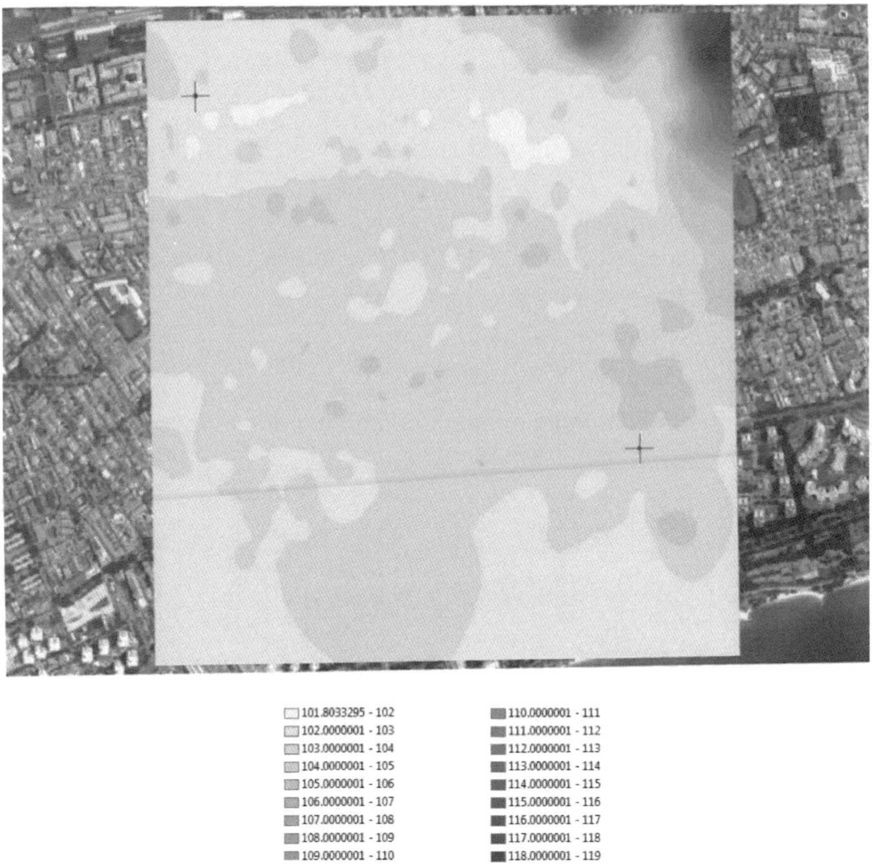


Fig. 3.4. Ground profile of selected site, hereafter referred to as "Baseline".

to the whole island in terms of land use and, subsequently, Capital Stock (see Section 3).

The land use of the selected site was estimated and every effort was undertaken to include as much detail as possible; however, information such as the total area of residential buildings or roads is approximate.

## 2.6 *Areas Inundated Under Different Scenarios*

As stated previously, ML is set at 1.652 m above Chart Datum which is set to 98.348 m, and the maximum high tide level can reach 3.3 m above the Chart Datum. The ground levels on the topographic map used are based on the ML; therefore during the maximum high tide, the level above ML is:

$$3.3\,\text{m}\,(\text{Max. Spring High Tide}) + 98.348\,(\text{Datum}) = 101.648\,\text{m}$$

(See Table 3.5)

Therefore this level forms the basis for the different sea level rise scenarios (Table 3.7), and the maximum sea level can be estimated. For sea level anomaly, the value chosen (0.6 m) represents an extreme condition, which nevertheless has been observed in Singapore. Based on all the above the maximum sea level was estimated using the SVY21 for all five scenarios.

Table 3.7. Sea level calculation for five sea level rise scenarios for Singapore.

| | Scenario 1 | Scenario 2 | Scenario 3 | Scenario 4 | Scenario 5 |
|---|---|---|---|---|---|
| Sea level Rise (m) | +0.24 (from Singapore Study) | +0.65 m (from Singapore Study) | +1.00 m (from Rahmstorf Study) | +0.65 m (from Singapore Study) | +1.00 m (from Rahmstorf Study) |
| High Tide level (Above ML) (m) | 100.048 | 100.048 | 100.048 | 100.048 | 100.048 |
| Max. Spring High Tide | 101.648 | 101.648 | 101.648 | 101.648 | 101.648 |
| Sea level Anomaly (m) | No | No | No | 0.6 | 0.6 |
| Maximum sea water level (m) | 101.888 | 102.298 | 102.648 | 102.848 | 103.248 |

Fig. 3.5. Ground profile of selected site showing inundated areas when Scenario 1 (101.888) takes place.

Using the profile from Figure 3.4 and the maximum water levels from (Table 3.7), five area inundation scenarios were created in ArcGIS (Figure 3.5 to Figure 3.9). The shaded outlined contours represent the extent of the area that will be inundated under the different scenarios. Land inundation presented in Figure 3.5 to Figure 3.9 (dark shaded contours) only shows areal extent and not depth. Therefore it is possible that some inundated areas are in fact only covered by, for e.g., 100 mm of water. Due to the added complexity, the depth of inundation was not calculated.

It can be seen that only when Scenarios 3, 4 or 5 are chosen is there an identifiable inundated area in the site. For Scenarios 3, 4 and 5 the total area of inundation was estimated. For Scenario 1, the area to be inundated is insignificant; and for Scenario 2, the area that is inundated is far away from the coast, therefore, it cannot be attributed to sea level rise, but is treated as a localised depression. Therefore the assumption is that for Scenarios 1 and 2 inundation is zero.

Fig. 3.6. Ground profile of selected site showing inundated areas when Scenario 2 (102.298) takes place.

For Scenarios 3, 4 and 5, only a selected portion of the total inundated area was selected (circled with dotted lines in Figures 3.10–3.12) as the remaining inundated areas were located away from the coast and were isolated, with no clear flow path connecting them to the coastline. In contrast, the areas by the coastline were assumed to flood as visual inspections indicate that they are low lying. The approximate area to be inundated for Scenario 5 is shown in Figure 3.12.

In total, the area that will be inundated for scenario 5 is approximately 120 ha. For Scenarios 3 and 4, the inundated areas are approximately 30 and 60 hectares respectively. Therefore, based on the analysis above, there is a significant area (approximately 32% of the total site) for Scenario 5 that will be inundated. This percentage drops to 16% and 8% for Scenarios 4 and 3 respectively.

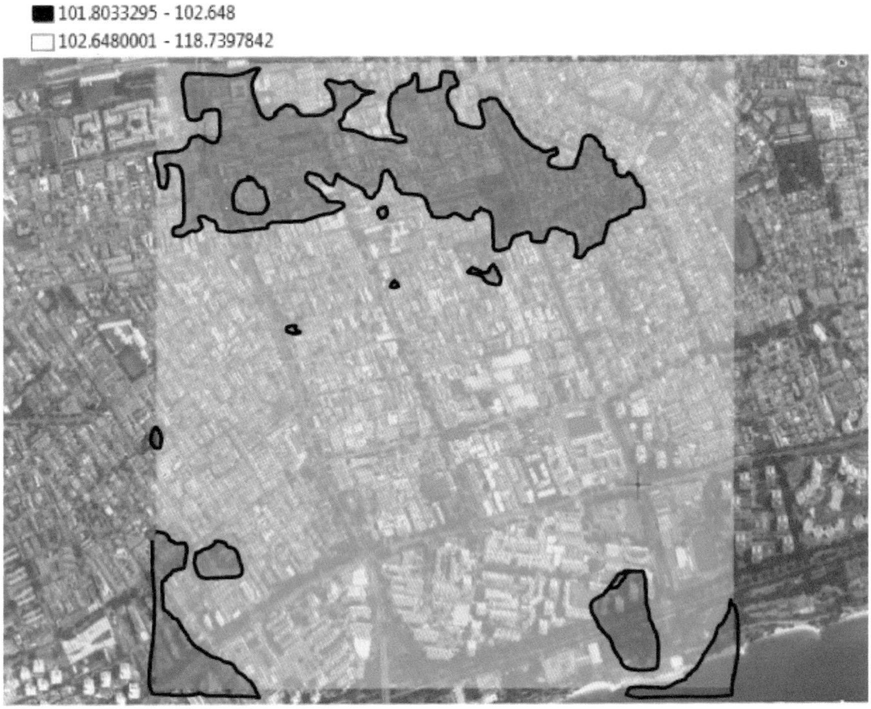


Fig. 3.7. Ground profile of selected site showing inundated areas when Scenario 3 (102.648) takes place.

## Capital Stock and Cost Estimations

Following the selection of the seawall design and the identification of the inundated land, the Capital Stock lost due to inundation in the selected site and the cost of the seawall construction to protect against inundation for 100 years were estimated and then compared.

### 3.1 *Estimation of Capital Stock of Inundated Land*

Capital Stock or capital formation is used to describe net capital accumulation during a time period. Gross Capital formation consists of outlays on additions to the fixed assets of the economy plus net changes in the level of inventories (The World Bank, 2012). These fixed assets comprise land improvement works such as drains, plant, machinery, construction of roads and railway lines and also schools, offices, hospitals residential, industrial

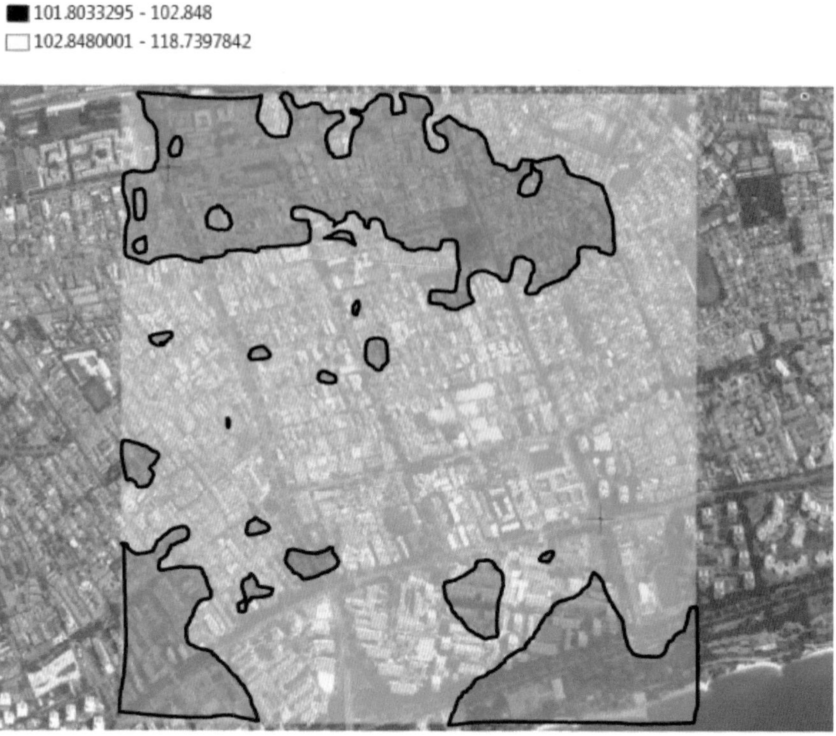


Fig. 3.8. Ground profile of selected site showing inundated areas when Scenario 4 (102.848) takes place.

and commercial buildings. A country uses capital stock coupled with labour to provide services and produce goods; an increase in this capital stock is known as capital formation.

To estimate the capital stock of Singapore, the Perpetual Inventory Method was used. The data on Gross Capital Formation (GCF) based on a constant Local Currency Unit (constant LCU) were used for this estimation (The World Bank, 2012).

The data were transformed into natural logarithms, i.e., Ln(GCF) and plotted against time (from 1975 until 2011 — data availability for Singapore) (Table 3.9 and Figure 3.13).

The trend line has the following equation type: Ln (GFC) $= \alpha 1 + \gamma$ (T). The coefficient $\gamma$ is needed to determine the size of Capital Stock for the initial year of the time series. From the trend line, equation $\gamma = 0.0455$.

The initial year of the time series is 1975. The size of Capital Stock (hereafter abbreviated as KM) for this year (KM1975) was estimated from

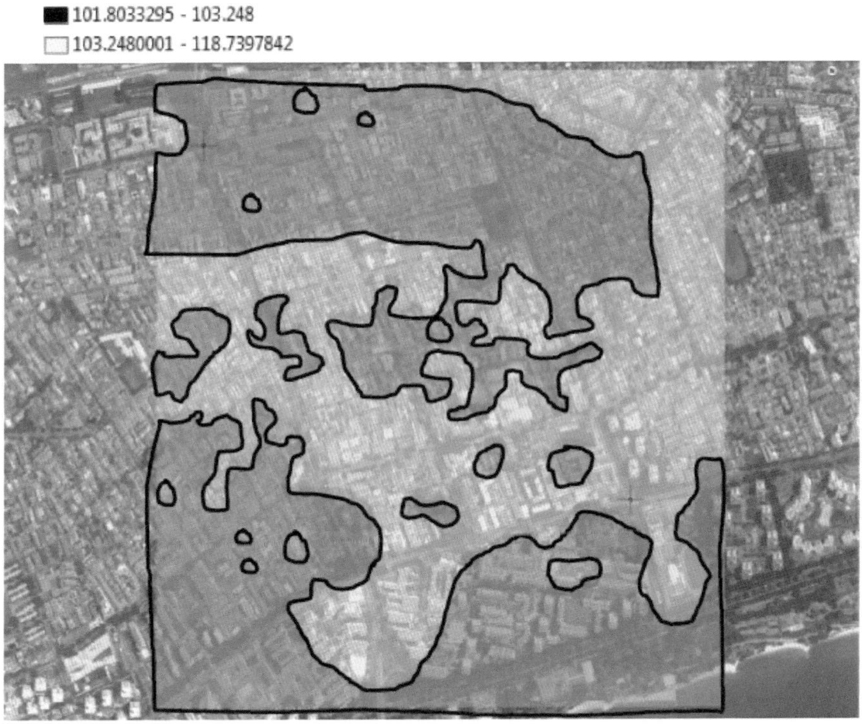


Fig. 3.9. Ground profile of selected site showing inundated areas when Scenario 5 (103.248) takes place.

the GCF value from the data set. This is defined as follows:

$$KM_{1975} = GFC_{1975}/\gamma + \delta$$

where $\gamma = 0.0455$ and $\delta$ is the rate at which capital stock depreciates, which is equal to 1/30, assuming that the life span of the plant and the machinery is 30 yrs = 0.033.

So the value of the Capital Stock for 1975 can be estimated as:

$$KM_{1975} = SGD\$\,2.58418E + 11$$

The size of Capital Stock for Singapore for the subsequent years was estimated by adopting the following formula (Table 3.10):

$$KM_T = KM_{T-1} + GCF_T - (\delta^* KM_{T-1})$$

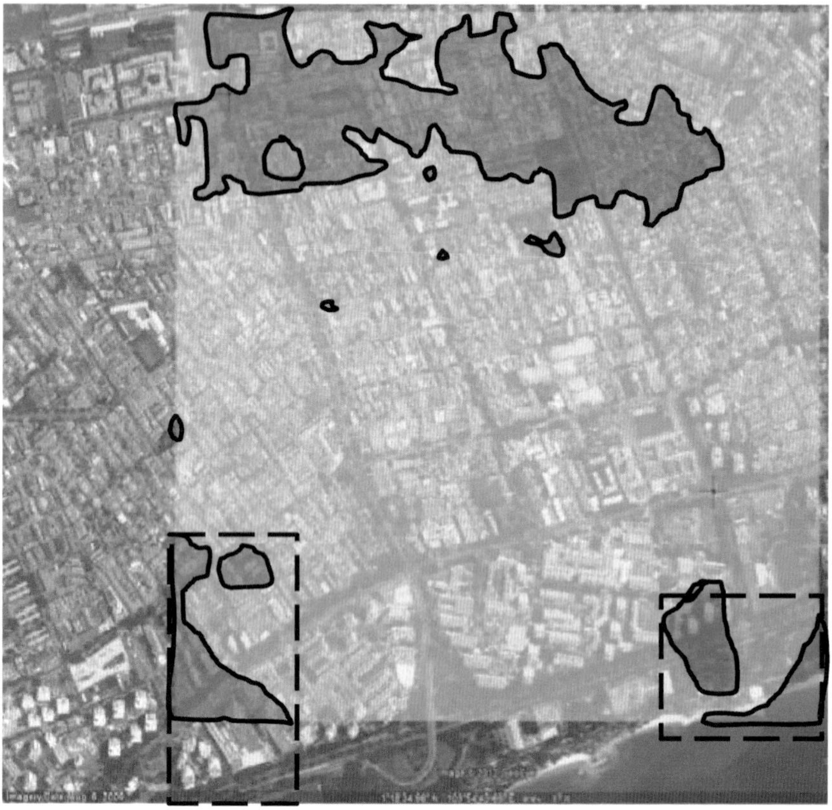

Fig. 3.10. Selected inundated area from Scenario 3 (102.648).

### 3.2 *Modified Seawall Description and Cost*

The modified seawall design is comprised of three main elements: Granite boulders, mortar and the concrete tiles which are mounted on the seawall. At the time of the study, there were no available designs of this seawall and all the concrete tiles were "retrofitted" on existing seawalls instead. Therefore for the cost estimation of such construction, it was assumed that a standard seawall will be constructed followed by retrofitting of concrete tiles. It is expected that in the future these types of modified seawalls will already be fitted with the concrete tiles or are constructed using other materials.

To estimate the construction of the standard seawall, the rates from the Punggol–Serangoon Dam/Reservoir project were used as a guide *only*.

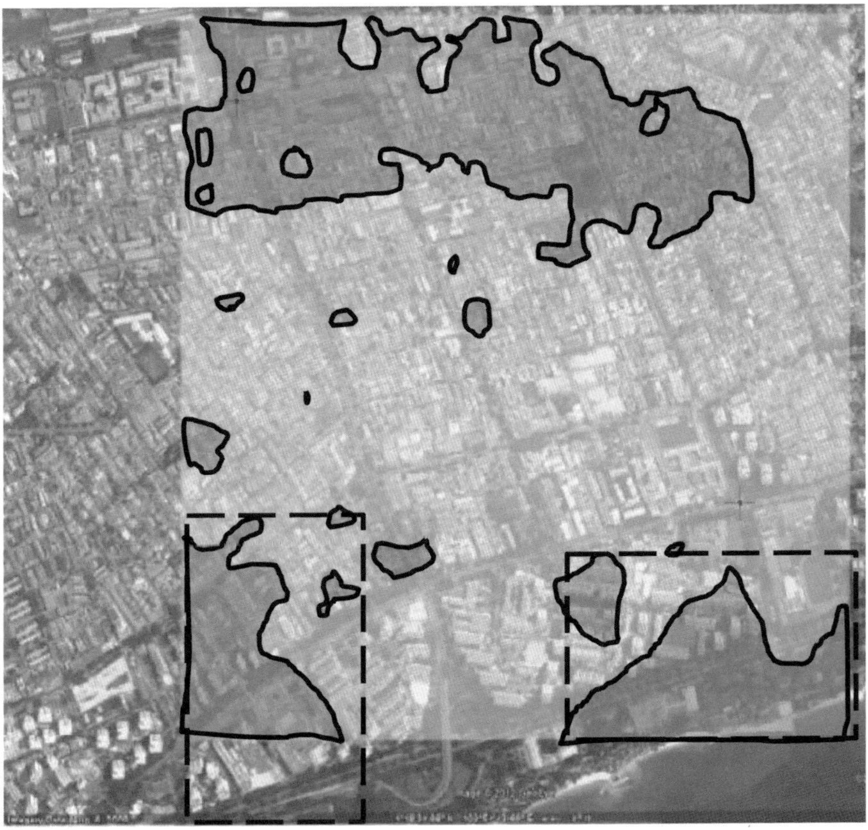

Fig. 3.11. Selected inundated area from Scenario 4 (102.648).

The project involved the transformation of Sungei Serangoon and Sungei Punggol from a river to a freshwater reservoir. A seawall or bund was used to "block" the river flow. The height of the embankment in the project was approximately 8–10 m, which is relatively high compared to the height of the seawall that may be needed for the study site. However, since the physical configuration of the study site coastline is not known, the embankment in Sungei Punggol was chosen as a conservative approximation.

The original rates were estimated in 2006; therefore they were adjusted to match 2012 prices. Prices were appreciated based on average inflation for Singapore from January 2006 until September 2012, which was estimated to be 3.31% per year (Department of Statistics, 2012). The costs estimates are presented in Table 3.11.

Fig. 3.12. Selected inundated area from Scenario 5 (103.248).

In order to establish the cost of installation of the concrete biodiversity-enhancing tiles, the staff from the Department of Biological Studies involved in the project provided basic information about the concrete tile installation (Table 3.12). The work is still in the experimental stage, therefore, the shape or dimensions of the concrete tiles are likely to be modified in the future. This will influence the cost but not dramatically.

Based on the information from Table 3.12, the construction and installation costs for the concrete tiles were estimated (Table 3.13).

From Table 3.13, the cost of the concrete tile manufacturing and installation is very low compared to the construction of the seawall (only 0.2% of the total cost). When the costs for the seawall and the tile retrofitting are added up, it will reach a total (in 2012) of SGD 53,246,048.13.

Table 3.8. Gross Capital Formation (GCF) data based on constant LCU for Singapore.

| | | | | | | | | | | |
|---|---|---|---|---|---|---|---|---|---|---|
| **1975** | **1976** | **1977** | **1978** | **1979** | **1980** | **1981** | **1982** | **1983** | **1984** | **1985** |
| 1.176E+10 | 1.231E+10 | 1.186E+10 | 1.33E+10 | 1.535E+10 | 1.726E+10 | 1.802E+10 | 2.006E+10 | 2.2E+10 | 2.381E+10 | 2.117E+10 |
| **1986** | **1987** | **1988** | **1989** | **1990** | **1991** | **1992** | **1993** | **1994** | **1995** | **1996** |
| 1.967E+10 | 2.129E+10 | 2.072E+10 | 2.296E+10 | 2.692E+10 | 2.705E+10 | 2.93E+10 | 3.433E+10 | 3.42E+10 | 3.909E+10 | 4.248E+10 |
| **1997** | **1998** | **1999** | **2000** | **2001** | **2002** | **2003** | **2004** | **2005** | **2006** | **2007** |
| 5.076E+10 | 3.835E+10 | 4.264E+10 | 5.356E+10 | 4.244E+10 | 3.967E+10 | 2.77E+10 | 4.187E+10 | 4.17E+10 | 4.898E+10 | 5.704E+10 |
| **2008** | **2009** | **2010** | **2011** | | | | | | | |
| 7.268E+10 | 5.805E+10 | 6.143E+10 | 6.737E+10 | | | | | | | |

Table 3.9. Natural logarithms of Gross Capital Formation (GCF) data.

| | | | | | | | | | | |
|---|---|---|---|---|---|---|---|---|---|---|
| **1975** | **1976** | **1977** | **1978** | **1979** | **1980** | **1981** | **1982** | **1983** | **1984** | **1985** |
| 23.19 | 23.23 | 23.20 | 23.31 | 23.45 | 23.57 | 23.61 | 23.72 | 23.81 | 23.89 | 23.78 |
| **1986** | **1987** | **1988** | **1989** | **1990** | **1991** | **1992** | **1993** | **1994** | **1995** | **1996** |
| 23.70 | 23.78 | 23.75 | 23.86 | 24.02 | 24.02 | 24.10 | 24.26 | 24.26 | 24.39 | 24.47 |
| **1997** | **1998** | **1999** | **2000** | **2001** | **2002** | **2003** | **2004** | **2005** | **2006** | **2007** |
| 24.65 | 24.37 | 24.48 | 24.70 | 24.47 | 24.40 | 24.04 | 24.46 | 24.45 | 24.61 | 24.77 |
| **2008** | **2009** | **2010** | **2011** | | | | | | | |
| 25.01 | 24.78 | 24.84 | 24.93 | | | | | | | |

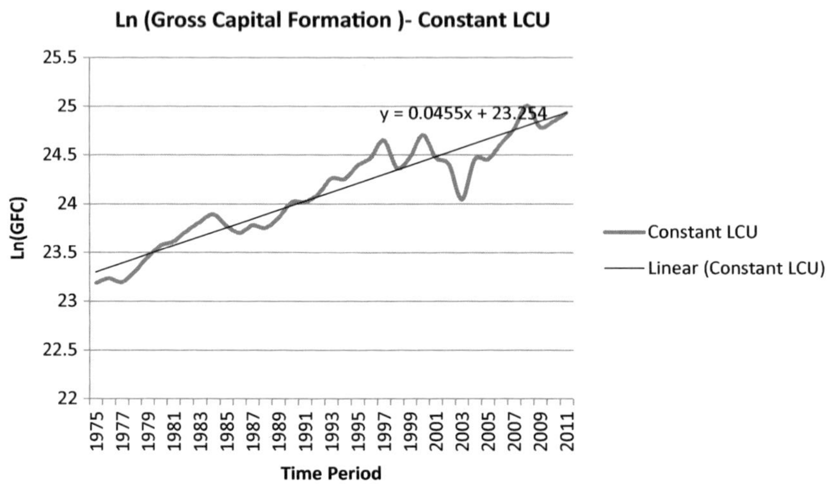


Fig. 3.13. Plot of logarithmic GFC data vs time and trend line.

To assess the cost of the seawall through the whole of the 100 year period, the Net Present Value (NPV) for the seawall was estimated. The first estimation assumed that the seawall lifespan is 50 years (Huaren & Swann, 2001); therefore, it shall be re-constructed at year 50, whilst the second estimation assumed no re-construction.

The main assumptions for the NPV calculations were:

a) Maintenance of the seawall is annual and the rate is SGD 7,015/km/year (Turner *et al.*, 2007); therefore, SGD 14,030 for 2 km
b) No increase due to inflation throughout the 100 year span
c) The declining discount rate used is: 3.5% for years 1–30, 3% for years 31–7, 5 and 2.5% for years 76–100 (Turner *et al.*, 2007)

Table 3.10. Estimated size of Singapores capital stock (in Singapore dollars).

| | | | | | | | | | | |
|---|---|---|---|---|---|---|---|---|---|---|
| **1975** | **1976** | **1977** | **1978** | **1979** | **1980** | **1981** | **1982** | **1983** | **1984** | **1985** |
| 2.58E+11 | 2.62E+11 | 2.65E+11 | 2.69E+11 | 2.76E+11 | 2.84E+11 | 2.92E+11 | 3.029E+11 | 3.14E+11 | 3.28E+11 | 3.38E+11 |
| **1986** | **1987** | **1988** | **1989** | **1990** | **1991** | **1992** | **1993** | **1994** | **1995** | **1996** |
| 3.56E+11 | 3.56E+11 | 3.65E+11 | 3.76E+11 | 3.90E+11 | 4.04E+11 | 4.20E+11 | 4.40E+11 | 4.60E+11 | 4.83E+11 | 5.10E+11 |
| **1997** | **1998** | **1999** | **2000** | **2001** | **2002** | **2003** | **2004** | **2005** | **2006** | **2007** |
| 5.44E+11 | 5.64E+11 | 5.88E+11 | 6.22E+11 | 6.43E+11 | 6.61E+11 | 6.67E+11 | 6.87E+11 | 7.060E+11 | 7.31.E+11 | 7.64E+11 |
| **2008** | **2009** | **2010** | **2011** | | | | | | | |
| 8.11E+11 | 8.42E+11 | 8.75E+11 | 9.13E+11 | | | | | | | |

Table 3.11. Estimation of seawall construction.

| Item no | Description | Unit | Amount ($) 2006 | Amount ($) 2012 | Amount ($) 2012 |
|---|---|---|---|---|---|
| | **DAM ENBANKMENT** | | Per linear m of wall | Per linear m of wall | For total length of 2000 m |
| 1 | Sand filling for dam embankment above the sand-key (sand replacement due to dredging not included) | Sum | $14,685.59 | $15,140.84 | $30,281,688.45 |
| | **Slope and Toe Protection at Seaward Side** | | | | |
| 2 | Class A stones | Sum | $4,659.40 | $4,803.84 | $9,607,673.43 |
| 3 | Class B stones | Sum | $161.55 | $166.55 | $333,106.73 |
| 4 | Class C stones | Sum | $709.65 | $731.65 | $1,463,307.67 |
| 5 | Geofabric | Sum | $1,012.02 | $1,043.39 | $2,086,781.49 |
| 6 | Armour rock properly pitched and set in cement mortar | Sum | $569.35 | $587.00 | $1,173,999.70 |
| | **Slope and Toe Protection** | | | | |
| 7 | Class A stones | Sum | $1,460.74 | $1,506.02 | $3,012,047.75 |
| 8 | Class B stones | Sum | $279.47 | $288.13 | $576,263.39 |
| 9 | Class C stones | Sum | $477.38 | $492.18 | $984,361.31 |
| 10 | Geofabric | Sum | $1,016.21 | $1,047.71 | $2,095,423.15 |
| 11 | Armour rock properly pitched and set in cement mortar | Sum | $747.39 | $770.56 | $1,541,120.05 |
| 12 | Total | Sum | $25,778.75 | $26,577.89 | **$53,155,773.13** |

Table 3.12. Construction information on concrete biodiversity enhancing Tiles.

| Item no | Description of item | Dimensions/ Material |
|---|---|---|
| 1 | Tile Dimensions | Circular 600 mm diameter × 40 mm deep |
| 2 | Tile Material | Concrete |
| 3 | Tile Weight | 25kg |
| 4 | Amount of Tiles needed | 25 tiles per 50 $m^2$ |
| 5 | Time required to install 1 tile | 10 minutes |
| 6 | Way of mounting | Using cement |
| 7 | No of people needed/tile | 2 people |

Table 3.13. Estimation of concrete tile construction and installation of concrete biodiversity enhancing tiles.

| Item No | Description | Unit | Diam. | Thickness | Quantity | Rate | Amount (SGD) | Remarks |
|---|---|---|---|---|---|---|---|---|
| | **CONCRETE TILE CONSTRUCTION AND INSTALLATION FOR BIODIVERSITY IMPROVEMENT** | | | | | | | |
| 1 | Concrete disc | m3 | 0.6 | 0.04 | 0.02 $m^3$ of concrete | $182.00 | $3.64 | Assumes it is mass concrete Grade 50. Rates taken based on I-Enviroconcrete 50 with 7% S-Fume Extreas MC special DM, S4 Slump (Tremie) 160–210 mm |
| 2 | Labour Cost (2Nos) | hr | | | 0.34 | $11.50 | $3.91 | Assumes that 2 workers spend 10 minutes to lay 1 tile |
| | **Cost to Supply and Install one Concrete Tile** | | | | | | **$7.55** | |
| 3 | Mould | | | | 1 | | **$3,000.00** | Assume that mould is required to produce the pre-fabricated concrete tiles |
| 4 | Number of tiles needed per 50 $m^2$ of seawall | | | | 25 tiles | | | |
| 5 | Number of tiles needed for a 2000 m long × 10 m wide (20000 $m^2$) seawall | | | | 10000 | | | |
| 6 | Total cost for all tiles construction and installation of 10000 tiles including mould and 15% for unforeseen costs | | | | | | **$90,275.00** | |

Based on the NPV analysis, the construction of the seawall will cost SGD 117,397,867.06 and SGD 53,665,768.55 respectively, which reveals that the design life of the seawall is very significant as the cost of re-construction is very high.

### 3.3 *Comparison of Capital Stock Lost Due to Inundation versus Modified Seawall Cost*

In sections 3.1 and 3.2 the Capital Stock of Singapore and the cost for the construction of the seawall were estimated. The Capital Stock lost due to inundation can be simply estimated by using the following analogy for different inundation scenarios:

$$Area\ inundated\ (see\ section\ 2.6)/Singapore\ area = x$$
$$Capital\ Stock\ lost\ in\ inundated\ areas = x^{*}Capital\ Stock\ for\ Singapore\ 2011$$

This approach assumes that Capital Stock of the inundated area is a percentage of the total Capital Stock for Singapore based on the ratio of the inundated area over the total land area of Singapore. These ratios are presented in Table 3.14 for the inundations in Scenarios 3, 4 and 5; inundation for Scenarios 1 and 2 was considered to be zero. The amount of Capital Stock likely to be lost due to inundation is shown in Table 3.15.

The area of Singapore has been estimated at 714.3 km$^2$ (Department of Statistics Singapore, 2012).

Therefore, based on the tables above, the loss of capital stock for each of the three inundation scenarios is greater than the cost of constructing the seawall (SGD 117,397,867.06 and SGD 53,665,768.55, with and without reconstruction, respectively), as presented in Table 3.16. Therefore it can be argued that for scenarios 3, 4 and 5, the construction of a modified seawall can be justified in economic terms.

Table 3.14. Ratio of area lost by inundation over total land area of Singapore.

| | Inundation scenario 1 | Inundation scenario 2 | Inundation scenario 3 | Inundation scenario 4 | Inundation Scenario 5 |
|---|---|---|---|---|---|
| Inundated area | 0 | Estimated 0 | 0.3 km$^2$ | 0.6 km$^2$ | 1.2 km$^2$ |
| Singapore area | 714.3 km$^2$ | 714.3 km$^2$ | 714.3 km$^2$ | 714.3 km$^2$ | 714.3 km$^2$ |
| Inundated area/ Singapore area | | | 0.00042 km$^2$ | 0.00084 km$^2$ | 0.00168 km$^2$ |

Table 3.15. Estimation of capital stock lost losses for different Inundation scenarios.

| | Inundation Scenario 3 | Inundation Scenario 4 | Inundation Scenario 5 |
|---|---|---|---|
| Capital stock of Singapore in 2011 | \$9.13E+11 | \$9.13E+11 | \$9.13E+11 |
| Inundated area/Singapore area | 0.00042 $km^2$ | 0.00084 $km^2$ | 0.00168 $km^2$ |
| Estimated loss of capital stock in inundated areas | \$412,668,547 | \$825,337,095 | \$1,572,070,656 |

Table 3.16. Capital stock losses versus estimated construction and maintenance of seawall.

| | Inundation scenario 3 | Inundation scenario 4 | Inundation scenario 5 |
|---|---|---|---|
| Estimated maximum cost of constructing seawall (no reconstruction) | \$53,665,768.55 | \$53,665,768.55 | \$53,665,768.55 |
| Estimated maximum cost of constructing seawall (with reconstruction) | \$117,397,867.06 | \$117,397,867.06 | \$117,397,867.06 |
| Estimated loss of capital stock in inundated areas | \$412,668,547 | \$825,337,095 | \$1,572,070,656 |

## Conclusions

From the analysis undertaken, it was shown that the construction of the modified seawall to protect against inundation is generally much cheaper compared with the losses in Capital Stock for extreme sea level rise scenarios. In fact, only if the inundated area lost was less than 9 ha would the Capital Stock losses become less than the cost of the seawall construction.

This is in agreement with the study conducted in 2005, which found that across ten coastal sites representing market land, protection against inundation from sea level rise until year 2100 was the lowest cost strategy (Ng & Mendelsohn, 2005). The main difference is that they used land market value, rather than Capital Stock loss, for their comparison. Therefore, it can be concluded that due to the increased industrialisation, population density and scarcity of land, it appears that total protection of the Singapore coastline against sea level rise can be an option worth considering.

The above argument can be further reinforced if loss of jobs, rise in unemployment and reduction of productivity is also considered together with losses in capital stock, which can also arise if inundation due to sea level rise takes place. Therefore, taking them into account makes the option of total protection an even more attractive proposal.

In this study a few different methods of protection against sea level rise were presented. A biodiversity enhanced seawall can provide protection for the valuable land of the Singapore coastline. As land is limited, if better protection can be offered with a seawall that occupies less space than natural defences such as wetlands, and the Capital Stock losses outweigh its ecosystem service value, then the choice to construct a seawall is a logical one.

## Limitations/Assumptions & Recommendations

In this study, no valuation of the inevitable environmental impact of the seawall construction was undertaken. In addition, there are other impacts such as aesthetic and economic impacts, as the construction of this enhanced seawall may limit visitorship to East Coast Park, restrict certain activities and affect businesses, at least in the short term.

Also, the estimation of the capital stock lost due to inundation was based on the ratio below:

$$Area\ inundated/Singapore\ area = x$$

$$Capital\ stock\ lost\ in\ inundated\ areas = x * Capital\ Stock\ for\ Singapore\ 2011$$

This simplified analogy may be overestimating the capital stock, as it is mainly a residential area with no major industrial or manufacturing uses and no major infrastructure such as hospitals, rail transport systems, etc. On the other hand, the costs for the construction of the enhanced seawall can be considered to be conservative as well. The height of the sea wall is high (8–10 m) and it has not been adjusted based on the different sea level rise scenarios. However, as both the Capital Stock and the cost for the construction of the sea wall are conservative, the comparison can still be considered appropriate.

In addition, in this study, it was assumed that a "traditional" seawall will first be constructed, and then retrofitted with biodiversity enhancing tiles later. However, it is more likely that the seawall will be redesigned to improve biodiversity and then constructed. Thus, it is suggested that the cost of this enhanced biodiversity seawall should be re-calculated when more information on the actual design is available.

Another issue that was not explored in this study is the effects of the construction of the seawall on the local coastal processes. The seawall may protect the selected area, however, it can lead to increased erosion in other areas due to changes in local hydrodynamics. Therefore before the construction of a seawall, a thorough analysis of the alteration in local hydrodynamic conditions needs to be undertaken and the effects need to be identified.

When other sea level rise protection techniques, especially those that include the use of ecosystem engineering species, are explored, the economic valuation should account for the value of ecosystem services that the protection measures can offer. It is therefore suggested that for all sea level protection methods, the estimation of the loss or increase of ecosystem services or other services that may contribute to the wellbeing of the citizens should be included in any future work. This may shift policy in favour of more environmentally-friendly protection designs, even when the economic valuation is not favourable. Additionally, inputs such as $CO_2$ capture shall be incorporated into the valuation of schemes such as managed realignment or construction of wetlands. Finally, finding ways to calculate the Capital Stock in specific areas rather than the whole country will assist in making the Capital Stock estimation more accurate and targeted for the area of interest.

## References

Abel, N., Gorddard, R., Harman, B., Leitch, A., Langridge, J., Ryan, A. and Heyenga, S. (2011). "Sea Level Rise, Coastal Development and Planned Retreat: Analytical Framework, Governance Principles and an Australian Case Study", *Environmental Science & Policy*, **14**(3), 279–288.

Alongi, M.D. (2008). "Mangrove forests: Resilience, Protection from Tsunamis, and Responses to Global Climate Change", *Estuarine, Coastal and Shelf Science*, **76**, 1–13.

Amos, J. (2012). "Satellites Trace Sea Level Change", *British Broadcast Corporation (BBC)*. [Online] Available at http://www.bbc.co.uk/news/science-environment-19702450, [Accessed September 2012].

Bas, W., Borsje, B.K., van Wesenbeeck, F.D., Paalvast, P., Bouma, T.J., van Katwijk, M.M. and de Vries, M.B. (2011). "How Ecological Engineering can Serve in Coastal Protection", *Ecological Engineering*, **37**(2011), 113–122.

Bos, A.R., Bouma, T.J., Geertje, L.J., de Kort, M. and van Katwijk, M. (2007). "Ecosystem Engineering by Annual Intertidal Seagrass Beds: Sediment Accretion and Modification", *Estuarine, Coastal and Shelf Science*, **74**, 344–348.

Chadwick, A. and Morfet, J. (1993). *Hydraulics in Civil amd Environemntal Engineering, 3rd ed.*, E & FN Spon/Routledge, London, pp. 526–527.

Chang, C.W., Deputy Director, Public Utilities Board, Singapore (2006). "SINGAPORE'S APPROACH TO CLIMATE CHANGE", Presentation to the International Water Association (IWA). [Online] Available at: http://www.iwahq.org/contentsuite/upload/iwa/Document/Atlanta2-06.pdf, [Accessed August 2012].

Church, J.A., Gregory, J.M. and Huybrechts, P. (2001). "Changes in Sea Level" in Houghton, J.T. (ed.), *Climate Change 2001; The Scientific Basis*, Cambridge University Press, Cambridge, pp. 639–694.

Church, J.A., White, N.J. and Hunter J.R. (2006). "Sea level Rise at Tropical Pacific and Indian Ocean islands", *Global and Planetary Change*, **53**(2006), 155–168.

Church, J.A., White, N.J., Coleman, R., Lambeck, K. and Mitrovica, J.X. (2006). "Estimate of the Regional Distribution of Sea Level Rise over the 1950–2000 period", *Journal of Climate*, **17**, 2609–2625.

Department of Statistics Singapore. *Time Series on CPI (2009=100) & Inflation Rate.* [Online] Available at: http://www.singstat.gov.sg/stats/themes/economy/hist/cpi.html [Accessed September 2012].

Department of Statistics Singapore. *Key Annual Indicators (2012) Land Area (Sq km) (for year 2011).* [Online] Available at: http://www.singstat.gov.sg/stats/keyind.html, [Accessed August 2012].

Environment Agency (EA), United Kingdom. (2012). *Medmerry Managed Realignment Scheme (MMRS).* [Online] Available at: http://www.environmentagency.gov.uk/homeandleisure/floods/109062.aspx, [Accessed September 2012].

Gacia, E. and Duarte, C.M. (2001). "Sediment Retention by a Mediterranean Posidonia Oceanica Meadow: The Balance between Deposition and Resuspension", *Estuarine, Coastal and Shelf Science*, **52**(2001), 505–514.

Hong, J. (2012). "Turning Sea Walls into 'Housing' Units for Marine Life", *The Straits Times*, 06 June. [Online] Available at: http://nusbiodiversity.wordpress.com/2012/06/06/lynettes-sea-wall-project-in-the-news-straits-times/, [Accessed September 2012].

Huaren, D. and Swann, L.H. (2001). "Design and Construction of a Seawall on Soft Marine clay" in: *Soft Soil Engineering*. Lee, C.F., Lau, C.K., Ng, C.W.W., Kwong, A.K., Pang, P.L.R., Yin, J.-H. & Yue, Z.Q. (Eds.), *Proceedings of the Third International Conference on Soft Soil Engineering*, 6–8

December, Hong Kong. Swets and Zeitlinger B.V., Lisse, The Netherlands, pp. 185–193.

Intergovernmental Panel on Climate Change (IPCC) (2007). *Fourth Assessment Report: Climate Change, Working Group I, Section 5.5.2.*

Jevrejeva, S., Moore, J.C., Grinsted, A. and Woodworth, P.L. (2008). "Recent Global Sea Level Acceleration Started over 200 Years Ago?", *Geophysical Research Letters*, **35**, L08715.

Marcos, M., Jordà, G., Gomis, D. and Pérez, B. (2011). "Changes in Storm Surges in Southern Europe from a Regional Model under Climate Change Scenarios", *Global and Planetary Change*, **77**, 116–128.

Marine Biodiversity Research Group, Tropical Marine Science Institute, National University of Singapore. (2012). *Urban Development in the Coastal Zone: Garden Cities in the Sea (GCS).* [Online] Available at: http://www.tmsi.nus.edu.sg/marine-biodiversity, [Accessed October 2012].

Maritime and Port Authority of Singapore (2011). "Tide Tables Year 2012", Hydrographic Department Maritime and Port Authority of Singapore, Singapore, pp. 7.

McGranahan, G., Balk, D. and Anderson, B. (2007). "The Rising Tide: Assessing the Risks of Climate Change and Human Settlements in Low Elevation Coastal Zones", *Environment & Urbanization*, **19**(1), 17–37.

National Environmental Agency (2010). *Singapore's Second National Communication.* [Online] Available at: http://www.nea.gov.sg/docs/default-source/weather-and-climate/second-nc.pdf?sfvrsn=2 [Accessed September 2012].

Ng, W.S. and Mendelsohn, R. (2005). "The Impact of Sea Level Rise on Singapore", *Environment and Development Economics*, **10**, 201–215.

One Map (2012). *Singapore Online Cityguide.* [Online] Available at: http://www.onemap.sg/index.html [Accessed July 2012].

Pethick, J. (2001). "Coastal Management and Sea level Rise", *Catena*, **42**(2), 307–322.

Rahmstorf, S. (2007). "A Semi-Empirical Approach to Projecting Future Sea level Rise", *Science*, **315**, 368.

Rupp-Armstrong, S. and Nicholls R.J. (2007). "Coastal and Estuarine Retreat: A Comparison of the Application of Managed Realignment in England and Germany", *Journal of Coastal Research*, **23**(6), 1418–1430.

Shepherd, A., Ivins, E.R., Geruo, A, Barletta, R.V., Bentley. M.J., Bettadpur, S., Briggs, K.H., Bromwich, D.H., Forsberg, R., Galin, N., Horwath, M., Jacobs, S., Joughin, I., King, M.A., Lenaerts, T.M.J., Li, J., Ligtenberg, S.R.M., Luckman, A., Luthcke, S.B., McMillan, M., Meister, R., Milne, G., Mouginot, J., Muir, A., Nicolas, J.P, Paden, J., Payne, A.J., Pritchard, H., Rignot, E., Rott, H., Sørensen, L.S., Scambos, T.A., Scheuchl, B., Schrama, E.J.O, Smith, B., Sundal, A.V., van Angelen, J.H., van de Berg, W.J., van den Broeke, M.R., Vaughan, D.J., Velicogna, I., Wahr, J., Whitehouse,

P.L., Wingham, D.J., Yi, D., Young, D., Zwally, H.J. (2012). "A Reconciled Estimate of Ice-Sheet Mass Balance", *Science*, **338**(6111), 1183.

Street Directory. *Streetmap Singapore Maps*. [Online] Available at: http://www.streetdirectory.com.sg/, [Accessed July 2012].

Titus, G.J. (2009). "Insitutional Barriers" in *U.S. Climate Change Science Program*, Report by the U.S. Climate Change Science Program (MACUCCSP) and the Subcommittee on Global Change Research. U.S. Environmental Protection Agency, Washington D.C., pp. 163–176.

Titus, G.J., Anderson, E.K., Cahoon, D.R, Gill, S., Thieler, R.E. and Williams, J.S. (2009). *Coastal Sensitivity to Sea level Rise: A Focus on the Mid-Atlantic Region, Climate Change Science Program* (CCSP), Report by the U.S. Climate Change Science Program and the Subcommittee on Global Change Research. U.S. Environmental Protection Agency, Washington D.C. [Online] Available at http://www.climatescience.gov/Library/sap/sap4-1/final-report/default.htm [Accessed September 2012].

Tkalich, P., Vethamony, P., Babu, M.T. and Pokratath, R. (2009). "Seasonal Sea Level Variability and Anomalies in the Singapore Strait", *Proceedings of International Conference in Ocean Engineering*, ICOE 2009, IIT Madras, pp. 874–880.

Todd, P. (2009). "Hantu Island an Experimental Site for Costal Protection Design. Designing Innovative Coastal Protection using Ecosystem-based Approaches", *The Hantu Bloggers*. [Online] Available at: http://www.pulauhantu.org/hantu-island-an-experimental-site-for-costal-protection-design/ [Accessed August 2012].

Tol, R.S.J. (2009). "Economics of Sea Level Rise", *The Journal of Economic Perspectives*, **23**(2), 29–51.

Trading Economics. *Singapore Inflation Rate* 1962–2015. [Online]. Available at: http://www.tradingeconomics.com/singapore/inflation-cpi [Accessed September 2012].

Trisirisatayawong, I., Naeije, M., Simons, W. and Fenoglio, L. (2011). "Sea Level Change in the Gulf of Thailand from GPS-Corrected Tide Gauge Data and Multi-Satellite Altimetry", *Global and Planetary Change*, **76**(3–4), 137–151.

Turner, R. K., Burgess, D., Hadley, D., Coombes, E. and Jackson N.A. (2007). "A Cost-Benefit Appraisal of Coastal Managed Realignment Policy", *Global Environmental Change*, **17**(3–4), 397–407.

WEATHERwise Singapore (2012). "Singapore Yearly Mean Temperature", *National Environment Agency of Singapore (NEA)*. [Online] Available at: http://app2.nea.gov.sg/data/cmsresource/20090721544571208250.pdf [Accessed August 2012].

Wong, P.P. (1992). "Impact of a Sea Level Rise on the Coasts of Singapore: Preliminary Observations", *Journal of Southeast Asian Earth Sciences*, **7**(1), 65–70.

World Bank. (2012). *World Development Indicators, Databank.* [Online] Available at: http://data.worldbank.org/data-catalog, *World Development Indicators, Data, Data Catalog.* [Online] Available at: http://data.worldbank.org/indicator/NE.GDL.TOTL.KN, [Accessed September 2012].

Zanuttigh, B. (2011). "Coastal flood protection: What Perspective in a Changing Climate? The THESEUS Approach", *Environmental Science & Policy*, **14**(7), 845–863.

# 4

# WATER SCARCITY TO WATER SECURITY: HOW CAN ASIAN CITIES ACHIEVE A SUSTAINABLE TRANSITION?

MAITREYEE MUKHERJEE, DENNIS WICHELNS
and NAMRATA CHINDARKAR

Unprecedented demographic growth imposes substantial demands on the water supply and other utility services. Urban water supply systems fail to meet the ever increasing demand, due to various constraints such as inadequate water sources, high loss due to leakage, poor maintenance of the system network and, most importantly, inadequate revenue generation by water utility boards. These constraints result in substantial gaps between the demand and supply of water of sufficient quality and quantity. Lagging economies, financial constraints, political unwillingness, lack of farsightedness and absence of motivation further aggravate water supply issues. This chapter aims to identify probable solutions for cities, through comparative case studies, to ensure a sustainable recovery from water scarcity to water security.

## 1. Urban Water Conservation and Management: Why is it Necessary?

The world population is anticipated to grow from 6.9 billion in 2010 to 9.1 billion in 2050, of which nearly 6.3 billion would live in urban areas (UN World Water Development Report, WWDR4, 2012). Presently, 85% of world human settlements are in the driest half of the planet. An estimated 783 million people lack access to clean water, while 2.5 billion lack adequate sanitation facilities in their homes (UN Water Facts and Figures, 2013). Although urban areas have better services for basic utilities than their rural counterparts, they are still struggling to keep pace with the growing demand of fast rising populations. More water is being withdrawn on a daily basis than can be returned to the water sources by precipitation, runoff and recharge. With limited supplies to support future growth and development, cities are investing fortunes to develop additional water supplies.

The three water dependent sectors that drive the modern world are agriculture, domestic use, and industry. When compared to other sectors, urban water use accounts for less than 20% of total country-based water distribution (even for countries such as the United States), yet urban domestic water management is still important, because cities are the economic and cultural focal points of any nation, and they are home to most of the population. Most cities depend on water from external sources; hence it is in their interest to become self-sufficient. Efficient water use among households has great potential to effectively reduce a city's water demand and has proven to be beneficial to consumers, water utilities, environment and the economy as a whole (Maddaus, 1999).

In this chapter, we focus on urban domestic water conservation and management. We draw upon case studies from across Asia, examine their successes and failures, and highlight key takeaways to achieve urban water sustainability. Looking forward, sustainable development and continued success in the water sector will be achieved only if cities can control the main threats to urban water management — namely excessive population growth, unrestrained exploitation of natural resources, water scarcity (physical or economic), diminishing water quality, climate change issues, lowering ground water tables, infra-structural, institutional and social problems (Biswas *et al.*, 2010).

## 2. Research Objectives and Methodology

When cities fail to upgrade their infrastructure to provide minimum service to increasing populations, they move further behind in the development process. Lagging economies, financial constraints, political unwillingness, lack of farsightedness and absence of motivation further aggravate their developmental failures. Since cities are the centre point of culture, trade, livelihoods and civilizations, their populations will continue increasing. Thus, it is necessary to find ways to safeguard a sustainable future, in terms of proper provision of basic utilities such as water, food, shelter, employment and health facilities.

In this chapter we will:

- Discuss the interplaying factors in urban water demand-supply interactions and efficient water management.
- Explore and analyse, within the Asian context, cases of successful and failed urban water management.
- Suggest policy directives based on key takeaways from the case studies.

We adopt the comparative case studies approach using qualitative data from secondary sources. We first examine the main features and modes of operation of water utility systems in selected better performing Asian cities. We then examine water management challenges and failed policies in other Asian cities. And finally, drawing upon these analyses, we identify key takeaways and suggest plausible policy directives.

## 3. Economic Framework behind Water Management

The normative concept of water as a human necessity has meant that water is often oversupplied but under-priced, leading to unsustainability of provision (Brooks, 2005). Yet another problem of under-pricing water is that it leads to water being "undervalued", resulting in over-exploitation. The distinction between price and the economic concept of value was provided by Adam Smith in his *Wealth of Nations* (Smith, 1776). He argued that value has two meanings — value in use and value in exchange. Value in use refers to the utility of a particular object, while value in exchange refers to its purchasing power. Even though the 'value in exchange' of water is low because it is either under-priced or has no price, its value in use is significantly greater, as it is a necessity (Hanemann, 2006). When consumers pay a very low or no price for water, the true utility and therefore the true value of water is undermined. Undervaluing water further causes individuals and households to have no incentive to use water conservatively, as the marginal cost of an additional unit of water is either zero or negligible.

Gleick (2004) notes that in spite of increasing concern regarding water issues, internationally, economic support for all types of water projects are marginal and diminishing. However, Marra and Alabaster (2008) propose that MDG (Millenium Development Goal) targets for water and sanitation may be attained in urban and peri-urban regions through a new shift of policy perspective. In particular, utility services may be made accessible to groups of households, rather than individuals. This will require less investment, while generating successful outcomes.

In this regard, Smith (2001) observes that nations such as South Africa, have applied ideas of distributive justice in their adoption of public policies that ensure access to public utilities to all, both rich and poor. Maiese (2004) describes two approaches for ensuring this: one focuses on outcomes (performances) and another concentrates on the procedural steps needed to achieve these outcomes (policy interventions). Fair processes always provide dependable information, transparency in decision-making and inclusion of opinions of affected parties. This may pave the way for public acceptance and enable smooth operation of services. Therefore, in the case of water utility systems, two notions of fairness arise. First, when water is taken as a commodity, which requires investments and infrastructure for provision, fairness in policy formulation is necessary for sustainable operations of the services. Second, when water is regarded as a basic human right, the focus shifts towards the principle of distributive justice, requiring fair performance to assure universal access. We have considered both these concepts of fairness, while examining and suggesting policy recommendations.

A common strategy for coping with increasing demand for water is to enhance supply. The supply strategy is effective and economically efficient only when water resources are not constrained. However, in the context of global water availability indices, water resources are indeed limited in many cities. Nevertheless, instead of focusing on controlling demand, the water supply agencies generally give prominence to supply enhancement (Raj, 2013). By doing so, and in the absence of sufficient revenue generation, these agencies get stuck in a "low-level equilibrium trap", which is marked by low levels of service and poor and unreliable quality of water supply (Raj, 2013). We analyse the water demand-supply problems by emphasizing water demand management through revised volumetric pricing models, prepayment metering for poor communities and other awareness programs. While proposing demand-side management policies, affordability for the very poor must be kept in mind. Two important concepts are equity and sustainability. Wichelns (2013) defines water equity as the notion whereby all members of a community have access to an affordable supply of water to satisfy their basic needs and support livelihoods. The equity criterion implies that water cannot be allocated purely through markets. However, there is also the sustainability criterion which Wichelns (2013) defines as generating sufficient revenues to offset the operation, maintenance, and fixed costs of water delivery. This can be linked to the "low-level equilibrium trap" wherein water suppliers who do not satisfy the sustainability criterion are likely to provide poor and inefficient delivery.

Two pivotal economic concepts that influence any water-pricing policy are the ability to pay (ATP) and willingness to pay (WTP). ATP is the maximum monetary value that a household can pay for a unit of water if it were not free or subsidized — or in other words, if they were to purchase water in a market. Quite often it has been noticed that in the absence of sufficient supply of water from a public provider, people are forced to buy from private vendors at high prices. Thus, it may be assumed that if ATP (ability to pay) is higher than the cost of production of water for the utility boards, there would be sufficient WTP (willingness to pay). Hence, if regular supply is assured at reasonably fair prices, even poor households might be convinced to pay for delivery services.

## 4. Case Studies from Asia — Lessons to Learn

In this section, we discuss the primary criteria for successful provision and management of urban water. The model (described below) enables us to understand the interaction between components necessary for smooth performance of an urban water purveyor in our case studies. Though it might be argued that each city has its unique people, culture, expectations and limitations, in our opinion, the three pillars for any development are 1) the willingness of authorities, 2) integrated co-ordination and management of various agencies, and 3) clear business objectives.

According to the three spheres of sustainable development, the primary components influencing the process of development are social, environmental, and economic targets. As illustrated in Figure 4.1, there is continuous interaction and inter-dependence between these sectors, making it necessary

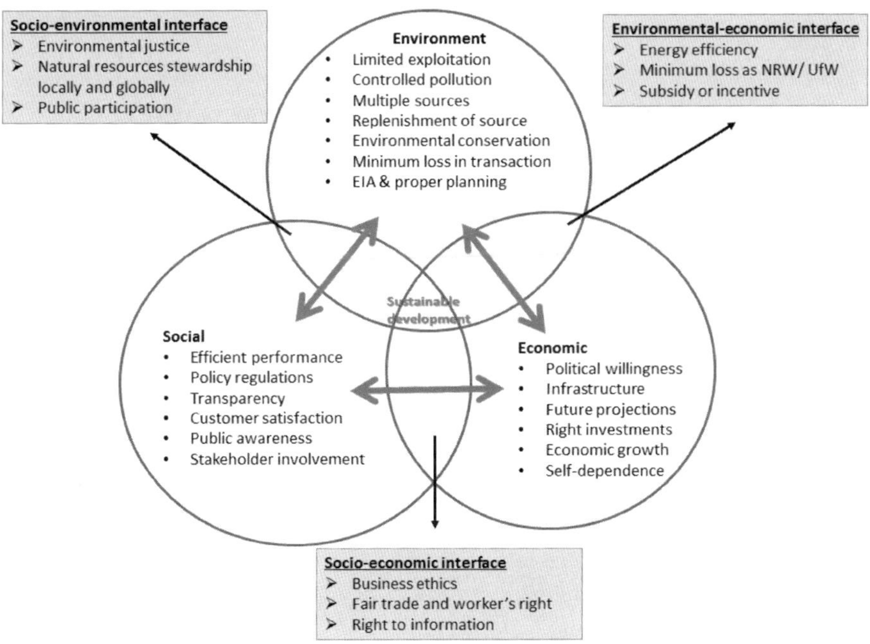


Fig. 4.1. Schematic compilation of necessary factors for efficient performance standards and long-term sustainability of urban water management (Authors' compilation, as adapted from Morse, 2010).

to maintain environmentally bearable, economically viable, and socially equitable interfaces among them. Using this model, we examine selected Asian cities and highlight the underlying reasons owing to which they developed either successful or failed water delivery systems.

## 4.1 *Asian Successes in Urban Water Management*

*Phnom Penh, Cambodia: Transformation from no administration to an institution of good governance*

Until 1992, the Phnom Penh Water Supply Authority (PPWSA — responsible for supplying clean water to the city) was plagued with corruption, inefficiency, bureaucracy, meagre salaries and lack of motivation. PPWSA had been reduced to a dysfunctional institution, both financially and from a managerial perspective (Biswas *et al.*, 2010). As a result, the city suffered from poor service and delivery of a sub-standard quality of water. In 1992, only about 20% of city residents received water through piped connections at very low pressure (0.2 bar) for 10 hours per day. In addition, there was an absence of proper metered connections resulting in nearly 72% of unaccounted-for-water and poor quality of water as reflected by high diarrhoeal incidents. However, between 1993–2008, PPWSA underwent radical development; with consequential improvement of service delivery, profit generation, un-interrupted water supply with good pressure and proper connections for all households, along with subsidised rates for the poor. This extraordinary success was achieved through a combination of strict legislation that allowed the utility financial and operational autonomy, government backing, sound leadership, professionalism, integrity, commitment, public awareness, community participation, transparency, and accountability (Chan *et al.*, 2012). Their journey included carefully involving donor agencies, international aid; providing autonomy to the water utility board; controlling unaccounted for water; recovering cost through proper pricing; metering all connections; establishing an accurate database; ensuring proper bill collection ratios; providing subsidised water for poor; and slowly gaining financial stability through sustainable ventures and inclusive policies (Das *et al.*, 2010).

Today, PPWSA provides a 24-hour, clean water supply to more than 90% of the city's households. Its annual water production has increased by 437%; water pressure has improved by 1260%; the distribution network has improved by 557%; customer base by 662%; and the percentage of metered connections by 5255% — compared to 1992 statistics (Biswas *et al.*, 2010).

Table 4.1. Performance indicators of PPWSA.

| Performance indicators | 1993 | 2008 | 2009 |
|---|---|---|---|
| Staff/1000 connections | 22 | 4 | 3.2 |
| Production capacity, $m^3$/day | 53,660 | 234,284 | 246,864 |
| Water quality | — | WHO | WHO |
| Coverage area | 20% | 90% | 90% |
| Supply duration | 10 | 24 | 24 |
| Supply pressure | 0.2 | — | 3 |
| Supply network, km | 288 | 1,556 | 1,660 |
| Number of connections | 26,881 | — | 191,092 |
| Number of subsidized connections | — | — | 17,657 |
| Unaccounted for water | 72% | 6.19% | 5.94% |
| Billing ratio | 28% | 93% | — |
| Collection ratio | 48% | >99% | 99.9% |
| Total revenue, million Riels | 700 | 100,317 | 105,780 |
| Total expenses, million Riels | 1400 | 70,431 | 72,467 |
| Operating Profits, million Riels | 700 | 29,882 | 33,313 |

*Source*: Water Diplomacy, 2013; Phnom Penh Water Supply Authority, 2011.

Unaccounted-for-water, through leaked pipelines and illegal connections has declined from 72% to only 6.9%. From a meagre number of 101, presently, nearly 26,778 poor households are connected to the utility system. This achievement (from 1993 to 2009; Table 4.1) by the utility has been acclaimed worldwide, and PPWSA has received many international awards, including the Ramon Magsaysay Award for Government Service in 2006, and the Stockholm Industry Water Award in 2010 for their outstanding performance.

*Tokyo Waterworks: A marriage of responsibility, technology and transparency*

Tokyo's water supply mainly comes from upstream catchment areas in the Tama River basin. The Bureau of Tokyo Waterworks has been conserving 21,631 hectares of primary forest (consisting of natural forests, planted forests, public and private forests) for more than 100 years to sustain its natural ecological services as water catchment and purification zone for the Tama River and Ogouchi reservoir.

The Bureau of Waterworks is dedicated to ensuring environmental conservation and sound business management. With a 12.8 million customer base, Tokyo Waterworks is dedicated toward best services, commitment, faith, and loyal relationships. With 6 branch offices and 26 service centres, the bureau is continuously available for customer complaints. It has formed the Tama Waterworks Bureau Environment Plan 2010–2012, embracing ISO 14001 standards, in order to formulate and implement its action plans. This is done to encourage environment-friendly water projects following PDCA (Plan Do Check Act) principles for future endeavours. To maintain transparency, it discloses its Environmental Report every financial year since 2000.

Japan being an earthquake prone country, the bureau has implemented measures such as waterworks facilities earthquake reinforcement, backup facilities and establishment of emergency measures for resilience building. Continuous research is conducted to determine better earthquake resistant structures, pipes and storage facilities. In the case of disruption of supply, immediate measures are taken to provide relief to customers (Water Supply in Tokyo, 2012).

The Tokyo Waterworks sustains its production, operation and maintenance from the collection of adequate water revenue. New construction and expansion projects are funded by external loans, government subsidies and the company's own funds. Tokyo Metropolitan Assembly has authority to regulate the waterworks' annual budget, tax revisions and account settlements, but actual implementation of these decisions is carried out by local bodies and agencies (Takizawa *et al.*, 2005; Water supply in Tokyo, 2012).

In November 2006, the utility board launched the "Tokyo Waterworks Principal Vision STEP II — Safe Waterworks, our pride to the world" (Tokyo Water Management Plan, 2013). The bureau also has specific centres for technical and managerial training, as well as Research & Development sections for continuous improvement in knowledge and technology (Murakumi *et al.*, 2006). To facilitate exchange of knowledge and experience, the waterworks bureau sends its staff overseas for training. The bureau also receives trainees for the purpose of sharing their knowledge, according to requests from JICA (Japan International Cooperation Agency).

### *Singapore: Security in Diversity and Turning Vulnerability into Strategic Advantage*

Driven by the vision of water sustainability, Singapore has created a robust, diversified and secured water supply from its four national taps (Tortajada, 2006):

- Harvested rain water from local catchment areas
- Imported water from Johor, Malaysia

- Reclamation of used water, known as NEWater
- Desalinated water from the sea

The Singapore Public Utilities Board (PUB) is the national water agency responsible for collection, production, distribution and reclamation of water in Singapore. Rainwater is harvested through rivers, streams, canals and drains, and stored in 17 reservoirs spread across the city state. These reservoirs are connected through pipelines that channel excess water from one reservoir to another, thus maximising storage utility (Luan, 2010). The collected water is then piped to water treatment centres for purification and stored in covered reservoirs before distribution. Used water from consumers is again collected by PUB via the Deep Tunnel Sewerage System (DTSS) and transported to water reclamation plants. The treated water may be disposed to the ocean or further purified to generate NEWater and mixed into natural reservoirs to allow naturalisation. At present, NEWater meets nearly 30% of the country's water demand (NEWater: The 3rd National Tap; PUB website). NEWater is also useful as high grade pure water for many industrial processes.

In addition to playing an important role in daily lives, water has been used innovatively to beautify the country's landscape and enhance its citizen's quality of life. The reservoirs are used for water sports and the parks around them harbour a variety of biodiversity. By involving people as important stakeholders in preserving the nation's water resources through reasonable pricing and awareness campaigns, PUB has succeeded in realizing their vision of 'Water for all' and 'Conserve, Value, Enjoy'. The Singapore story of water management is another example of success achieved through competent governance, prudent fiscal management, innovation, clear vision, goal setting, and active involvement of people.

### *Hong Kong: Innovative Solutions to Manage Limited Resources*

Hong Kong is considered to be one of the world's most densely populated cities. Growing prosperity, trade, and development, along with population density, have resulted in increased demand for water and energy. In spite of having insufficient natural water resources, Hong Kong has utilized technology and space to collect rainwater through a semi-artificial drainage system that includes several artificial reservoirs across the territory. Hong Kong was one of the pioneers in constructing freshwater reservoirs by damming off the sea. Plover Cove (1973) and High Island reservoirs

(1978) are two notable examples. Nearly one-third of Hong Kong's land area has been transformed into water gathering grounds, including both direct catchment areas and indirect drainage systems that divert run-off and seepage into storage systems. Hong Kong's man-made catchment area, within urban premises, is probably the largest in the world. However, development of extensive water collection ground is limited by the high cost of collection and the lack of clean water seepage areas (Man *et al.*, 2011).

Hong Kong built and used desalinisation plants with 66 million $m^3$ capacity, for a short time but abandoned (1991) this technology long ago due to its high cost of production, and the availability of Dongjiang water at lower prices. But, interestingly, sea water is used widely in Hong Kong for toilet flushing, through a separate, parallel pipeline. This conserves nearly 22% of freshwater consumption, saving about 200 million $m^3$ of Dongjiang water, valued at about HK$ 600 million in year 2000 (Chen, 2001; Hong Kong: The Facts, 2013).

However, only 20% of its annual demand can be met by Hong Kong's internal water catchment capacity (Hong Kong: The Facts, 2013). The remainder is obtained by importing water from the East Pearl River or Dongjiang river basin of Guangdong province, China. Since the 1960s, Hong Kong's dependence on Guangdong province for water has increased manifold. While China has ensured the major share of Dongjiang river water to Hong Kong (often compromising other contesting cities such as Shenzen and the Daya Bay area), Hong Kong has generated large economic returns for the water it received (Chen, 2001).

The cost of imported water paid by Hong Kong is one of the highest in the world. The cost has increased from HK\$3.09/$m^3$ (1999) to HK\$8/$m^3$ by 2013[1] (South China Morning Post, 2013). This is about 8 to 9 times more expensive than that paid by Singapore for purchasing water from Johor, Malaysia. After 2006, Hong Kong and Guangdong have agreed on a yearly lump sum payment for the annual water requirements, in which the annual cost for purchasing water in 2012, 2013 and 2014 are HK$3,538 million, HK$3,743 million, and HK$3,959 million, respectively, for a target supply of 820 million $m^3$ [LC Paper No. CB(1)734/12-13(04)][2]. The water fees

[1] 1 USD = 7.75 HK$ (as on 06/11/2014).

[2] Water agreements between Singapore and Malaysia, are comparable to the agreement between Hong Kong and Guangdong. However, the nature, terms and conditions are quite different. Singapore (agreements of 1961, 1962, 1999) is allowed to draw 1.14 million $m^3$ per day from the Johor river in Malaysia. For this, Singapore bears the cost of building

paid by Hong Kong have supported notable industrial and economic growth in Guangdang Province. Hence this mutually-benefitting relationship has proved effective for both China and Hong Kong. Thus, as Hong Kong's economic growth continues in future, there should be water available from Guangdong Province, at a price found to be agreeable by both parties.

### 4.1.1 *Key takeaways*

From the four case studies, we identify the first principle of successful water delivery systems to be clearly laying out the goals of environmental and financial sustainability, equal distribution, and provision of good quality water. In the face of increasing demand from an expanding population, urban managers should not digress from **environmental targets** of limited exploitation, reduced pollution of sources, diversification of sources to enable distribution of demand pressures, conservation of forests and natural watershed areas, and use of proper impact assessment. Simultaneously, **economic goals** should be maintained through competent financial management, minimal transaction losses in the form of NRW (Non-Revenue Water)/UfW (Unaccounted-for-water), proper pricing, proper revenue collection, along with regular monitoring of the system. **Social responsibility** can be carried out effectively through stakeholder involvement, public awareness and participation, performance efficiency checks, customer satisfaction, transparency, right to information, and labour productivity. For financial and functional stability of the water delivery system, competent policy regulations at regular intervals, along with provision of incentives for customers' responsible behaviour is an absolute necessity. Environmental, economic and social stability requires the existence of robust regulatory institutions that would certify good governance and appropriate policy implementation, both in private as well as public sectors. **Good governance** should recognise the corrosive effects of corruption on public morale, economic efficiency, political stability

---

infrastructure, operation and maintenance of dams, pipelines, treatment plants, and equipment. They pay for the water, in addition to paying rent for the land used. The agreement will expire in 2061. Malaysia is a reliable water provider, but it has used this issue to create political pressure in Singapore. Hence, Singapore has developed strategies and technology to achieve water supply self-sufficiency in the future. Such a constraint does not exist for Hong Kong, as it is under Chinese jurisdiction. China has agreed to supply abundant water, in exchange for high tariff rates. Hong Kong also pays for the infrastructure needed for its water supply, while China bears the operational and maintenance costs.

and sustainable development in general (Caiden, 2007); hence water sector managements should work on the principle of 'zero tolerance to corruption'. In order to facilitate smooth developmental achievements, state water board authorities should adopt democratic and transparent decision-making procedures along with financial accountability.

**Political will and administrative determination** also plays pivotal roles in achieving good governance. However, political will does not always lead to competent governance, as the presence of multi-party democracy in many countries tends towards implementation of populist policies and governance (Yadav, 1996). For example, providing autonomy to the water sector institutes, competent water tariff structures, legalising connections, metering, and restricting pro-poor policies may seem unattractive to populist government authorities, who are more interested to retain their power after a 5-year term. The four cities discussed in this section have one thing in common — they are run by a single strong party rule. Strong party presence has ensured a clear-cut vision, long-term strategic planning, execution, and realization of end-effects. In contrast, the presence of a strong democracy often gives rise to multiple political motives, agendas and weak decisions, as often sighted in other Asian cities. Even the ruling party lacks the willingness to take risks in strong policies (in water as well as other sectors) fearing unpopularity among voters. In addition, the change of power at frequent intervals results in delayed developmental progress (Feng, 1997; Moser, 1999).

In short, sustainability requires the integrated involvement of communities (the main stakeholders), planners, managers, scientific researchers, workers, NGOs and responsible government authorities. The process takes place through a cascade of steps where each tier ensures active interaction and smooth functionality. Figure 4.1 captures this flow of inter-dependence among various stakeholders in the society for achieving common sustainable end results.

## 4.2 *Plight of Other Asian Cities: Consequences of Over-Exploitation, Mismanagement and Political Indifference*

In contrast to the cities discussed in the previous section, other Asian cities have failed to achieve the necessary level of water service delivery and continue to face challenges. Although these cities are experiencing economic growth, the standards of living, condition of public services and public administrative processes are not keeping pace. With increasing population, the per capita availability of water resources has decreased, resulting in increased competition, exploitation, surface water pollution, groundwater contamination and diminishing quality of watersheds. Excessive extraction of groundwater has led to land subsidence, seepage of salinity into freshwater bodies and resultant losses to urban economies.

Urbanisation also has other side-effects, such as the concentration of urban poor, increased unemployment, rising disparity of income, lack of basic service availability, artificial water scarcity, and the need to import of water from distant basins. Water import from outside sources involves large energy expenditures, which raises the cost of water production and distribution. These increased costs are relayed to consumers in the form of higher water prices. In many cases, the absence of adequate tariff formulation and revenue collection has severe outcomes for water purveyors in the form of heavy losses, lack of maintenance, sub-standard service quality, inability to repay loans and ultimately, disincentive for improvements.

With a population of about 20 million, **Beijing** is among the world's fastest growing cities. Twenty-five years of drought and exploitation have strained the city's fresh-water resources. While more than 200 rivers and streams are depicted in its official map, most of these have receded or dried up. The same is true for many reservoirs built since 1950. Beijing, once known for its abundant sweet water springs, now relies heavily on imported water from outside provinces and other groundwater resources (Probe International Beijing Group, 2008).

Beijing's decision to ensure free water supply to the capital, even as its rivers and reservoirs are receding, has had adverse consequences on the agricultural sector. In the past decade, the State Council and Beijing authorities formulated several policy solutions which include (Zhang *et al.*, 2007; Pan, 2004):

- "emergency water transfers" from existing and proposed fresh-water reservoirs at Hebei Province (2008) and the Yangtze River in Hubei Province (2010) — the 'south-to-north' water diversion projects costing up to US$12 billion;
- restrictions on surface and groundwater use in upstream Hebei province in order to increase river flow to Beijing;
- extraction of karst groundwater from depths of 1,000 metres or more from Beijing's surrounding districts;
- introduction of water-saving irrigation methods and water-saving equipment for households;

- recycling industrial water and trying to reduce the industrial demand for fresh water.

Instead of trying to limit the city's over-use of water and making it self-reliant, the government decided to import more water from outside provinces to meet their increasing demand and flush out its polluted waterways, thereby putting pressure on those external watersheds. Such short-term policy intervention cannot be a sustainable solution. Long distance importation of water is expensive and jeopardises the system itself. Projections created by the water supply board of Beijing show an increasing trend of water use by 3.4%, while the population increased by only 1%. The municipality is forced to pump out nearly 3 billion $m^3$ of groundwater annually to keep pace with increasing demand. Like most Asian cities, the policy focus is on supply-side management; neglecting the control of consumer demands by encouraging water conservation (Zhang *et al.*, 2007).

**Bangkok, Jakarta, Hanoi, Kuala Lumpur and Manila** are major national capitals in Southeast Asia. Although different in cultural aspects, they share physical, environmental, socio-economic and demographic similarities. Lying within the tropics, these megacities enjoy average temperatures within the range of 25–30°C, and an average 1500–2000 mm of annual precipitation, leading to heavy run-off and occasional floods during the rainy season. In spite of abundant water resources, blocked waterways or drainage systems, scarce infrastructure for providing clean water, and heavy pollution has led to artificial water scarcity. More often, relatively few households from richer sections receive clean, piped water at subsidised rates, while poorer residents are forced to buy expensive water from private vendors. In addition, the over-exploitation of groundwater leads to contamination, depletion, wide-spread flooding, land subsidence and salinization of freshwater aquifers (Babel, 2010; Albuquerque, 2013; Duc, 2013). Among these ASEAN megacities, effects of un-regulated or poorly planned privatisation in **Jakarta** as well as **Manila** are worth mentioning.

In 1997, the Jakarta service area was divided between two private companies following poor service and heavy losses. Thames Water (UK), along with local company Kekarpola Airindo (KATI), Suez (France), and Garuda Dipta Semesta (GDS), were awarded the rights to develop the infrastructure, and operate and maintain the water-sewerage system in the city of Jakarta, with a concession period of 25 years. But these contracts started with some unconventional agreements. These were awarded without proper competitive bidding, owing to close business links with (then) President Suharto. Moreover, the introduction of water tariffs as the market value of water charged, protected private players against revenue risks (Wu *et al.*, 2013). Unfortunately, just after this privatisation drive, the Asian recession of 1997 started, and the value of the Indonesian Rupiah crashed. Simultaneously, anti-movements against President Suharto led to a power change in the country, leading to a renegotiation of the contracts. The foreign counterparts incurred heavy losses and so, exited; selling out most of their shares to local companies. In 2001, the Jakarta Water Supply Regulatory Body (JWSRB) was formed to supervise concessions and tariff rates on behalf of the government. In addition, PAM Jaya[3] was responsible for monitoring the proper functioning of private agencies. After privatisation, Jakarta's water price was one of the highest in Asia — in spite of substandard service, poor coverage and intermittent supply (Wu *et al.*, 2013). From 2006–2009, owing to extensive customer dissatisfaction, protests, and reduced target achievements by the respective companies, the government froze the tariff rates. The cumulative effects of the economic crisis, decline of the Indonesian rupiah, miscalculated exchange rates, unconventional personal favours drawn by family members of the political supremo and freezing of tariff rates led to heavy losses incurred by the companies (Bakker *et al.*, 2006). By 2012, the low performance of water purveyors, selling out and exit of initial companies, lack of expertise among new companies, and repeated negotiations, devastated the water supply sector. The latest survey by the Jakarta Water Supply Regulatory Body (JWSRB) revealed that the average annual consumer growth after privatisation has been much less than that before privatisation (Wu *et al.*, 2013). A few other studies also show the same observation: where, in 1997, before privatisation only 42% of residents had access to piped water; the coverage still remained below 50% in 2010 — ten years after privatisation (UNICEF Indonesia, 2012; UNDP Jakarta Indonesia Case study (Water), 2012).

In the case of Manila, the privatisation of the water services in Manila in 1997 (around the same time as Jakarta) has led to mixed results, with Manila Waters (a joint partnership of United Utilities, Bechtel & Ayala, local enterprises) serving the east and Maynilad (with international partners Suez and Benpres) serving the western part of the city. Both contracts were given a 25-year concession. In contrast to Jakarta, Manila

[3]PAM Jaya = Perusahaan Air Minum Jakarta Raya, Jakarta's municipal water authority.

had established a Regulatory office (RO) in the beginning to supervise concession agreements, target achievements as well as tariff rates. From 1997, Manila Waters improved their infrastructure and services consistently, along with expanding its operation to other Philippino cities such as Laguna, Boracay and more. In 2005, Manila Waters was registered on the Philippine Stock Exchange owing to the strengthening of the company's credible financial status (Wu *et al.*, 2013). In comparison, Maynilad failed to provide competent service, went bankrupt and was ultimately taken over by government owned MWSS (Manila Waterworks and Sewerage Systems, a government corporation overlooking water supply and sewerage service in Metro Manilla). In 2006, Maynilad's shares were rebid and the water service rights were won over by DMCI-MPIC (the consortium of DMCI Holdings Inc. and Metro Pacific Investment Corporation, a local enterprise). Since 2010, Maynilad claims to have achieved good infrastructure, extensive service coverage (above 95%), competent tariff structure, and adequate revenue collection leading to ensured financial stability. Apart from a few favourable economic factors, this turn-about was mainly possible due to a focus on improving labour productivity, restructuring workforce morale and atmosphere along with minimisation of NRW (from 66% to 43% by 2012) and compulsory metering (Jensen *et al.*, in press). However, in spite of developments in its water-sewerage system, Manila which is prone to destructive typhoons, heavy rainfall and flooding, and has yet to develop resilience for flood control and service disruption measures (ADB Evaluation Study, 2008).

The water sector in **India** is also far from satisfactory. According to a recent World Bank report, India is the largest groundwater user in the world, drawing more than a quarter of the global reserve annually. Groundwater supplies are sustaining more than 60% of irrigated agriculture and 85% of potable water demand in the country. Urban areas are also becoming more dependent on groundwater as their primary water source. Groundwater is an essential buffer against the unpredictable monsoon rains. A 2004 nationwide groundwater assessment states that 29% of the country's aquifer regions are in a semi-critical, critical or over-exploited state, and the level of groundwater is fast depleting in most densely populated and economically viable areas. It is feared that if the current status continues, in two decades, India will face severe groundwater shortages, leading to serious impacts on agricultural harvests, long-term food security, economic standards and the general livelihoods of people (The World Bank, 2012). Furthermore, climate change and its various consequences would put even

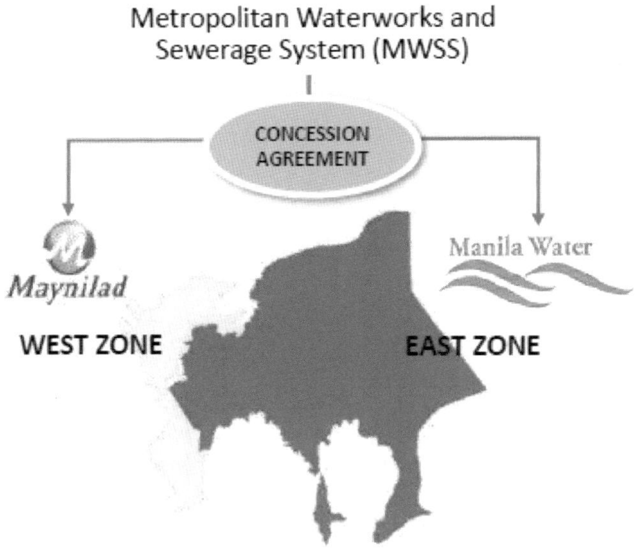


Fig. 4.2. Concession agreements in east and west parts of Manila in 1997 (Oreta, 2011).

more pressure on water resources in Indian cities, along with receding sources and excessive pollution.

### 4.2.1 *Key takeaways*

**Mismanagement of environmental capital**, such as watersheds, forests, lakes, reservoirs, and river catchment areas often have negative social consequences. Pollution, surface water degradation, depletion of water sources, and encroachments has led to artificial scarcity and substantial economic losses. Such losses can also result from inappropriate or misguided investments. For example, Beijing, instead of diverting large volumes of water from outside provinces, might consider its own supply and demand management capacities, and try to strengthen them. Formulating policies following the water catchment conservation model of Tokyo might be beneficial for long term water security. Investing in water treatment plants can also brighten economic and environmental prospects. This will not only help clean up the cities' water system, but also assist in restoring groundwater levels. The technology for generating NEWater from grey water is being taken up in many cities, such as Singapore, to supplement and replenish natural water resources.

**Inadequate water storage and improper wastewater treatment facilities**, common in Asian cities, are critically aggravating the adverse effects of urbanisation on ecosystems. In Singapore, in spite of having limited space and water resources, the city-state has constructed reservoirs as rainwater catchment areas. These reservoirs, apart from storing rainwater, also receive treated used water or NEWater to supplement the country's water supply. Creation of fresh water reservoirs to collect fresh water from rivers and rainwater run-off by constructing sea-walls is another technology used in Singapore and Hong Kong. In contrast, natural assets such as Manila Bay and Laguna Lake in the Philippines are being lost by extensive pollution and land reclamation to support more commercial and tourist developments. Apart from their fresh water storage capacity, these sensitive ecosystems should be valued for their numerous ecological services.

**Privatisation of public utilities** is yet another issue requiring careful attention. Jakarta and Manila have experienced notable failure in water utilities due to privatisation. These cases clearly point out that privatization in water utilities should be done mandatorily by competitive bidding and proper regulatory mechanisms. Moreover, private operators should not be given a total free-hand in operation or decision-making in important sectors like water. Instead, a better way would be that the government themselves develop and own the infrastructures, then slowly allow private players to operate in parts, while maintaining a stringent regulatory system under its jurisdiction. Public–private partnerships or PPP models, as practised in Singapore, are gaining popularity in the water sector. Infrastructure for water utilities, capacity building, operation, and management of facilities should be planned and regulated with strict policy guidelines.

**Inadequate water pricing and scarce distributions of household connections** are two pressing concerns. All water utilities incur expenses for raw water extraction, treatment and distribution. These costs of production must be recovered to prevent financial losses that can lead to inadequate service provision and poor service quality. Most countries consider water to be an essential commodity, a birth right for citizens, and therefore supply it at highly subsidised prices. However, policies based on this premise tends to undermine the use value of water (Smith, 1776; Hanemann, 2006). Over time, failing to charge for water can result in poor quality of service, sub-standard water quality, and high rates of leakage in systems owing to poor maintenance. This phenomenon forces residents to depend on private sources, invest in purification systems, and thus, indirectly, spend more from their monthly budget for water services. Water pricing is a politically sensitive issue, as this may lead to loss in popularity among voters. Moreover, before pricing a utility, its service should be improved to a level that encourages consumers to pay for the service. Hence, water purveyors must first ensure a continuous, clean water supply, and proper connections for each and every household; and then through proper public consultation and campaigns, introduce an increasing block-rate tariff structure, where households consuming less water are subsidised by those who consume more. Phnom Penh provides a good example in creating a sustainable water service system by ensuring good performance and proper cost recovery.

Lastly, to establish or restore good service, **political will, administrative determination, bold decision-making ability, along with demand and supply side management of water utility services,** are essential factors. In our view, the above mentioned factors are the major loopholes in urban water planning and management of many Asian cities that are posing as an invisible barrier to sustainable water security.

## 5. Conclusion: Towards Interdisciplinary Approaches, Cross-Sectorial Initiatives and Definitive Targets for Sustainable Urban Water Management

Over the past three decades, many Asian cities have been growing rapidly, undergoing functional and physical transformation into international economic hubs. It's most spectacular consequence is the emergence of megacities, characterised by intensive urbanisation, stretching up to 100 km into adjacent suburban hinterlands. This phenomenon has created new challenges for urban managers, engineers and planners, because while the cities have grown beyond their capacity of resource or facility provision, their infrastructure, technology, modernisation and planning for the future have not kept pace. Due to the lack of planning and implementation of necessary policies, urbanisation has gone hand in hand with urban poverty, resource depletion and environmental degradation, particularly in urban hydrological systems, such as natural drainage, run-off, groundwater quality, sedimentation, water scarcity and unavailability of clean, safe water. Even when Asian cities have witnessed economic 'miracles', urban water supply has remained well below satisfactory standards, increasing the burden of disease and health expenses for residents.

The reasons for these failures are numerous and systemic. One is the general pattern of "top-down" exercise of political authority, which excludes the needs of many residents in its routine governance and planning in the public domain. Often, water departments at municipal levels lack the power or resources to effectively manage regional environmental problems. Although problems are better understood and managed more appropriately at the local level, local governments in many Asian countries lack political, financial, technical or economic capability. Coupled with this is the universal problem of corruption and unaccountability. In addition, the lack of awareness and understanding regarding the gravity of the issue leads civil society to continue their unsustainable ways of water extraction, usage and pollution. Inadequacies in environmental legislation and missing incentives for cleaner environmental services mean lower costs for enterprises (both public and private), but enhanced health risks for citizens. Unfortunately, many governments tend to avoid these issues, focusing more on economic or financial development, while public water utilities and the environmental or eco-system value of water degrades due to continued negligence and inaction.

In the Asian context, cities such as Singapore, Hong Kong and Tokyo have gained exemplary achievement in providing safe water to all, by restoring watershed quality and developing new technologies. Even Phnom Penh, Cambodia, has improved notably from a non-functional system into a successful, transparent and accountable enterprise within one decade. Determination, will power, clear vision, competent management and dedicated staff helped to make the difference. On the other hand, the primary constraints in other Asian cities seem to be the mismanagement of environmental water-resource assets, degradation of eco-system services, lack of modern technology, age-old infrastructure, poor quality of service provision, inadequate pricing, and privatisation of public water utilities. Populist policies and political instabilities in some countries hamper long-term goal setting. In addition, inadequate funding and complicated time-consuming administrative processes slow down developmental projects (such as purification plants, sewage treatment, pipeline network coverage, renovation works and so on) for water sectors.

Even in advanced countries, development does not come easily. Fast industrialisation in western nations led to quick degradation of their environment, water and air sheds, and human health. When this degeneration reached its saturation limit, communities thought of rationalising their activities and restoring lost environmental sources. Asian cities are now going through a similar journey. Considering Smith's approach (Smith, 1776) of use and exchange value of water, it might be suggested that as long as water is abundant, it has no exchange (commodity) value. But as exploitation, degradation of source and pollution reduces the use value of water, making it scarce; the commodity value of water is recognised. In cities with limited water resources such as Hong Kong and Singapore, water has an exchange value which is set by competitive market forces. On the contrary, in cities where surface or groundwater is still sufficiently available, irrespective of its quality, even its use value is not considered and consumers refuse to pay for their basic necessities. Such careless policies are unsustainable in the long run and can result in severe water insecurity. With exploitation, when water sources are degraded and depleted, a city is forced to divert water from outside sources at higher costs. Hence to maintain the sustainability of water for the future, its use value should be recovered in order to re-invest this price for restoration and maintenance of the water source itself. This way, in the long term, the city will avoid making water an exchangeable commodity for a high price (buying or transporting water from distant basins).

We will conclude by summarising the following steps we feel are necessary to form a sustainable Integrated Urban Water Management Plan. These steps are illustrated in Figure 4.3. They can be broadly categorised into three phases (three coloured boxes):

**Urban institutional occupation in water management** — Institutions involved in water management include the water supply board, public health department, environment protection agency and other related agencies. Each institution has specific jurisdiction areas which are interconnected in terms of their end result towards the community or environment.

Introduction of competent policies and long term clear cut goals is the first effective step. The water and sanitation institutions need to take care of its operation, maintenance and expansion. Capacity building, construction of infrastructure, cutting-edge research to establish new technologies and management guidelines need to be implemented. Protection of water supply sources (overlap with environment protection agency) while improving their quality would bring high dividends in the future. Networks and pipelines should be maintained properly to reduce water losses by leakage. Sanitation and water treatment facilities need to be developed to minimise pollution, contamination, damage and risk of disease (intersection with public health department).

**Emergency services for water supply** — Water purveyors should also be prepared for emergency situations such as storms, disruptions of service, drought or flood conditions and other hazards. In addition, they should prepare consumers for such situations. Emergency preparation and requirements vary regionally. For example, in Japan, creation of earthquake-resistant infrastructure is critical. In the Philippines or Bangladesh, investing in storm protection capacities is essential. Many coastal areas in Thailand face land subsidence and sea-water infiltration. Water purveyors in such cities would require more investment to protect their inland fresh water from encroaching salinity.

In cases of excessive groundwater extraction, such as in Bangalore or Hanoi, alternative sources including rain-water harvesting, reuse of grey water and non-potable uses of recycled water should be encouraged. Pollution of water resources is another common problem faced by most water managers. This not only degrades water sources, but also poses serious health threats for communities. Thus maintaining catchment areas and retaining water quality [total dissolved solids, TDS] within normal standards should also be considered with equal importance and prioritised under emergency services.

**Institutional regulation, implementation and supervision** — Successful water purveyors need some level of decision making autonomy.

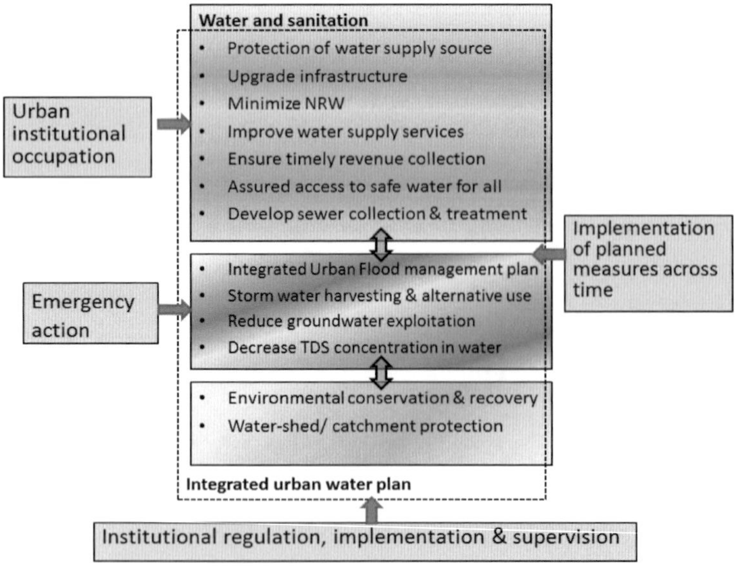


Fig. 4.3. Necessary steps of integrated urban water management and planning.

Fostering regular interaction between local and central government, institutional heads and representatives of civil society helps to bring in transparency and faith among various stakeholders. Creation of interdisciplinary networks (between water engineers, administrators, policy makers, ground-level workers and consumers) for urban water management will help strengthen institutional frameworks, incorporate regulatory policies effectively, and improve governance. Apart from good governance, both local and central institutions should also emphasise the protection and restoration of environmental capital to ensure the steady flow of ecological services.

Ultimately, the effective coordination of various interfaces between administrative, institutional, and managerial sectors is necessary to achieve a sustainable transition from water scarcity to water security. As shown in Figure 4.3, these are all inter-related and inter-dependent. Therefore an integrated approach should percolate all the way from policy making to implementation and maintenance of good quality service.

## References

ADB Evaluation Study (2008). ADB Assistance to Water Supply Services in Metro Manila. Available online: http://www.oecd.org/countries/philippines/42229254.pdf (Accessed 6th Jan. 2014).

Albuquerque, C. (2013). Thailand: Striking contrasts in access to safe drinking water and sanitation — UN expert urges establishment of independent water and sanitation regulator. United Nations Human Rights. Available online: http://www.ohchr.org/en/NewsEvents/Pages/DisplayNews.aspx?NewsID=12975&LangID=E (Accessed 17th Nov. 2013).

Babel, M. S., Rivas, A. A. and Kallidaikuruchi, S., (2010). "Municipal Water Supply and Management in Bangkok: Achievements and lessons", *International Journal of Water Resources Development*, **26**(2), 193–217.

Bakker, K., Kooy, M., Shofiani, N. E. and Martijn, E. (2006). Disconnected: Poverty, Water Supply and Development in Jakarta, Indonesia. Available online: http://hdr.undp.org/en/reports/global/hdr2006/papers/bakker_et_al1.pdf (Accessed 17th Nov. 2013).

Biswas, A. K. and Tortajada, C. (2010). "Water Supply of Phnom Penh: An Example of Good Governance", *Water Resources Development*, **26**(2), 157–172.

Brooks, D. B. (2005). "An Operational Definition of Water Demand Management", *International Journal of Water Resources Development*, **22**(4), 521–528.

Caiden, G. E. (2007). "Corruption and governance", *Combating Corruption, Encouraging Ethics: A Practical Guide to Management Ethics*, 78–80.

Chan, E. S., Vermersch, M. and Vaughan, P. (2012). "Water supply in Phnom Penh: from devastation to sector leadership", *Water Utility Management International*, **7**(3), 09.

Chen, Y. D. (2001). "Sustainable Development and Management of Water Resources for Urban Water Supply in Hong Kong", *Water International*, **26** (1), 119–128.

Das, B., Chan, E. S., Visoth, C., Pangare, G. and Simpson, R., eds. (2010). Sharing the Reforms Process, Mekong Water Dialogue Publication No. 4, Gland, Switzerland: IUCN. Available online: http://cmsdata.iucn.org/downloads/mwd_ phase_1_final_report_1.pdf (Accessed 23 Sept. 2013).

Duc, L. V. (2012). Sanitation of Water Source and Treatment of Garbage and Wastes in Hanoi City. Statement at East Asian and Middle-South American Conference on Environmental Industry. Available online: http://www.mofa.go.jp/region/latin/fealac/pdfs/4-6_vietnam.pdf (Accessed 20th Nov. 2013).

Feng, Y. (1997). "Democracy, political stability and economic growth", *British Journal of Political Science*, **27**(03), 391–418.

Gleick, P. H. (2004). "The millennium development goals for water: crucial objectives, inadequate commitments" in *The World's Water: The Biennial Report on Fresh Water Sources, 2004–2005.* Gleick, P. H. (ed.). Island Press, Washington, DC.

Hanemann, W. M. (2006). "The Economic Conception of Water" in *Water Crisis: Myth or Reality?*, P. P. Rogers, M. R. Llamas and L. Martinez-Cortina (Eds.), London: Taylor & Francis.

Hong Kong: The Facts (2013). published by Information Services Department, Government of Hong Kong. Available online: http://www.gov.hk/en/about/abouthk/factsheets/docs/wp&g_supplies.pdf (Accessed 26 Sept. 2013).

Jensen, O., Leong, C. and Wu, X. (2014). *The revival of a water concession: the case of Maynilad*, unpublished.

LC Paper No. CB(1)734/12-13(04), March 2013. Available online: http://www.legco.gov.hk/yr12-13/english/panels/dev/papers/dev0326cb1-734-4-e.pdf (Accessed 23rd Oct. 2013).

Luan, I. O. B. (2010). "Singapore water management policies and practices", *Water Resources Development*, **26**. (1), 65–80.

Maddaus, W. O. and Alamo, C. A. (1999). "Realizing the benefits from water conservation", *Water Resources Update*, **114** (1), 2.

Maiese, M. (2004). "Procedural Justice" in *Beyond intractability.* Burgess, G. and Burgess, H. (eds). Conflict Research Consortium, University of Colorado, Boulder. Available online: www.beyondintractability.org/essay/procedural_justice/ (Accessed: 6th Sept. 2013).

Man, A., Ng, B., Shou, S. and Vickridge, I. (2011). "Sustainable Water Supply Infrastructure for Hong Kong (PDF)", *America Water Works Association Journal*, **103** (7), 57–62.

Manila Water website. Available online: http://www.manilawater.com/ Pages/Index.aspx (Accessed 20th Nov. 2013).

Mara, D. and Alabaster, G. (2008). "A new paradigm for low-cost urban water supplies and sanitation in developing countries", *Water Policy*, **10**(2), 119–129.

Morse, S. (2010). *Sustainability: a biological perspective.* Cambridge University Press, pp. 1–5.

Moser, R. G. (1999). "Electoral systems and the number of parties in postcommunist states", *World Politics*, **51**(3), 359–384.

Murakuni, S. and Dixon, J. A. (2006). "Water Resources Management in Japan: Policy, Institutional and Legal Issues", World Bank Analytical and Advisory Assistance (AAA) Program. Available online: http://siteresources.world bank.org/INTEAPREGTOPENVIRONMENT/Resources/ WRM_Japan_experience_EN.pdf (Accessed 21st Oct. 2013).

Mukherjee, M., Chindarkar, N. and Gronwall, J. (2014). "Non-Revenue Water (NRW) and Cost Recovery in Urban India: Case of Bangalore", *Water Policy*, accepted on 10th September, 2014, (in press).

Oreta, L. J. B. (2011). *Turning Risks into Opportunities: The Manila Water Approach.* Presentation from 2011 World Water Week in Stockholm. Available online: http://www.worldwaterweek.org/documents/WWW_PDF/2011/ Wednesday/K24/Water-Intensive-Companies/Turning-risks-into-opportuni ties-The-Manila-water-approach.pdf. (Accessed 29th Nov. 2013).

Pan, A. (2004). Water Management in Beijing; Beijing Water Authority. Available online: http://sustainabledevelopment.un.org/content/documents/ 3377pan.pdf (Accessed 10th Nov. 2013).

Phnom Penh Water Supply Authority (2011). Financial Statements for the Year Ended 31 December 2011. Available online: http://www.ppwsa.com.kh/ Administrative/finance/English/Annual-Report/doc/PPWSA_Financial_ Statements_for_the_year_ended_31_Dec_2011_English.pdf (Accessed 23 Sept. 2013).

Phnom Penh Water Supply Authority (2012). Financial Statements for the Year Ended 31 December 2012. Available online: http://www.ppwsa.com. kh/Administrative/finance/English/Annual-Report/doc/PPWSA_FS_The_ year_ended_31_Dec_2012%28EN%29.pdf (Accessed 25 Sept. 2013).

Probe International Beijing Group (2008). Beijing's Water Crisis; 1949–2008 Olympics. Available online: http://www.chinaheritagequarterly. org/016/_docs/BeijingWaterCrisis1949-2008.pdf (Accessed 10th Nov. 2013).

PUB website. Available online: http://www.pub.gov.sg/Pages/default.aspx (Accessed 17th Oct. 2013).

Raj, K. (2013). Sustainable Urban Habitats and Urban Water Supply: Accounting for Unaccounted for Water in Bangalore City, India. *Current Urban Studies*, **1**(04), 156.

Smith, A. (1776). *An Inquiry into the Nature and Causes of the Wealth of Nations.* Cosimo classics, current edition 2011.

Smith, L. (2001). "The urban political ecology of water services in Cape Town", *Urban Forum,* **12**(2), 204–224.

South China Morning Post, 22 March 2013. "Hong Kong's unsustainable water policies". Available online: http://www.scmp.com/comment/insight-opinion/article/1196453/hong-kongs-unsustainable-water-policies (Accessed 23rd Oct. 2013).

Takizawa, S., Iwasaki, C. and Oguma, K (2005). "Effects of water tariff structures on water demand in Tokyo metropolis", *Water Science and Technology: Water Supply*, IWA publishing online, **5**(6), 235–242.

The World Bank (2012). India Groundwater: A Valuable but Diminishing Resource. Available online: http://www.worldbank.org/en/news/feature/2012/03/06/india-groundwater-critical-diminishing (Accessed_20 Nov. 2013).

Tokyo Water Management Plan 2013. Available online: http://www.waterworks.metro.tokyo.jp/eng/plan/01.pdf?1 (Accessed 21st Oct. 2013).

Tortajada, C. (2006). "Water management in Singapore", *Water Resources Development,* **22**(2), 227–240.

UN Water Facts and Figures (2013). Available online: http://www.unwater.org/water-cooperation-2013/water-cooperation/facts-and-figures/en/ (Accessed 9th Oct. 2013).

UNDP, Jakarta Indonesia Case Study (Water) (2012). Available online: http://www.esc-pau.fr/ppp/documents/featured_projects/indonesia.pdf (Accessed 3rd June 2015).

UNICEF Indonesia, Issue Briefs (2012). Available online: http://www.unicef.org/indonesia/A8-_E_Issue_Brief_Water_Sanitation_REV.pdf (Accessed 3rd June 2015).

Water Diplomacy, Aquapedia Case study Database (2013). Drinking Water Supply in Phnom Penh, Cambodia. Available online: http://aquapedia.waterdiplomacy.org/wiki/index.php?title=Drinking_Water_Supply_in_Phnom_Penh,_Cambodia#The_Problem (Accessed 23 Sept. 2013).

Water Supply in Tokyo (2012). Available online: http://www.waterworks.metro.tokyo.jp/eng/supply/ (Accessed 21st Oct. 2013).

World Bank website. Project Appraisal Document on a proposed loan in the amount of US$ 275 million to the Land Bank of the Philippines with the guarantee of the Republic of the Philippines for the Metro Manila wastewater management project. pp. 1–3. Available online: http://www-wds.worldbank.org/external/default/WDSContentServer/WDSP/IB/2012/04/26/000350881_20120426091055/Rendered/PDF/596750PAD0P1130osed0402601200SIMULT.pdf (Accessed 20th Nov. 2013).

Wichelns, D. (2013). Enhancing the performance of water prices and tariff structures in achieving socially desirable outcomes. *International Journal of Water Resources Development*, **29**(3), 310–326.

WWDR4 (The 4th edition of the UN World Water Development Report) (2012). Available online: http://www.unesco.org/new/en/natural-sciences/environment/water/wwap/wwdr/wwdr4-2012/ (Accessed 9th Oct. 2013).

Wu, X. and Ching, L. (2013). "The French Model and water challenges in developing countries: Evidence from Jakarta and Manila", *Water Policy*, **32**(2), 103–112.

Yadav, Y. (1996). "Reconfiguration in Indian politics: State assembly elections, 1993–95", *Economic and Political Weekly*, **31**(2/3), 95–104.

Zhang, S., Deng, L., Yue, P. and Ciu, H. (2007). Study on Water Tariff Reform and Income Impacts in China's Metropolitan Areas: The Case of Beijing; The World Bank. Available online: http://siteresources.worldbank.org/INTEAPREGTOPENVIRONMENT/Resources/ReportWaterpricingBeijingENFinal.pdf (Accessed 10th Nov. 2013).

# 5

# PLASTIC BAGS IN SINGAPORE: A CRITICAL EXAMINATION OF ISSUES IN SUSTAINABILITY

Marra Lin Teasdale-Hensby and Lye Lin-Heng

When Singapore became an independent republic in 1965, it began a fast paced modernisation program which sparked economic growth and an increase in consumption which has led to issues of environmental sustainability. One of these issues is that this increase in consumption and consumerism has resulted in the overuse of plastic bags. This study uses an interpretive approach of secondary materials and consumer surveys to quantify the current issue of plastic bag usage on Singapore's environment, analyse potential impacts of alternatives, review current reduction programs and the effectiveness of reduction programs in other countries.

Plastic bags serve many practical purposes but their use is excessive. In 2006, Singapore was using approximately 2.5 billion plastic bags a year, which was equivalent to 1.7 bags per person each day, almost double the average number in other first world countries such as the United States. Due to their excessive use and public littering, they are finding their way into Singapore's green spaces and waterways, adversely impacting its wildlife and changing our ecosystems to a level that may take decades to understand. Governments from around the world are investing money and resources to try and minimize their negative effects through new legislation and it is clear by the impacts highlighted in this study that it is time for Singapore to do the same.

## 1. Introduction

When Singapore became an independent republic in 1965, it began a fast paced modernisation programme, which focused on establishing a manufacturing industry, public housing and a robust education system. In 1970 the population of Singapore was just over 2 million, and as of 2012, has grown to 5.3 million.[1] This rapid population growth has been paired with an equally dramatic and impressive economic growth with Singapore's economy growing, on average, nine percent each year.[2]

[1]Department of Statistics Singapore's website http://www.singstat.gov.sg/. Accessed on 30 Jan 2013.

[2]"World Economic Outlook Database, September 2006" International Monetary Fund. 7 May 2009.

With this growth comes an increased consumer appetite, matched with an increase in consumption and, as Gary Gardner, Director of Research for Worldwatch, has said, "Most of the environmental issues we see today can be linked to consumption".[3]

Plastic bag reduction has increasingly become one of the most popular government-imposed policies worldwide to reduce the environmental impact of consumers. Numerous sources estimate that over 1 trillion bags are used annually, which would be the equivalent to 1 million used every minute.[4]

Whether used as carrier bags or to manage household waste, all plastic bags have an environmental impact. From Canada to Bangladesh, governments are passing legislation to try and minimise this impact; and even though Singapore is seen by many as a model of sustainable development, there is currently no legislation to address this issue.

To date, measures to promote the reduction of single-use plastic bags in Singapore are virtually nonexistent. Some residents feel that this is not an issue as these bags do not end up in a landfill, but instead are incinerated with all other waste that is generated in Singapore. However, considering that the average Singaporean uses twice the amount of plastic bags than the average person in other developed countries, it is difficult to argue that their use is not excessive. What is unclear at this time, is the environmental impact that this overuse is having and what actions should be taken.

### *Academic Literature Review*

Previous research has been conducted by students at the National University of Singapore (NUS) and will be reviewed, as well as international academic sources that specifically look at the impact of fees on the consumer consumption of plastic bags.

A recent thesis written by Takeo Sugii (2008) at Tufts University, USA, focused on the reduction of plastic bags through policy changes. The study showed that reducing plastic bag consumption through new policy and fees not only reduced their direct environmental impact, but also influenced people's lifestyle choices and attitudes towards the environment. When consumers had to purchase plastic bags, or pay a fee for them, it forced consumers to make a choice to either pay the extra money, or to decline

[3] "As Consumerism Spreads, Earth Suffers, Study Says", National Geographic News. Hillary Mayell, 12 Jan 2004.
[4] http://www.reuseit.com/. Accessed on 04 Feb, 2013.

them all together, resulting in the conscious contemplation of the impact on the environment every time they bought goods.

A study conducted by Gupta (2011) from the Delhi School of Economics tested various policies and approaches to minimise the use of plastic bags in Delhi, India. Back in 2009, the Delhi government introduced a ban on the use of plastic bags; but as of 2011, 94 percent of consumers still continued to use them despite the ban. This was attributed to a lack of reinforcement, which led the study to look at other ways to control/reduce their use. They tested three approaches through field experiments which included (i) provision of information to consumers on the environmental impacts of plastic bags (ii) a cash-back scheme (giving cash back to consumers who used alternatives to plastic bags) and (iii) provision of substitutes for plastic bags. The results from this study showed that by using all three of these approaches, the proportion of consumers who brought their own bags increased from 4.6 percent to 17.7 percent. This study shows that in cases where a ban may not be possible, reductions can still be accomplished.

Although no other NUS students had researched sustainability issues with plastic bags, Jing (2011) looked at overall sustainability efforts in Singapore and found that many of them focused on energy efficiency, green technologies, and pollution control. As a result Jing reported that Singapore relied too much on the efficiency of resource use and environmentally-sound technologies, and that not enough attention was paid to reducing the overall level of consumption. This paper highlights the need for increased resource conservation, to encourage Singaporeans to consume less, not just more efficiently.

This message was also highlighted in a report written by Theng (2002), which looked at the incineration of waste in Singapore and the long term effects of emissions. It was argued that even though Singapore has legislation in place and technologies to reduce the amount of emissions from incineration, there will still be minimal levels released and these cumulative amounts must still be considered.

Due to the growth in consumerism in Singapore there is a definite need for a critical review of this issue. Based on this literature review there is a clear gap in extensive research pertaining to plastic bag usage in Singapore. This is why the following study is necessary.

> "Every individual, organization and company can make a difference to the environment in the choices and decisions that they make every day"
>
> (The Singapore Green Plan 2012, 2006 Edition)

## 2. Environmental Impacts of Plastic Bags Globally

It is estimated that more than 1 trillion plastic bags are used globally every year. This copious use of plastic bags has many environmentalists worried about the impact this is having on our wildlife as they make their way into waterways and nature areas. The littering of plastic bags also has another impact; the economic burden that is put on governments to remove them from littering our landscapes. Both of these impacts will be reviewed.

### *Impacts on Wildlife*

Many countries that use landfills to dispose of their waste are concerned that a single plastic bag can take up to 1,000 years to degrade. Their life span in landfills is not a concern for Singaporeans as the local waste stream leads to incineration. However this does become a concern when plastic bags do not make their way into the recycling or trash bin, as they are light weight and aerodynamic and easily transported by a gust of wind to our green spaces, waterways and coastal areas.

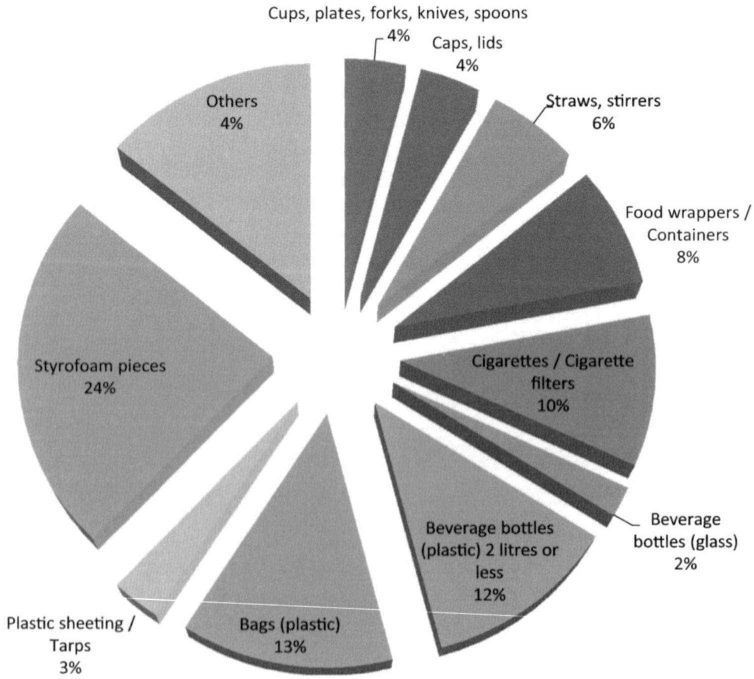


Fig. 5.1. Debris Collected during the 2012 ICC event in Singapore.

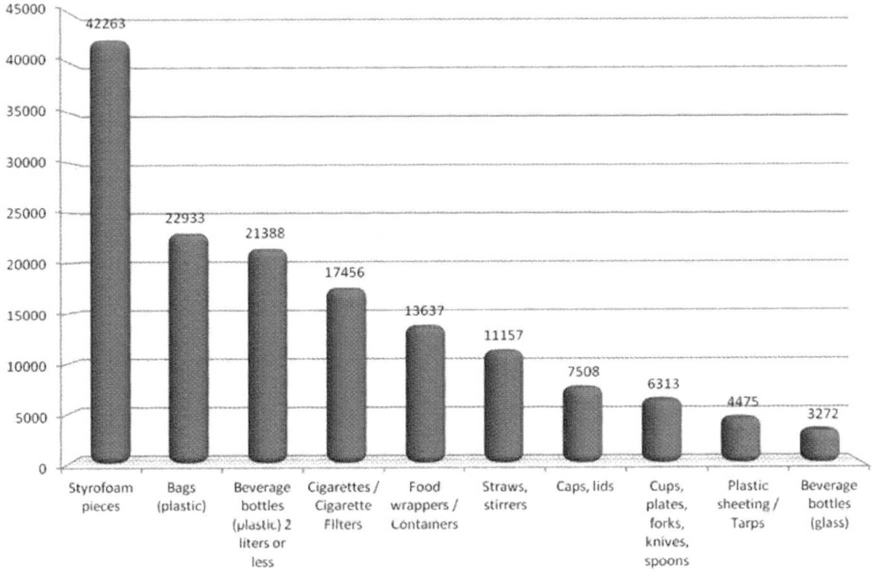


Fig. 5.2. Top 10 items collected during the 2012 ICC event in Singapore.

One programme, which aims to reduce their impact, is the International Coastal Cleanup (ICC) event. During the 2012 event, volunteers collected 173,574 items along Singapore's coastal areas. Thirteen percent of these items were plastic bags as shown in Figure 5.1. As shown in Figure 5.2, plastic bags were the second most common type of debris found after Styrofoam pieces.[5]

These results show that plastic bags are a very large percentage of litter found around the coastal areas. With 22,933 littering plastic bags collected from Singapore's coasts in just one day, it is clear that plastic bags pose a serious litter problem in Singapore. Although no formal studies have been done to determine the amount of plastic bags found littering other areas of Singapore, it can be assumed that similar results would be found as to the percentage of litter that is plastic bags in other areas of the island as well.

Every plastic bag that finds its way into our forests and green spaces threatens the natural progression of wildlife. One of the most common threats is animals mistaking them for food and either suffocating on them or ingesting them, resulting in the animal choking to death or blocking

[5]As per personal email communications with N. Sivasothi, Coordinator, International Coastal Cleanup Singapore. 13 March 2013.

their intestinal tract. Due to the slow break down rate, a single plastic bag has the potential to kill multiple animals. Some studies have estimated that one bag can kill one animal every three months through digestion or inhalation.[6] These bags are also very dangerous for marine life when they find their way into our waterways. In particular, marine hunting mammals can easily mistake the size, shape, and texture of a plastic bag for a meal and wind up cutting-off their airway by ingesting the bag.

> "The scale of contamination of the marine environment by plastic debris is vast. It is found floating in all the world's oceans, everywhere from polar regions to the equator"
>
> (*Plastic Debris in the World's Oceans*. 2009. A study by UNEP and Greenpeace)

### *Economic Impacts*

The plastic bags that we see littering our landscapes not only pose a threat to our wildlife, they have an economic burden as well. Governments spend large sums of money to clean them up as their visual impacts could result in community dissatisfaction and reduced tourism. They can also clog waterways by getting lodged in storm drains, disrupting flow and hindering rainwater drainage resulting in flooding or even contamination.

Singapore is fortunate that it has one of the cleanest landscapes in Asia due to many initiatives, including stiff littering penalties, strict enforcement against littering, and a highly developed cleaning regime. Although the National Environment Agency (NEA) was not able to share the amount of money spent on litter collection, a great amount of resources are dedicated to this task. To estimate the cost of plastic bag litter, studies conducted in California will be reviewed, as much effort has been made by their government to quantify this issue.

A Los Angeles County Staff Report on Plastic Bags (2007) had one section that looked specifically at the impacts of plastic bag litter. This report views plastic bag litter as a larger nuisance compared to other littered materials because it is usually white or brightly coloured (causing a significant eye sore when found in natural landscapes) and because it is so light and can become airborne very easily, making it difficult to collect. In Los Angeles County, the Department of Public Works, and the

[6]www.healthguidance.org "The Effects of Plastic Bags on Environment" by Jamey Wagner. Accessed 18 Feb 2013.

Flood Control District, spend US$18 million annually on activities such as street sweeping, catch basin cleanouts, overall litter prevention, and public education, in order to comply with the Federal Clean Water Act.

This study also found that communities within close proximity to landfills and solid waste processing areas have higher levels of plastic bag litter due to bags escaping from waste collection trucks as they travel to these sites. Local landfill and waste transfer operators estimate that they spend from US$1,500 to US$25,000 per month at each facility, on patrols to pick up littered plastic bags.[7]

Aside from plastic bag litter negatively impacting the visual appeal of landscapes, this report also looked at other effected areas of city maintenance. Plastic bags were found in the storm drain systems impacting storm water runoff; they even impacted city landscape maintenance by becoming tangled in the lawn mowing equipment. Although the exact percentage of plastic bags in the total litter stream in Los Angeles is unknown, as part of this report, they conducted a study on catch basin cleaning to quantify their impact. In June of 2004 a waste characterisation study was conducted by cleaning 30 storm drain catch basins. They found that plastic bags were 25 percent of the waste by weight and 19 percent by volume.

Although California has an anti-littering law in place that carries fines, it still failed to solve the issue of plastic bag litter. Anti-littering laws are difficult to enforce, as officers have to catch the person in the act. This is why reduction in the volume of plastic bags that enter the waste stream is the best way to fight against it and reduce the economic burden that plastic bag litter has on governments.

Aside from the impact on governments from litter around our terrestrial landscapes, there is also a significant economic impact from the litter that finds its way into our oceans. A study by McIlgorm, *et al.*, (2008) found that marine debris is costing Asia-Pacific Economic Cooperation (APEC) member economies, in which Singapore is a member, up to US$1.3 billion annually as it threatens the fishing, shipping and tourism industries. Previous research suggests that 6.4 million tons of debris find its way into the world's oceans every year, comprising of around 8 million individual items every day. Plastics comprise 60 to 80 percent of the total debris recorded.

[7]County of Los Angeles. 2007. An Overview of Carryout Bags in Los Angeles County. A Staff Report to the Los Angeles County Board of Supervisors. August 2007. CA, USA.

## 3. Plastic, Paper, and Reusable Bags: A Comparative Analysis of Impacts

If plastic bags are banned or their availability is significantly reduced, consumers will have to find alternatives such as paper bags or reusable bags made of fabrics such as cotton. This paper next looks into the environmental concerns surrounding plastic bags if they follow their intended lifecycle of production, use and proper disposal. It will also compare the high level impacts of alternatives available to consumers such as paper bags and reusable bags. The environmental impacts of all three will be reviewed and analysed, primarily through published Life Cycle Assessment (LCA) studies which focus on Singapore and other countries. Neither a full LCA nor a partial LCA was prepared for this report. This comparison will focus on the resource use and production stages of all three types of bags, as a study conducted by the Environment Agency in England and Wales (2011)[8] found these stages to be where the highest environmental impact occurs.

### *Plastic Bags*

According to the NEA website[9] the most commonly used types of plastic bags in supermarkets and convenience stores are High Density Polyethylene (HDPE) and Low Density Polyethylene (LDPE) plastic bags. Khoo *et al.* (2010)[10] conducted a study that compared the overall environmental impacts of plastic bags produced in Singapore to bio-based bags produced in the U.S. using crops such as sugar and corn. Bio-plastics are traditionally viewed as a better environmental alternative as they are made from renewable feedstock, without depleting non-renewable resources such as crude oil and natural gas. However, this study discusses these environmental benefits, and shows how they are debatable as the type of energy utilised throughout their entire lifecycle needs to be taken into account.

In this particular example, the energy to produce the bio-bags was supplied by the U.S. electricity mix which is coal powered. This made the environmental impacts of production of these bio-bags five times greater

---

[8] The Environment Agency (England & Wales), 2011. "Life cycle assessment of supermarket carrier bags: a review of the bags available in 2006". Environment Agency, Horizon House, Bristol, UK.

[9] http://app2.nea.gov.sg/ Singapore's National Environment Agency. Accessed 5 March 2013.

[10] Khoo, H., Tan, R., Chng, K. (2010). "Environmental impacts of conventional plastic and bio-based carrier based", *The International Journal of Life Cycle Assessment,* **15** (3), 284–293.

than regular plastic bags produced in Singapore. Additionally, the shipment of these bio-bags from the U.S. to Singapore increased their environmental impact from the emissions of volatile organic compounds (VOCs) from marine transportation. Even though our current analysis is not comparing plastic bags to bio-bags, this study is important as it highlights that the manufacturing of plastic bags in Singapore is local, which decreases their overall environmental impact compared to other studies based in the UK where plastic bags are shipped in from China.

A study commissioned by The Environment Agency in England & Wales (2011) conducted a lifecycle assessment on plastic bags available in the UK and suggests that plastic bags may not be the eco villain that they are portrayed to be. They found that one of the most significant environmental impacts of plastic bags in the UK was the transportation from China where they are produced. This research also found that over 40 percent of plastic bags were reused as bin liners in the UK, and that reusing a plastic bag as a bin liner had greater environmental benefits than recycling them.

Another finding showed that the end of life management of the plastic bags in the UK reduced the overall environmental impact. This was because the UK waste stream is sent to waste to energy incineration plants, which is the same process currently followed in Singapore. During incineration, energy is captured as heat, which is then used to boil water to produce steam that drives turbines to generate electricity. Hence, the petroleum used to make plastic bags is reused to generate power.[11]

Waste to energy incineration is a viable option for waste management in land-scarce Singapore as it reduces the volume of waste by as much as 90 percent, and also provides approximately 2 percent of Singapore's power. Still, there are environmental impacts to incinerating plastic bags. An impact assessment of waste management in Singapore conducted by Tan *et al.* (2006) found that the incineration of all materials causes harm to human health and the environment, especially the incineration of materials such as plastics. They also found that even though the incineration plants in Singapore produce energy, these benefits are outweighed by the air pollution which is produced. Even with modern pollution control technologies, only 90 percent of NOx and 99 percent of toxic metals and acid gases are removed. These pollutants contribute to climate change, acidification, and

[11]http://www.wildsingapore.com/news/20070506/070512-4.htm Article written on Wild Singapore, 2007. Accessed 6 March 2013.

eco toxicity. As pointed out in this study, incineration is still the best choice for Singapore due to land scarcity, however reduction of total waste generated is key to minimising these effects.

### *Paper Bags*

Since the backlash against plastic bags began, several studies have sought to compare the environmental effects of plastic and paper grocery bags. One of the first issues highlighted in the media about plastic bags was the fact that they can take up to 1,000 years to decompose, which is why paper bags were viewed as a better alternative. Although paper bags were initially marketed as a more responsible environmental choice, there is growing evidence that their production, use and disposal put a larger burden on natural resources. Millions of trees are cut down every year to manufacture paper bags and these bags require four times the amount of energy in production compared to the average plastic bag.[12]

Another drawback to paper bags is that they seem to have minimal reuse as they are not reused as bin liners and they are not as durable as plastic bags, being easy to split or tear. The study conducted by the Environment Agency in England and Wales (2011) looked at the number of times that an average paper bag would have to be reused to produce less global warming than an average plastic bag. Even if a plastic bag was not reused and put into the trash, you would still have to reuse the paper bag 3 times to produce less global warming. If a plastic bag was reused as a bin liner, then you would have to reuse the paper bag 7 times in order to produce less global warming.

#### *Reusable bags*

A current marketing trend for companies is to give away branded reusable bags at events and many grocery stores sell branded reusable bags at their checkout areas. The main issue with this trend is that not everyone who is given a reusable bag actually reuses it. Sometimes they go unused because they are of a cheap quality, or maybe because the person who accepted the free bag never gets into the habit of using it. Even if these bags are reused, are they being reused enough to make them a better environmental choice than a standard plastic bag?

---

[12]http://earth911.com/recycling/paper/brown-paper-bags/facts-about-brown-paper-bags/ "Facts about brown paper bags" Accessed 1 March 2013)

According to the NEA website, most reusable bags in Singapore are made from Polypropylene (PP). The study conducted by the Environment Agency in England and Wales (2011) found that a consumer would have to reuse the PP reusable bag 11 times and a cotton reusable bag 131 times in order to have less impact on global warming than a plastic bag thrown in the trash after one use. If you reused your plastic bag as a bin liner, then that same PP reusable bag would have to be reused 26 times and the cotton bag would have to be reused more than 327 times to have a lesser impact on global warming.

## 4. Current Plastic Bag Reduction Programmes in Singapore

### *Bring Your Own Bag Day (BYOBD)*

On February 11$^{th}$ 2006 the "Why waste plastic bags? Choose reusable bags!" campaign was launched by the NEA — the first national campaign to reduce the use of plastic bags in Singapore.

> "Our purpose is three-pronged: apart from discouraging wastefulness, we want to promote sustainability by moving away from the single-use and disposable mentality towards that of maximizing the utility of everything we use, and protecting the environment. In the urban context, plastic bags can end up as litter, and in the natural environment, plastics have a strong tendency to suffocate plant and marine life. Continuous education is key."
>
> (Mr. Yatin Premchand, Senior Manager, Singapore Environment Council)[13]

The goal was to encourage Singaporeans to bring their own bags or to use reusable bags when they were shopping. Table 5.1 shows the incentive schemes used by participating retailers that were referred to in the press release.[14]

The incentive scheme was not consistent, as all retailers chose how they promoted the programme individually. Despite the fact that this may have been confusing for consumers, retailers showed that since the launch of the campaign approximately 2 percent of their shoppers brought their own

[13]http://app2.nea.gov.sg/news NEA news release, "Why waste plastic bags? Choose reusable bags!", 11 February 2006.
[14]Email correspondence with Yang Hong (Ms), Manager (Waste Minimisation & Recycling), NEA from Jan 2013–March 2013.

Table 5.1. Summary of Retailers' Incentive Schemes. ("Why waste plastic bags? Choose reusable bags!")

| Retailer | Timeframe | Incentives to Customers |
|---|---|---|
| Carrefour | 11 Feb 18–2006 | The price of their reusable trolley bags will be reduced from $7.90 to $4.90 for a one-week period |
| | 11 Feb 10 Mar 2006 | The normal reusable bags will be sold at $0.95 for a one-month period |
| NTUC FairPrice | 09 Feb–08 Mar 2006 | Customers who bring their own bags to shop at NTUC outlets will be able to participate in the lucky draw. Prizes are some electronics items or vouchers. |
| | 09 Feb–08 Mar 2006 | Reusable bags (2 to 4 colour selection) will be put up at all outlets for customers to purchase at the affordable price of $1 |
| Prime | 11 Feb 18 Feb 2006 | Whereby shoppers will be given a token, if shoppers do not take Prime's plastic bags but bring their own bags or purchase reusable bags from Prime, instead |
| | 04 Feb–18 Mar 2006 | Shoppers who do not ask for plastic bags and brings his/her own bags for purchase (e.g. shopper can bring his/her own bag or buy reusable bags from Prime) can participate in a lucky draw. $50 worth of cookware to be won — 100 winners. |
| Cold Storage | 11 February, 2006 | free reusable bags will be given to each of the first 50 shoppers with purchase of $30 or above on a single receipt at every outlet of Cold Storage (35 outlets) |
| Giant | 11 February, 2006 | free reusable bags will be given to each of the first 50 shoppers with purchase of $30 or above on a single recept at every outlet of Giant (7 outlets). |
| IKEA | 11 Feb–17 Feb 2006 | Reusable bags will be sold at reduced prices of $1.00 for a one-week period. |

shopping bags as of June 2006.[15] This percentage was small, but showed that in just five months consumer habits began to change.

These positive results led the NEA to take the programme a step further and on 18 April 2007 they launched the Bring Your Own Bag Day (BYOBD) campaign. During the first Wednesday of every month, consumers were encouraged to bring their own plastic bags or reusable

[15]http://app2.nea.gov.sg/news NEA news release, "Make A Difference On World Environment Day — Choose Reusable Bags!", 2 June 2006.

bags. The other option was to voluntarily donate $0.10 for each plastic bag taken from participating retailers. The proceeds were collected and sent to the SEC to fund the BYOBD campaign and other environmental programmes.[16]

Surveys during the monthly BYOBD events revealed that on average 60 percent of shoppers participated either by bringing their own bags, by buying reusable ones or by making donations when they required plastic bags. Approximately two thirds of consumers surveyed also supported increasing the frequency of BYOBD events.[17]

As a result of the public support shown through these surveys, from June 4th 2008, BYOBD events were extended to every Wednesday. According to NEA staff, there is no data to show if this extension was a success as no further consumer surveys were conducted and no data was collected from participating retailers. When this programme was extended on June 4th 2008, the NEA also handed the programme over to the SEC. As of March 2013, details of the BYOBD programme can still be found on the NEA and SEC websites, including a list of participating retailers and reasons why plastic bag reduction is important for Singapore. However, as discovered through interviews with NEA staff, the BYOBD campaign was discontinued in 2010. When NEA staff was questioned as to the reasons for cancelling this programme, it was mentioned that monetary collections had decreased significantly and that the program was no longer self-sustaining.

According to an article published on Chanel News Asia in 2010[18] talks were taking place between the SEC and major retailers to implement a nationwide scheme. To date, no such program has been launched. As per interviews with members of the SEC[19] they are still in the process of conceptualising how to begin research for a possible new programme, even though it was first mentioned almost two years ago.

The demise of this programme, may have been caused by the voluntary aspect of the fee; or perhaps, from a lack of public education. One thing is for sure, that the failure of the BYOBD campaign should not stop other

[16]http://app2.nea.gov.sg/byobd.aspx BYOBD information page on the NEA website. Accessed 9 January 2013.
[17]Personal emails and phone interviews with Yang Hong (Ms), Manager (Waste Minimisation & Recycling), NEA from Jan 2013–March 2013.
[18]"New scheme planned to replace Bring Your Own Bag Day" by Saifulbahri Ismal, Channel News Asia. 23 April 2010.
[19]Phone interview with Vaidehi Shah, Project Manager at Singapore Environment Council, 8 March 2013.

similar programmes from being launched in Singapore. The following are two success stories of plastic bag reduction programmes in Singapore.

### *Ikea — Plastic Bag Reduction Programmes*

Since April 2007, Ikea Singapore stopped giving away free plastic bags, at their two locations in Singapore and customers were charged $0.05 for every plastic bag purchased. This was considered a bold move as they were the first retailer in Singapore to charge customers for plastic bags. Ikea saw a reduction of 5.34 million plastic bags in the first year of their programme.[20]

In January 2013, Ikea Singapore announced that it would be taking another step in reducing the usage of plastic bags by ceasing to sell them as of March 23rd 2013.[21]

> "This initiative is the next natural step for us to further reduce the use and consumption of disposable plastic bags in Singapore and at the same time support change in people's everyday behavior for a positive sustainable impact on the environment."
>
> (Mr. Christian Rojkjaer, Managing Director of Ikano Retail Asia)

With this latest initiative, consumers shopping at Ikea will be encouraged to bring their own bags, or to purchase one of Ikea's blue reusable shopping bags. According to their website, there are 2.5 billion plastic shopping bags used every year in Singapore which is equivalent to 2500 bags per family per year or seven bags a day. If each family uses one less, Singapore could save 50 million bags each year. These numbers demonstrate the staggering results that a few little changes in consumer behaviour can have on the overall environment.

### *NUS Food Centres — Green Canteens*

Green Canteens is one of the subcommittees of Students Against Violation of the Earth (NUSSU SAVE) at the National University of Singapore (NUS). Green Canteens launched a tax on plastic bags given out on the NUS campus in 2010. When students, teachers or staff request for a plastic

[20] Personal email communications with Marcus Tay Guan Hock, Sustainability Manager (Singapore), Ikea. 27 February, 2013.

[21] http://www.ikea.com/ms/en_SG/about_ikea/our_responsibility/national_community_involvment/index.html "IKEA to do away with disposable shopping bags". Published 17 Jan. 2013. Accessed 28 Feb 2013.

bag, they are required to drop $0.10 for each plastic bag taken into a coin collection box located next to the counter.[22]

This programme aims to discourage the irresponsible use of plastic bags on the NUS campus, and was implemented based on feedback collected from students. In 2009, NUSSU SAVE organised an "NUS VOTES EARTH" campaign aimed at engaging the NUS population to vote on ways to fight against climate change by making changes in their daily lives. According to Ms. Lee Tse-Sin, Project Director of Green Canteens, 71 percent out of roughly 10,000 poll participants voted in favour of phasing out free plastic bags on the NUS campus. All proceeds collected from this programme go to the NUS Sustainability Fund which funds environmental projects that benefit the larger Singapore community.

Since the Green Canteen plastic bag tax was implemented, there has been an 86 percent reduction in plastic bag use on the NUS campus. This proves that you do not have to be a retail giant such as Ikea to implement a successful plastic bag reduction programme. One contributing factor to this programme's success is that the NUS community supported its mission from day one as shown in the results of the "NUS Votes Earth" campaign. This demonstrates the importance of consumer surveys to show consumer buy-in when looking to implement a plastic bag reduction programme.

*From a consumer perspective — consumer survey results*

Current bag reduction programmes in Singapore have all been carried out in the absence of any legislative requirements pertaining to the use of plastic bags. As part of the current study, the potential for such legislation, or future government led initiatives will be explored. The Consumer Survey in Figure 5.3 aims to collect data from consumers on their opinions regarding plastic bags, their usage in Singapore, and whether consumers feel that a levy should be implemented to reduce their usage.

The surveys were carried out in February of 2013, and 270 surveys were completed. There were a total of 9 questions and a comments section. The first two questions determine a general profile of the consumers such as gender and age as shown in Figure 5.4 and Figure 5.5.

The next question in the survey captured consumers' perceptions of whether or not they believe there are too many plastic bags in the Singapore

[22]Email communications with Ms. Lee Tse-Sin, Project Director of Green Canteens. 27 February 2013.

| | | | | | |
|---|---|---|---|---|---|
| What is your age group | Teens | 30s | 50s | 70s | 90s |
| Male Female | 20s | 40s | 60s | 80s | |
| Do you believe there are too many plastic bags used in Singapore? | Yes | No | Undecided | | |
| What purposes do you **reuse** plastic bags for? | Purpose 1 | Purpose 2 | Purpose 3 | Purpose 4 | Purpose 5 |
| How often do you use reusable shopping bags when you **grocery shop**? | Every Time | Sometimes | Rarely | Never | |
| How often do you use reusable shopping bags when you shop for goods **other than** | Every Time | Sometimes | Rarely | Never | |
| What things deter you from using reusable shopping bags? | Too expensive to buy | I forget to take them when shopping | I don't like carrying them around with me | I need plastic bags for my trash bin @ home | For hygenic reasons |
| | I use them all the time | Other: | Other: | Other: | Other: |
| Do you think there should be some form of levy or fee on plastic bags to deter | Yes | No | Undecided | | |
| If there was a levy on plastic bags what amount do you feel would be reasonable to deter overuse of bags? | $0.05 | $0.10 | $0.15 | $0.20 | Other: |
| Other Comments: | | | | | |

Fig. 5.3. Consumer survey on plastic bags in Singapore.

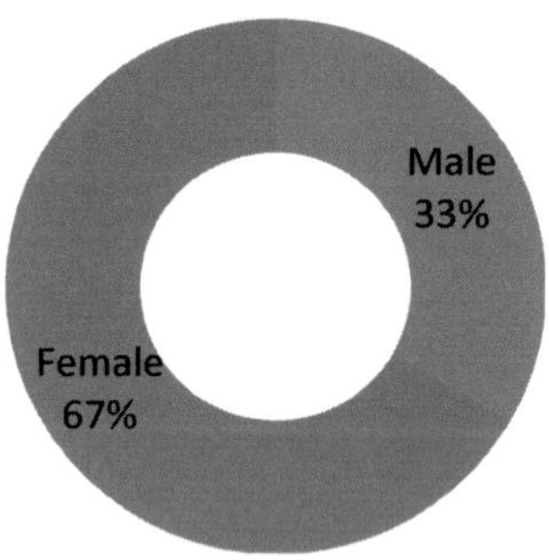


Fig. 5.4. Gender of respondents.

market. These results are of particular interest because they show that 87 percent of respondents feel that there are too many plastic bags being used in Singapore, and only 5 percent of respondents felt that there is no issue with the amount being used (as shown in Figure 5.6).

The next question on the survey looked at the reuse of plastic bags. This question did not provide examples of reuse options but instead consumers had to identify options with no help from the surveyors. Figure 5.7 shows the different reuse options as identified by the respondents.

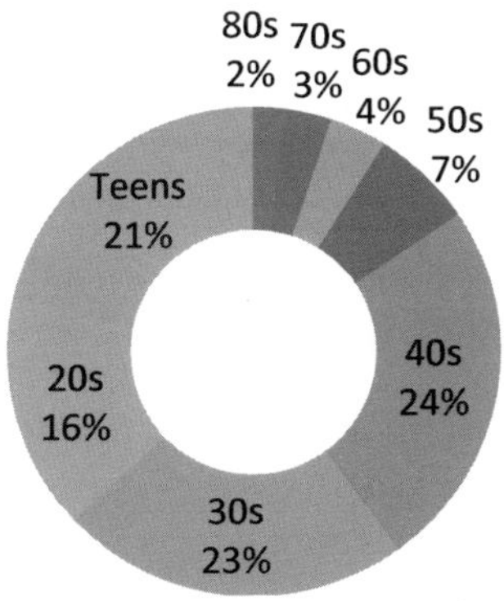


Fig. 5.5. Age of respondents.

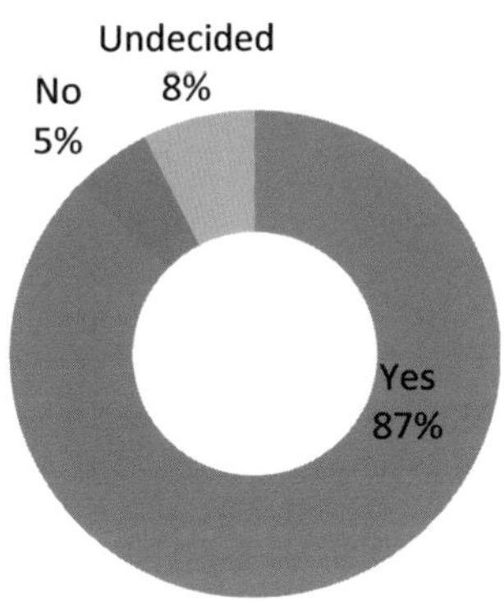


Fig. 5.6. Are there too many plastic bags used in Singapore?

The next three questions looked at the use of reusable bags by consumers in Singapore to determine if Singaporeans had integrated their use into their daily shopping habits and if not, what the reasons were.

As shown in Figure 5.8, only 19 percent of respondents said that they use reusable shopping bags every time that they grocery shop, increasing to 27 percent when asked about shopping for other goods.

The next question on the survey sheds some light as to why this may be, by asking respondents which factors deter them from using reusable shopping bags. Not all respondents answered this question and some chose more than one answer. The results are shown in Figure 5.9.

According to the results of this survey, the number one reason that respondents do not use reusable shopping bags is that they forget to bring

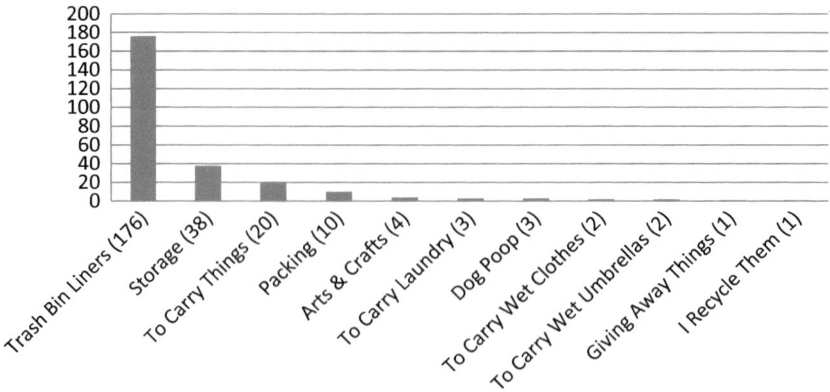


Fig. 5.7. What purposes do you reuse plastic bags for?

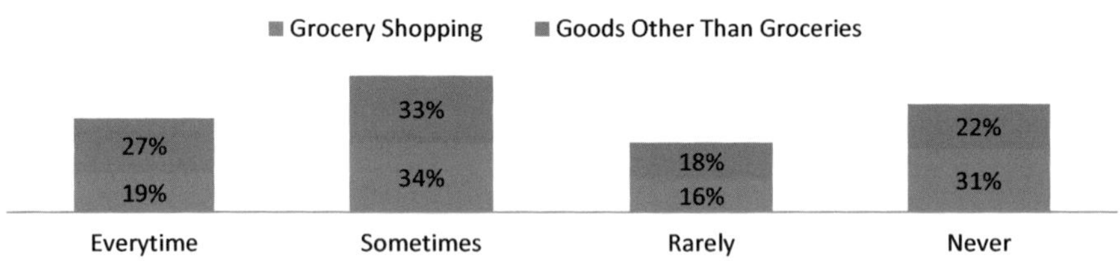


Fig. 5.8. How often do you use reusable shopping bags?

them when shopping. Using plastic shopping bags for bin liners at home was only chosen by 20 percent of the respondents. This would suggest that even though many people use plastic bags as bin liners at home, this does not deter them from using reusable bags.

The last two questions on the survey asked respondents if they felt that there should be a levy on plastic bags and if so, how much should that levy be? 71 percent of respondents chose "Yes", while only 22 percent answered "No" and 7 percent were "Undecided". The 192 respondents that answered "Yes" were asked to choose what they felt to be a reasonable dollar amount for a levy to deter over usage of plastic bags. Figure 5.10 shows the breakdown of their responses.

62 percent of respondents felt that a levy should be set at either \$0.05 or \$0.10, which would be in line with other countries' levies, which will be discussed in Section 6 of this paper. The final section of the survey allowed

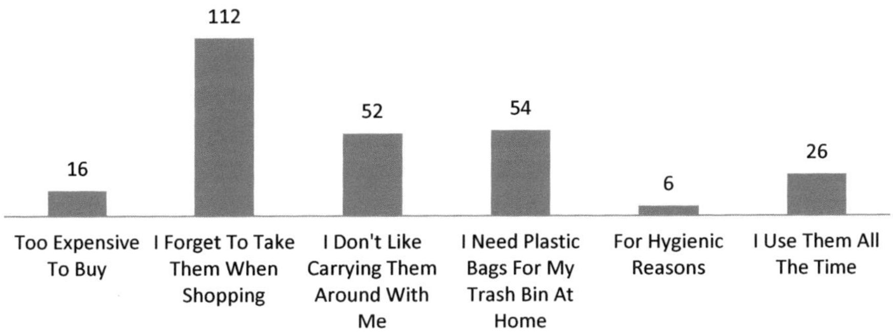


Fig. 5.9. What things deter you from using reusable shopping bags?

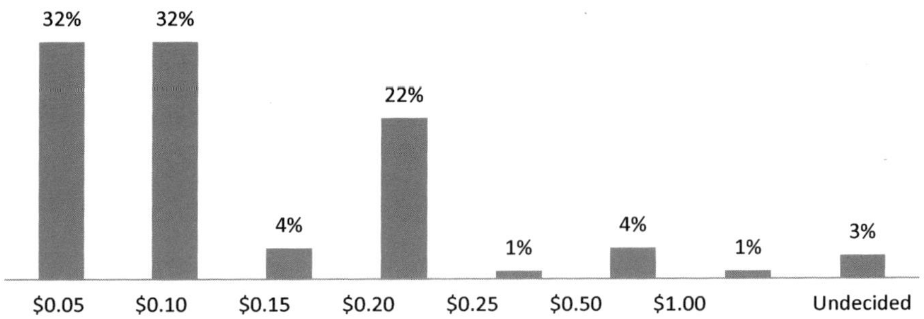


Fig. 5.10. What $ amount should a levy on plastic bags be in order to deter overuse?

for additional comments from the participants. Below are samples of what some of them had to say:

> "Usage of Plastic Bags should be reduced. The supermarkets should spread awareness about less usage of plastic bags. You can't get rid of plastic bags."

> "[Levy] rate [sic.] of plastic bags should be higher so people take reusable bags with them."

> "Encourage to not use plastic bags through campaign, not to provide plastic bags every Wednesday/Friday [sic.]."

> "Charge for plastic bags so people think twice."

> "If there is a fine, it should be extensive so that people feel the pinch."

"They should provide an incentive to not use bags rather than fine people."

This survey shows that out of the 270 consumers who were surveyed, 87 percent felt that there are too many plastic bags in Singapore, and 71 percent would support a levy on their use. Despite these findings only 19 percent of respondents said they used reusable bags every time that they grocery shop, and 27 percent used them every time when they shop for other goods. When asked what detered them from using reusable bags, the most common response was that they forgot to take them with them when shopping. Based on these responses there were no significant reasons why customers do not use reusable bags, and instead points to an issue with consumer behaviour of forgetting them.

## 5. Lessons Learned from Our Island

Several implications can be drawn from the three programmes highlighted. First looking back on the government-led initiatives, starting with the "Why waste plastic bags? Choose reusable bags!" in 2006. Based on this brief campaign, in only five months, retailer-led surveys showed that 2 percent of customers were using reusable bags. Although this percentage was small, it showed that the campaign was successful in motivating consumers to change their habits.

By April 2007, consumer surveys showed that 60 percent of consumers were participating in BYOBD, and that approximately two thirds of consumers supported the BYOBD events to be more frequent; resulting in the events being extended to every Wednesday. By 2010 the voluntary $0.10 donations had decreased so much that the programme was no longer sustaining the costs of operations and was discontinued. Due to the lack of data, it is difficult to determine the exact causes of the programme's deterioration. However when looking at other successful models of plastic bag reduction programmes in Singapore, some observations can be made.

The successful programmes run by Ikea and NUS both involved mandatory fees on plastic bags and consumer education programmes. These consumer education programmes were also consistent throughout their locations providing a clear message to consumers. While the BYOBD campaign had no mandatory consumer education at participating retailer locations and voluntary fees. This may have resulted in reduced consumer engagement, as it did not force consumers to stop and think about the issue in the same way that a mandatory fee did.

This is not to say that a nationwide program could not be successful. Looking at the Ikea and NUS programmes, it is clear that with proper structure, consumer education and dedication, consumer behaviours can be changed in Singapore. With 87 percent of the respondents in this study believing there are too many plastic bags in the Singapore market and 71 percent supporting a levy on their use, it is clear that even the public wants to see a nationwide programme. Hopefully the SEC will look at successful programmes in Singapore, such as the ones administered by Ikea and NUS, and draw from their experience. There is also a wealth of information to be drawn from successful nationwide programmes that other countries have launched.

## 6. Lessons Learned from Our Neighbours

It is clear that Singaporeans feel plastic bag usage is an issue and that the current programmes are not sufficient in addressing this. So we will look at some of the ways other countries have tackled the overuse of plastic bags and the lessons learned from their experiences. First, we look at Ireland, where the government imposed a tax on their use to reduce their consumption and to increase public awareness of littering. Then Hong Kong's plastic bag levy will be reviewed to highlight the lessons learnt from introducing such a programme into a developed, highly populated Asian market.

### *Ireland's Plastic Bag Tax*

The plastic bag tax introduced in Ireland in 2002 is a perfect example of how product taxes can influence consumer behaviour, provide environmental payoffs and do so at negligible costs to the government. This tax of € 0.15 per plastic bag was designed to change consumers' behaviour; to reduce their consumption of plastic bags and to increase public awareness of littering. The plastic bag tax was regulated under the Waste Management (Environmental Levy)(Plastic Bag) Regulations, which were passed in 2001.[23] Although there are many countries that have implemented a similar tax, Ireland's programme is seen as one of the most successful.

According to a paper written by Sugii (2008), two major factors led to this tax being implemented: shortage of landfills and damage to natural

[23]Irish Statute Book Webpage, http://www.irishstatutebook.ie/2001/en/si/0605.html. Accessed 11 March, 2013.

scenery. The Irish government decided to tackle the shortage of landfills by pursuing policies aimed at reducing the amount of waste produced and increasing recycling efforts — the plastic bag tax was one of those policies. The second issue, damage to natural scenery caused by plastic bags, was highlighted as a major issue through a national survey on Irish citizens' concerns for the environment in 2000 called "Attitudes and Actions".[24]

A study by Fehily Timoney & Company (1999) was initiated by the Department of Environment and Local Government (DoHLG) to look at the consumption and disposal of plastic bags, and alternative management measures; and to assess policy options such as a levy, deposit refund scheme, prohibition and public awareness campaigns.

As a result of this study, the Irish government introduced a levy that has clearly worked as the use of plastic bags dropped by more than 90 percent.[25] Concerns about their impact on natural scenery were also reduced as the national litter composition attributed to plastic bags dropped from 5 percent to 0.32 percent in 2002, and 0.22 percent in 2004.

Also, the overall costs of administration only amounted to 3 percent of revenues; enabling 97 percent of the money collected to be put into an environmental fund that benefited a variety of environmental programmes around Ireland.[26] The low operating cost was mostly due to the fact that it was possible to integrate reporting and collection into existing tax reporting systems. Enforcement of the programme also required little funds, as it was integrated into the daily responsibilities of local authorities who made sure the tax was charged properly.

The overall response from the public regarding this programme was positive. Convery *et al.* (2007) reported a strong perception among consumers surveyed, that the effect on the environment was positive; and many even reported feelings of guilt when they forgot their reusable bags when shopping. It is clearly a strong model for other countries to study if they are considering the implementation of a plastic bag reduction programme. However, some may argue that consumer behaviour and attitudes in Europe towards the environment differ from those

[24]The Litter Monitoring Body. (2003). The national litter pollution monitoring system. Retrieved from http://www.litter.ie/docs/DoEHLG%20System%27s%20Results%20Report%20Final%202002.pdf

[25]http://www.litter.ie/. Accessed 05 Feb 2013.

[26]Convery *et al.* (2007). "The most popular tax in Europe? Lessons from the Irish plastic bags levy", *Environ Resource Econ.*, **38**, 1–11.

in Asia. For this reason the plastic bag levy in Hong Kong will be reviewed.

### Hong Kong's Plastic Bag Levy

The overuse of plastic bags in Hong Kong is a major and visible environmental problem. A landfill survey conducted by the Environmental Protection Department found that approximately 8 billion plastic bags are disposed of in landfills every year. In 2007 when this survey was conducted, that number translated into more than three plastic bags per person per day.[27] Shortly after the results of this study were released, the government proposed a levy on plastic bags in May 2007.

One of the first steps taken in Hong Kong was to launch a public consultation. According to the Public Consultation Report on the proposal for an environmental levy on plastic shopping bags (2007), the main objective of the proposal was to "reduce the indiscriminate use of plastic shopping bags through an economic disincentive".[28] The approach that the government proposed was a phased-in introduction of the levy. The first phase would apply a $0.50HKD ($0.06 USD) levy on each plastic bag distributed at large retailers, including chain or large supermarkets, convenience stores and personal health and beauty stores.

A public opinion survey was conducted in June and July of 2007 with 1,102 telephone interviews being completed. Almost 90 percent of respondents agreed that there was room to reduce the use of plastic bags in their daily lives and 66 percent supported the introduction of an environmental levy on plastic bags. Due to the high levels of support received from the public consultation the Product Eco-responsibility Ordinance (PERO) (Cap. 603) was enacted in July 2008. The Ordinance provided a legal basis for introducing producer responsibility schemes, with the environmental levy on plastic shopping bags as the first scheme under the Ordinance. The Product Eco-responsibility (Plastic Shopping Bags) Regulation that further set out the implementation details of the environmental levy scheme, was approved by the Legislative Council in April 2009. The actual charging of

---

[27]Environmental Protection Department, www.epd.gov.hk, the Government of Hong Kong's Website. Accessed 18 Feb 2013.

[28]"Public Consultation Report on the Proposal on An Environmental Levy on Plastic Shopping Bags (2007)". Environmental Protection Department, The Government of Hong Kong. Retrieved from http://www.epd.gov.hk/epd/english/environmentinhk/waste/prob_solutions/files/Consultation_Report_Levy_Eng.pdf

the environmental levy was implemented on 7 July 2009 and was the first mandatory producer responsibility scheme in Hong Kong.[29]

Within the first year of the programme, the number of plastic bags distributed by registered retailers had reduced by 90 percent.[30] In 2011, their internal Central Policy Unit conducted another telephone survey targeting consumers and found that 75 percent of respondents no longer accepted plastic bags from retail outlets, and that nearly 80 percent felt that the levy had helped them to develop the habit of bringing their own bag when shopping.[31]

There were a total of 40 retailers who operated 3100 registered outlets under phase one of the levy scheme. Table 5.2 shows the obligations of retailers registered under phase one of the programme.

The second phase of the levy scheme that the government proposed in 2011 was to extend the levy to small and medium enterprises (SMEs).

Table 5.2. Obligations of registered retailers in Hong Kong.[32]

| |
|---|
| A registered retailer shall fulfil the following requirements as set out in the Ordinance[33] as well as the Regulation |
| **To charge customer** — Registered retailers are required to charge a customer an amount of not less than 50 cents for each plastic shopping bag provided directly or indirectly to a customer. |
| **To display the Certificate of Registration** — Registered retailers shall ensure that a certificate of registration is displayed in a prominent position of the registered retail outlet to which the certificate relates. |
| **To submit quarterly returns and pay the levy** — Registered retailers shall submit a return on a quarterly basis in respect of each registered retail outlet, and pay to the Government the same total amount of levies stated in the return. |
| **To keep records** — Registered retailers must ensure that records, invoices, receipts, delivery notes, or any other documents that contain sufficient details to enable the Government to readily verify the number of plastic shopping bags provided and the charge for those bags, are kept. |

[29]HK Environmental Protection Department, "Environmental levy on Plastic Shopping Bags", http://www.epd.gov.hk/epd/english/environmentinhk/waste/prob_solutions/env_levy.html. Accessed on 11 March 2013.
[30]Hong Kong's Information Services Department, "First anniversary of plastic bag levy sees encouraging results", 4 July 2010.
[31]Consultation Document on the Extension of the Environmental Levy Scheme on Plastic Shopping Bags (2011). Accessed through www.epd.gov.hk. Accessed on 22 Jan 2013.
[32]Environmental Protection Department, Hong Kong. http://www.epd.gov.hk/epd/psb/en/environmental_admin.html. Accessed 22 Jan 2013.
[33]Product Eco-responsibility Ordinance (PERO).

This next phase would have to take into account their abilities to comply with phase one's administrative and statutory requirements of the levy (as listed in Table 5.2). Also, many SMEs may operate in the informal economy; which means that they currently pay no tax, are not monitored by the government, and are not included in the gross national product. Informal economies usually refer to small cash businesses, which the Asian market has no shortage of. Regulating their compliance is difficult, if not impossible, as they do not exist on paper. Hong Kong has still not launched phase 2 of the levy scheme to include SMEs, but are currently undertaking the preparatory work for the extension.

The success from phase one proves that a levy on plastic bags created a direct economic disincentive to encourage consumers to reduce their usage. With a 90 percent decrease within the first year, it is clear that consumers responded to the levy scheme in a positive manner and that these programmes are not only successful in a European market, but that markets in Asia can have the same success. It also highlights some issues with applying an environmental levy to products sold in Asia, which is figuring out how to apply it to SMEs and informal economies.

## 7. Recommendations

Singaporeans use approximately 2.5 billion plastic bags a year, which is equivalent to 1.7 bags per person each day.[34] One of the main justifications for the high use of plastic bags in Singapore is that they are reused as bin liners, which reduces their environmental impacts when compared to the alternatives. According to the study conducted by the Environment Agency in England and Wales (2011), if a plastic bag is reused as a bin liner you would have to use a paper bag 7 times; a Polypropylene (PP) reusable bag 26 times; and a cotton reusable bag 327 times, to have a lesser impact on global warming than the reused plastic bag. However, based on each person using 1.7 plastic bags a day, it is doubtful that every plastic bag is being used a second time as a bin liner.

It is clear by looking at the results of the consumer survey that residents of Singapore believe that the issue of plastic bag usage needs to be addressed. Out of the 270 consumers who were surveyed, 87 percent of them felt that there are too many plastic bags in Singapore, and 71 percent would support a levy on their use. Despite these findings, only 19 percent

[34] "New move to cut use of plastic bags" by Sarah Ng The Straits Times: 12 Feb 2006.

of respondents said they used reusable bags every time they grocery shop, and 27 percent used them every time when they shop for other goods. When asked what deters them from using reusable bags the most common response was that they forgot to take them out when shopping, which highlights an issue with consumer behaviour; of not making lifestyle choices which positively affect the environment.

The thesis written by Sugii (2008) showed that by implementing new policy and fees on plastic bags, not only was consumption reduced, but people's lifestyle choices and attitudes towards the environment were also influenced. When consumers had to purchase or pay a fee for plastic bags, it forced consumers to make a choice to either pay the extra money or to decline them all together, resulting in the conscious contemplation of the impact on the environment every time they bought goods.

Singapore may not have many of the issues associated with plastic bags ending up in landfill, but the 2012 International Coastal Cleanup (ICC) event in Singapore showed that plastic bags accounted for 13 percent of rubbish collected around coastlines, showing that plastic bags are still having a negative effect on the health of Singapore's overall environment. Reduction in the quantity of plastic bags used is the most important factor when minimising their impact on the environment.

Consumer feedback, high rates of usage, and plastic bags' high contribution to litter, all support the need for Singapore to develop a comprehensive strategy to combat excessive usage of plastic bags. Current programmes run by Ikea and NUS show how mandatory fees and consumer education can be successful in Singapore. Lessons can also be learned from the BYOBD programme being discontinued. Due to lack of data it is difficult to say the exact reasons this programme was discontinued, but the lack of mandatory fees and consumer education were most likely contributing factors.

Based on the research and data collected in this study, a levy on plastic bags is recommended. The case studies reviewed on Ireland's and Hong Kong's programmes provide valuable examples of how a program could be applied in Singapore.

Based on the estimate that each Singaporean uses 1.7 plastic bags a day, which amounts to the equivalent of 11.7 per week; if Singapore were to implement a similar programme as those implemented in Ireland and Hong Kong and reach a 90 percent reduction, each Singaporean would still use almost 1.2 plastic bags a week. Arguably enough to use as bin liners when you consider the number of people in the average household (approximately, 3.51).[35] This would translate into 4.2 plastic bags still being used each week by the average Singaporean family, and available to be used as bin liners, while other purchases could be carried using reusable bags.

However, before a plastic bag reduction programme could be successfully implemented in Singapore, there are a number of areas that require further research. Although the consumer survey conducted for this study provides a good first look into consumer perceptions, only 270 responses were recorded. A larger study should be conducted in various areas of the city to capture a more complete view of consumer perceptions. Additionally, the current survey only measures the buy-in of consumers, and does not measure the buy-in of other stakeholders such as retailers, manufacturers, government etc.

### *Concluding Remarks*

> "They are free in most instances, and cheap as dirt when they come at a price. Add to that qualities such as waterproof, durable, light, flexible, strong, as well as being ideal for lugging groceries, and containing rubbish, and it certainly looks like the plastic bag is here to stay. But therein lies the problem. Our apparent love affair with the old bag has reached a stage of insufferable dependency"
>
> (*Element05*, Issue 2)[36]

The plastic bag issue is complex with no easy answers. It seems that Singaporeans love their plastic bags and that their over-consumption will be a hard habit to break. Whether used as a carrier bag or to manage household waste, all plastic bags have an environmental impact. They have been part of our society for over 40 years, and with estimates of usage reaching 1 trillion per year globally, it's hard to imagine our world without them. They fit easily into many people's fast-paced lives due to their convenience, and in Singapore, serve secondary purposes such as trash bin liners. However, the environmental impacts that have been highlighted in this paper cannot be ignored. Due to their excessive use and public littering, they are finding their way into our green spaces and waterways. They are impacting countless animals every day, and changing our ecosystems to a level that may take decades to truly understand.

---

[35] Department of Statistics Singapore, http://www.singstat.gov.sg/statistics/latest_data.html#13. Accessed 15 March 2013.

[36] *Elements05*, A publication of the Singapore Environment Council, in partnership with the Singapore Green Business Alliance. July–Sept 2005, Issue 2.

Even though there are a few programmes currently aimed at reducing plastic bag usage in Singapore, there are no government-led, legislated initiatives. By considering the negative environmental impacts they have, it is clear that a more proactive, wide-reaching approach is necessary. As Jing (2011) found, overall sustainability efforts in Singapore are focused on energy efficiency, green technologies, and pollution control, and not enough attention is paid to reducing the overall level of consumption. There seems to be a common misconception that consumers are doing their part to protect the environment by choosing more eco-efficient consumables, even if the level of consumption remains high. This highlights that Singaporeans need to consume less, not just more efficiently.

The SEC has voiced that a new nationwide plastic bag reduction programme is still a top priority. This shows that they believe there is still an issue with plastic bag usage that needs to be addressed with the implementation of a new government-led initiative. The consumer surveys carried out for this study also supports this view. With 87 percent of respondents believing there are too many plastic bags in the Singapore market, and 71 percent supporting a levy on their use, it is clear that the public wants to see action taken to reduce their presence in the market. With minimal numbers reporting consistent use of reusable bags, it is clear that a nationwide programme is necessary to motivate consumers to change their habits. Implementing a programme that will make consumers stop and think about how their decisions effect the environment will be the first step in facilitating green consumerism in Singapore.

## References


Allsopp, M., Walters, A., Santillo, D., Johnston, P. (2009). *Plastic Debris in the World's Oceans*, United National Environment Program & Greenpeace.

Barnes, D. (2002). "Invasions by marine life on plastic debris", *Nature*, **416**, 808–809.

Burritt R.L., Lewis, H., James, K. (2005). "Analysing the Effectiveness of an Environmental Voluntary Agreement: The Case of the Australian National Packaging Covenant", *The Handbook of Environmental Voluntary Agreement.*

California Integrated Waste Management Board (2009). *California 2008 Statewide Waste Characterization Study.* Prepared by Cascadia Consulting Group. August 2009. CA, USA.

Consultation Document on the Extension of the Environmental Levy Scheme on Plastic Shopping Bags (2011). Can be accessed through www.epd.gov.hk.

Convery, F., McDonnell, S., Ferreira, S. (2007). "The most popular tax in Europe? Lessons from the Irish plastic bags levy", *Environ Resource Econ.*, **38**, 1–11.

County of Los Angeles (2007). "An Overview of Carryout Bags in Los Angeles County", *A Staff Report to the Los Angeles County Board of Supervisors.* August 2007. CA, USA.

Drury Research on Behalf of the Department of the Environment and Local Government (2000) *Attitudes and actions — a national survey on the environment.* Drury, Dublin.

*Elements05.* A publication of the Singapore Environment Council, in partnership with the Singapore Green Business Alliance, July–Sept 2005, Issue 2.

Gupta, K. (2011). *Consumer Responses to Incentives to Reduce Plastic Bag Use: Evidence from a Field Experiment in Urban India*, Delhi School of Economics, New Delhi, India.

ICF International (2010). *White Paper on the Methodology for Analyzing the Cost of Litter Clean-up Efforts*, San Francisco, CA.

Jing, J.-X. (2011). *Sustainability in Singapore: An Ecological Footprint Perspective*, National University of Singapore, Singapore.

Khoo, H., Tan, R., Chng, K. (2010). "Environmental impacts of conventional plastic and bio-based carrier based", *The International Journal of Life Cycle Assessment*, **15**(3), 284–293.

Klick, J., Write, J. D. (2012). *Grocery Bag Bans and Foodborne Illness.* University of Penn, Inst. for Law & Econ. Research Paper No. 13-2.

Lee, K.-Y. (2000). *From Third World to First: The Singapore Story 1965–2000, Memoirs of Lee Kuan Yew*, New York: Harper Collins Inc.

McIlgorm, A., Campbell H. F., Rule, M. J. (2008). *Understanding the Economic Benefits and Costs of Controlling Marine Debris in the APEC region*, Univ. of New England & Southern Cross University, New South Wales, Australia.

Public Consultation Report on the Proposal on An Environmental Levy on Plastic Shopping Bags (2007). Environmental Protection Department, The Government of Hong Kong. Retrieved from http://www.epd.gov.hk/epd/english/environmentinhk/waste/prob_solutions/files/Consultation_Report_Levy_Eng.pdf.

Sugii, T. (2008). *Plastic Bag Reduction: Policies to Reduce Environmental Impact*, Tufts University, Massachusetts, USA.

Timoney, F., & Company (1999). *Consultancy study on plastic bags.* Dublin Department of Environment and Local Government.

Tan, R., Khoo, H. (2006). *Impact Assessment of Waste Management Options in Singapore.* Air & Waste Management Association. **56**, 244–254.

The Environment Agency (England & Wales) (2011). *Life cycle assessment of supermarket carrier bags: a review of the bags available in 2006*, Environment Agency, Horizon House, Bristol, UK.

The Litter Monitoring Body. (2003). *The national litter pollution monitoring system.* Retrieved from http://www.litter.ie/docs/DoEHLG%20System%27s%20Results%20Report%20Final%202002.pdf

*The Singapore Green Plan 2012* (2006 Edition): The Ministry of the Environment, and Water Resources, Singapore.

Theng, A. P. (2002). *Incineration Does Not Undermine the Importance of Recycling*, National University of Singapore, Singapore.

"World Economic Outlook Database, September 2006", International Monetary Fund, 7 May 2009. Retrieved from http://www.imf.org/external/pubs/ft/weo/2006/02/data/index.aspx

## News Articles

**(i) Chanel News Asia**

23 April 2010, "New scheme planned to replace Bring Your Own Bag Day", by Saifulbahri Ismal.

**(ii) Hong Kong's Information Services Department** — www.news.gov.hk

1 July 2010, "First anniversary of plastic bag levy sees encouraging results", by HK Government.

**(iii) Inter Press Service News Agency**

29 Jan 2002, "BANGLADESH: Ban on Plastic Bags Changes Shopping Habits", by IPS Correspondents.

**(iv) Journalism and Media Studies Centre — The University of Hong Kong**

12 May 2008, "Green Group Finds HK's Plastic Bag Problem Suffocating", by Elmy Lung.

**(v) National Environment Agency News Releases** — http://app2.nea.gov.sg/news

11 February 2006, "Why waste plastic bags? Choose reusable bags!".
2 June 2006, "Make A Difference On World Environment Day — Choose Reusable Bags!".

**(vi) The San Francisco Chronicle**

21 Sept 2012, "New SF checkout bag law takes effect Oct. 1", by Stacy Finz.

**(vii) The Straits Times**

12 Feb 2006, "New move to cut use of plastic bags", by Sarah Ng.

**(viii) The Telegraph**

27 Nov 2007, "The plastic bag debate", by Rebecca Hoskings and David Tyson.

**(ix) National Geographic News**

12 Jan, 2004, "As Consumerism Spreads Earth Suffers Study Says", by Hillary Mayell.

## Websites and Key Descriptions

http://app2.nea.gov.sg/byobd.aspx (BYOBD information page on the NEA website, accessed 9 January 2013)

http://www.reuseit.com/, (Reuse it.com company website, accessed on 04 Feb 2013)

http://www.singstat.gov.sg/ (Department of Statistics Singapore's website, accessed on 30 Jan 2013)

http://www.europen.be/ (The European Organization for Packaging and the Environment's website, accessed on 13 Feb, 2013) http://app2.nea.gov.sg/ (National Environment Agency's website, accessed 06 Feb 2013)

www.plastics.ca. ("A Microbiological Study of Reusable Bags and 'First or single-use' Plastic Bags" (2009). A study funded by EPIC in Toronto Canada, accessed on 15 Feb 2013)

http://www.epa.gov/ (United States Environmental Protection Agency's website, accessed on 07 Feb 2013)

www.healthguidance.org ("The Effects of Plastic Bags on Environment", by Jamey Wagner, accessed 18 Feb 2013)

www.epa.ie (Irelands Environmental Protection Agency website, accessed 02 Feb 2013)

http://app2.nea.gov.sg/topics_wastestats.aspx (NEA Recycling Statistics 2011, accessed on 26 Feb 2013)

www.dcenr.gov.ie ("Dempsey Shares Expertise on Plastic Bag Levy with San Francisco Environment Officials", press release from Department of Communications, Marine and Natural Resources website, 11 March 2006, accessed 05 Feb 2013)

www.epd.gov.hk (Environmental Protection Department, The Government of Hong Kong's Website, accessed 18 Feb 2013)

www.theplasticfreetimes.com/bangladesh-plastic-bag-ban, (*Plastic Free Times*, accessed 22 February 2013)

http://www.irinnews.org/Report/92072/BANGLADESH-Plastics-proliferate-despite-ban IRIN — Humanitarian news and analysis, "Bangladesh: Plastic proliferate despite ban", accessed 22 Feb 2013)

http://www.gigantic-idea.com/blog/?p=118 (*Gigantic Idea Studio Blog*, 22 Jan, 2013, by Nancy Roberts, accessed 24 Feb 2013) http://cascade.uoregon.edu/fall2012/expert/expert-article/ (University of Oregon, *Paper or Plastic? The Answer may Surprise you.* Fall 2012, by Matt Cooper, accessed 28 Jan 2013)

http://www.wildsingapore.com/news/20070506/070512-4.htm ("Wild News" on *Wild Singapore*, 2007, accessed 6 March 2013)

http://earth911.com/recycling/paper/brown-paper-bags/facts-about-brown-paper-bags/ (*Facts about brown paper bags*, accessed 1 March 2013)

http://www.ikea.com/ms/en_SG/about_ikea/our_responsibility/national_community_involvment/index.html (Ikea press release, "IKEA to do away with disposable shopping bags". Published 17 Jan, 2013, accessed 28 Feb 2013)

http://www.irishstatutebook.ie/2001/en/si/0605.html (Irish Statute Book Webpage, accessed 11 March 2013)

http://www.epa.ie/irelandsenvironment/waste/ (Ireland's Environmental Protection Agency webpage, accessed 11 March 2013)

http://www.epd.gov.hk/epd/english/environmentinhk/waste/prob_solutions/env_levy.html (HK Environmental Protection Department, Environmental levy on Plastic Shopping Bags, accessed on 11 March 2013)

http://www.singstat.gov.sg/statistics/latest_data.html#13 (Department of Statistics Singapore, accessed 15 March 2013)

http://www.tradingeconomics.com/bangladesh/gdp-per-capita (Trading Economics Website — World Bank, accessed 17 March 2013)

# 6

# ENVIRONMENTAL IMPACT ASSESSMENT LAWS OF MALAYSIA AND HONG KONG: LESSONS FOR SINGAPORE

NIDHI MEHRA and LYE LIN-HENG

A law mandating Environmental Impact Assessment (EIA) requires development projects to bring environmental considerations to the forefront. It aims at avoiding or minimising adverse impacts to the environment, commencing from the planning stage of a project, and is part of good environmental governance and management. Although Singapore has excellent land use planning regulations and good environmental management practices, it does not have an EIA law. The rapid economic development of Singapore has increased stresses on its environment, and with public concern about the environment rising, it is imperative to examine the need for a formal EIA law and draw lessons from the experiences of similar Asian countries. This study traces the origins of the EIA law in the United States, and compares the EIA laws in Hong Kong and Malaysia. It finds that Singapore would benefit enormously if it formulated an unambiguous, firm and procedures-specific law for EIA. The necessity for the exploration of alternatives including questioning the need for the development project itself, and increased public participation is emphasised in the recommendations for Singapore's EIA law.

## 1. Introduction

"*The environment is where we live; and development is what we all do. The two are inseparable.*"

— Gro Brundtland, *Our Common Future* (1987).

### 1.1 *Environmental Impact Assessment*

The core objective of environmental management (EM) is to ensure the sustenance of the natural environment for future generations. Sustainable development is an integral part of EM. One essential tool of EM and decision-making is the Environmental Impact Assessment (EIA). The United Nations' Economic and Social Commission for Asia and the Pacific defines environmental impact assessment as "*an assessment of a planned activity on the environment*" (UNESCAP, 2003).

EIA is a tool that identifies the future consequences of a proposed action, and its aim is to achieve better environmental quality, social equity and sustainable development in the society. It is important to note that an EIA is not limited to environmental impacts but is inclusive of social and health impacts and economic analysis. EIA is a systematic and integrative process in which public consultation and participation play a fundamental role.

### 1.2 *The Evolution of EIA Laws*

(i) *The USA's National Environmental Policy Act (NEPA), 1969*

The 1960s and 1970s witnessed an increased environmental consciousness in American minds. Environmental accidents and incidents like the Santa Barbara oil spill, the Cuyahoga river fires; and the use of defoliation tactics by Americans in the Vietnam War, attracted domestic and international attention. The startling revelation by Rachel Carson's book, *Silent Spring*, in 1992 — that the indiscriminate use of synthetic pesticides, particularly dichlorodiphenyltrichloroethane (DDT) was damaging the ecological system; in particular, endangering the life cycle of birds — inspired an upwelling of public concern. These events resulted in the passage of the National Environment Policy Act (NEPA) in 1969. NEPA viewed the environment with a 'systems' perspective, legally mandating the use of "precaution" for large projects and "*opened up decision-making processes to affected members of the public, environmental interest groups and interested government agencies*" (Craik, 2008). Section 102 (2) (c) of NEPA has had a profound influence on policies in United States and abroad, as it mandated the preparation of environmental impact statements predicting the adverse and beneficial impacts on the environment before the execution of a wide range of projects, plans and programs (Hildebrand and Cannon, 1993). NEPA has evolved over the years and influenced environmental planning and decision-making, it is the "most frequently copied", and "*most cited law*" of all domestic U.S. laws worldwide (Rodgers Jr., 1993).

(ii) *The Development of EIA Laws*

NEPA sowed the seeds of EIA law internationally. Following the Stockholm Conference on the Human Environment in 1972[1] and the Rio Conference

[1]Stockholm Conference on the Human Environment, 1972: Elements pointing to EIA are found in Principles 13 and 14 of the Stockholm Declaration. See http://www.unep.

on Environment and Development (the Earth Summit) twenty years later in 1992,[2] numerous international treaties and instruments[3] now include EIA principles. Principle 17 of the Rio Declaration and Agenda 21, Earth's action plan for sustainable development at all levels of governance, calls for the adoption of EIA. In the context of the environmental infrastructure of countries, Article 7.41 (b) of Agenda 21 clearly states, "all countries should *ensure that relevant decisions are preceded by environmental impact assessments and also take into account the costs of any ecological consequences*". The need for greater public participation is emphasized in Agenda 21, Article 23.2.[4]

EIA processes are now adopted in more than seventy-five jurisdictions, in all political systems; and have been established in common law, civil law, and socialist traditions (Robinson, 1992).

(iii) *Singapore and (lack of) EIA law*

Singapore is an active member in regional and international environmental forums; Singapore presented its inaugural "Singapore Green Plan"[5] in 1992 (updated in 2012) at the Earth Summit, developed the Singapore Index on Cities' Biodiversity[6] and is party to many Multilateral Environmental Agreements (MEA).[7] However, Singapore does not have an Environmental Impact Assessment (EIA) law. In its absence,

---

org/Documents.Multilingual/Default.asp?documentid=97&articleid=1503 (Accessed on November 5, 2013)

[2]See Craik, 2008, 'Table of International Instruments' for a list of international treaties and instruments that have adopted EIA principles.

[3]United Nations Conference on Environment and Development, 1992 (Earth Summit) See Principle 17 http://www.unesco.org/education/nfsunesco/pdf/RIO_E.PDF (Accessed on February 27, 2013)

[4]See Agenda 21 Section 23.2 at http://www.unep.org/Documents.Multilingual/Default.asp?DocumentID= 52&ArticleID=71 (Accessed on February 27, 2013)

[5]The Singapore Green Plan, issued in 1992, was Singapore's first formal environmental blueprint to ensure that Singapore achieves economic development through sound environmental management without compromising the needs of future generations. See http://app. mewr.gov.sg/web/Contents/Contents.aspx?ContId=1342 (Accessed on February 4, 2013)

[6]The Singapore Index or Cities Biodiversity Index measures biodiversity in cities and is a monitoring tool. It was presented at the tenth Conference of Parties of Convention on Biological Diversity at Nagoya. See http://www.cbd.int/authorities/gettinginvolved/cbi.shtml (Accessed on November 5, 2013)

[7]See "Ministry of Foreign Affairs" for a list of environmental international treaties and MEAs that Singapore has ratified or acceded to. Accessed on February 5, 2013 from http://www.mfa.gov.sg/content/mfa/international_issues/environment.html

Singapore has steadily lost areas of great biodiversity and serves as an example of what other Asian countries can lose (Brook *et al.*, 2003).

### 1.3 *Hong Kong and EIA Law*

Hong Kong Special Administrative Region (HKSAR), and Singapore are highly urbanised Southeast Asian countries known for their meteoric economic success in the 1980s and 1990s (Chau, 1997). HKSAR faces greater pollution risks compared to Singapore, due to population pressure and its proximity to China's highly industrialised southern cities. HKSAR has robust EIA and Strategic Impact Assessment (SEA) laws while Singapore has no legal mandate for EIAs and all EIAs are performed ad hoc. Given the obvious similarities, Singapore may make a more informed decision on the EIA by analysing HKSAR's experience with enforcement of its EIA laws.

### 1.4 *Malaysia and EIA Law*

Singapore is sited at the tip of the Malayan Peninsula. The two countries have a common history of colonisation by the British and Japanese occupation, similar ethnic immigration patterns, and shared languages. Malaysia is also a supplier of fresh water to Singapore as Singapore is fresh water deficient. Sharing maritime borders, Malaysia is a very important neighbour for Singapore strategically.

The environment in Malaysia suffers from air pollution, trans-boundary haze, and large-scale conversion of forestland for commercial crop plantations, soil erosion and increased urban waste. EIA law in Malaysia exists but often fails to serve its real purpose. Lacunas exist in its application, as transparency is not mandated in the EIA law. With similarities and trans-boundary issues between Singapore and Malaysia, Singapore has many lessons to learn from Malaysia's experience with EIA law.

## 2. The Hong Kong EIA Experience

Hong Kong (Fig. 6.1) is a metropolis with a population of almost 7.26 million people cramped on an island of 1104 $km^2$ (Population Census data, HK, 2014). Although the average population density of HKSAR is 6500/$km^2$, in its densest district, Kwun Tong, the density can reach

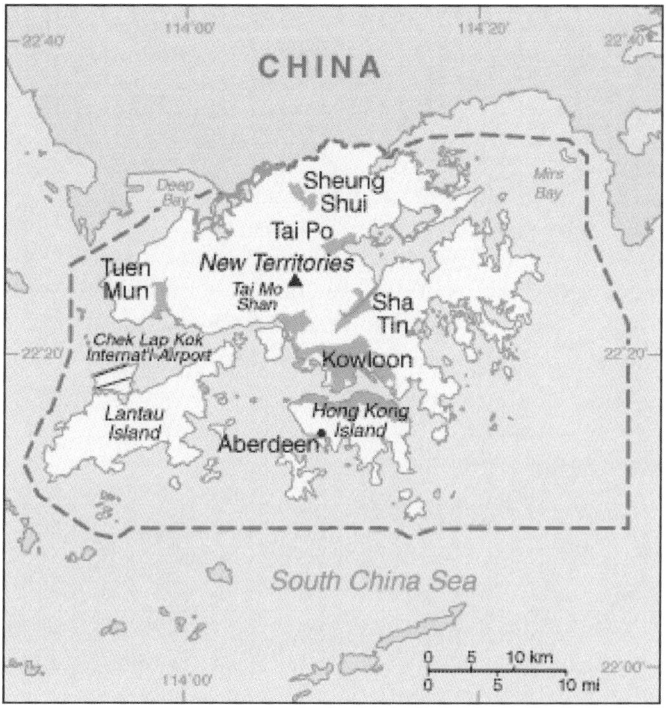


Fig. 6.1. Map of HKSAR (Central Intelligence Agency, 2013).

50,000 people/km$^2$. The vertical city has 1,268 skyscrapers, more than any city in the World.[8]

## 2.1 *HKSAR Basic Law*

Hong Kong consists of Hong Kong Island, Kowloon and the New Territories. The British government transferred power to the People's Republic of China (PRC) on July 1, 1997. China created a Special Administrative Region of Hong Kong (HKSAR) under Article 31 of the Constitution of PRC[9] on resuming sovereignty of Hong Kong.

On transfer of sovereignty, a Joint Declaration was made between the British governments and PRC that bound the Central People's government

[8]Data accessed from http://www.emporis.com/statistics/most-skyscraper-cities-world wide. Emporis is a global provider of building information.

[9]http://en.people.cn/constitution/constitution.html (Accessed February 12, 2013)

of PRC to adhere to policies under the declaration for 50 years until 2047. The Joint Declaration is stipulated in "The Basic Law of HKSAR",[10] which is HKSAR's constitutional document. Under the Basic Law, the Judiciary consists of the Court of Final Appeal, High Court (Court of Appeal and the Court of the First Instance), district courts, magistrates' courts, and other special courts (CIA, HK, 2013). 'The Basic Law' mentions the term 'environment' thrice, in articles, 7, 97, and 119. Article 119 is of greater importance as it entrusts the government of HKSAR to formulate policies to promote and co-ordinate the development of various trades along with the protection of the environment. This article provides the HKSAR courts with the discretion to issue mandatory orders to the government to fulfil its responsibility of paying adequate regard towards the protection of the environment (Fong, 2012).

## 2.2 *Hong Kong's Environment*

Nearly 40% of the land area (about 24 country parks) in HKSAR is protected for conservation and nature under the Country Parks Ordinance of 1976.[11] HKSAR has rich biodiversity with over 400 types of native trees, 510 bird species, 50 types of mammals, 180 types of fresh water fish and over 1000 kinds of marine fish. It is also home to the endangered Chinese White Dolphin and the Chinese Pangolin. HKSAR also has over 100 types of stony corals (BrandHK.gov, 2012). Consisting of more than 260 islands, HKSAR has an extensive coastline (over 700 km) with different geological formations. Coastlines are impacted by urbanisation and land reclamation, yet they remain important habitats for migratory birds, mangroves and associated fauna. Mai Po Marsh and Inner Deep Bay are identified as wetlands of international importance under the Ramsar Convention on Wetlands.[12] HKSAR has four marine parks and one marine reserve under the Marine Parks Ordinance of 1995 (Harris, 2012; Warren-Rhodes and Koenig, 2001).

[10]http://www.basiclaw.gov.hk/en/facts/index.html#3_1 (Accessed February 12, 2013)
[11]http://www.legislation.gov.hk/blis_pdf.nsf/4f0db701c6c25d4a4825755c00352e35/CFD5745BF7B8EB75482575EE004CF43B/$FILE/CAP_208_e_b5.pdf (Accessed February 12, 2013)
[12]The Ramsar Convention on Wetlands is the only international Treaty that is for a particular ecosystem. *It provides a framework for national action and international cooperation for the conservation and wise use of wetlands and their resources.* http://www.ramsar.org/cda/en/ramsar-home/main/ramsar/1_4000_0__ (Accessed on November 9, 2012)

### 2.2.1 *Pollution*

Air pollution is a big problem for HKSAR. It has severely affected public health and HKSAR's reputation for commerce and tourism (Pak, 2013). Direct health costs of air pollution in HKSAR are quite high: approximately HK$839 Million in 2008 (Harris, 2012). The leading cause of suspended particulate matter in the air here is trans-boundary pollution from neighbouring China. However, road transport and a busy port also contribute to almost one-third of the total air pollution. High buildings in densely populated areas have restricted the flow of air through what is called the 'deep canyon effect'; polluted air gets trapped between 'deep, narrow, canyon-like spaces between tall buildings' and air pollution in the city persists.

Like most other island cities, the water supply in HKSAR is not adequate for its residents. Despite having large reservoirs as water catchments, HKSAR imports 70–80% of its water from Guangdong, China. Sewage treating facilities were created here only after the 1980s. The marine environment in HKSAR is still quite polluted and has had a history of 'harmful algal blooms' or 'red tides' due to excessive organic pollution (Harris, 2012).

## 2.3 *EIA in HKSAR*

### 2.3.1 *History*

Today, HKSAR stands out as a country with exemplary EIA and SEA laws and administrative framework (World Bank, 2006). To address increasing pollution, a landmark report on HKSAR's environment was published in 1977 that proposed comprehensive administrative and legal framework for the protection of HKSAR's environment (Hills and Barron, 1997). Thereafter, the EIA was introduced in HKSAR in 1979 through administrative means, and selected projects underwent the process (Wood and Coppell, 1999). In 1989, on continued environmental degradation, the government issued a white paper on pollution (Planning, Environment and Lands Branch, 1989). The paper categorically stated "*The White Paper acknowledges that the main reason why our environment is now in an unsatisfactory state is that in the past the Government and the community made choices which gave too little emphasis to the needs of the environment*"; the paper called for strengthening the EIA system as an "*anticipatory control measure.*" Nevertheless, the administrative implementation system continued with wider application on policies and projects until 1998, due to the non-statutory status of the EIA process.

### 2.3.2 *Environmental Impact Assessment Ordinance (EIAO)*

Hong Kong's EIAO, chapter 499 was enacted on February 4 1997 and came into force on April 1, 1998. The construction, operation and decommissioning of designated projects listed in Schedule 2[13] of the EIAO requires an evaluation of the EIA report and an 'environmental permit' (EP) before the project can be implemented (EPD, 2003). The process seeks to serve two competing public interests of environmental protection and ensuring that major projects are brought to fruition in a timely and efficient manner (Lahiri, 2006).

EIA principles were introduced in policy formulation (Strategic Environmental Assessment) only after a 1997 "Study on Sustainable Development for the 21st Century" (SUSDEV21)[14] emphasised the need to integrate environmental, social and economic factors in decision-making (Ng and Obbard, 2005).

### 2.3.3 *Definitions*

The principal definitions under EIAO are:

"***environment***" *means the components of the earth; and includes-*

*(a) land, water, air and all layers of the atmosphere;*
*(b) all organic and inorganic matter and living organisms; and*
*(c) the interacting natural systems that include any of the things referred to in subparagraph (i) or (ii);*

"***environmental impact***" *for a designated project, means-*

*(a) an on-site or off-site change that the project may cause in the environment;*

[13]Schedule 2 projects include (I) major projects and (II) decommissioning projects. Source: Section 2: Designated projects requiring environmental permits, Cap 499, Environmental Impact Assessment Ordinance http://legislation.gov.hk/eng/home.htm (Accessed on March 7, 2013)

[14]SUSDEV21: "The Study on Sustainable Development for the 21st Century" aimed at introducing the concept of sustainability into decision making, was commissioned by HKSAR government in 1997. http://www.pland.gov.hk/pland_en/p_study/comp_s/susdev/ex_summary/sum_eng.pdf (Accessed on March 6, 2013)

*(b) an effect of the change on-*

*(i) the well-being of people, flora, fauna and ecosystems;*
*(ii) physical and cultural heritage;*
*(iii) a structure, site or other thing that is of historical or archaeological significance;*

*(c) an on-site or off-site effect on any of the things referred to in paragraph (b) from activities carried on for the project;*
*(d) a change to the project that the environment may cause, whether the change or effect occurs within or outside the site of the project.*

Source: Cap 499 Sched 1: Interpretation EIAO 1998.

2.3.4 *Authorities/Regulators*

The Director of the Environmental Protection Department (EPD) is responsible for the review and approval of the EIA reports and Environmental Permit (EP). He maintains a register (open for public view without charge) containing study briefs, EIA reports, EP applications, his decisions, etc. (S15 and Schedule 1: Interpretation, EIAO).

The Advisory Council on the Environment (ACE) is the principal advisory body for environmental protection and nature conservation. Its main tasks are to keep track of, and advise the government on, policies regarding pollution and the environment in HKSAR. A sub-committee on EIA under ACE reviews EIA reports and comments on Strategic Environmental Assessments (SEA) (EIAO guidance Note No. 1/2002).

The Secretary for the Environment issues the 'Technical Memorandum' (TM); setting out principles, procedures, guidelines, requirements and criteria for technical content of the EIA reports, report requirements, Eps, etc. The Legislative Council (LegCo) has the right to repeal the TM if LegCo passes a resolution against it (S16, EIAO). The Secretary, in consultation with ACE, sets out regulations regarding qualifications of people carrying out EIAs, appeals, fees, etc. (S 32, Regulations, EIAO).

An Appeals Board is set up to hear and resolve the disagreements between project proponents and the decisions of the Director regarding the preparation of the EIA study brief, permission to directly apply for Environmental Permit, or approval of EIA report or EP (S17, EIAO). At least three people constitute the Appeals Board. The Chief Executive appoints a qualified District Judge as the Chairman and Deputy Chairman of the Board. The contents of the TM are not questioned in appeals.

### 2.3.5 *The EIAO Process*

The main features of the statutory EIA process are:

- Both public and private sector projects designated under Schedule 2 and 3[15] require an EP or an EIA report.
- For a designated Schedule 2, construction, operation and decommissioning project, under Sections 5 to 8 of EIAO, the authorised project proponent needs to:
  - (a) Apply for an EIA study brief which sets out the Terms of Reference (TOR)
  - (b) Proceed to EIA study
  - (c) Submit the EIA report for approval under the Ordinance or proceed to request for an EP directly. For a designated Schedule 3 project, under Sections 6 to 8, the EIA report needs to be approved.

Abaza *et al.* (2004) define Strategic Environmental Assessment (SEA) as "*a formal, systematic process to analyse and address the environmental effects of policies, plans and programmes and other strategic initiatives*". While the EIA is restricted to a particular project's impacts on the environment, the objectives of the SEA process are to provide for a high level of protection of the environment and to promote sustainable development by integration of all interests.

The simplified EIA process is as shown below in Fig. 6.2:

- The public and the Advisory Council on the Environment (ACE) have the opportunity to comment in the EIA process on three occasions as shown by the arrows in Fig. 6.2.
- Statutory time limits imposed on each phase of the EIA evaluation process are shown in Fig. 6.3 below:
- There are clear penalties for offences relating to Environmental Permits (EPs) and interference with respect to the enforcement of the legislation. A project that starts construction without a proper EP carries criminal liability of a maximum fine of HK$2 million (USD 129,016[16]) and

[15]Schedule 3 projects: urban development projects with an area greater than 20 Ha or involving population greater than 100,000, and redevelopment projects — all require Environmental Impact Assessment. See http://legislation.gov.hk/eng/home.htm (Accessed on March 7, 2013) and (Wong, 2011)

[16]1.00 HKD = 0.129017 USD. Retrieved April 27, 2015 from http://www.xe.com/currencyconverter/convert/?Amount=1&From=HKD&To=USD

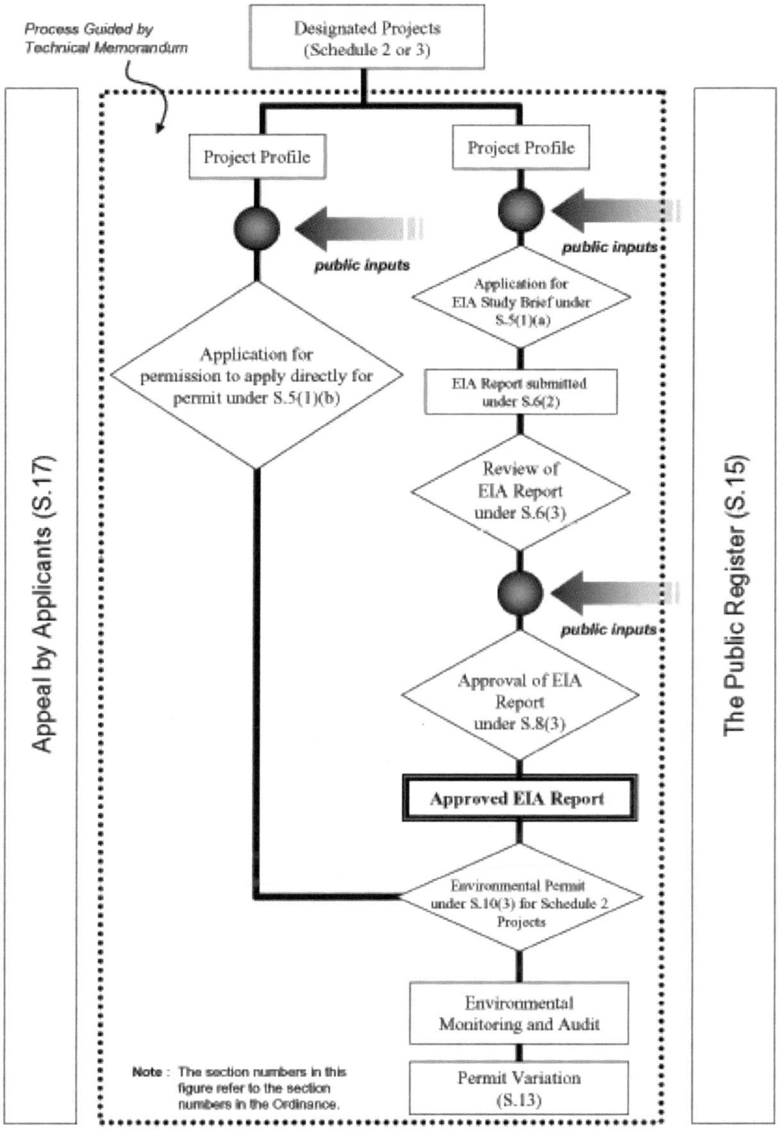


Fig. 6.2. The statutory EIAO process (Hong Kong) (A guide to the EIA process, 2013).

imprisonment of six months upon the first conviction. The liability is increased to a maximum fine of HK$5 million (USD 645,082) and imprisonment of two years on the second or subsequent convictions. If the offense continues, an additional fine of HK$10,000 (USD 1,290)

| | The Authority | The Public | ACE |
|---|---|---|---|
| 1. Application for an EIA Study Brief | 45 days | 14 days | 14 days |
| 2. Application for approval to apply directly for Environmental Permit | 45 days | 14 days | 14 days |
| 3. Review of EIA Report | 60 days | | |
| 4. Public Exhibiton of EIA Reports for comments | | 30 days | 60 days |
| 5. Request for information in response to comments | 14 days | | |
| 6. Approval of EIA Report | 30 days | | |
| 7. Application for Environmental Permit | 30 days | | |

Fig. 6.3. HKSAR statutory time limits (Wong, 2011).

per day is imposed (S.26 (1) EIAO). For comparison, the annual per capita gross domestic product of Hong Kong for 2014 was HK$ 310,113 (USD 40,010) (National Income and Balance of Payments, 2014).

### 2.3.6 *Performance of the EIAO Process in HKSAR*

One of the criticisms of the EIA legislation system is that it hinders the pace of development projects (Price and Ho, 2011). Contrary to this belief, only seven EIA reports were rejected between 2011 and 1998 since EIAO was implemented in HKSAR (Cheung, 2011; Cheung and Wong, 2011). The EIAO website http://www.epd.gov.hk/eia/index.html maintains an online register of EIA study briefs and reports and decisions taken on a yearly basis. We can ascertain the number of EIA study briefs applied for, EIA reports that were submitted, the subsequent decisions on the EIA reports and EPs issued since 1998. This allows us to understand the practical applications of the EIAO in HKSAR.

#### A. Statistical Study

Graphs 6.1–6.3 below show the summary statistics regarding the EIA applications made and approved for all projects from 1998 to 2012.

Graph 6.1 shows the distribution between approved, rejected and withdrawn applications of EIA reports before public inspection. This graph

Table 6.1. Application statistics EIAO HKSAR 1998 to 2012 (Monthly update of applications under environmental impact assessment ordinance, 2013).

| Application type | Received | Issued/ granted | Considered suitable for public inspection of EIA report | Approved EIA report | Rejected or considered not meeting TM and/or SB requirements | Required EIA submission for EP variation sought |
|---|---|---|---|---|---|---|
| Study Brief (*9 applications withdrawn) | 255 | 246 | N/A | N/A | N/A | N/A |
| Permission to Apply Directly for Environmental Permit (*13 applications withdrawn) | 222 | 204 | N/A | N/A | 4 | N/A |
| Approval of EIA Report (*28 applications withdrawn) | 206 | N/A | 177 | 168 | 7 | N/A |
| Environmental Permit (*14 applications withdrawn) | 444 | 428 | N/A | N/A | 2 | N/A |
| Variation of Environmental Permit (*26 applications withdrawn) | 387 | 343 | N/A | N/A | N/A | 15 |
| Further Environmental Permit (*4 applications withdrawn) | 143 | 138 | N/A | N/A | 1 | N/A |
| Surrender of Environmental Permit (*9 surrenders withdrawn) | 81 | N/A | N/A | N/A | N/A | N/A |
| Total | 1738 | 1359 | 177 | 168 | 14 | 15 |

Source: http://www.epd.gov.hk/eia/english/whatnew/monthly_update_applications.pdf (Accessed on 8 February 2013)

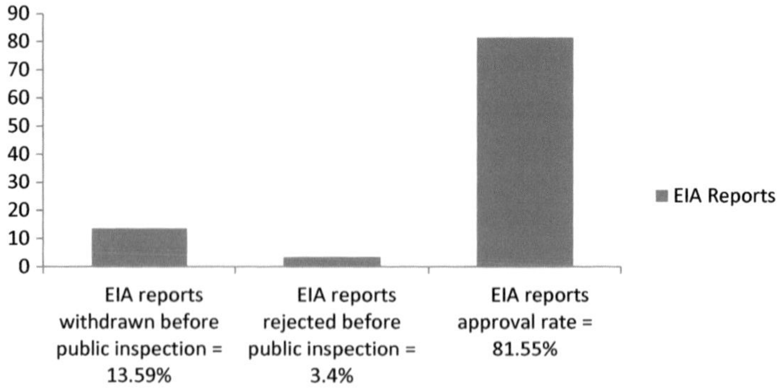


Graph 6.1. EIA reports applications statistics before public inspection.

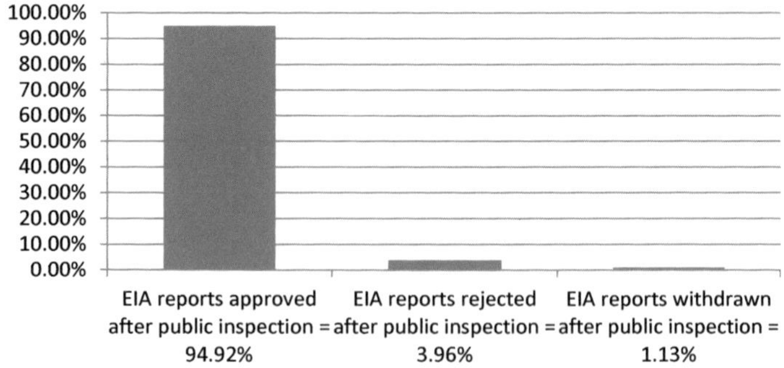


Graph 6.2. EIA reports applications statistics after public inspection.

indicates that the Director approved 81.55% of EIA reports received. Of the remaining 18.45% of EIA reports, 13.59% were withdrawn by the project proponents themselves and 3.4% were rejected. The withdrawn applications were probably those that did not meet the technical requirements as specified in the TM or were not feasible given the mitigating measures required due to the project's impacts.

Graph 6.2 shows the distribution between approved, rejected and withdrawn applications of EIA reports after public inspection. This graph indicates that after public inspection, 94.92% of the EIA reports were approved. The project proponents withdrew 3.96% themselves and only about 1% were rejected. Thus, most of the withdrawals and rejections

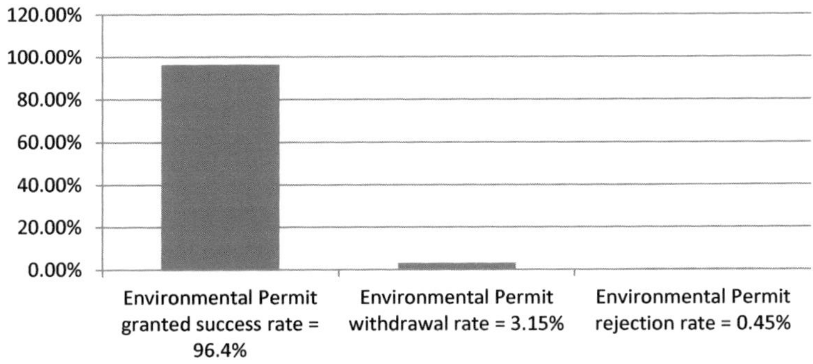


Graph 6.3. Environmental permit applications statistics.

occurred before public inspection. This suggests that public objections were not the primary cause for rejections of EIA reports.

Graph 6.3 shows the distribution regarding approval, withdrawals and rejections of EPs. The approval rate for EPs is 96.4%; withdrawal rate is 3.15%; and a small rejection rate of 0.45%. The EP applications are inclusive of the direct applications for EPs without needing the EIA report. If we assume that the projects that go through EIA are rejected at the EIA report approval stage itself, then the rejection rate here primarily represents the rejection of those projects that did not go through the EIA report evaluation phase.

That EIA laws' promotion of sustainable development, and how they are not against economic development, cannot be fully appreciated in these statistics alone, though they are certainly indicative. The following case study illustrates the above points qualitatively.

B. *Case Study* (*Hong Kong*)

Lok Ma Chau Spur Line project

The Lok Ma Chau spur line project (Project) was commissioned by the HKSAR government in 1999 to connect the Sheung Shui station in HKSAR to a new border crossing in mainland China, in Lok Ma Chau. Kowloon–Canton Railway Corporation (KCRC) was responsible for the construction of a 7.3 km double-tracked rail line, the railway terminal at Lok Ma Chau and associated footbridge, etc. The Project was designated as the most urgent infrastructure for public transport (Lo, 2002). The railway line passed through two ecologically sensitive areas in the New Territories in HKSAR. (Josh, 2008, IAIA):

- Long Valley: This 33 Ha agricultural fresh water wetland area has no special legal conservation protection. It has over 200 bird species, of which 29 are of conservation importance.
- Deep Bay Wetlands Buffer Area: The fishponds in the Deep Bay Buffer Area serve as a buffer region to maintain the integrity of the Deep Bay Wetlands, which is a 'Ramsar Convention', protected region. The fishpond habitats are contiguous with the Deep Bay Wetlands and are deemed as 'wetland conservation area' by the Town Planning Board.

The Project was a Schedule 2 designated project and thus required an EIA report and an EP before the work could start.

The EIA process started with KCRC submitting the Project profile to the Director under section 5(1) (a) of EIAO in December 1998. The Director considered comments from the Advisory Council on the Environment (ACE) and the public in drawing up the study brief under section 5(6) of EIAO and section 3 of the Technical Memorandum (TM) (*KCRC* v *Director EPD, No. 2 of 2000*). Information regarding the study brief was provided to ACE and the project was advertised. After verification of the EIA report from KCRC, the Director of EPD opened the report for public inspection in August 2000.

The Director received 225 comments opposing the project. On further deliberations by ACE, the project proponents acknowledged that the proposed central route through Long Valley was the worst choice ecologically compared to the other options considered, and will lead to ecological damage and fragmentation of the wetlands. There was also doubt among ACE members regarding the success of the proposed wetland being created in lieu of the one being impacted. Considering the opinions of the public, ACE and the Director of Agriculture, Fisheries and Conservation, on the 700-metre railway track passing through Long Valley, the Director rejected the EIA report in October 2000 under section 8 of EIAO.

KCRC appealed to the Appeals Board against the Director's decision and sought an order to approve the EIA report with conditions and grant the permit with conditions. The Appeals Board unanimously decided to reject KCRC's appeal and stated that *"the underlying aim of the impact assessment process is to ensure that no designated project is undertaken unless it is shown to be environmentally acceptable"*. The Appeals Board believed that the Project could be environmentally acceptable if the feasibility of mitigation and construction methods could be improved and re-evaluated. The Project proponent improved upon their initial plan during the appeal. KCRC amended the report, included alternatives, their assessments and results, and conclusions based on the new proposals given during the appeals.

The project eventually utilised a system of tunnels and viaducts, ensuring the conservation and enhancement of 37 hectares of the wetlands. After the amended EIA report went through the process of public scrutiny and reassessment, the Director of EPD approved the amended report with conditions on 22$^{nd}$ March 2002. This project is now cited as an example of what an efficient EIA system can achieve in terms of strategic infrastructure development and environmental protection. (Josh, 2008).

The environmental scrutiny took close to $2\frac{1}{4}$ years and at least HK$ 46 million was spent on wetland conservation (ibid). However, the consideration of alternatives resulted in a far more sustainable project than it would have been without the EIA process in place (Zhao, 2009).

Large projects often have too much might in terms of financial backing and political support; environmental regulations and the EIA legislation in particular act as a 'check and balance mechanism' to curb unrestrained development and to promote the application of the 'precautionary principle'.

## 3. The Malaysian EIA Experience

Malaysia consists of two distinct physical parts (East and West Malaysia) with thirteen states and three federal territories (Map: Fig. 6.4). Malaysia is one of the twelve mega-diverse countries in the World in terms of biodiversity. It has a population of 29.6 million (Department of Statistics, Malaysia, 2013). The average population density of the Federation is 86 persons/km$^2$ but rises to 6,891 persons/km$^2$ in the highly populated capital city of Kuala Lumpur.

The judiciary consists of the Court of Appeal and two High Courts of co-ordinate jurisdiction and status: the High Court of Malaya and High Court of Sabah and Sarawak. Malaysia adopts a common law system along with religious Syariah Courts (Mottershead, 2002). The Malaysian Federal Constitution is the supreme law of the land and there is no mention or provision for protection of Environment in the Constitution or State legislature (Kanniah, 1999; Chan, 1998; Mohammad, 2011). Each State has its own State Constitution.

Fig. 6.4. Map of Malaysia (Source: CIA world factbook).

### 3.1 *Environment*

Approximately 63% of the land area consist of rainforests (of which 11% are primary rainforests) and over 1.2 million acres of mangrove forests. About 80% of Eastern Malaysia's once rich bio-diverse rainforests, have been felled in Sarawak (Marshall Cavendish, 2009). There are approximately 45 State Parks[17] and six sites designated as "wetlands of international importance" under the Ramsar Convention.[18] There are two sites of natural heritage: Kinabalu Park at Sabah, and Gunung Mulu National Park in Sarawak, and two sites of cultural importance under the World Heritage Convention[19,20] (UNESCO, Malaysia). Malaysia is home to some 15,500 species of higher plants, 746 birds, 300 mammals, 379 reptiles, 198 amphibians, and 368

[17]National Parks in Malaysia are de facto State Parks. http://en.wikipedia.org/wiki/List_of_ national_parks_of_Malaysia (Accessed on March 15, 2013)

[18]The 6 sites are: Kuching Wetlands National Park, Lower Kinabatangan–Segama Wetlands, Pulau Kukup, Sungai Pulai, Tanjung Piai and Tasek Bera. http://www.ramsar.org/cda/en/ramsar-pubs-annolist-annotated-ramsar-16529/main/ramsar/1-30-168%5E16529_4000_0__ (Accessed on March 15, 2013)

[19]The 1972 World Heritage Convention links "*together in a single document the concepts of nature conservation and the preservation of cultural properties.*" http://whc.unesco.org/en/convention/#Brief-History (Accessed on March 15, 2013)

[20]New Straits Times "Malaysia now has four world heritage sites", July 2, 2012. http://www.nst.com.my/latest/malaysia-now-has-four-world-heritage-sites-1.101377 (Accessed on March 15, 2013)

species of fish (Mongobay, Malaysia). There are also 286 species of mammals in Malaysia, of which 27 are endemic and 51 are threatened.

### 3.1.1 *Pollution*

Over 250,000 km$^2$ of land area in Malaysia and Indonesia is covered with peat deposits, approximately containing 70 GT of carbon. The South East Asian region suffers from biomass and peat land burning that has increased considerably in the last 20–30 years (Salinas *et al.*, 2012). Many parts of Malaysia suffer the consequences of trans-boundary haze due to biomass burning in the Kalimantan region, with health costs, production losses, tourism-related losses and the cost of avertive action losses running in the billions in 1997–1998[21] (Awang *et al.*, 2000), 2005 and 2011. Heavy vehicular traffic in urban areas and Malaysia's dependence on fossil fuels (36.5% coal and 55.9% natural gas) for power plants also contributes to air pollution (Ali *et al.*, 2012).

The rapid industrialisation of Malaysia with activities such as quarrying and land clearing have contributed most considerably to water pollution in the nation. Adding to it is urban sewage, animal wastes, and wastes from manufacturing and agro-based industries. These together are the major contributors to the pollution of waters and the eutrophication of aquatic ecosystems in the country (Ibid. and Gasparatos, 2011). The Straits of Malacca are very busy shipping lanes. Land reclamation activities with shipping activities, discharges from oil spills, etc. contribute to marine pollution in the country.

## 3.2 *Evolution of EIA*

Consideration for EIA legislation in Malaysia began after Perak, the second largest state in Peninsular Malaysia, and Kuala Lumpur suffered severe floods in 1967 and 1970. As flooding was a major concern for Malaysia, the World Bank assessed the situation and emphasised the need for EIA for any major development project in its report. Post Stockholm Conference, a comprehensive legislation for *prevention, abatement, control of pollution and enhancement of the environment,* the Environmental Quality Act of 1974 — Act 127 (EQA) — was enacted in Malaysia and came into force on 15 April 1975.

[21]Economic losses in 1997–1998 for the Malaysia–Singapore region was $9.3 billion. http://news.mongabay.com/2005/0804-ap.html (Accessed on March 16, 2013)

Thereafter, an Environment Division under the Ministry of Local Government was formed in 1975.[22] The Minister for Environment (Minister) appointed a Director General of Environmental Quality (DG) in the Environment Division to administer the EQA. An Environmental Quality Council (EQC) was established to advice the Minister on environmental matters and the EQA. EQC comprised members from various government ministries with representation from the states of Sarawak and Sabah, industry, academia, and registered non-governmental environmental societies.

### 3.2.1 *Section 34A of EQA*

Section 34A came into effect in April 1988 and its Schedule lists the Prescribed Activities[23] that require an EIA. The Minister has the authority to specify any activity as 'prescribed activity' if a large environmental impact is expected. Even though American and Canadian EIA processes served as models for EIA legislation development, the Malaysian authorities decided that preliminary EIAs would be sufficient for most projects and a detailed EIA would be required only for projects with serious impacts.

Three projects underwent informal EIAs in 1980 and 29 projects (mostly hydroelectric) between 1981 and 1984 (Kanniah, 1999). Between 1988 and 1995, 1705 formal EIAs were conducted (Vun and Latiff, 1999).

Subsequent amendments to the EQA in 1996 (Act A953) and in 2012 (Act 1441) require EIA reports to be prepared by qualified environmental consultants registered with the Department of Environment (DOE) to counter the unsatisfactory quality of the majority of EIA reports (Hezri and Nordin Hasan, 2006). Thus, the development of EIA law in Malaysia underwent phases of inspiration by international experience, learning from international agencies, following foreign ideas, and finally to evolving the law to suit country-specific circumstances.

### 3.2.2 *Definitions*

The important definitions in the EQA are given below:

> "**environment**" means the physical factors of the surroundings of the human beings including land, water, atmosphere, climate, sound, odour,

[22]Source: Background and History from DOE website http://www.doe.gov.my/portal/jabatan/background-history/ (Accessed on March 15, 2013)

[23]Prescribed Activities are based on size of the project and the number of people, the project would affect. Source: Schedule, P.U.(A) 362/87 Accessed on March 19, 2013 from Lawnet Malaysia Database

taste, the biological factors of animals and plants and the social factor of aesthetics;

The Assessment division in DOE defines "**Environmental Impact Assessment**" as a study to identify, predict, evaluate and communicate information about the impacts on the environment of a proposed project and to detail out the mitigating measures prior to project approval and implementation.[24]

### 3.2.3 *Authorities/Regulators*

The DG of DOE maintains a list of qualified consultants for producing quality EIA reports; sets guidelines for EIAs; examines, evaluates, approves or disapproves the report; and follows up on compliance with the conditions of approval of the report. Preliminary EIA Reports are processed and concluded by the DOE State Offices. Detailed EIA Reports involving more than one state are processed and concluded at the DOE Headquarters.

Under section 35(e) of the EQA, there is a provision for an Appeal Board comprising three members. The Chief Justice appoints the Chairman and Deputy Chairman from among advocates or solicitors of Malaysian High Courts who have been practising for at least seven years.

## 3.3 *The EIA Process*

The EIA process in Malaysia (see Fig. 6.5) begins with ensuring that the project complies with the rules laid down in the land use planning procedures prior to the EIA study.[25]

### 3.3.1 *Preliminary EIA*

The Preliminary EIA procedure is explained in Fig. 6.6. The Preliminary EIA report needs to be evaluated within five weeks from the date of submission. Public engagement is required in the guidelines, but it is upon the project proponent to decide the form of the engagement.

### 3.3.2 *Terms of Reference (TOR) Approval*

A list of seventeen activities that require detailed EIA are in Appendix 4 of the Handbook.[26] After the screening and consultation process, for detailed EIAs, the project proponent must set the TOR (see Fig. 6.7).

---

[24]http://eia.doe.gov.my/portal/?page_id=236 (Accessed on March 19, 2013)
[25]Project Concept, EIA, DOE. Accessed on march 19, 2013 from http://eia.doe.gov.my/portal/?page_id=735
[26]See http://eia.doe.gov.my/portal/?p=725 (Accessed on March 19, 2013)

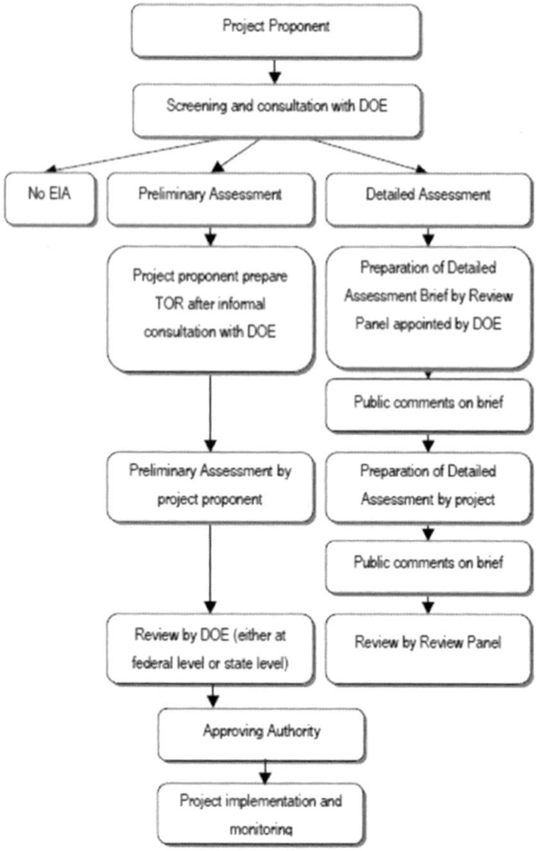


Fig. 6.5. EIA process in Malaysia (Stærdahl *et al.*, 2004).

### 3.3.3 *Detailed EIA*

Based on the TOR, detailed EIA briefs need to be advertised and displayed publicly at public libraries, State DOE offices and on the DOE website for 30 days to enable public comment. An independent expert panel also reviews the report (Yusof, 2012). The detailed EIA reviewing time is 5 weeks — but based on DOE/Notice 8/2012, in order to expedite faster reviewing times for activities expected to have lower environmental impact, the reviewing time is reduced to 3 weeks (see Fig. 6.8).[27]

[27]The activities with reduced EIA review time are: Housing Development covering an area of 50 hectares or more; Construction of New Townships; Municipal Sewage: Construction Of Wastewater Treatment Plant; Chemical Activities — where production

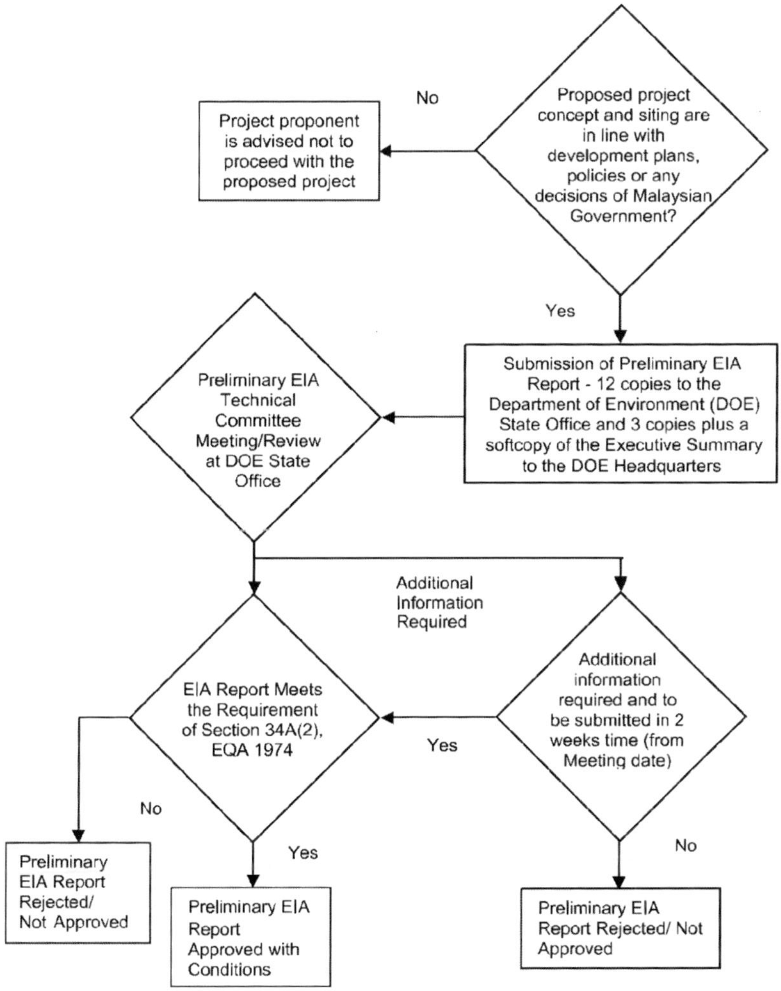


Fig. 6.6. Procedure for a preliminary EIA (DOE Malaysia, accessed on March 19, 2013).

The DG can approve the report with conditions attached. In case of non-compliance, the maximum fine is MYR 500,000 (Malaysian ringgit)

---

capacity of each product or of combined products is greater than 100 tonnes/day; Construction of Recovery Plant (Off-Site) and Construction of Storage Facility (Off-Site). Source: http://www.doe.gov.my/eia/wp-content/uploads/2012/11/Notice-8-2012-3-weeks-EIA-processing5.pdf

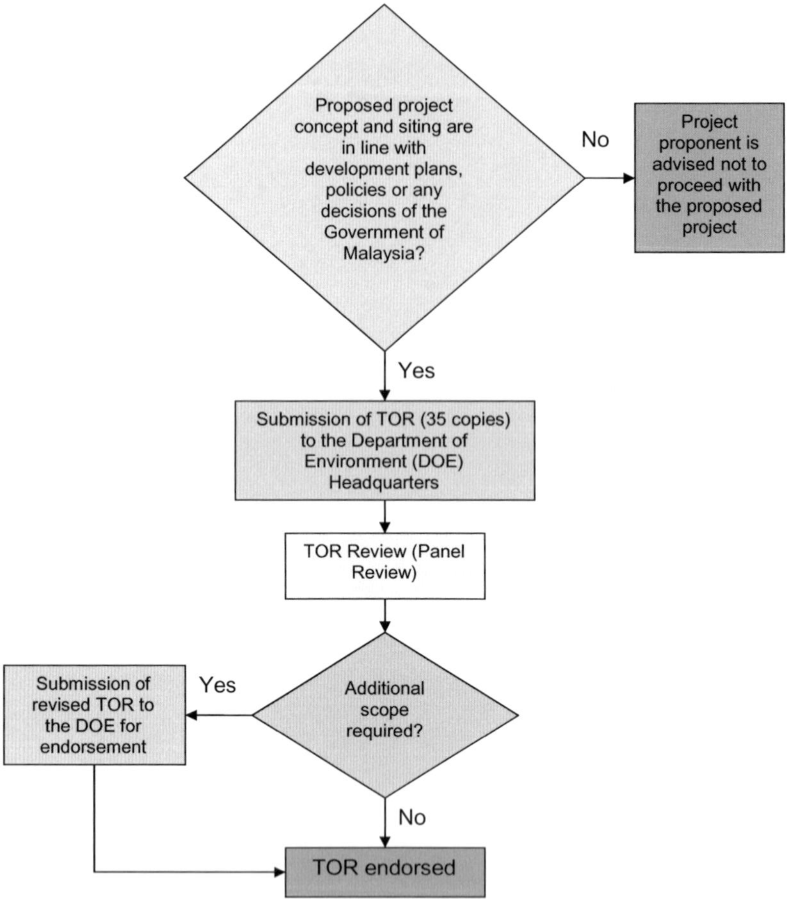


Fig. 6.7. TOR approval process (Appendix 5, EIA report, accessed on March 19, 2013).

(~USD 140,762)[28] or a maximum imprisonment of five years, and a further fine of MYR 1000 (~USD 282) for continuation of offense every day. In accordance with section 34AA of the EQA, the DG may even issue a stop work order if the project has been initiated without approval from the DG or if the project implementer is not meeting the conditions in the

[28]Currency Exchange Rate as of April 28, 2015: 1 Ringgit = 0.28 USD (Source: http://www.xe.com/currencyconverter/convert/?Amount=1&From=MYR&To=USD)

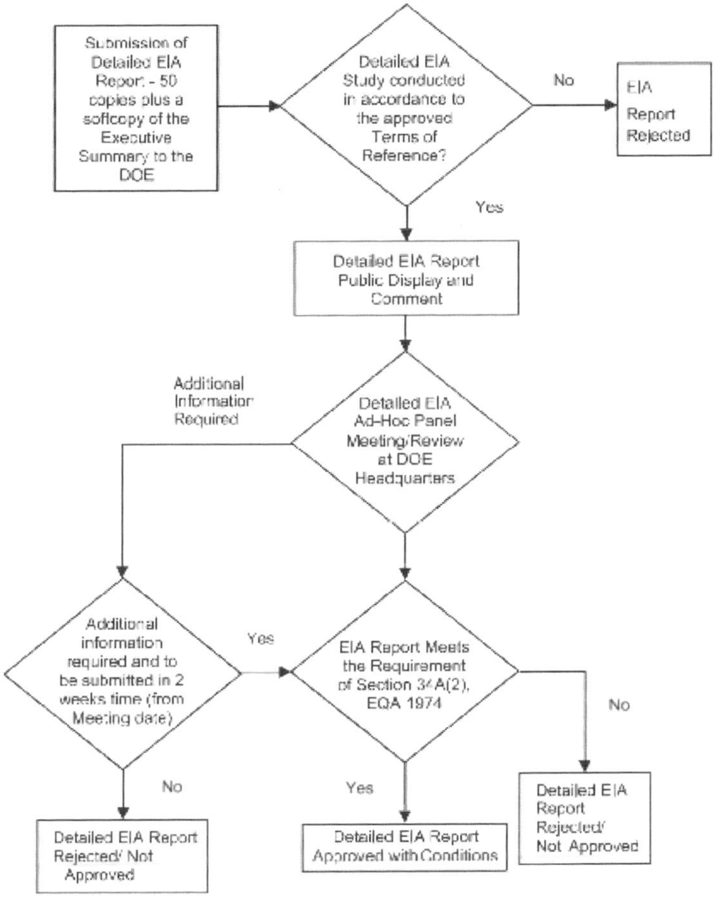


Fig. 6.8. Procedure for detailed EIA reports (DOE Malaysia. Accessed on March 19).

approval. These penalties can be compared against the annual per capita gross domestic product in Malaysia, which is MYR 37,363 (~USD 10,538) (World Bank data, 2014)

### 3.3.4 *Implementation of the EIA Law in Malaysia*

It is observed that in the case of Malaysia, more can be done in the realms of public participation, impact management and mitigation plans (Briffett, 1999). This point is illustrated through the following case study:

### 3.3.5 *Case Study (Malaysia)*

New Pantai Highway

The 'New Pantai Highway' was planned in 1996 to provide access to Kuala Lumpur from Subang Jaya and Bandar Sunway, and to avoid traffic jams on the Federal Highway Route 2 during rush hour.[29] The Project proponent was Maxtro Engineering. The project mainly involved land extension of existing roads and new alignment of some portions of the road. The main issues in this case were land acquisition and relocation of the affected residents.

Since DOE did not deem the predicted impacts significant, approval of only a Preliminary EIA report was sufficient for the implementation of the project. In Preliminary EIA reports, public participation is not mandated under EQA — though it is encouraged by the DOE and the method of the participation is left to the project proponent.

Maxtro Engineering issued the Preliminary EIA report in December 1996. Public participation during the preparation of the report took place through interviews using questionnaires. The analysis was scattered, based on only 185 questionnaires from 11 different localities in the vicinity. Consequently, the impacted local population was exposed to incomplete information, which in turn led to protests. The affected people refused to move from the project site (Leong Kar Yen, 2001). Despite the standoff, the highway was constructed. This is representative of the real EIA process in Malaysia where it would appear that little attention is paid to include the public's views in decision-making.

Under the Ninth Schedule of the Constitution of Malaysia, 1957, land acquisition, forestry, agriculture, etc. is under the State List; and hence States have the right to formulate laws to regulate their environments. This federal structure and its application in recent years in Malaysia has weakened the federal EIA provision in EQA. The Environmental Quality (Prescribed Activities) (Environmental Impact Assessment) Order 1987 was amended retroactively in 1995 to exclude certain prescribed activities from the original list for Sabah and Sarawak, and that excluded the public participation rights in these states for certain activities.

Since EIAs in Malaysia are initiated after land use planning, the land for the project has already been purchased (Kanniah, 1999). Hence, alternative

[29]New Pantai Expressway: accessed on March 30 2013 from http://en.wikipedia.org/wiki/New_Pantai_Expressway

locations are not taken seriously in the EIA process. The website by DOE is not transparent and statistics are not available. The EQA is lacking in providing an adequate voice to the people and in considering alternatives during the planning process of the development projects — all of which are the basic purposes of an EIA.

## 4. Singapore

Singapore is a 714.3 $km^2$ (Yearbook of Statistics, 2012) city-state where greenery, parks and air conditioners abound. Singapore was a tropical rainforest with tigers, leopards and sambur deer when Sir Stamford Raffles discovered it in 1819 (Sanson, 1992). Today, the bustling metropolis has a high population density of 7,257 people/$km^2$ (Yearbook of Statistics, 2012). Since its independence, Singapore has achieved better liveable standards for its society despite its small size and resource paucity.

The Judiciary in Singapore comprises the Supreme Court (composed of the High Court and the Court of Appeal) and the subordinate courts (comprising the District Courts, Magistrates' Courts, Family Court, Juvenile Court, Coroner's Court and the Small Claims Tribunal[30]). Environmental law in Singapore comprises Statutory Laws passed by Parliament, Common Law principles of tort law, and soft laws such as Codes of Practice and Guidelines issued by relevant Government authorities.

### 4.1 *Environment*

Consisting of more than 60 islands, Singapore legally protects representatives of 'key indigenous ecosystems': the Bukit Timah Nature Reserve (lowland dipterocarp forest), the Central Catchment Nature Reserve (including freshwater swamp forest), Sungei Buloh Wetland Reserve (mangroves), and Labrador Nature Reserve (coastal hill forest). In addition, one of the reserves, the Sungei Buloh Wetland, holds the distinction of being an ASEAN Heritage Park,' (NParks, 2009).

Singapore submitted a 'Green Plan' at Rio in 1992 and the revised versions have committed 5% of its land area to conservation. Singapore has designated 3500 hectares of its total 70,400 hectares (2012) as conserved areas, which is approximately 4.5% of the total (Neo, 2007). Due to continuous reclamation of land and heavy industrialisation, there has been

[30]Part VIII Section 94 (1), Constitution of the Republic of Singapore http://statutes.agc.gov.sg/aol/home.w3p (Accessed March 20, 2013)

extensive loss of biodiversity, habitat, coastal mangroves and mudflats, as most inland rivers are dammed to create water-storage reservoirs. "All remaining nature areas, including designated natural reserves and national parks remain under constant threat due to intense development pressures and lack of virgin building land" (Briffett and Mackee, 2002).

### 4.1.1 *Pollution*

Rapid industrialisation, high population density, loss of nature areas and trans-boundary haze from burning of forests in Indonesia, make air quality in Singapore vulnerable. Singapore is a water scarce country, and the alteration and channelisation of almost all its natural watercourses has taken place (Hesp, 1995). Making water security a priority in Singapore by its first Prime Minister has resulted in Singapore becoming a hub for water innovation and leader in water-related technologies. However, its marine environment has suffered, as Singapore's port is among the busiest in the world, and land reclamation and industrialisation takes its toll.

### 4.1.2 *Environmental Agencies/Regulators*

There is no umbrella ministry for the environment in Singapore.

1. The Ministry of Environment and Water Resources (MEWR), is responsible for pollution control, public health and water. There are two statutory boards under MEWR:[31]
   - National Environment Agency (NEA) for pollution control and public health and
   - Public Utilities Board (PUB) for supply and quality of water
2. Under the Ministry of National Development (MND), there are five statutory boards:[32]
   - Urban Redevelopment Authority (URA) for land use planning,
   - Housing Development Board (HDB) for public housing
   - National Parks Board (NParks), for conservation and management of nature areas

---

[31] Statutory Boards of Ministry of Environment and Water Resources. Accessed from http://app.mewr.gov.sg/web/Contents/Contents.aspx?Id=189

[32] Ministry of National Development Statutory Boards. Accessed from http://app.mnd.gov.sg/AboutUs/Introduction.aspx

- Agri-Food and Veterinary Authority (AVA) for safeguarding the health of animals and plants, and
- Building and Construction Authority (BCA) for development of the 'built environment' in Singapore

3. The Maritime and Port Authority (MPA) manages marine pollution in Singapore.[33]

The National Environment Agency (NEA) is responsible for Environmental Protection in Singapore. The Minister for the Environment and Water Resources under Section 3(1) of the Environmental Protection and Management Act (Chapter 94A) (EPMA) appoints the Director General of NEA (DG).

## 4.2 *Singapore and (lack of) EIA Law*

There is no mention of environment or any provision relating to the environment in Singapore's Constitution (1965). There is no law mandating EIAs for development projects in Singapore.

During the first rapid phase of industrialisation and development post-independence before 1989, virtually no EIA was performed. However, since 1989, EIAs are conducted for both public and private projects on an ad-hoc basis (Hesp, 1995). Khairul (2009) labels Singapore's few, subject-specific EIA requirements as 'Scattered EIA practice' based on the provisions of assessment studies or pollution control studies from different statutory boards. These requirements are evaluated further in the following segments.

### 4.2.1 *The EIA in Singapore*

- <u>Sections 26 and 36 of the EPMA</u>

The closest to an EIA requirement in Singapore are Sections 26 and 36 of the EPMA (Lye, 2008). Under Section 26(1) of EPMA, the DG of NEA by notice in writing may require the 'owner or occupier of hazardous installations' to carry out impact analysis studies. "Hazardous installations", are premises "used or intended to be used to carry out activities involving the storage, handling and use of hazardous substances".

[33]MPA website. Accessed on April 1, 2013 from http://www.mpa.gov.sg/sites/global_navigation/about_mpa/about_mpa.page

"Impact analysis studies" require:

(a) identification of all possible potential hazards that may threaten the health or safety of any person, or cause pollution of the environment;
(b) estimation of the frequency or probability of occurrence of such potential hazards as identified in paragraph (*a*);
(c) quantification of the consequences and risk levels of such potential hazards as identified in paragraph (*a*);
(d) evaluation of the effects of potential fires or other disasters including the potential for release of toxic materials or toxic combustion products and the potential for release of contaminated fire-fighting water into the environment; and
(e) identification of all necessary preventive measures to avoid and control the hazards identified in paragraph (*a*) and formulation of a programme to implement the measures.

The maximum penalty for contravening S.26 (5) is SGD 20,000 (USD 15,029). The general maximum penalty for offences involving hazardous substances is SGD 50,000 (~USD 37,573) or imprisonment of maximum two years or both under S.27 of EPMA. In comparison, the annual per capita gross domestic product in Singapore is SGD 72,667 (~USD 55,182.5) (World Bank Data, 2013).[34]

Section 36 of EPMA states that the DG may require any person intending to carry out any activity that the DG deems pollutive to the environment to conduct a "study on environmental pollution control"; to submit to the DG for approval, and to implement measures to prevent, reduce or control pollution. The DG may issue guidelines for conducting such study and may require any modifications or additions in such measures as the DG regards as necessary and even require further study on the matter.

The penalty for this section is not specified. The general penalty for offences under EPMA for which no specific penalty is specified is SGD 20,000 (~USD 15029)[35] for the first offence and a further fine not exceeding SGD 1,000 (~USD 752) for every day that the offence continues. A fine not exceeding SGD 50,000 (~USD 37,573) is imposed on a second offence and a continuing fine of SGD 2,000 (~USD 1,503) per day is stated (S.67).

---

[34]http://www.mom.gov.sg/Documents/statistics-publications/yearbook12/mrsd_2012 YearBook.pdf (Accessed March 20, 2013)
[35]1 SGD = 0.75 USD based on currency conversion on April 28, 2015 (Source: http:// www.xe.com/currencyconverter/convert/?Amount=1&From=SGD&To=USD

- Singapore Standard (SS) 593: 2013 — Code of Practice for Pollution Control

SS593 states that developers of special industries using large quantities of hazardous chemicals may be required to conduct:

(i) Quantitative Risk Assessment (QRA) study and
(ii) Pollution Control (PC) study

The QRA specified calls for identification and quantification of hazards and risks, recommends measures to keeps risks to a low level and facilitate development of an 'Emergency Response Plan'.

The PC study calls for identifying sources of pollution and measures to reduce pollution and mitigate adverse effects of pollution.

- BCA Green Mark for Infrastructure

The Building Control Authority (BCA)'s Green Mark is a green building rating system approved by NEA; it evaluates a building/project for its environmental impact and performance. Under the BCA Green Mark Scheme, only infrastructure projects like roads, barrages, bridges, etc. mention impact assessment studies as evaluation criteria.

Under Project Management assessment criteria, Environmental Risk Analysis (ERA) is required and is awarded points under the scheme. ERA ensures that planners and designers incorporate environmental criteria in their designs at an early stage.

> *The risk analysis should include relevant impacts such as, but not limited to, soil contamination, air pollution, noise health effects, and ecology impacts, including endangered species, geological hazards, and water pollution.*

(a) Paragraph 4.4 of BCA Green Mark for Infrastructure (version 1) guide states:[36]

> *The feedback and inputs from the consultation should be used to improve the design and construction process. If any negative impacts are unavoidable, mitigation measures should be implemented to reduce these impacts as far as possible. These shall be communicated to the community and interest groups in a timely manner.*

[36]Retrieved from http://www.bca.gov.sg/greenmark/others/GM_Infra_V1.pdf

Paragraph 4.5 calls for follow up consultations during construction phase and as required later.

### 4.2.2 *Weaknesses of the Present System*

The weaknesses in the system in Singapore can be summarised as follows:

- "The nearest approximations of the EIA in Singapore are contained in sections 26 and 36 of the EPMA" (Lye, 2008a). EPMA S.26 and 36 call for impact assessments at the discretion of the Director General of Pollution Control. S.26 applies only to "hazardous installations" while specifications of S.26 aim more at prevention/mitigation of hazards. There is clarity regarding penalties but there is ambiguity in the requirements regarding timelines and specifications. Even if the impact assessment is carried out before the implementation of the project, these studies do not form part of the decision-making process. The maximum penalty for an offence under S.26 is only SGD 20,000 (~USD 15,029)[37] when the general maximum penalty involving hazardous substances is SGD 50,000 (~USD 37,573). It can be inferred that S.26 does not incur the highest penalty possible under this section of EPMA.
- S.36 is broader in scope than S.26. S.36 can require any person who *intends* to carry out a pollutive activity to conduct a Pollution Control study. This implies that the S.36 requirement is specifically for projects before implementation. However, the term 'any activity' also makes the requirement highly ambiguous and only a basis of 'ad-hoc EIA studies' performed in Singapore. Would a proposed golf course, for example, warrant an S.36 study? This is left to the discretion of the DG of Pollution Control. Essential elements of an EIA study like ecological and societal impacts, public information and participation, the consideration of alternatives, and proper screening requirements, are woefully missing.
- SS 593 calls for a QRA for industries handling hazardous chemicals but it is unclear if S.26 of EPMA is referred to; S.26 requirements are for hazardous substances and it makes no mention of the QRA. The PC study mentioned in the SS matches with the requirements of S.36 of EPMA. However, these requirements only point to identification and mitigation of impacts caused by the project and not an environmentally conscious planning process.

[37] 1 SGD = 0.75 USD based on currency conversion on April 28, 2015 Data accessed from http://www.xe.com/currencyconverter/convert/?Amount=1&From=SGD&To=USD

- BCA's Green Mark guidelines for infrastructure are the only provisions that consider the steps of pre-assessment, impact assessment, risk analysis and public consultation. However, these are only guidelines applicable to infrastructure projects and are not legally enforceable.

### 4.3 *Does Singapore Need an EIA Law?*

Professor Tommy Koh, Ambassador at Large for Singapore in a *Straits Times*' article (Koh, 2012) makes the introduction of an EIA law in Singapore, his first wish in his "wish list". In 2002, Malaysia objected to Singapore's land reclamation activities in Pulau Tekong and the Tuas extension, claiming harm was caused to the marine environment, as well as damage to jetties and reduced catch for fishermen in the Straits of Johor (Koh and Lin, 2006).Malaysia invoked Article 286 of the 1982 UN Convention on the Law of the Sea (UNCLOS), and initiated arbitration under Annex VII of the UNCLOS. In 2005, through effective arbitration, the case was settled owing to international consulting and research organisation DHI-Water and Environment's rigorous and voluminous report on the impacts of land reclamation by Singapore during the arbitration. DHI's EIA report proved that Singapore's actions were not significantly impacting the marine environment. Singapore would have been better prepared for such a situation if it were ready with its EIA prior to the start of the project.

Some projects that have generated considerable controversy, calling into question the need for EIA laws, include:-

- Proposed golf course at Lower Peirce Reservoir, 1992.

  In 1992, shortly after Singapore's participation at the Rio Conference, plans were announced to build a golf course on part of the Lower Peirce Reservoir, an area that is legally protected as part of our nature reserves. The government declined to reveal its EIA to the public, whereupon the Nature Society of Singapore[38] conducted and published its own EIA, demonstrating the environmental losses that would occur if the project were allowed without assessing its impacts (Wee, 1992). The government decided not to proceed.

[38]The Nature Society (Singapore) is a non-government, non-profit organisation dedicated to the appreciation, conservation, study and enjoyment of the natural heritage in Singapore.

- Chek Jawa[39] is a cape located on the south-eastern tip of Pulau Ubin, an island off the north-eastern coast of the main island of Singapore. It is unique with some six eco-systems — sandy beach, rocky beach, sea grass lagoon, coral rubble, mangroves and coastal forest — in one place (Agnarsson *et al.*, 2007; Chou, 2006). Chek Jawa was slated for reclamation in 2001 by the Singapore Government (Lye, 2008a) and had it not been for the efforts of volunteer conservationists and the National Parks Board, this unique site would have been lost (Wee and Hale, 2008; Hobson, 2006). An EIA done by the Urban Redevelopment Authority was inadequate. The government stopped reclamation at the last minute with assurance to keep it as is, for as long as possible (Soh and Yuen, 2006). The lack of mandatory EIA laws in Singapore make places like Chek Jawa vulnerable to development as Chek Jawa's status is not legally conserved and new developments do not need an EIA.

- Bukit Brown Cemetery.

  In 2011, the government proposed the widening of a road that would cut into the Bukit Brown Cemetery; a 230-hectare cemetery with more than 100,000 graves,[40] which claims to be the largest Chinese cemetery outside China. It is covered with secondary forests and is host to 90 migrant and resident birds with 13 nationally threated species. In 1991, it was designated a Tree Conservation Area.[41] The new road was announced without a proper EIA and some 5,000 graves have now been removed. More homes are expected to be built in the area. In the absence of a well-crafted EIA law, Singapore may lose another area of natural and heritage importance.

Singapore's Population White Paper released in 2013[42] plans for a population of 6.9 million in 2030. This caused considerable consternation in Parliament. Nominated MP Faizah Jamal highlighted the need for a

[39]Conserving marine biodiversity and the role of individuals: A case study of Chek Jawa, Pulau Ubin, Singapore by Teh Tiong Saand K. Raju Accessed from http://www.globalislands.net/userfiles/Singapore_4.pdf
[40]See Nature Society of Singapore http://www.nss.org.sg/ and Singapore Heritage Society http://www.singaporeheritage.org/ for their position papers on Bukit Brown
[41]See more at All Things Bukit Brown, Heritage, Habitat, History. Accessed from http://bukitbrown.com/main/; See also http://remembersingapore.org/bukit-brown-cemetery/
[42]See http://202.157.171.46/whitepaper/downloads/population-white-paper.pdf (Retrieved February 14, 2013)

compulsory EIA in Parliament. An EIA process open to public scrutiny is to facilitate "*openness, transparency and public accountability*" in planning processes (Chan, 2013). The exercise of predicting, planning and preparing for 2030 has regrettably, failed to call for mandatory EIAs and this is inadequate from the perspective of sound environmental planning and management.

## 5. Conclusion

The driver for the evolution of EIA law in the United States was the American society, as it felt increasingly frustrated with environmental pollution and disasters. An environmental law with elements of the precautionary principle was thus established. This study explored the formulation and implementation of EIA law in Malaysia and Hong Kong, countries that are close to Singapore in terms of geography and economy, in the hope that Singapore may learn from their experiences and pass an appropriately crafted EIA law.

### 5.1 *Main Findings*

This study has found that:

- In Hong Kong, the EIA law is independent and enforceable. In Malaysia, the laws are nestled within the larger pollution control framework legislation and enforcement is compromised due to the Federal structure of the Constitution.
- The coverage of EIA laws in HKSAR is comprehensive, while there are lacunas in the coverage of EIA law in Malaysia that allow project developers to avoid the provisions if they wish to do so.
- HKSAR and Malaysia both do not have consideration of alternatives in the statutory process — and the EIA process guidelines mandate the consideration of alternatives. While the no-project scenario is a requirement in the HKSAR system, it is often inconsequential — a weakness of the HKSAR system. In Malaysia, due to the confusion with the land zoning laws, etc., the consideration of alternatives, including not proceeding with the project, sis rare.
- Public Participation has been included as an important part of the statutory EIA process in HKSAR; there are clear specifications on the participation methodologies and timings for this process in HKSAR. In Malaysia, however, proper public participation is severely limited as it

involves only one level of participation. This has weakened the Malaysia EIA law substantially.

### 5.2 *Recommendations for Singapore*

An analysis of Singapore's laws underlines its inadequacy in mandating an EIA for development projects that may adversely impact the environment in this city-state. Sections 26 and 36 of the EPMA are inadequate as they are narrowly confined to pollutive impacts. While development is essential, sustainable development has been emphasised by Singapore government. A firm, clear EIA law will aid sustainable development in the city and enhance the present environmental management system. The recommendations for Singapore's EIA law emphasise clarity of processes, transparency, and public participation. There are many lessons that Singapore can draw from the comparison of EIA laws in Malaysia and HKSAR.

## References

Abaza, H., Bisset, R., and Sadler, B. (2004). *Environmental impact assessment and strategic environmental assessment: towards an integrated approach.* London: Earthprint.

Agnarsson, I., Wayne, P., Maddison, and Avilés, L. (2007). "The phylogeny of the social Anelosimus spiders (Araneae: Theridiidae) inferred from six molecular loci and morphology", *Molecular Phylogenetics and Evolution*, **43**(3), 833–851.

Ali, R., Daut, I., and Taib, S. (2012). A review on existing and future energy sources for electrical power generation in Malaysia", *Renewable and Sustainable Energy Reviews*, **16**(6), 4047–4055.

Awang, M.B., Jaafar, A.B., Abdullah, A.M., Ismail, M.B., Hassan, M.N., Abdullah, R., Johan, S., and Noor, H. (2000). "Air quality in Malaysia: Impacts, management issues and future challenges", *Respirology*, **5**(2), 183–196.

Birnie, P.W., Boyle, A.E., and Redgwell, C. (2009). *International Law and the Environment* 3rd ed. New York: Oxford University Press.

BrandHK.gov, May 2012, accessed on February 10, 2013 from http://www.brandhk.gov.hk/en/facts/factsheets/pdf/05_green_hongkong_en.pdf

Briffett, C. (1999). "Environmental impact assessment in Southeast Asia: fact and fiction?", *GeoJournal*, **49**(3), 333–338.

Briffett, C., and Mackee, J. (2002). "Environmental assessment in Singapore: an enigma wrapped up in a mystery!", *Impact Assessment and Project Appraisal*, **20**(2), 113–125.

Brook, B.W., Sodhi, N.S., and Ng, P.K.L. (2003). "Catastrophic extinctions follow deforestation in Singapore", *Nature*, **424**, 420–423.

Brundtland, G.H. (1987). *Report of the World Commission on Environment and Development: "Our Common Future"*. United Nations.

Census and Statistics Department Hong Kong; Hong Kong in Figures Accessed February 9, 2013 from http://www.statistics.gov.hk/pub/B10100062012AN12E0100.pdf

CIA, Central Intelligence Agency, Malaysia; The World Factbook. Accessed on March 14, 2013 from https://www.cia.gov/library/publications/the-world-factbook/geos/my.html

CIA, HK, Central Intelligence Agency Hong Kong; The World Factbook. Accessed on February 5, 2013 from https://www.cia.gov/library/publications/the-world-factbook/geos/hk.html

Chan, R. (2013). "Assess impact on nature: NMP Faizah Jamal", *The Straits Times*, February 7, 2013. Accessed February 7, 2013 from http://www.straitstimes.com/microsites/parliament/story/assess-impact-nature-nmp-faizah-jamal-20130207

Chan, W.-H. (1998). "Environmental Protection in Malaysia: Lessons from the Bakun Hydroelectric Project Litigation", *Environmental Liability*, **6**, 11–17.

Chau, L.C. (1997). "Hong Kong: a unique case of development", *Lessons from East Asia*, **4**(2.4), 35.

Cheung, J. (2009). "*China Builds Bridge to Link Southern Cities*", *The Wall Street Journal*, Asia News, December 15, 2009. Accessed on March 11, 2013 from http://online.wsj.com/article/SB126088119555891949.html?mod=googlenews_wsj#articleTabs%3Darticle

Cheung, C.F. and Wong, O. (2011). "Green law has limited impact Approval for major works is almost a foregone conclusion despite requirements of 1998 act", July 19, 2011 *South China Morning Post.* Accessed on February 13, 2013 from Factiva database.

Chou, L.M. (2006). "Marine habitats in one of the world's busiest harbours" in *The Environment in Asia Pacific Harbours*, pp. 377–391. Netherlands: Springer.

Craik, N. (2008). *The International Law of Environmental Impact Assessment: Process, Substance and Integration*, Vol. 58. Cambridge: Cambridge university press.

Cross, F.B. (1996). "Paradoxical perils of the precautionary principle", *Wash & Lee L. Rev.*, **53**, 851–1571.

Department of Environment, Environmental Quality Report, 2011 Malaysia. Accessed on March 14, 2013 from http://www.doe.gov.my/portal/publication-2/

Department of Statistics Malaysia, Population Clock. Accessed on March 14, 2013 from http://www.statistics.gov.my/index.php?r=home/index&menu_id=b2ROaWpITmQ5NnAvMHVmRjRkZzlBQT09

DOE/Notice 8/2012 Reviewing of EIA report within 3 weeks (working days) by the Department of the environment DOE for selected prescribed activities. Accessed on March 19, 2013 from http://eia.doe.gov.my/portal/wp-content/uploads/2012/11/Notice-8-2012-3-weeks-EIA-processing5.pdf

EIAO guidance Note No. 1/2002; A guide to the EIA Ordinance. Accessed on March 7, 2013 from http://www.epd.gov.hk/eia/english/guid/ordinance/guide1-3.html#3

EPD, Environmental Assessment and Noise Division, HKSAR Government, '*The Operation of Environmental Impact Assessment Ordinance From 1998 to 2001*' (2003). Accessed on March 6, 2013 from http://ebook.lib.hku.hk/HKG/B35848893.pdf

Fong, A. (2012) "Protecting Hong Kong's Natural Resources: Air Pollution and the Basic Law." *Available at SSRN 2170670* (2012).

Food and Agriculture Organization, United Nations, Malaysia, Accessed on March 16, 2013 from http://www.fao.org/forestry/country/57478/en/mys/

Gasparatos, A., Stromberg, P., and Takeuchi, K. (2011). "Biofuels, ecosystem services and human wellbeing: Putting biofuels in the ecosystem services narrative", *Agriculture, Ecosystems & Environment*, **142**(3), 111–128.

Hesp, P.A. (1995). "The Environmental Impact Assessment process in Singapore with particular respect to coastal environments and the role of NGOs", *Journal of Coastal Conservation*, **1**(2), 135–144.

Harris, P.G.E. (2012). *Environmental Policy and Sustainable Development in China*. The Policy Press.

Hezri, A.A., and Hasan, M.N. (2006). "Towards sustainable development? The evolution of environmental policy in Malaysia", *Natural Resources Forum*, **30**(1), 37–50.

Hildebrand, S.G., and Cannon, J.B. (1993). *Environmental Analysis The NEPA Experience*, CRC.

Hills, P., and Barron, W. (1997). "Hong Kong: the challenge of sustainability", *Land Use Policy*, **14**(1), 41–53.

Hobson, K. (2006). "Enacting environmental justice in Singapore: Performative justice and the Green Volunteer Network", *Geoforum*, **37**(5), 671–681.

Josh, L.A.M. (2008). "The Art and Science of EIA in Achieving Sustainability In a Hong Kong Transport Development Project". IAIA.

Kanniah, R. (1999). *An analysis of the laws relating to environmental impact assessment in Malaysia with specific reference to federal-state jurisdiction.* A dissertation submitted to the University of Malaya in accordance with the requirements for the degree of LL.M in the Faculty of Law, 1999.

KCRC v Director EPD, Environmental Impact Assessment Appeal Board No. 2 of 2000. Accessed on February 25, 2013 from http://www.epd.gov.hk/eia/english/content/files/appealboard_2000.html

Khairul, A.A. (2009). *Environmental impact assessment in Singapore: is it absent, or just hard to find.* An undergraduate honours dissertation submitted to the Department of Geography, supervised by Dr Harvey Neo Choong Tiong, National University of Singapore, Singapore 2009.

Koh, T. (2013). "Green Thoughts inspired by Stockholm and Rio", *The Straits Times*, June 6, 2012. Accessed February 15, 2013 from http://www.spp.nus.edu.sg/ips/docs/pub/pa_tk_ST_Green%20thoughts%20 inspired%20by%20Stockholm%20and%20Rio_160612_full%20version.pdf

Koh, T., and Lin, J. (2006). "The Land Reclamation Case: Thoughts and Reflections", *SYBIL*, **10**, 1.

Shourav, L. (2006). *Court of Final Appeal Rules on Environmental Impact Assessment Process*, November 13 2006. Accessed on March 9, 2013 from http://www.internationallawoffice.com/newsletters/detail.aspx?g=1e8e5161-0a69-db11-a275-001143e35d55

Leung, K.Y. (2013). "Residents protest against highway developer for reneging on deal", *Malaysiakini*, October 21, 2001. Accessed March 30, 2013.

Lo, K.-I. (2002). "Sustainable railway infrastructure development: lessons from the Sheung Shui to Lok Ma Chau Spur line".

Lye, L.H. (2008a). "A Fine City in a Garden-Environmental Law and Governance in Singapore", *Singapore Journal of Legal Studies*, 68–117.

Lye, Lin-Heng Singapore, *International encyclopaedia of laws* General Editor Blanpain, Roger Kluwer law international, 2008 Singapore-1 to Singapore-127

Marshall Cavendish Corporation (2009). *World and Its Peoples*, Vol. 13. Marshall Cavendish.

Mohammad, N. (2011). "Environmental Law and Policy Practices in Malaysia: An Empirical Study", *Australian Journal of Basic and Applied Sciences*, **5**(9), 1248–1260.

Mongobay Malaysia, Accessed on March 15, 2013 from http://rainforests.mongabay.com/20malaysia.htm

Monthly Manufacturing Statistics Malaysia, August 2012, Department of Statistics, Malaysia. Accessed on March 30, 2013 from http://www.statistics.gov.my/portal/index.php?option=com_content&view=article&id=1756&Itemid=153&lang=en

Mottershead, T. (2002). *Environmental Law and Enforcement in the Asia-Pacific rim*. Sweet & Maxwell Asia.

National Income and Balance of Payments, Census and Statistics Department, The Government of Hong Kong Special Administrative Region, 2014. Accessed on April 27, 2015 from http://www.censtatd.gov.hk/hkstat/sub/so50.jsp

Neo, H. (2007). "Challenging the developmental state: Nature conservation in Singapore", *Asia Pacific Viewpoint*, **48**(2), 186–199.

Ng, K.L., and Obbard, J.P. (2005). "Strategic Environmental Assessment in Hong Kong", *Environment international*, **31**(4), 483–492.

NParks, A Holistic Approach 2009. Accessed November 18, 2013 from http://www.nparks.gov.sg/cms/index.php?option=com_content&view=article&id=162&Itemid=129

Pak, J. (2013). "Pollution threatens Hong Kong's competitiveness", *BBC News Hong Kong*, 16 January 2013. Accessed on February 5, 2013 from http://www.bbc.co.uk/news/business-21038495

PELB, Planning, Environment and Lands Branch (1989), *Pollution in Hong Kong — A Time to Act*, White Paper. Accessed on February 8, 2013 from http://www.epd.gov.hk/epd/english/resources_pub/policy/files/White_Paper-A_time_to_act.pdf

Population Census data 2014 HK; Population Census Summary Results. Accessed April 28, 2015 from http://www.censtatd.gov.hk/home/

Price, R., and Ho, J. (2011). "Air pollution in Hong Kong: The failure of judicial review and the slight promise of recent cases", *Singapore Journal of Legal Studies*, 394–411.

Rio Declaration on Environment and Development (1992). The United Nations Conference on Environment and Development. Accessed on November 5, 2012 from http://www.unep.org/Documents.Multilingual/Default.asp?documentid=78&articleid=1163

Robinson, N.A. (1992). "International trends in environmental impact assessment", *Boston College Environmental Affairs Law Review*, **19**.

Rodgers Jr., W.H. (1993). "The Seven Statutory Wonders of US Environmental Law: Origins and Morphology", *Loy. LAL Rev.*, **27**, 1009.

Salinas, S.V., Chew, B.N., Miettinen, J., Campbell, J.R., Welton, E.J., Reid, J.S., Yu, L.E., and Liew, S.C. (2012). "Physical and optical characteristics of the October 2010 haze event over Singapore: A photometric and lidar analysis", *Atmospheric Research*,.

Sanson, V. (1992). *Gardens and Parks of Singapore*. Oxford University Press.

Singapore Constitution 1965, Constitution of the Republic of Singapore. Accessed on April 1, 2013 from Attorney General's Chambers http://statutes.agc.gov.sg/

Soh, E.Y., and Yuen, B. (2006). "Government-aided participation in planning Singapore", *Cities*, **23**(1), 30–43.

Stærdahl, J., Schroll, H., Zakaria, Z., Abdullah, M., Dewar, N., and Panich, N. (2004). "Environmental Impact Assessment in Malaysia, South Africa, Thailand, and Denmark: Background, layout, context, public participation and environmental scope", *The Journal of Transdisciplinary Environmental Studies*, **3**(1).

UNESCO Malaysia World Heritage Sites, 'The States Parties' Malaysia. Accessed on March 15, 2013 from http://whc.unesco.org/en/statesparties/my

UNESCAP United Nations' Economic and Social Commission for Asia and the Pacific, Virtual Conference: Examples of different definitions of EIA 2003. Accessed February 4, 2013 from http://www.unescap.org/drpad/vc/index.htm

Vun, L.W., and Latiff, A. (1999). "Preliminary ecological input assessment and environmental impact assessment for coastal resort development in Malaysia." *Impact Assessment and Project Appraisal*, **17**(2), 133–140.

Warren-Rhodes, K., and Koenig, A. (2001). "Escalating trends in the urban metabolism of Hong Kong: 1971–1997", *AMBIO: A Journal of the Human Environment*, **30**(7). 429–438.

Wee, Y.C. (1992). "Proposed Golf Course at Lower Peirce Reservoir: An Environmental Impact Assessment", Singapore: Nature Society (Singapore).

Wee, Y.C., and Hale, R. (2008). "The Nature Society (Singapore) and the struggle to conserve Singapore's nature areas", *Nature in Singapore*, **1**, 41–49.

Wood, C., and Coppell, L. (1999). "An evaluation of the Hong Kong environmental impact assessment system", *Impact Assessment and Project Appraisal*, **17**(1), 21–31.

World Bank Data (per capita GDP). Accessed on April 28, 2015 from http://data.worldbank.org/indicator/NY.GDP.PCAP.CD

World Bank Environmental Impact Assessment Regulations and Strategic Environmental Assessment Requirements: Practices and Lessons Learned in East and Southeast Asia, April 2006. Accessed on October 23, 2012 from http://siteresources.worldbank.org/INTEAPREGTOPENVIRONMENT/Resources/EIA&SEA-regional-review.pdf

Wong H.M., (2011). Environmental Protection Department, HKSAR Government; *Sustainable Environmental Management in Urban Asia, Environmental Impact Assessment, Hong Kong Experience*, MEM 10th Anniversary Conference, December 2011.

Yearbook of Statistics (2012). Department of Statistics Singapore. Accessed March 24 from http://www.singstat.gov.sg/publications/publications_and_papers/reference/yearbook_2012/yos2012.pdf

Yusof, N.M. (2013). "Public Participation approaches and strategies under the EIA process in Malaysia". AECEN Regional Workshop 'Strengthening Public Participation on Environmental Management', 10–11 July 2012 Accessed on March 19, 2013 from http://www.aecen.org/sites/default/files/public_participation_in_eia_in_malaysia.pdf

Zhao, Y. (2009). "Assessing the environmental impact of projects: a critique of the EIA legal regime in China", *Nat. Resources J.*, **49**, 485.

7

# SINGAPORE'S FRESH PORK SUPPLY: FOOD SECURITY AND ENVIRONMENTAL IMPACTS

LIM CHIEN-FANG and VICTOR R. SAVAGE

An overview of Singapore's pork supply since the 1970s showed the importance of food safety in the larger framework of food security. Singapore's pork supply went through three phases of changes with the source extending further at each stage. With the increase in food miles, the greenhouse gas emitted during transport of the fresh pork supply rose from 0.8 million to nearly 34 million tonnes. Singapore's fresh pork market is dominated by Australian supply. The need for source diversification and considerations for Singapore's ability to sustain the costly imports are evaluated in anticipation of volatile environmental and economic conditions in the future.

## 1. Introduction

Singapore is the third most densely populated country in the world[1] and imports over 90% of its food supply (AVA). Fresh pork is a popular food item in Singapore for which the supply has undergone noteworthy transitions in the past few decades. Singapore was self-sufficient in fresh pork production up to the mid-1980s (Gursansky and Peacock, 2000) with more than 600 pig farms (Chou, 1998). Today, there are no pig farms in Singapore; all of the fresh pork in Singapore is imported. The transition of the fresh pork supply offers valuable insights into Singapore's food security, its pork consumer footprint and its environmental impacts from domestic output to international reliance.

### *Nature of Pork Supply in Singapore*

Singapore's spatial context of its pork supplies reflect three phases:

- Before 1985: Domestic pork production
- 1980–1990s: Pork imports from Malaysia
- Since 1999: Chilled pork imports from Australia

[1]Based on data from United Nations Population Division, "World Population by continents and countries — Nations Online Project". Available online at http://www.nationsonline.org/oneworld/world_population.htm. Accessed on.

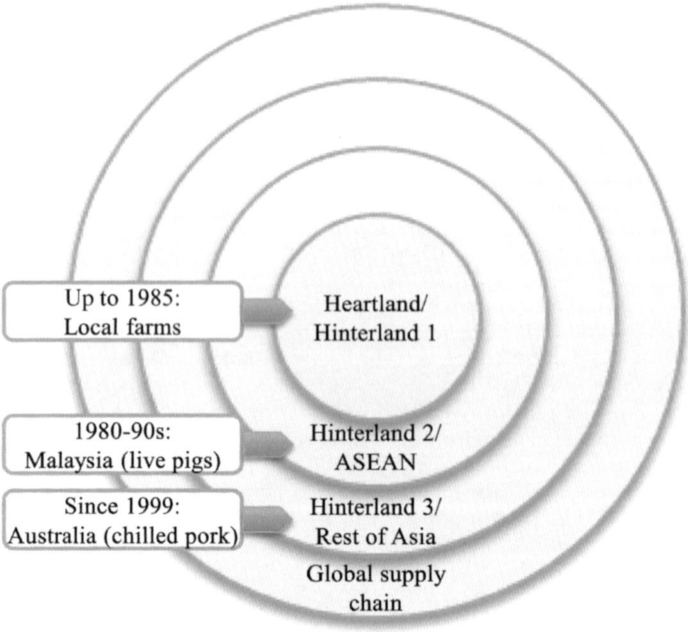


Fig. 7.1. Singapore's growing "hinterland" with respect to fresh pork supply (adapted from Teng and Escaler, 2010).

For most of its history, Singapore relied on domestic pig farms for its pork consumption. However, pig farming is a resource-intensive and a highly-polluting industry. When Singapore gained independence and began moving to its vision of being a garden city, the government decided to completely ban local pork production due to both environmental issues and land scarcity. Singapore then turned to Malaysia for imports of live pigs. Two developments then led to Singapore's expanding hinterland of its food sources. Firstly, the Nipah virus outbreak in Malaysia in 1999 led the Singapore government to stop imports of fresh pork from Malaysia. The government has become more pro-active in ensuring food safety given that an outbreak of any degree could have rapid implications for citizens in the compact city and the environment. Secondly, Singapore could depend on an expanding hinterland due to advances in transport technology that enabled pork to be brought in from countries located much further away.

## 2. An Overview of Singapore's Fresh Pork Industry

### *Singapore Pork Consumption*

In 2011, the world produced 109 million metric tons of pork (Best, 2012). Pork is mostly consumed in the country of origin, as such major producers are likely to be major consumers themselves. The top three largest exporters are the United States, the European Union (EU) and Canada, which are considerably far from Singapore (Table 7.1). Large producers within ASEAN, such as Vietnam and Philippines, consume more than they produce. Thus, there is little opportunity for Singapore to import from these two neighbouring countries.

Pork is very much favoured by the Chinese majority (74% of the total population) in Singapore (Department of Statistics, Singapore, 2012a). Pork imports regularly make up around 30% of all meat imported into Singapore (Table 7.2). By import volume, pork is second only to chicken.

The popularity of pork is also reflected in the per capita consumption. As shown in Table 7.3, a Singaporean consumes an average of 20 kg of pork per year. In 1999, the per capita consumption dropped sharply from 22.7 kg in 1998 to 15.2 kg in 1999. This decrease was due to the Nipah virus scare (Tey *et al.*, 2009). The figures quickly picked up, returning to per capita consumption of 18 kg in 2000. The health-consciousness of Singaporeans can be seen with the corresponding sharp increase in the consumption of chicken and fruits in 1999. Due to fear of contracting the disease, Singaporeans turned to consuming chicken and ate more fruits to boost their immunity and 'cleanse' the body of toxins (Tey *et al.*, 2009).

Tey *et al.*'s (2009) study of Singapore's Household Expenditure conducted in 2002/03 showed that fresh pork makes up the highest share of household expenditure on meat in Singapore. An updated survey in 2007/08 shows that food takes up 7.6% of the household expenses. Of these, chilled pork continues to take up the highest expenditure spent on meat produce (at 7% of the food expenses, Table 7.4).

#### *Current sources for Singapore's pork*

Singapore presently imports about 1,000 pigs daily from Pulau Bulan in Indonesia. This makes up about 20% of Singapore's total pork consumption (Ramesh, 2009). However, the production facilities in Indonesia are owned by the KMP Group, Singapore (Fraser, 2000). This helps Singapore

Table 7.1. Pork selected countries summary (2011) (USDA Livestock & Poultry, 2012) 1,000 metric tons (Carcass Weight Equivalent).

| | (A) Production | (B) Total domestic consumption | (C) Total imports | (D) Total exports | (B)–(A)* |
|---|---|---|---|---|---|
| China | 49,500 | 50,004 | 758 | 244 | 504 |
| EU-27 | 22,750 | 20,564 | | 2204 | −2,186 |
| United States | 10,332 | 8,339 | 364 | 2356 | −1,993 |
| Brazil | 3,227 | 2,644 | | 584 | −583 |
| Russia | 1,995 | 2,940 | 946 | | 945 |
| Vietnam | 1,960 | 1,995 | | 8 | 35 |
| Canada | 1,770 | | 204 | 1197 | |
| Japan | 1,267 | 2,522 | 1254 | | 1,255 |
| Philippines | 1,260 | 1,358 | | | 98 |
| Mexico | 1,182 | 1,690 | 594 | 86 | 508 |
| South Korea | 837 | 1,487 | 640 | | 650 |
| Taiwan | | 894 | | | |
| Hong Kong | | | 432 | | |
| Australia | | | 175 | 41 | |
| Belarus | | | 125 | 78 | |
| Ukraine | | | 119 | 17 | |
| Chile | | | | 139 | |
| Others | 5,582 | 6,849 | 984 | 28 | |
| **Total** | **101,662** | **101,286** | **6,595** | **6,982** | |

Greyed-out cells indicate where data is not available.
*Positive figures indicate that the country requires imports to meet its needs.

safeguard the safety standards and continuity of supply. As at 12 February 2013, there are 24 countries which are approved to export pork to Singapore (AVA, 2013). The following tables and charts show the figures and trends of Singapore pork import from 2009–2011.

The majority of Singapore's pork comes in frozen form. There has been a slight steady decrease in the consumption of fresh/chilled pork. A longer

Table 7.2. Singapore food imports from 2002 to 2011 (AVA).

| Item | | 2002 | 2003 | 2004 | 2005 | 2006 | 2007 | 2008 | 2009 | 2010 | 2011 | % Difference 2011 vs. 2002 |
|---|---|---|---|---|---|---|---|---|---|---|---|---|
| Chicken | Tonne | 130,293 | 152,532 | 126,173 | 140,675 | 134,326 | 156,835 | 163,811 | 160,693 | 169,221 | 180,051 | 38% |
| | $'000 | 155,780 | 298,214 | 275,306 | 347,157 | 307,180 | 405,234 | 492,572 | 414,590 | 461,737 | 536,311 | 244% |
| | Tonne | 16,766 | 15,721 | 10,821 | 14,842 | 12,665 | 14,777 | 15,049 | 13,561 | 14,514 | 13,532 | −19% |
| Ducks | $'000 | 48,178 | 44,698 | 36,080 | 44,703 | 44,071 | 54,806 | 64,680 | 56,638 | 57,994 | 55,825 | 16% |
| **Pork** | **Tonne** | **75,925** | **83,267** | **87,487** | **83,448** | **89,017** | **95,423** | **93,260** | **96,268** | **103,368** | **101,221** | **33%** |
| | **$'000** | **272,005** | **291,609** | **320,148** | **322,236** | **343,909** | **359,690** | **387,262** | **414,507** | **451,645** | **426,132** | **57%** |
| | Tonne | 15,697 | 19,545 | 18,470 | 18,691 | 24,385 | 27,780 | 28,848 | 27,897 | 31,872 | 36,659 | 134% |
| Beef | $'000 | 73,239 | 85,884 | 94,673 | 98,961 | 118,800 | 145,140 | 174,845 | 160,213 | 192,639 | 238,044 | 225% |
| | Tonne | 8,480 | 8,556 | 8,602 | 8,032 | 8,648 | 9,489 | 10,014 | 10,740 | 9,280 | 9,260 | 9% |
| Mutton | $'000 | 35,649 | 39,964 | 45,437 | 42,872 | 44,315 | 51,191 | 52,850 | 58,245 | 68,175 | 74,689 | 110% |
| | Tonne | 179,665 | 163,648 | 167,302 | 168,217 | 169,966 | 163,549 | 158,608 | 154,291 | 153,195 | 145,678 | −19% |
| Fish | $'000 | 649,713 | 632,634 | 684,848 | 746,687 | 705,219 | 694,063 | 746,151 | 691,856 | 766,624 | 800,019 | 23% |
| | Tonne | 390,424 | 391,683 | 399,268 | 458,588 | 434,851 | 456,036 | 462,767 | 486,184 | 489,638 | 497,069 | 27% |
| Vegetables | $'000 | 302,096 | 307,494 | 312,275 | 360,517 | 362,329 | 401,959 | 431,569 | 449,133 | 498,671 | 510,160 | 69% |
| | Tonne | 423,562 | 393,107 | 378,442 | 371,509 | 365,091 | 358,177 | 361,766 | 365,300 | 361,816 | 369,277 | −13% |
| Fruits | $'000 | 411,794 | 404,103 | 386,407 | 384,188 | 400,556 | 419,028 | 448,926 | 454,345 | 490,758 | 519,340 | 26% |
| | Tonne | 49,121 | 52,432 | 44,256 | 52,456 | 55,819 | 60,859 | 67,019 | 69,880 | 73,226 | 72,635 | 48% |
| Hen eggs | $'000 | 66,991 | 71,272 | 75,283 | 90,330 | 82,375 | 99,612 | 125,064 | 124,789 | 132,042 | 145,444 | 117% |

Table 7.3. Per capita consumption of food commodities in Singapore, 1997–2011.

| | Chicken (kg) | **Pork (kg)** | Beef (kg) | Duck (kg) | Mutton (kg) | Fish (kg) | Vegetables (kg) | Fruits (kg) | Eggs (pcs) |
|---|---|---|---|---|---|---|---|---|---|
| 1997 | 33.5 | **23.8** | 3.5 | 3.3 | 2.4 | 27.4 | 71.7 | 82.5 | 277.8 |
| 1998 | 31.4 | **22.7** | 2.9 | 2.9 | 2.5 | 25.3 | 75.8 | 80.3 | 272.2 |
| **1999** | **38.4** | **15.2*** | 3.7 | 3.7 | 2.4 | 25.3 | 79.4 | **88.6** | 279.8 |
| 2000 | 36.9 | **18** | 3.4 | 3.3 | 2.4 | 24.3 | 79.9 | 84.2 | 271.9 |
| 2001 | 34.7 | **17.3** | 3.3 | 2.8 | 2.3 | 23.3 | 82.3 | 81.8 | 277 |
| 2002 | 35.8 | **18.1** | 3.7 | 4 | 2 | 23.9 | 83.6 | 86.8 | 281.6 |
| 2003 | 38 | **20.1** | 4.7 | 3.8 | 2.1 | 23.4 | 84.8 | 83.7 | 307.5 |
| 2004 | 31.8 | **20.8** | 4.3 | 2.6 | 2 | 23 | 85.5 | 80.9 | 267.8 |
| 2005 | 32.3 | **19.3** | 3.7 | 3.5 | 1.8 | 21.2 | 84.1 | 79.3 | 285.3 |
| 2006 | 29.2 | **20.1** | 3.8 | 2.9 | 1.9 | 20.8 | 84.2 | 75.6 | 291.4 |
| 2007 | 33 | **20.6** | 4.3 | 3.2 | 2 | 19.5 | 85.7 | 72.3 | 301.6 |
| 2008 | 32 | **19** | 4 | 3 | 2 | 23 | 95 | 70 | 300 |
| 2009 | 31 | **19** | 4 | 3 | 2 | 23 | 94 | 69 | 300 |
| 2010 | 32 | **20** | 4 | 3 | 2 | 21 | 96 | 67 | 309 |
| 2011 | 33 | **19** | 5 | 3 | 2 | 21 | 95 | 66 | 308 |

*Source*: AVA, Singapore.
*Notes*: All calculations are based on total population. Total population comprises Singapore residents (citizens and permanent residents) and foreigners staying in Singapore for at least one year.
*Decrease in intake of pork due to Nipah Virus Scare.

period of data is required to ascertain this trend (Table 7.6, Figure 7.2). Nonetheless, the most striking fact presented by this data is Australia's dominance of the fresh/chilled pork market in Singapore in almost every product category (Table 7.5).

Most fresh/chilled pork arrives in Singapore in the form of carcasses and half-carcasses. These are purchased by food processors such as abattoirs. The non-carcasses are often sold to retailers and food services. The reverse is true for frozen pork, as the majority of the import are not in the form of carcasses.

*Singapore's chilled pork market*

The export of Australia fresh/chilled pork into Singapore is an important trade for both countries. Singapore is the key market and accounts for

Table 7.4. Singapore household expenditure survey 2007/08 (Tey, *et al.*, 2009).

| | Percentage of | |
|---|---|---|
| | Total food expenses | Meat expenses |
| Bread and Cereals | 20% | |
| Meat | 17% | |
| **Chilled pork** | | **38%** |
| **Frozen pork** | | **0%** |
| Chilled beef | | 20% |
| Frozen beef | | 0% |
| Chilled mutton | | 0% |
| Frozen mutton | | 0% |
| Chilled poultry | | 23% |
| Frozen poultry | | 8% |
| Other chilled or frozen meat | | — |
| Bacon, ham and sausages | | 8% |
| Other processed meat products | | 15% |
| Fish and Seafood | 17% | |
| Milk, Cheese and Eggs | 11% | |
| Oils and Fats | 3% | |
| Fruits | 11% | |
| Vegetables | 13% | |
| Sugar, Jam, Honey, Chocolate and Confectionery | 4% | |
| Food Products | 4% | |

more than 50% of Australia's total pork export (Figure 7.3). Singapore contributes to 46.4% of Australia's Moving Annual Total (MAT) June 2010 export volume market share, and 59% of MAT June 2010 export value share (APL). 13 establishments in Australia are approved by AVA for the export of chilled pork into Singapore.

While it is desirable to create multiple sources for the same food item, certain supplying markets would gain economy of scale through mass marketing or the setting up of integrated supply chains. Therefore, it is not surprising that while many countries are approved for pork export to Singapore, Australia dominates the market.

*Singapore's frozen pork market*

Since the ban on Malaysian pork was imposed in 1999 after the Nipah virus outbreak, the lower end food service market for pork shifted to using frozen products. The processed pork industry also increased their imports of frozen pork for production of food items such as barbecued pork, sausages, dumplings, etc. (Stanton, Emms and Sia, 2010).

Table 7.5. Australia's share of Singapore's pork imports by quantity (IE Singapore Statlink).[2]

| Category | Product | Source | 2009 Quantity | 2009 % | 2010 Quantity | 2010 % | 2011 Quantity | 2011 % |
|---|---|---|---|---|---|---|---|---|
| Live | Other live swine below 50kg (NMB) | World | | | | | 36.00 | |
| | Other live swine at least 50kg (NMB) | World | 284,884 | | 290,549 | | *See note* | |
| Frozen | Other meat of swine frozen (TNE) | World | 51,165 | 100.0 | 57,644 | 100.0 | 52,925 | 100.0 |
| | | Australia | 759 | 1.5 | 609 | 1.1 | 858 | 1.6 |
| | Other edible offal of swine frozen (TNE) | World | 5,914 | 100.0 | 6,884 | 100.0 | 6,509 | 100.0 |
| | | Australia | 370 | 6.3 | 398 | 5.8 | 466 | 7.2 |
| | Carcasses & half carcasses of swine frozen (TNE) | World | 723 | 100.0 | 1,195 | 100.0 | 1,605 | 100.0 |
| | | Australia | 4 | 0.6 | | — | | — |
| | Livers of swine frozen (TNE) | World | 1,324 | 100.0 | 1,270 | 100.0 | 1,319 | 100.0 |
| | | Australia | 106 | 8.0 | 217 | 17.1 | 319 | 24.2 |
| | Hams shoulders & cuts thereof of swine | World | 824 | 100.0 | 1,067 | 100.0 | 1,046 | 100.0 |
| | | Australia | 64 | 7.8 | 69 | 6.5 | 191 | 18.3 |
| Fresh/ Chilled | Carcasses & half carcasses of swine fresh or chilled (TNE) | World | 12,018 | 100.0 | 9,659 | 100.0 | 7,575 | 100.0 |
| | | Australia | 12,018 | **100.0** | 9,649 | **99.9** | 7,575 | **100.0** |
| | Other meat of swine fresh or chilled | World | 2,197 | 100.0 | 2,474 | 100.0 | 2,600 | 100.0 |
| | | Australia | 2,038 | **92.8** | 2,274 | **91.9** | 2,431 | **93.5** |
| | Edible offal of swine fresh or chilled (TNE) | World | 622 | 100.0 | 561 | 100.0 | 494 | 100.0 |
| | | Australia | 620 | **99.7** | 560 | **99.8** | 492 | **99.6** |
| | Hams shoulders & cuts thereof of swine | World | 317 | 100.0 | 322 | 100.0 | 244 | 100.0 |
| | | Australia | 229 | **72.2** | 297 | **92.2** | 243 | **99.6** |

[2] An enquiry made to AVA confirmed that Singapore was still importing live pigs from Pulau Bulan, Indonesia, in 2011 and 2012.

Table 7.6. Singapore's pork import by quantity (IE Singapore Statlink).

| | Prepared | Fresh/Chilled | Frozen |
|---|---|---|---|
| 2009 | 3,621 | 15,154 | 59,950 |
| 2010 | 3,722 | 13,016 | 68,060 |
| 2011 | 4,402 | 10,913 | 63,404 |

*Excluding approximately 290,000 live swine of at least 50 kg.

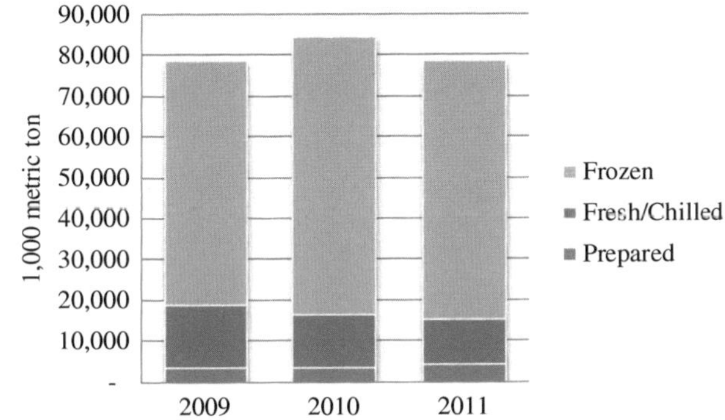


Fig. 7.2. Singapore pork imports (trends by quantity) (IE Singapore Statlink).

Consumers may prefer fresh pork but the frozen pork market is important for food security in Singapore. High quality fresh pork supply is limited around the region. Air freighted chilled pork is costly and needs careful management to avoid unsanitary conditions. To create a more diverse food source and encourage alternatives to chilled meat, AVA has been running a Frozen Meat campaign.

## *Greenhouse Gas Emissions from Pork Production*

Fiala (2008) noted the global increase in beef, chicken and pig consumption (Table 7.7). By 2030, pork has the potential to emit 586 million tonnes of greenhouse gases — half the emissions by beef (1,164 million tonnes), although beef emits nearly four times more greenhouse gases than pork.

The growing concern of climate change will also have "more or less severe impacts on all components of food security: food production and availability, stability of food supplies, access to food and food utilization." Climate change will affect all four dimensions of food security: food availability, food accessibility, food utilization and food systems stability.

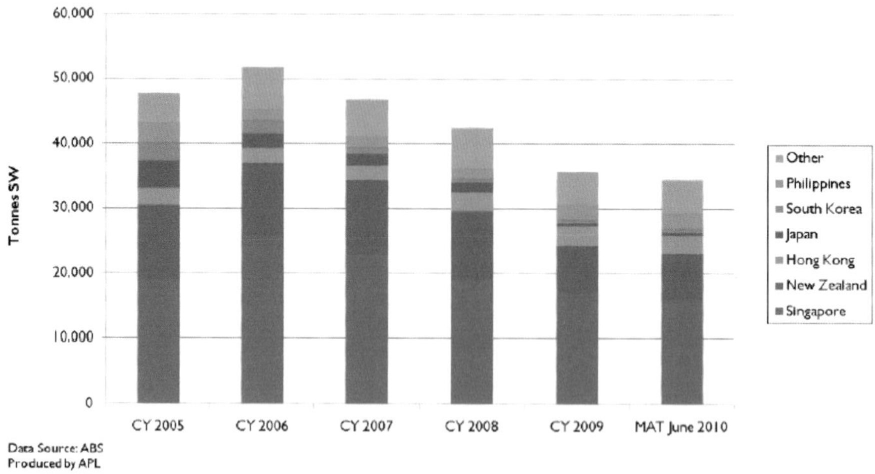


Fig. 7.3. Australian Farmed Pigmeat Exports by Country. Reproduced from: Submission to the Productivity Commission Inquiry into Bilateral and Regional Trade Agreements, 16 Sep 2010 (APL).

Table 7.7. Estimate of total greenhouse gases in million tonnes of $CO_2$ equivalent for each commodity. (Fiala, 2008).

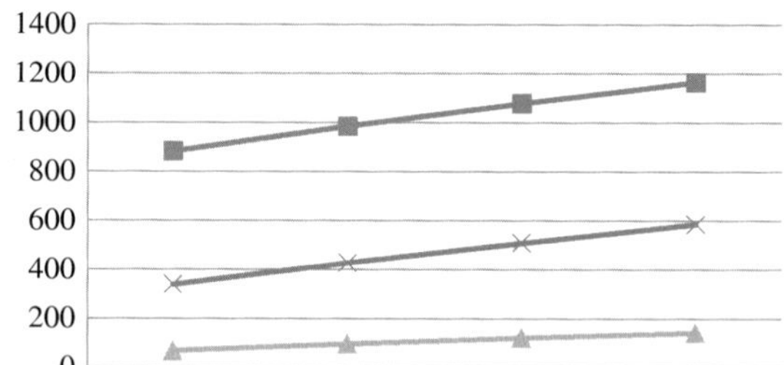


| | 2000 | 2010 | 2020 | 2030 |
|---|---|---|---|---|
| Beef | 882 | 984 | 1077 | 1164 |
| Chicken | 67 | 96 | 121 | 142 |
| Pig | 338 | 427 | 509 | 586 |

It will have an impact on human health, livelihood assets, food production and distribution channels, as well as changing purchasing power and market flows. Its impacts will be both short term, resulting from more frequent and more intense extreme weather events; and long term, caused by changing temperatures and precipitation patterns (FAO).

Adopting a higher emission rate per unit, a more recent review by Steinfeld (2011) has estimated that pork production contributes about 793

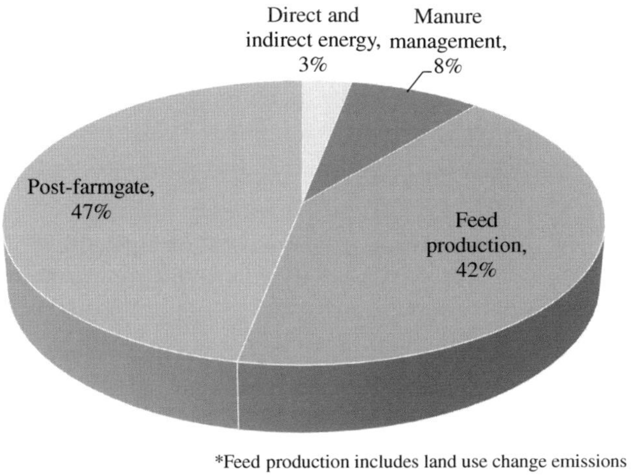


Fig. 7.4. Proportion of pork's contribution to greenhouse gas emissions. Reproduced from BPEX (2012).

million tonnes of $CO_2e$. This is 86% higher than Fiala (2008)'s projection for 2010 and makes up approximately 1.6% of global greenhouse gas emissions. Correspondingly, Steinfeld's new estimate gives a relatively higher impact of about 6.9 kg $CO_2e$/kg carcase weight as compared to Fiala's 3.8 kg $CO_2e$/kg (Table 7.8). Figure 7.4 showed the proportion of greenhouse gas emission contributed by different components of pork production according to Steinfeld (2011). While the exact percentage differs from other studies owing to the different locality, the main contributing components are fairly similar.

The pig industry seemed to have taken the environmental concerns seriously. The International Meat Secretariat prepared a report entitled "Pigs and the Environment" which highlights the global pig industry's effects on the environment (BPEX, 2012). The report addresses their responsibilities to mitigate these effects and provides details on the changes that some major producers have started to make. The industry saw these as important endeavours as they are coming under pressure from environmental restrictions, reduced public acceptance of swine units, and increasing lack of suitable farm land.

## 3. Three Phases of Singapore's Fresh Pork Supply

The changes undergone by the Singapore pork supply can be categorically described as three phases moving from local, to regional (ASEAN), then global.

### A) *Singapore Pork Industry: Natural Footprint*

Since its independence in 1965, the growth of the Singapore economy and its population has led to greater competition for land for various uses. As part of the nation building efforts, the late Mr Lee Kuan Yew (then Prime Minister) had the vision of transforming Singapore into a Garden City.

The numerous polluting farms in Singapore were obstacles to realising this vision. One of the initiatives implemented in Singapore towards becoming a Garden City was the 10 year clean-up programme of the Singapore River and Kallang Basin. The programme started in 1977 at a cost of S$200 million. These waterways were vital to the economic development of Singapore as the commercial hub with numerous financial institutions and commercial activities located herein. In the 1970s, the Singapore River and Kallang Basin were heavily polluted and both carried highly offensive odour; these hampered development plans for the nation (at that time) (Chou, 1998).

The pig farms were one of the main sources of pollution for the Singapore River and Kallang Basin. There were more than 600 pig farms located along the waterways. The Kallang River (which drains into the Kallang Basin) was particularly overwhelmed with pig wastes (Chou, 1998). In 1979, pork was the most popular meat consumed in Singapore. It made up 47% of the total 16,086 tonnes of meat consumed. The farms enabled Singapore to be self-sufficient in pork supply (Cheng, 1982).

Recognising that even small-scale localised pig farming can cause serious pollution, the government started to enlarge them into capital-intensive specialised farms that were restricted to certain designated localities (Cheng, 1982). This intensive production more than exceeded Singapore's needs for pork, producing 104% of the total needs (AVA). Nevertheless, self-sufficiency of pork was shelved aside for a cleaner environment. The government eventually decided that a highly polluting industry such as pig farming would have to be eradicated from the Singapore landscape altogether. This meant Singapore needed to explore new avenues of food security. The government imported 2,000 live pigs from Thailand towards the end of 1981 to see how the public would respond (AVA).

Subsequently in 1985, pig farming was totally banned in Singapore to devote the scarce land to higher economic activity and eliminate heavy costs associated with managing the waste (Gursansky and Peacock, 2000). From this point onwards, the environmental impacts from the pork supply to Singaporeans moved from its own land onto foreign soil.

Pork consumption continued to be high after the eradication of pig farming in Singapore. In 1989, Singaporeans consumed 1.1 million pigs. The main supply came from Malaysia, where 2,000 to 3,000 pigs were brought into Singapore daily by truck. The rest of the supply came from Indonesia by ship (Neo, 1992). This arrangement remained fairly stable until the deadly Nipah virus outbreak occurred in Malaysia in 1999.

### B) *Enter the Malaysian Pig Farming Industry*

During the 1990s, Singapore was entirely reliant on the region for its fresh pork supplies. Live pigs were transported for slaughter in Singapore abattoirs. Malaysia had about 85% of the market share at that time. Indonesia provided the remaining 15% of Singapore's pork (Neo, 1992). Thus, Singapore's pork environmental impact was specifically related to Malaysia and Indonesia.

Malaysia and Indonesia are both Muslim countries where the pork industries are facing problems due to strict government controls and religious pressure. Indonesia has the largest Muslim population of any country in the world and 61.3% of Malaysians are Muslim.[3] Pork is just one of many food and drinks that are prohibited in Islam. Nurdeng (2009) outlined that these prohibited items are 'injurious to one's physical health but also affect the character and spiritual faculties of man' and is one of the main causes that leads man to corrupt deeds (Nurdeng, 2009).

Near the time Singapore started importing live pigs from Malaysia, laws governing pig farms in Malaysia changed during the 1980s to 1990s. Four states in Malaysia (Johor, Terengganu, Negri Sembilan and Malacca) enacted local laws to stymie the growth of the pig industry (such as refusing licensing, rejecting appeals and arresting offenders of the law). The resulting socio-political tension took a toll on the Malaccan pig industry. The racial and religious divide compounded the problem. It was a far cry from the 1950s when the industry was flourishing under market-driven conditions. Back then, there was wide support, as pig farming was seen to spur rural activities. The authorities had even developed marketing strategies to benefit the pig farmers (Neo, 2009).

---

[3]Data from US Department of State. Background Note: Malaysia March 2, 2012. See http://www.state.gov/r/pa/ei/bgn/2777.htm

i) *Effects of the Nipah virus outbreak on Malaysia supply of fresh pork*

Ironically, the laws that pushed for the relocation of pig farms were a major factor for the cause of the Nipah virus outbreak in 1999. It brought on ecological impacts that led to major health issues. The Malaysia Nipah virus outbreak could be partially attributed to the habitat destruction caused by clearing of land to make way for plantations, and the continual slash-and-burn activities which forced animal species to come closer together and enabled the pathogens to spread. Abrupt orders to shut down and relocate pig farms without adequate time for careful planning made matters worse.

The outbreak was severe enough to reduce Malaysia's pork market share in Singapore to zero overnight with a permanent ban on all Malaysia pork import into Singapore that was in place until 2015. The Nipah virus outbreak affected 265 humans, including 105 deaths (Chua, 2003). One of the deaths was an abattoir worker in Singapore. It was this death that led Singapore to impose a full ban on all varieties of pork products from Malaysia. As Singapore was highly dependent on Malaysian pig supply prior to the disease outbreak, this pig related death had an immediate and decisive impact on Singapore's general food security and specifically to pig imports.

The Nipah virus outbreak was "pivotal to Singapore opening to sourcing of pork from more countries. If the disease did not hit Singapore, Australian pork will be niche rather than a dominant player [*sic.*]."[4]

ii) *Obstacles for the renewal of Singapore–Malaysia pork trade*

Singapore's ban on Malaysian pork products has been upheld for the last 14 years from 1999 to 2013. While deadly, after-effects of the Nipah virus outbreak could not be the only reason for this extended ban. For instance, Singaporeans are not issued with travel advisory cautions not to consume pork in Malaysia. Malaysia did not have any further Nipah virus outbreaks since 1999, and their population remains safe from consumption of local pork products. Malaysia also continues to be a major supplier of food to Singapore. As of 2007, Singapore was buying almost every other major food produce Malaysia has to offer, except pork (Cheong, 2007).

Due to the disease outbreak, unpleasantness in the past dealings and religious sensitivity, Singapore was motivated to source elsewhere for the

[4]Ms Quek H.M. (International Manager, Marketing, Singapore, APL) email interview dated 22 November 2010.

much desired pork. It is unlikely that in the 14 years that have passed, Malaysia has not been able to improve its pork industry standards to meet Singapore's requirements. It seems that both sides have an unspoken agreement to not restore the pork trade which would revive the Malaysian pig industry.

While the Malaysian authorities cited poor environmental conditions, in reality, it is the religious (Muslim) factor that has determined the federal policies against pig farmers. The Singapore ban on pigs from Malaysia has ironically helped the Malaysian authorities to marginalise and reduce pig farming. Along with the religious tensions, there are ethnic implications. The prosperity of the pig industry benefits mainly the Chinese population who owns these farms. Since 1970, Malaysia has a long-standing "bumiputera" policy introduced as the New Economic Policy. It is essentially "economic discrimination in favour of ethnic Malays and indigenous groups" (The Economist, 2008).

The Malaysian pig farmers have little clout to open pork sales to Singapore given that all pork exports received no encouragement from the government. Neo (2009) found that Malaysia's pig industry is subjected to instability due to the social infrastructure. His study confirmed that the Islamic and racial sentiments played a strong role in shaping the industry. When the pig industry started, it was widely encouraged as it was recognised as an economic venture that will improve the living standards of the people. However, the businesses were mainly run by the Chinese minority. As pork is considered "haram", the racially-dominant Malays have stayed away from the enterprise. As the pig industry expanded, the economic wealth was not shared by all, leaving the Malaysian Malay-Muslim community out. The increase visibility of pig farms quickly became a sensitive issue, given the predominant Malay population.

### C) *Australian Pork Breakthrough*

The Singapore ban on Malaysian pork supplies gave opportunity for Australian pork to enter the city state. In total, Australia exports 178 other agribusiness and food commodities to Singapore.[5] Table 7.8 shows some of the higher value items. Many of Australia's exports (beef, lamb, milk, cheese, etc.) also dominate the Singapore market. Singapore imported

---

[5]From data provided by Mrs Toh Guek Hong, Senior Business Development Manager, Food and Beverage from the Australian Trade Commission (email dated 25 Oct 2010).

Table 7.8. Overview of Singapore's food imports (2009).[6]

| Product | Total (S$'000) | Australia's share (S$'000) | Market share |
|---|---|---|---|
| Beef fresh chilled frozen | 147,219 | 65,169 | 44% (No. 1) |
| Lamb fresh chilled frozen | 57,533 | 49,754 | 86% (No. 1) |
| Pork fresh and chilled | 87,328 | 85,694 | 98% (No. 1) |
| Milk & cream | 701,530 | 175,422 | 25% (No. 2) |
| Butter | 84,929 | 21,370 | 25% (No. 2) |
| Cheese & curd | 78,069 | 29,018 | 37% (No. 1) |
| Vegetables fresh chilled etc. | 464,882 | 31,748 | 7% (No. 3) |
| Fruits & nuts fresh dried | 586,757 | 59,291 | 10% (No. 5) |
| **Grand Total (Food)*** | | | **S$ 8,083,413,000** |
| **Australia's Total*** | | | **S$ 867,668,000 (No. 3 supplier, 10.7%)** |

*The figure includes other major food categories.

S$8 billion worth of food products in 2009 and Australia is its third largest supplier, taking 10.7% of the total market share.

While Singapore makes up a small percentage (2–3%) of Australia's total export under this category, pork accounts for 8–12% of the Australian food items exported to Singapore. Singapore is a significant market for the Australian pork industry as it accounts for more than 50% of the total pork export (as seen in Figure 7.6 earlier). Before the Global Financial Crisis in 2008, this export was fetching A$82 million dollars a year. Table 7.9 below shows Singapore's share of Australia's total export for commodities under agribusiness and food.

The value of Australian pork export to Singapore has seen some decrease over the years. This could be owing to competition from other countries, or a result of foreign exchange rate fluctuations or lower consumption due to the 2008–2009 Global Financial Crisis. However, the overall agribusiness and food export from Australia to Singapore has been growing in terms of dollar value.

[6]Figures provided by Mrs G.H. Toh, Senior Business Development Manager, Food and Beverage from the Australian Trade Commission (email dated 21 Oct 2010). Source: IE Singapore.

[7]Figures provided by Mrs G.H. Toh, Senior Business Development Manager, Food and Beverage from the Australian Trade Commission (email dated 25 Oct 2010).

Table 7.9. Australia export statistics to Singapore, commodity: Agribusiness and food.[7]

| | All commodities exported under agribusiness & food (A$'000) | | Meat of swine (pork), fresh, chilled or frozen exported to Singapore: Value (A$'000) and percentage of total export to Singapore |
|---|---|---|---|
| | World total | Singapore total and as a percentage of world's total | |
| 2005–06 | 23,334,966 | 622,611 (2.67%) | 74,417 (11.95%) |
| 2006–07 | 22,787,954 | 643,060 (2.82%) | 82,295 (12.80%) |
| 2007–08 | 23,460,951 | 712,315 (3.04%) | 71,455 (10.03%) |
| 2008–09 | 27,883,712 | 759,749 (2.72%) | 70,459 (9.27%) |
| 2009–10 | 23,637,426 | 738,992 (3.13%) | 63,949 (8.65%) |

Although the ban on Malaysian live pigs import due to the Nipah virus outbreak opened the opportunity for Australian chilled pork to enter the market, Australian pork's success in Singapore has not been as easy as it seemed. The Australians had been planning the import protocol for two years before their pork could take off.[8] Australia had tried to establish a larger pork market in Singapore in the 1980s and 1990. The early attempts were not successful. Australian chilled pork was even banned in 1997 after the supply chain failed to meet the health requirements.[9] Nonetheless, this initial venture helped seal Australia's later success, as the Pig Research and Development Corporation (PRDC) and the Australian Quarantine and Inspection Service (AQIS) had worked with the Singapore Primary Production Department (PRD)[10] to begin to develop export protocols for Australian chilled pork much earlier on (Gursansky and Peacock, 2000).

In the mid-1990s, Singapore was importing A$200,000 worth of chilled pork a year from Australia. This has now increased 320 times to A$64 million in 2009–2010 (Table 7.11). By April 1999, the export protocols were fully established, which meant that the chilled pork trade could take off immediately (Gursansky and Peacock, 2000). The Australian pig industry worked with AQIS, Austrade and AVA in Singapore to begin distribution and building the brand of its product 'AIRPORK' — air flown fresh

[8]Mrs Toh G.H., Senior Business Development Manager, Food and Beverage, Australian Trade Commission (Singapore). Interview in person (19 Oct 2010).

[9]See 'Airpork: Pigs can fly' at http://www.ferret.com.au/n/Airpork-pigs-can-fly-n69 0345

[10]PRD was restructured into a statutory board and renamed the Agri-Food and Veterinary Authority (AVA), on 1 April 2000.

pork. [11] Since then, Australian pork dominated the Singapore chilled pork market.

### i) *Factors for Australia pork's success: local and foreign influences*

Numerous factors account for Australia's pig industry's eventual success, almost in the opposite imagery of the Malaysian pig industry. In 1990, the Australian pork industry enjoyed protection from AQIS's blanket quarantine on raw pork imports due to concerns over potential disease threats. This policy was subsequently revised and relaxed so that Australia could remain in the Cairns Free Trade Group within the World Trade Organisation (Ronan *et al.*, 2001).

The proportion of pork as part of total meat consumption in Australia went up from 11% to 19% from 1986–1996 (Brewis-Weston and Lewis, 1996). This increased domestic demand was very well met by the improvement in productivity. In 1994/95, Australia exported less than 1% of its total production (7,852 tonnes) but had to import 2,966 tonnes due to trade agreements it had with other nations.

In 1996, the Australian pork market was heading towards saturation. Australian researchers had analysed Singapore's potential as an export market and found that there was a rapid increase in Singapore's pork consumption as a result of higher personal income and increase in meat consumption as a whole. Australia could provide Singapore with the desired high quality lean cuts and could tap onto Singapore as a bridge into other Southeast Asian markets (Brewis-Weston and Lewis, 1996).

There were obstacles for entry into the Singapore market. At that time, Australian pork had higher production costs compared to other sources. Freight, marketing costs and inspection costs affected the competitiveness of the Australian pork in the Singapore market.

By 1998, Australia's domestic pork prices had collapsed due to low world prices and rising imports. At that time, the best efforts by leading pig exporters made little progress in Japan or elsewhere (Ronan *et al.*, 2001). A robust national framework for its food production system was setup. All pig farms in Australia were approved by AQIS to export their

[11] Letter by Kathleen Plowman, General Manager (Policy) of Australian pork addressed to Dr John Carter Secretary Foreign Affairs Sub-committee Joint Standing Committee on Foreign Affairs, Defence and Trade with reference to Inquiry into Australia's Relationship with ASEAN dated 19 January 2009. See http://aph.gov.au/house/committee/jfadt/asean1/subs/sub%2045.pdf

meat. The slaughtering would take place at AQIS-approved abattoirs prior to export.[12] The Australian pork industry is self-regulating and they work with AVA and AQIS on various issues. Some of this covers research activities to achieve standards that are higher than what AVA has stipulated.

The export protocols were fully established by April 1999. In less than five months, the rapid growth of Australia's market share of the chilled pork market in Singapore reached a stable peak in September 1999. Singaporeans have a strong preference for meat to be fresh or chilled but not frozen. North America and Europe were too far away for the chilled pork to still have an acceptable shelf-life. Australia's "clean and green image" and the high health status of the Australian herd were their greatest advantage over other pork exporting countries at this critical juncture when fear of disease featured highly.

ii) *Fulfilling consumers' food aspirations*

The main reason for the dismal trade volume of Australian pork in Singapore during the mid-1990s was due to the taste of (then) typical Australian pork. Singapore consumers thought that it tasted 'tainted', 'milky' or even 'beefy'. Therefore, after securing an entry opportunity into the Singapore fresh pork market in 1999, the Australian Trade Commission worked to ensure that the taste suited Singaporean consumers. Research showed that Singaporeans preferred gilts (young female pigs) and disliked the taste of non-castrated boars. The Australian farmers thus worked to modify their product offering to suit the local taste.

After resolving the taste issue, they worked to persuade the consumers that air flown pork would be as fresh as live ones slaughtered locally. At that time, the freshness of the product was in doubt as it was so geographically far — beyond Singapore's hinterland. The Confederation of Australian Pork Exporters (CAPE) had created the Australian Pork Industry Quality programme to improve the export standards to meet consumers demand. By being willing to service the demanding preferences of the Singaporean market, thereby fulfilling their food aspirations, Australia has maintained its dominance of the Singapore chilled pork market (Stanton, Emms, and Sia, 2010).

[12]Ms Quek H.M. (International Manager, Marketing, Singapore, APL) email interview dated 22 November 2010.

iii) *Well-established organizational infrastructure*

The Australian pig industry had great support from the government and could organise themselves to focus on setting up an export market. This was one area where the Malaysian pig industry had been crippled. In 2001, the Confederation of Australian Pork Exporters (CAPE) established a market and research entity named the Australian Pork Limited (APL). They invested A$3.5 million into research and strategizing to maintain the market lead in order to prevent any possibility of the Malaysian pig industry bouncing back after eradicating the virus (Stanton, Emms and Sia, 2010). APL was to oversee the entire industry and coordinate the national strategy. "The organisation aims to enhance opportunities for the sustainable growth of the Australian pork industry by delivering integrated marketing, innovation and policy services along the pork industry supply chain." APL functions both at domestic and international level.

To better serve its largest export market, APL has appointed one country-specific Marketing Manager — only in Singapore (Australian Pork). Ms H.M. Quek commented that "Singapore is a significant country of pork export for Australia with about 50% share of total exports."[13] APL has a multi-faceted role in Singapore that includes marketing management, product quality and research, trade relationship and government relationship.

iv) *Drawing on the 'clean and green' image*

Much could be lauded for the Australian pig industry's environmental efforts. Unlike the haphazard management of farming land in Malaysia, APL has worked out the *National Environmental Sustainability Strategy for the Pork Industry*. The strategy framework outlined seven major environmental sustainability challenges faced by the Australian pork industry. They include:

1. availability and use of land, water, energy and efficient transport;
2. security of feed grain supply for piggeries;
3. sustainable reuse of by-product nutrients to enhance soils and catchment health;
4. carbon abatement — effectively managing gaseous emissions;

[13]H.M. Quek (International Manager, Marketing, Singapore, APL) email interview dated 22 November 2010.

5. climate change — managing for the expected climate variability and drought;
6. environmental regulation; and
7. community interactions — amenity impacts and community perception of industry.

Born out of a pork health scare, Australia's stronghold on Singapore's fresh pork market owed greatly to Australia's reputation of being a clean and green producer (Gursansky and Peacock, 2000). All tourists to Australia have personally experienced the strict border control that has been put in place to prevent invasion of alien species and communicable diseases.

At the moment, Australian pork is marketed on its key attributes as being "From Australia, Healthier, More Nutritious, Fresh Chilled and Tastiest."[14] Besides producing for exports, Australians themselves have begun to demand products that are produced in an environmentally friendly way. They are also concerned about the use of chemicals in the food production.

Due to Australia's own consumer awareness and desire for environmental excellence, the Australian pork industry has also taken steps to improve its environmental performance. This is despite the lack of pressure from Singapore — its largest export market. When asked if the importance placed on environmental performance of Australian pork was a marketing strategy in Singapore, Ms H.M. Quek replied that "Environmental Management and the Welfare of Pigs, on their own merits, are currently not ranked high on the Singapore consumer's agenda in determining purchase. Similarly, the same can be said of the Singapore Government in permitting sources of supply to be approved for import into Singapore as the key criteria for approval of imports is food safety."

Ms H.M. Quek was of the opinion that the "wholesome quality for Australian pork which is unparalled by competition" was due to the Australian pork industry's embrace of pig welfare and environmental management in production. She sees Singaporean consumers as "relatively less concerned about (environmental and animal welfare) issues in discriminating their choice of purchase" because Singapore is not a country where "agriculture plays a big part in an economy" and "consumers of such countries are more aware of such issues."

[14]Ms H.M. Quek (International Manager, Marketing, Singapore, APL) email interview dated 22 November 2010.

The Australia pork industry believes that pork has the potential to be the most emission-friendly animal protein source of the future. They are aiming to reduce the current 3.1–5.5 kg $CO_2e$/kg hot standard carcass weight (HCSW) of pork to just 1 kg $CO_2e$/kg HCSW. So far, improved effluent treatment systems have driven this down to 2.3 kg $CO_2e$/kg HSCW. The Co-operative Research Centre for High Integrity Australian Pork aims to balance the economic, social and environmental needs of the Australian pork industry through innovative and collaborative research through the entire value supply chain (BPEX, 2012).

One of the key contributions from APL in their pursuit of environmental sustainability for all the piggeries has been its publication of a comprehensive National Environmental Guidelines for Piggeries (NEGP). The first edition was published in 2004. In 2010, they released a second edition.

Ms Quek commented that this is "one of the key documents APL now promotes." Both the 2004 and 2010 version "provides a general framework for managing the environmental issues associated with piggeries, and the guidelines have been tailored to industry circumstances and conditions most commonly encountered."[15] The NEGP is prepared with inputs from "all industry stakeholders, in particular, state government departments and environmental authorities, the research community and producers from all major pig-producing states."

The NEGP 2010 starts with the planning principles and desired environmental outcomes. The major contents are on site selection, cleaner production, effluent and solid waste collection and treatment systems, environmental risk assessment and management plans, and monitoring and assessment of sustainability. To be a useful guide for pig farmers, the NEGP was designed to a) be comprehensive, covering the major topics and contain the vital information; b) be applicable and easy to use even by farmers who did not have formal training on environmental management; c) address the environmental issues in a way that made sense to the pig farmers; and d) provide solutions for the social issues that are often most difficult to manage, such as complaints by neighbours on the odour and noise pollution coming from the pig farms.

v) *Strong trade relationship and close proximity*

Countries would opt to import goods from neighbours to avoid high transport costs and complications. Distance is particularly omnipresent in

[15]Ms Quek H.M. (International Manager, Marketing, Singapore, APL) email interview dated 22 November 2010.

food trade due to issues of perishability, greenhouse gas emission and food miles. Hence, food has traditionally been sourced from the hinterland. It is likely that trade between Singapore and Australia flourished due to a lack of similar quality products in the ASEAN region. The growing affluence of Singapore also means that desires for more premium products could be fulfilled. In the case of pork, Australia is also nearer to Singapore than other major pork exporting countries such as the EU, United States and Canada.

Bilateral trade agreement is a common tool used to drive trade for both sides. The Australia–Singapore trade story is no exception. Historically, having a long-term agreement significantly enhanced meat trade between countries (Koo *et al.*, 1994). Australian pork's success in Singapore has been underscored by the inking of the Singapore–Australia Free Trade Agreement (SAFTA), which came into force on 28 July 2003. This has paved the way for a stronger bilateral economic partnership. Australia sees Singapore as a 'natural first export market' for Australian products where Singapore acts as a gateway for Australia to enter the growing markets in ASEAN, India and China (Austrade).

## 4. Environmental Cost and Governance

### A) *Environmental Cost of Singapore's Pork Supply*

The analysis of the greenhouse gas emission will take a conservative approach, limited to three components — emissions per capita, production and transport. A single, most plausible and shortest route was chosen to calculate the transport distance. Only the Malaysian–Singapore and Australian–Singapore routes were analysed as they were the major supplier at each phase.

#### *The pork industry: greenhouse gas emission contributors*

Due to the higher volume consumed and weighty greenhouse gas emission at production, pork ranks high on the list of the worst climate change offenders among all meat in Singapore on a per capita basis. Following the data from Table 7.3 and 7.8, the amount of greenhouse gases emitted per capita in 2011 from the consumption of pork is 72.2 kg. This is just slightly lower than the 74 kg emitted by beef. However, beef emits 3.9 times more greenhouse gas than pork. Singaporeans consumed 14 kg more of pork than they did beef, and this drew the per capita comparison of greenhouse gas emission closer between the two products.

Table 7.10. Greenhouse gas emissions from pork produced for import into Singapore.

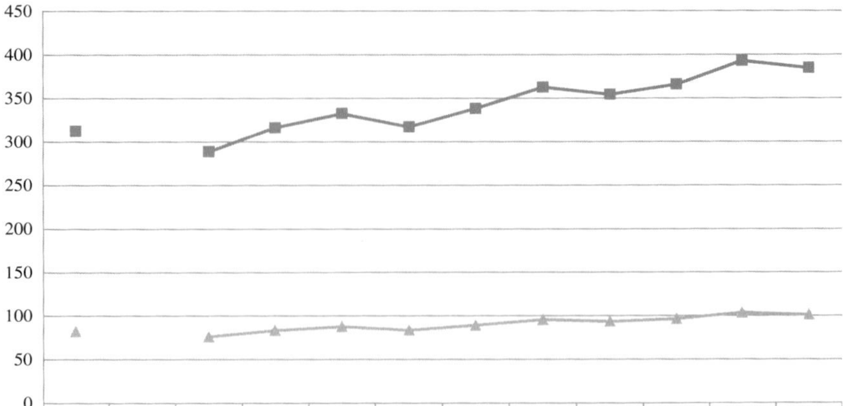


| | 1992 | | 2002 | 2003 | 2004 | 2005 | 2006 | 2007 | 2008 | 2009 | 2010 | 2011 |
|---|---|---|---|---|---|---|---|---|---|---|---|---|
| A | 312 | | 289 | 316 | 332 | 317 | 338 | 363 | 354 | 366 | 393 | 385 |
| B | 82 | | 76 | 83 | 87 | 83 | 89 | 95 | 93 | 96 | 103 | 101 |

A. = kg $CO_2$e based on 3.8 kg $CO_2$e/kg pork produced (Table 7.3).
B. = Tonnes imported (*Sources*: 1992 — Brewis-Weston and Lewis (1996); 2002–2011 — Table 7.7).

Along with the rise in consumption level of pork in Singapore, the greenhouse gas emitted from pork production has steadily been on the increase (Table 7.10).

The greenhouse gas emission from the extending hinterland predictably became greater as pork supply in Singapore underwent each of the three phases.

When the pigs were farmed locally in Singapore, the greenhouse gas emission was negligible. The transport of thousands of live pigs daily from Malaysia meant greater food miles and higher emission of greenhouse gases. In 1992, Singapore imported 61,669 tonnes of pork from Malaysia (Brewis-Weston and Lewis, 1996). Measuring conservatively, at 168g $CO_2$e of carbon emissions per tonne-kilometre, the freight of live pigs from Malaysia to Singapore using lorries would have amounted to 855,561 kg $CO_2$e.[16]

[16]Based on data from:
Dalgaard R, Halberg N and Hermansen JE 'Danish pork production. An environmental assessment', *DJF Animal Science*, n. 82 2007, see http://www.lcafood.dk/djfhus82ny.pdf
Distance between Singapore and Johor: 82.58 km. See http://www.distancefromto.net/between/Singapore/Johor

Air freight of pork from Australia has the greatest carbon emission. Each shipment of pork from Australia to Singapore would take six to eight hours.

— On average, an air cargo emits 500g of greenhouse gas per tonne-kilometre.[17]
— New South Wales has the highest number of pig farms in Australia.
— The international airport in Sydney, New South Wales is 6,305 km away from Singapore by flight.
— The greenhouse gas emission for one tonne of pork transported by air from Sydney to Singapore would amount to 3,152.50 kg.
— With the 10,741 tonnes of chilled pork exported by Australia to Singapore in 2011 (Table 7.5), this would have amounted to 33.86 million tonnes of greenhouse gas.

Air freight of chilled pork from Australia to Singapore alone has emitted more greenhouse gases than the generation of energy for entire countries such as Bahrain (30.69 million tonnes), Ecuador (24.43 million tonnes) and Croatia (23.43 million tonnes) in 2010.[18] The staggering food miles travelled by air freight pork is critical in understanding Singapore's pork environmental impact.

### B) *Singapore Government's Management of Food Security*

Importing over 90% of the food required from all over the world exposes Singapore to almost every single risk that is out there. Climate change concerns affect Singapore as a food importing country as much as it affects the producing country. The production levels are affected by changing climatic conditions, even for pork, as pigs need plant-based feed and a steady supply of fresh water.

Singapore's reach into Australia for air-flown pork is a serious climate change issue given the enormous amount of greenhouse gas emitted. Tougher international guidelines have surfaced regarding air transport such as EU's attempt to impose a cap on greenhouse gas emissions from flights going through EU airports. Singapore can ill afford to carry on business as

[17]Based on data from http://timeforchange.org/co2-emissions-shipping-goods
[18]See ranking from U.S. Energy Information Administration. *International Energy Statistics.* http://www.eia.gov/cfapps/ipdbproject/iedindex3.cfm?tid=90&pid=44&aid=8&cid=regions&syid=2008&eyid=2008&unit=MMTCD

usual as these international measures will affect food prices. Pork is almost a staple in the Singaporean Chinese diet. Escalating pork prices will stir up ill sentiments within the population.

Aside from food security concerns (availability, accessibility and use), the Singapore government is also the *de facto* guardian of food safety. Food importers and consumers take the cue from the government regarding the wholesomeness of the food items.

i) *Rigorous food import system and regulations*

The Agri-Food and Veterinary Authority of Singapore (AVA) is the primary body in charge of food matters in Singapore. AVA's vision is to have "safe food, healthy animals and plants for Singapore; trusted and respected globally." With AVA's sterling performance, Singapore has one of the lowest incidences of food-borne disease outbreaks in the world, despite importing over 90% of the food consumed. This is an impressive achievement given that about 3 million deaths occur globally out of the 1.5 billion cases of food-borne diseases that are reported (AVA).

AVA has a systematic way to safeguard food imports. The source country needs to first receive accreditation. This is followed by the individual establishments in that country. It is a practical way of ensuring that there can be sufficient checks and balances in the source country without Singapore spending excessive effort and cost. If the AVA-equivalent in the source country does not meet AVA's standards, much of the audits would fall on AVA who would have to factor in such expenses in their annual budget.

Licensing is also required before anyone can begin to import food. Importers have to comply with relevant legislation governed by AVA. For meat and fish, that would be the Wholesome Meat and Fish Act, Sale of Food Act and Food Regulations. "The Wholesome Meat and Fish Act is an Act to regulate the slaughtering of animals and the processing, packing, inspection, import, distribution, sale, transhipment and export of meat products and fish products and for matters connected therewith."

In evaluating whether or not a country can export its products to Singapore, AVA's considerations centre on the animal health and safety legislation and services that are in place. Individual establishments within the countries have to submit more details and be subjected to inspection visits before they can export to Singapore.

Singaporean importers can only import meat and meat-products from AVA-approved sources. Each consignment needs to be accompanied by a health certificate issued by a veterinary authority of the exporting country, certifying that they have complied with Singapore's safety requirements. An import permit or Cargo Clearance Permit is required regardless of the item or mode of transport. All consignments must be inspected and passed by AVA before they are sold, distributed or used. During inspection, importers are required to show relevant documents such as invoices or health certificates. Even after the goods (pork products) leave the port, AVA has the authority to issue food alerts and recall items that are subsequently found to be unsafe.

ii) *Strategies for food security: Diversification, substitution and securing at source*

Despite being a high income country, Singaporeans also feel the effects of high inflation on food prices. During the food crisis of 2007–2008, the prices of imported food increased by 12.1% between December 2006 and 2007 (Ramesh and Perry, 2008). One of Singapore's key strategies in ensuring food security has been to diversify its food sources. Local production of eggs, leafy vegetables and fish to higher levels of self-sufficiently has been promoted.[19] In attempting to diversify external sources, AVA aims to have no more than 25% of the supply of any item coming from a single source.

Other than diversification, Singapore has two other main strategies to ensure food security: 1) Promoting substitute products, e.g. frozen pork cuts instead of chilled pork meat; and 2) Securing food at source by investing in these foreign enterprises. These can take the form of contract farming or the creation of a designated 'food reduction and processing zone' situated in another country.

## 5. Conclusion

Singapore's expanding catchment of food supply has been sustainable due to its strong transport links and strong economic power. Although AVA does not specifically look into environmental performances of the pig farms in its source countries, its demand on high quality and safe food at an

[19]Speech by Dr Mohamad Maliki bin Osman, Senior Parliamentary Secretary for National Development and Defence at the opening ceremony of the International Conference on Asian Food Security (ICAFS 2011) on 10 August 2011 at the Grand Copthorne Waterfront Hotel, 10 Aug 2011, 09:35 AM.

affordable price could essentially only be provided by an environmentally-sound farm. An unclean farm with uncontrolled costs is unlikely to meet AVA's criteria.

The main challenge to Singapore's pork supply is diversity. Any disease outbreak among Australian pigs or disruption in air travel would destroy Singapore's supply. Currently, Singapore is still bringing in live pigs from Indonesia. AVA has also approved pork imports from 24 countries and would be able to call upon them to deliver the products should the markets in Australia be disrupted. This is evidence of Singapore's food security spreading its liabilities, though the viability remains to be tested. It is not known how Singaporeans will take to the taste of these alternative imports and whether the prices would be acceptable.

Ideologically, it is wise to restrict a supplying country's dominance to less than 25% as per AVA's goal, but this means that Singapore would be importing more of the same product from the second-best source. Inspection cost would go up as they would have to audit more production systems. Importers would also sacrifice any potential economy of scale from engaging a big supplying nation. It is also not feasible for AVA to meddle with market forces to ensure that each supplying country does not gain more than 25% market share.

Even without an economic downturn, the pork supply to Singapore may be under threat as other countries in the region prosper. Countries such as China, Vietnam and Philippines will turn towards importing pork when they undergo the same transition as Singapore. Australia has always seen Singapore as a gateway to these markets. With the far larger populations in these neighbouring countries, their market would be more attractive to the Australian farmers. At the very least, this will drive up the cost of pork coming into Singapore.

Tougher environmental conditions and controls will push food prices up. Global concerns over climate change and the high environmental cost of air freight will also affect the pricing of air cargo. The emergence of disease outbreak affecting livestock will require Singapore to be increasingly vigilant and incur expenses to inspect imports. Climate change will also affect the yield in the producing countries. Singapore cannot escape from the repercussions of this phenomenon. Droughts will affect crops, which will affect the amount of feed available for the pigs. Drier conditions in Australia will also subject the pigs to diseases and/or raise the cost of production.

To overcome these challenges, Singapore can move away from just being the consumer to be an important player in the food supply chain. One solution is to become the regional food processing and distribution hub. By maximising its potential as the geographical hub within the region, Singapore can serve to secure the region's food security and thus benefit itself.

Singapore needs to consider further diversification of its food sources. Closer ties with other pork supplying nations should be developed even if the supply from Australia is in a steady state. The frozen pork campaign is a step in the right direction and should be further promoted. Frozen pork can be brought in through shipping. Shipping is cheaper and produces less greenhouse gas emission and thus, Singapore would reduce its carbon footprint while making pork available to the population at a lower cost.

The impacts of climate change on Singapore's food security have been largely ignored. The stringent health and safety requirements for imported food indirectly require producing countries to be more vigilant in managing their farms. However, the post-production environmental harm needs to be addressed by Singapore as the importing country. This is not only because Singapore should strive towards practising better environmental stewardship, but also because international legislation and guidelines will affect the affordability of the products to Singaporeans. Creating greater awareness amongst Singaporean of the food miles for each product through initiatives such as carbon labelling, is a positive move that can be taken. This is important as internationally levied carbon taxes on greenhouse gas emission due to air freight may be introduced and passed down to consumers. Providing such information at this stage will ease consumers into considering food items that are less harmful to the environment.

## *Post-Study Update*

With the intention to diversify the source of pork, AVA has recently audited and approved one Malaysian establishment in Sarawak, East Malaysia, for the import of frozen pork (Boh, 2015). From February to May 2015, nine tonnes of frozen pork have been imported into Singapore from this Malaysian establishment. Although the figures are low for a start, further imports of frozen pork from Malaysia is highly beneficial to Singapore's carbon footprint for pork consumption on two fronts — the distance travelled is very short, and frozen pork does not require air freight.

## References

Agri-Food and Veterinary Authority of Singapore. See http://www.ava.gov.sg/
Austrade. See http://www.austrade.gov.au/Singapore-profile/default.aspx
Australian Pork. See http://www.australianpork.com.au/

Best, P. (2012). Global pig production: Trends shaping the pork industry (13 Jun 12) See http://www.wattagnet.com/Global_pig_production__Trends_shaping_the_pork_industry.html

Boh, S. (2015). "First batch of frozen pork from Malaysia since 1999", *The Straits Times*, 18 May 15. See http://www.straitstimes.com/news/singapore/more-singapore-stories/story/long-ban-pork-malaysia-lifted-ava-20150518

Brewis-Weston, S.G., Lewis, P.E.T. (1996). "The Singapore Pork Market: a Preliminary Assessment of the Potential for Australian exports", *Review of Marketing and Agricultural Economics*, **64**(1), 101–110.

Cheng, L.K. (1982). "Fresh Food Supplies in Singapore", *Geojournal Supplementary Issue*, **4**, 61–72.

Cheong, L. (2007). "Food supply for Singapore; Status and Challenges", Presentation delivered at GTA Seminar 2007, 31 March, in Singapore. See www.gta.org.sg/resources/Seminar2007/lesliecheong.pdf

Chou, L.M. (1998). "The cleaning of Singapore River and the Kallang Basin: approaches, methods, investments and benefits", *Ocean & Coastal Management*, **38**, 133–145.

Chua, K.B. (2003). "Nipah virus outbreak in Malaysia", *Journal of Clinical Virology*, **26**, 265–275.

Department of Statistics Singapore (2012). Population Trends 2011. See http://www.singstat.gov.sg/pubn/popn/population2011.pdf, p.3

FAO (2008). Climate change and food security: a framework document. See http://www.fao.org/forestry/15538-079b31d45081fe9c3dbc6ff34de4807e4.pdf

Fiala, N. (2008). "Meeting the demand: An estimation of potential future greenhouse gas emissions from meat production", *Ecological Economics*, pp. 412–419.

Fraser, C. (2000). "Bunge to Sell Stake in Indonesian Pork Operations", *Meat and Poultry Online*. See http://www.meatandpoultryonline.com/doc.mvc/Bunge-to-Sell-Stake-in-Indonesian-Pork-Operat-0001

Gursansky, B.C.G., Peacock, A. J. (2000). "Meeting the Needs of a New Market: Australian Chilled Pork to Singapore Australasian", *Journal of Animal Sciences*, **13** Supplement, 86–95.

IE Singapore Statlink. See https://statlink.iesingapore.gov.sg/

Koo, W.W., Karemera, D., Taylor, R. (1994). "A gravity model analysis of meat trade policies", *Agricultural Economics*, **10**, 81–88.

Neo, B.S. (1992). "The implementation of an electronic market for pig trading in Singapore", *Journal of Strategic Information Systems*, **1**(5), 278–288.

Neo, H. (2009). "Institutions, cultural politics and the destabilizing Malaysian pig industry", *Geoforum*, **40**, 260–268.

Nurdeng, D. (2009). "Lawful and unlawful foods in Islamic law focus on Islamic medical and ethical aspects", *International Food Research Journal*, **16**, 469–478.

Ramesh, S., Perry. M. (2008). "Singapore's Food Inflation Remains Low by International Standards", *Channel News Asia*, 3 Feb 2008. See http://www.channelnewsasia.com/stories/singaporelocalnews/view/326673/1/.html

Ramesh, S. (2009). "H1N1 detected in live pigs imported into Singapore", *Channel News Asia*, 03 September 2009. See http://www.channelnewsasia.com/stories/singaporelocalnews/view/1002648/1/.html

Ronan, G., Langberg, J., Moore, M. (2001). "Evaluating the Export Growth Strategy of the Australian Pork Industry", Paper presented to the 45th Annual Conference of the Australian Agricultural and Resource Economics Society, Adelaide, South Australia, 22–25th January, 2001.

Stanton, Emms, Sia (2010). *Competitive Industry Report on the Singapore Market for Imported Meat and Poultry. A Guide for Canadian Exporters* (*Abbreviated). Prepared for: The High Commission of Canada in Singapore & Office of Southeast Asia Regional Agri-Food Trade Commissioner Agriculture and Agri-Food Canada. See http://www.ats.agr.gc.ca/ase/5691-eng.htm

Steinfeld, H. (2011). "Food supply and pork production — a critical review", FAO, World Pork Conference, Bonn — 7th October 2011.

Teng, P., Escaler, M. (2010). "The Case for Urban Food Security: A Singapore Perspective", *NTS Perspectives RSIS Centre for Non-Traditional Security (NTS) Studies*, **2010**(4), 9.

Tey ,Y.S., Suryani, D., Emmy, F.A., Illisriyani, I. (2009). "Food consumption and expenditures in Singapore: implications to Malaysia's agricultural exports", *International Food Research Journal*, **16**, 119–126.

(2008). "Race restrictions in Malaysia", *The Economist*, 24 Apr 2008. See http://www.economist.com/node/13562131

USDA Livestock and Poultry (2012). World Market and Trade Circular Archives, Apr 2012. See http://www.fas.usda.gov/dlp/circular/2012/livestock_0412.pdf

# Part 3

# Green Business

# 8

# SUSTAINABLE & RESPONSIBLE: THE NEW FACE OF INDIAN BUSINESSES

HIMADRI MAHAJAN and AUDREY CHIA

With the world's population expected to reach 9 billion by 2050, the ever mounting challenges, ranging from social inequity to ecological degradation, are becoming more and more apparent, and so does the necessity to grow, create employment, and alleviate poverty. As these challenges continue to mount, economies like India are struggling to sustain their developmental realisations. For a developing economy like India, there is need for a transformational approach, which can re-define the growth trajectory and propel the nation towards sustainable and inclusive growth. Appreciating the opportunity to serve the nation's objective, the Indian business sector is at the vanguard of making this paradigm shift. Leadership companies are looking at the integration of the "*Triple Bottom Line*" and creating sustainable value as their "*corporate responsibility*". Leveraging on their enterprise potentials, namely creativity, innovation, financial and human resources, Indian businesses are facilitating societal change effectively. Ample evidence indicate that Corporate India is adopting a holistic approach towards converting risks into business opportunities by being sustainable and responsible to the core.

## Introduction: The Challenges of Development

It is increasingly being realised that global economic models have failed miserably to pursue the growth objectives and deliver a secure and sustainable future. There is no doubt that the global economic models have created significant material wealth; however, this has happened with a deferred social and environmental cost. The quest for higher and faster economic growth has created tremendous social inequalities and apathy towards natural capital.

The challenges — like climate change, loss of biodiversity, unprecedented rate of natural resource depletion, stern water stress, and food and energy security — are overwhelming societies, governments and businesses and forcing the world to confront unforeseen threats.

A fundamental transformation is required to pull 50% of the global population out of poverty and ensure a decent living for nine billion individuals within the ecological capacity of this planet. The transformational approach is required to redefine the growth trajectory which would propel nations towards low carbon, green economies and creating sustainable and inclusive societies.

## What Actually Went Wrong?

In the past, available resources to generate financial capital, which definitely improved human well-being, created employment and enabled the economy to cope with recessionary trends, thus, becoming the "*preferred yardstick*" of development. Conversely, Gross Domestic Product (GDP)[1] does not capture many aspects of human well-being; for instance quality of life, health, education, quality and quantity of natural resources as depicted in Figure 8.1.

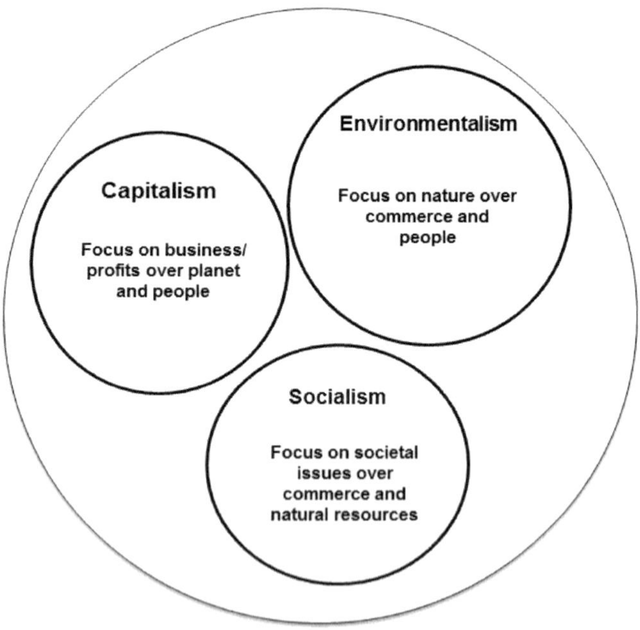


Fig. 8.1. Beyond the extremes. Adapted from the Reading Material of Business Leaders Programme, 2011 conducted by Prof Stuart L Hart, Cornell University, and organized by CII-ITC Centre of Excellence for Sustainable Development (CII-ITC CESD), India.

[1]Gross Domestic Product, "GDP is the market value of all officially recognized final goods and services produced within a country in a given period of time. GDP per capita is often considered an indicator of a country's standard of living". See http://en.wikipedia.org/wiki/Gross_domestic_product, last accessed on 9 January 2013.

The natural resources are primarily "public goods", which have no markets and no prices, and therefore, are rarely detected by our economic compass. The environmental assets are economic assets, whether or not they enter the market place, however, the conventional measures of national economic wealth and performance fail to reflect the "natural capital" stocks, contributing to their economic invisibility and misuse. Failure to incorporate the value of natural capital stock into economic decision making has led to the perpetuation of investments and activities causing inefficient use and a continuous degradation of natural capital, which is the foundation of our economies. Similarly, social capital is also a matter of concern as poverty, quality of life, education and health are inseparable societal issues.

## Sustainable Development as a Driver of Change

Challenging times are provoking governments and businesses to factor ecological and social aspects into growth targets, re-draw plans, re-work strategies and solutions to remain afloat. The Economics of Ecosystems and Biodiversity (TEEB) suggests that with a '***business-as-usual***' scenario, the world would have to face severe consequences and costs of inaction by 2050.[2] Unless Governments and businesses become sustainable to the core and adopt low-carbon growth policies, they shall cease to remain competitive.

Fortunately, there is a growing sense that the mounting sustainability crises are interconnected,[3] and this calls for the integration of all the three capitals/pillars to create an equitable society with positive environmental footprint (Figure 8.2).

In the wake of the social and ecological crises, the expectations which a society has of its government and businesses have increased many fold. The prevailing financial and ecological crises have already diminished the world's resilience, while increasing tensions across borders and causing social distress. This suggests that no stakeholder is able to cope with these global challenges alone.[4]

---

[2]TEEB (2010), "The Economics of Ecosystems and Biodiversity: Mainstreaming the Economics of Nature: A synthesis of the approach, conclusions and recommendations of TEEB" accessed on 11 December 2012.
[3]CII-ITC Centre of Excellence for Sustainable Development (2009), 4th Sustainability Summit: Asia 2009, Theme Paper.
[4]World Economic Forum (2011), "Redefining the Future of Growth: The New Sustainability Champions" by World Economic Forum in collaboration with The Boston Consulting Group, 2011.

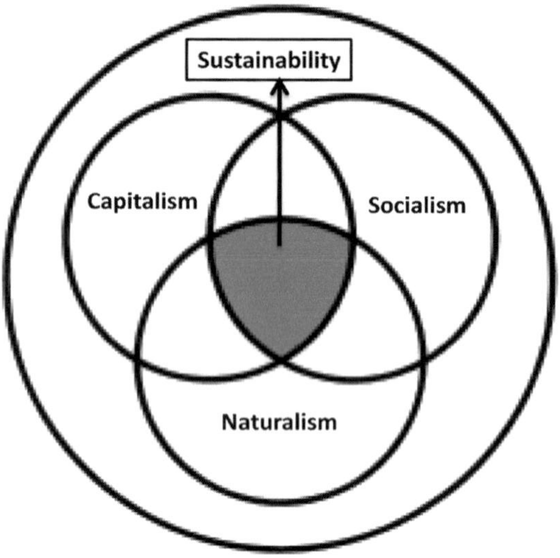


Fig. 8.2. Philosophical integration. Adapted from the Reading Material of Business Leaders Programme, 2011 conducted by Prof Stuart L Hart, Cornell University, and organized by CII-ITC Centre of Excellence for Sustainable Development (CII-ITC CESD), India.

Since no stakeholder is able to cope with these global challenges alone,[5] cooperation on an unprecedented scale between governments, business and society is needed, so that there is an enhanced understanding of their respective rights and responsibilities. While the Governments have the responsibility to formalise and formulate the policy regime, the role of businesses is significant in transforming the policies into actions by integrating sustainability into core business strategies and decision making.

## Role of Businesses in Sustainable Development

Businesses are the vital economic organ of society and are uniquely placed to play a constructive role in creating sustainable value for its stakeholders by leveraging on their creativity, innovation, financial and managerial resources.

[5]World Economic Forum (2011), "Redefining the Future of Growth: The New Sustainability Champions" by World Economic Forum in collaboration with The Boston Consulting Group, 2011.

### *The Concept of Triple Bottom Line*[6]

In the organisational context, the concept of sustainable development is best defined in terms of the integration of the economic, social and environmental bottom lines, commonly termed the ***Triple Bottom Line***[7] (TBL) approach (Figure 8.3), which means businesses are to be responsible for their impacts on people and the planet, beyond just making profits.[8] The mainstreaming of TBL as a way of conducting business is often described as "*Corporate Social Responsibility*" (henceforth CSR).

The economic bottom line includes the manufactured and financial capital, which involves wealth creation, increasing shareholder value, and profitability. Social capital is reflected in the social bottom line, involving a rise in quality of life for all strata of society with the help of economic growth. The environmental bottom line concerns natural capital and the challenges in conserving the natural resources.

Historically, businesses have merely focused on increasing their economic capital and enhancing shareholder value, taking the economic bottom line as a key business driver. From the business standpoint, it is imperative to have economic viability, without which the business cannot continue

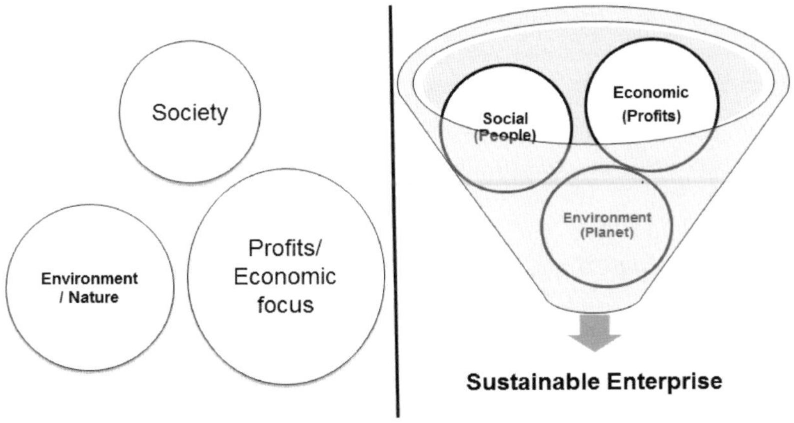


Fig. 8.3. Triple Bottom Line Concept.

[6]The term was coined by John Elkington in 1994.
[7]Brazilian Institute of Corporate Governance (IGBC), "Sustainability Guide for Companies: An Overview for Directors and Senior Executives", Corporate Governance Handbooks, 2009.
[8]CATALYST Consortium (July 2002), "What is Corporate Social Responsibility? — 8 Questions and Answers", USAID, Washington, 2002.

to exist. The prevailing global crisis prioritises alignment with the vision of sustainability, requiring a paradigm shift in the way businesses are conducted.

Companies can actually make meaningful contributions, if they future-proof against risks, recognise opportunities, and collaborate to innovate and facilitate development for all. This requires long-term strategic focus and solutions to convert these threats into opportunity, in order to create sustainable value.

Integrating sustainability helps businesses to be competitive, attracts shareholders, increases resilience and stability towards newer shocks, and enhances stakeholder value. The business case for sustainability also provides the next step in the business curve by embracing new forms of entrepreneurship. Companies that transform to embrace sustainability aspects to create sustainable value have a strong potential of creating both short-term and long-term value, improving benchmarking in the highly competitive business world, and increasing their resilience towards future risks and challenges.

There are various aspects of the TBL approach which have been seen as success factors within the market system. A few of these success factors or performance aspects are now discussed (Figure 8.4).

### *Eco-efficiency*

Eco-efficiency is considered as the basic contribution of businesses to sustainable development, and forms an integral part of corporate social and environmental responsibility. It improves the production process, while reducing resource use and waste generation with more value and lesser impacts, providing numerous business avenues. Various strategies have been used by businesses and governments to reduce the environmental impacts. A few strategies include:

- Pollution Prevention — focuses on pollution reduction and its environmental impact through reduction in material used at source and eliminating the waste within the process, but does not include generation of waste. Firms are developing ways to use practices like de-materialisation, material substitution to reduce environmental risks and impacts.[9]

[9] "Canadian Centre for Pollution Prevention", http://www.c2p2online.com/, last accessed on 4 January 2013.

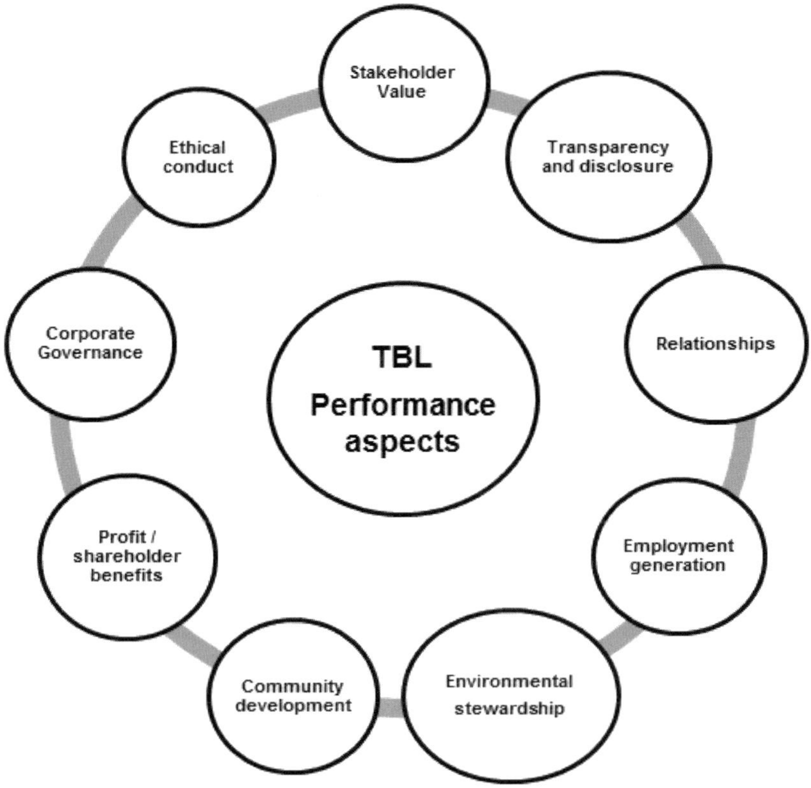


Fig. 8.4. TBL Performance Aspects. Adapted from Sustainability Management: Improve Brand & Profitability via TBL & CSR by Anand Subramaniam, June 2009.

- Waste Minimisation — USEPA defines it as "reduction to the extent feasible, of hazardous waste that is generated or subsequently treated, sorted, or disposed of".[10] Source reduction is the process of reducing the contaminants at the source, which would otherwise have entered the waste stream or been released into the environment without treatment. For example, General Motor's new model of involving a single supplier resulted in reduction of chemical waste, as compared to the earlier strategy of focusing on a group of suppliers, which used to reduce the profit ratio due to resource wastage. This modification in approach

[10]UNEPA official website, "Energy Recovery", see: http://www.epa.gov/epawaste/hazard/wastemin/index.htm (Accessed on 2 December 2014).

also helped in reducing the consumption of resources and resulting degradation of environment.

- Industrial ecology — relates to industrial systems designed using natural systems as a model for sustainability, and that emulate the way materials are recycled in an efficient and sustainable manner, to manage material and energy flow in production. The key objective is the transformation of the linear nature of the industrial system into a cyclical closed loop, wherein waste from one industry can be utilised as raw material for another. It minimises the impacts and maximises benefits by material recycling and reuse, resource use, and embedded energy optimisation.

  For instance, the eco-industrial park in Kalundborg, 100 km west of Copenhagen in Denmark, was an attempt at integrating a system which minimised resource use and waste generation. It had not resulted from a purposed plan, but had evolved as a result of mutual co-operation which developed between four neighbouring industries and the Kalundborg municipality. The partnerships developed are illustrated in Figure 8.5.

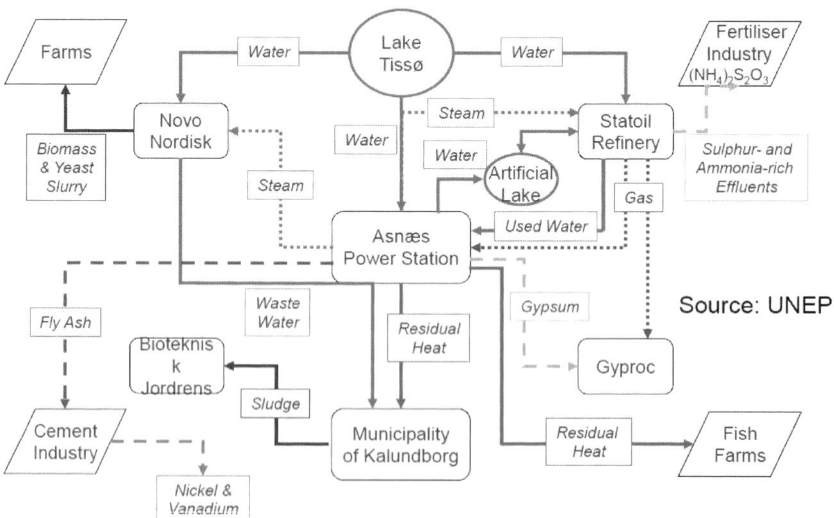


Fig. 8.5. Industrial Symbiosis of Kalundborg (Courtesy of Dr Kua Harn-Wei, National University of Singapore).

Source: C.G. Francis, adapted from Christensen (1999).

Thus, industrial ecology is occasionally referred to as the "Science of Sustainability".[11] Simply put, industrial ecology promotes broad scientific solutions to the sustainability issues pertaining to businesses. There are various system tools like Life Cycle Assessment and Carbon Footprinting, and strategies like Life Cycle Design (LCD) and Design for Environment (DfE).

Other aspects are:

*Stakeholder engagement*

The increased concern over corporate environmental and social responsibility has brought new perspectives of interaction between businesses and society. It is broadly recognised that the solutions to global problems cannot be found by a single stakeholder alone, therefore, there is a need to involve more and more stakeholders to devise sustainable solutions.[12] As an outcome, today's businesses recognise that stakeholder engagement is a crucial ingredient for sustainable business and social responsiveness.

Stakeholder engagement involves a range of benefits for businesses, including the "*social license to operate*", as it represents them as being socially responsible and builds up a trust level.[13] It also helps in managing the reputational risks and emphasises on transparency and reporting. Consideration of stakeholders' concerns and feedback proves valuable in assessing the outcomes of a particular business strategy. This may help in improved decision making and promoting learning and innovation.

Some standards and frameworks such as AA1000 Stakeholder Engagement Standard and GRI Reporting Guidelines provide direction on how to integrate the principles of materiality, stakeholder inclusiveness, and completeness into management practices.[14]

International Finance Corporation emphasises that the stakeholder engagement process is required to be iterative; which may involve planning, being specific in content, feedback, documentation, and disclosure. The

[11]http://industrialecology.blogspot.com/

[12]Confederation of India Industry (CII) and Indian Institute of Corporate Affairs, Ministry of Corporate Affairs (2010), "Stakeholder Engagement Handbook" by, Government of India.

[13]Confederation of India Industry (CII) and Indian Institute of Corporate Affairs, Ministry of Corporate Affairs (2010), "Stakeholder Engagement Handbook" by, Government of India.

[14]GRI Sustainability Reporting Guidelines.

Fig. 8.6. Stakeholder Engagement Process. Adapted from International Finance Corporation's Stakeholder Engagement: A Good Practice Handbook for Companies Doing Business in Emerging Markets.

stages recognised in the process (Figure 8.6) are: (1) Stakeholder Mapping; (2) Definition of objectives; (3) Identification of issues; (4) Stakeholder Prioritisation; (5) Assessment, maintenance and strengthening of capacity required; (6) Engagement Process Designing, and (7) Action, monitoring and reviewing and reporting back.

Companies are going beyond engagement, to the level which involves partnering with stakeholders to leverage on each other's strengths, skills and endeavours, experiences and learning.

### *Transparency, disclosure and sustainability reporting*

Companies are facing unprecedented pressure from their stakeholders to be responsible, accountable and transparent on their values, strategy and performance on aspects of sustainable development (Figure 8.7).

The financial markets are looking forward to a more transparent communication, depicting not only the financial but also the social and environmental performance of organisations. It requires reporting on intangible assets such as natural capital, stakeholders' partnership, human and intellectual capital, reputation, etc.

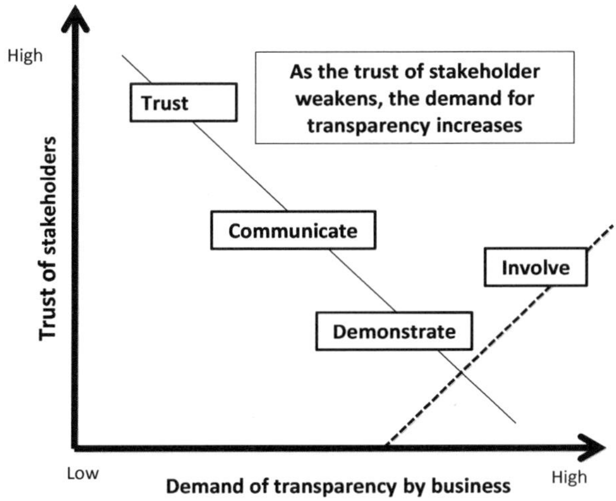


Fig. 8.7. Stakeholder demand of transparency. Adapted from "CII-ITC Sustainability Awards: Assessors Training Programme" Manual, 2009.

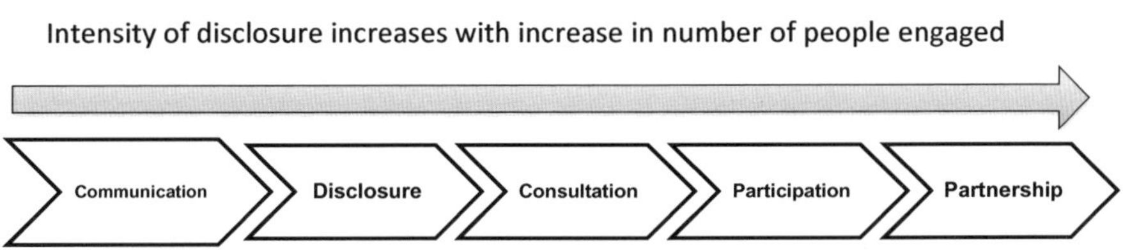


Fig. 8.8. Stakeholder Participation and extent of disclosure. Adapted from IFC's Stakeholder Engagement: A Good Practice Handbook for Companies Doing Business in Emerging Markets.

The extent of disclosure depends upon the level of stakeholder engagement as depicted in Fig. 8.8. Considering this, businesses have responded by publishing Sustainability Reports in addition to their financial disclosures. Sustainability reports play a key role in depicting a wide range of sustainability issues and business activities. Good governance, accountability, financial markets, innovation and learning are observed as drivers for sustainability reporting.

Reporting is an excellent approach of being transparent and communicating corporate sustainability values, practices and performance. This helps in building trust between the company and various stakeholders.

Reporting reflects that the company is striving to achieve their sustainability goals and meeting their stakeholders' expectations. This helps companies to recognise new alliances and dialogues that drive shareholder value.

There is no set framework of sustainability reporting as "one-size-fits-all" does not work here. Various frameworks, codes of conduct and guidelines such as the UN Global Compact, OECD Guidelines, GRI Sustainability Reporting Guidelines, etc. are performing as drivers of corporate disclosure and transparency. A few disclosure parameters by GRI have been illustrated in Figures 8.9 and 8.10.

Standards like AA1000 Assurance Standard (AA1000AS) are used to strengthen the reliability of the organisation's sustainability report, and is used by external auditing bodies to assure the same.[15]

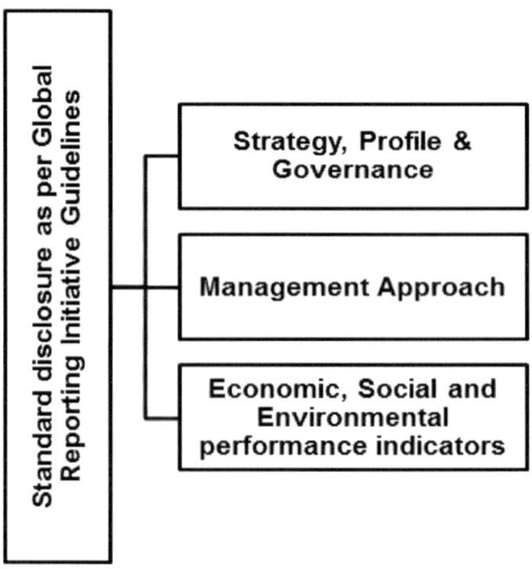


Fig. 8.9. Standard Disclosure as per GRI Reporting Guidelines. Adapted from the GRI G3.1 Reporting Guidelines.

### *Business Case for Sustainability*

Businesses around the world have started prioritizing and focusing on societal concerns as part of their business responsibility (Table 8.1). This

[15]http://www.proveandimprove.org/tools/aa1000AS.php; last accessed on 7 January 2013.

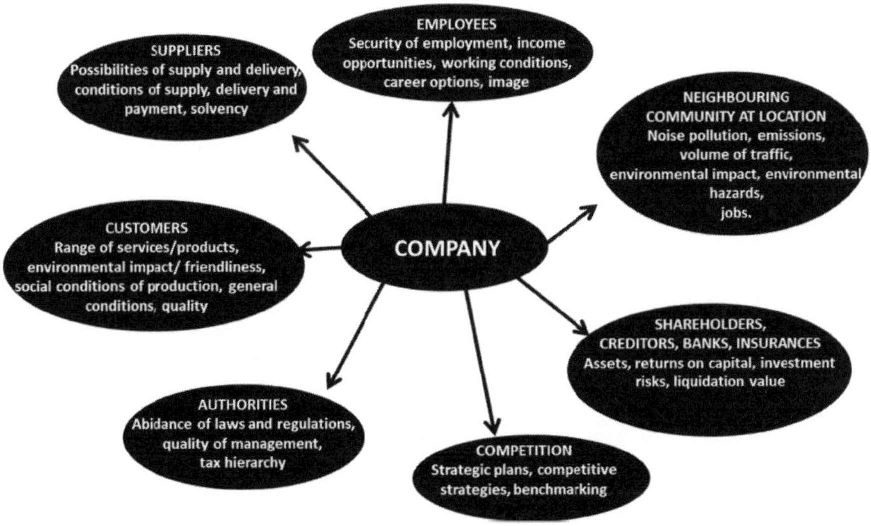


Fig. 8.10. Aspects of Information Disclosure. Adapted from the GRI G3.1 Reporting Guidelines.

Table 8.1. Sustainability revolutions.

| Aspect | Old paradigm | New paradigm |
|---|---|---|
| Markets | Compliance | Competition |
| Values | Business | Societal |
| Transparency | Closed | Open |
| Life Cycle Technology | Product | Function |
| Partnerships | Isolation | Co-operation |
| Time-Frame | Short-term | Long term |
| Corporate Governance | Advisory | Accountable |

Source: Adopted from Reading Material, "CII-ITC Sustainability Awards: Assessors Training Programme", 2009

has eventually resulted in new and manifold business opportunities and advantages:

*Increased efficiency and better production*

Eco-efficiency helps to improve products and services, simultaneously reducing resource use, waste and GHG emissions throughout the value chain. It means decoupling production with natural resource use, which may result in various new business opportunities.

*Reduced risks and production cost*

Closing the production loop and increasing the durability of the product not only advances product quality and profitability, but also decreases the production cost and risks associated.

*Legitimacy, reputation enhancement and brand image management*

Businesses that are committed to sustainable development principles to build a better future will be recognised by the emerging markets. This helps in enhancing the reputation and brand image.

*Leadership, learn to change and innovate*

Sustainability leadership fosters learning towards change and stimulating social and environmental values, while focusing on the economic bottom line and market reality. It capacitates businesses to innovate, reposition and benchmark in a highly competitive business environment through "creative destruction".

For instance, Procter & Gamble, in addressing micronutrient deficiency and hidden hunger in developing nations, introduced a product called "NutriDelight". It also formed an agreement with international bodies and local governments on generating awareness on issues of hidden hunger. This highly-nutritious, cost-effective product showed positive results leading to taller and brighter children.

*Creating market for all*

Sustainability offers an opportunity to tap on the base of the pyramid through innovation and societal development. In the case of business, it is very clear, as it promotes new forms of partnerships and dialogues with the community.

The World Business Council for Sustainable Development (WBCSD) in "The business case for sustainable development" illustrates ten building blocks (Table 8.2) for businesses to realise their sustainability goals and contribute to the formation of a sustainable society.

Table 8.2. Business case for sustainable development.

| Building blocks | Business case for sustainable development |
|---|---|
| Market | Markets help in attaining the goals of sustainability through supporting efficiency and innovation. |
| Right Framework | Governments would have to set up a right framework for businesses to pave a way for their support in forging towards sustainable development. |
| Eco-efficiency | Eco-efficient practices and processes improve the environmental performance and financial results. |
| CSR | It is a concept that re-defined corporate responsibility to serve needs of its stakeholders. CSR underlines the role of governments and businesses in actualizing sustainable growth. |
| Learning to change | TBL approach requires change in thinking throughout the corporation. Businesses have to appreciate the need to consider social and environmental justice. |
| From dialogue to partnership | There is a need to go beyond stakeholder dialogues and move to the next step to partner with them to combine skills, which results in increasing the credibility of actions. |
| Informing and providing consumer choice | In case of free markets, where the consumers have choices, businesses have a significant responsibility to inform and provide consumers with the environmental information of that choice. |
| Innovation | In order to move up the sustainability curve, the company would have to innovate new products and services and processes that are more eco-efficient. Also, companies should engage their stakeholders in the process for social acceptance of innovation. |
| Reflecting the worth of Environment and natural resources | Businesses can only protect what they value and they can only manage what they measure. Hence, proper economic evaluation of natural capital can serve as a solution to the global resource crisis. |
| Making markets work for all | The biggest obstacle to sustainable growth is poverty. Well-constructed markets can alleviate poverty and provide the poor with the opportunity to earn their livelihoods, the key motive of any economy. Governments cannot accomplish the task alone, without the assistance of the private sector, but they can and must certainly set up a framework for this to happen. |

Source: WBCSD's Sustainable Development Reporting: Striking a balance.

Businesses across the world have started recognising the need to find out the social and environmental impacts associated with their activities, integrating these facets into their core business strategies and operating sensitively with sustainable development as a priority. They are addressing the needs and expectations of stakeholders by executing engagement strategies and developing new partnerships.

Companies that do not act in response to the changing sustainability trends will cease to exist. Hence, to remain afloat, manage business risk, and remain competitive, companies would have to reorganise and reinvent themselves. Sustainability is becoming a norm of the business world, which goes beyond greening and involves social well-being. Businesses are required to strategise processes and practices that can transform the threats into opportunities and contribute towards sustainable and inclusive growth, while creating economic benefits.

## Sustainable Development — What Does it Mean to a Country Like India?

For India, the need for sustainable development cannot be emphasised enough. Although the economic reforms of 1990s made India's economy one of the growing economies in the world, sustaining a high GDP to improve the quality of life and giving adequate attention to inclusive growth, while preventing the ecosystems from degradation, remains at the heart of India's paradox.

With over 400 m people living in multi-dimensional poverty, surviving on less than US$1 a day,[16] India has over 50 million malnourished children and the highest rate of maternal deaths during childbirth in the world.[17] It is estimated that India would have over 41% of its population living in cities by 2030, leading to a phenomenon called "*urbanisation of poverty*". A survey declared, "*India has moved up the ranks, but is still the poorest among the G-20*".[18] India held a 134th position out of 187 countries in the Human Development Index.[19]

India remains at a threshold of critical environmental catastrophe with almost half of its total 600 districts water-stressed, and with climate change intensifying the crisis. As estimated by the Caron Disclosure Project,[20] the

[16]ITC Limited (2012), Developmental Challenges for India, Reports and Accounts 2012.
[17]Business Standard (2014) "Poverty, child, maternal deaths high in India: UN report", http://www.business-standard.com/article/current-affairs/poverty-child-maternal-deaths-high-in-india-un-report-114070800186_1.html, last accessed on 11 June 2015.
[18]"Economic Survey 2012–2013. "India poorest among G20 nations despite growth", See http://indiatoday.intoday.in/story/economic-survey-2012-13-india-growth-poverty/1/177886.html, last accessed on 10 January 2013.
[19]http://genderindex.org/country/india, last accessed on 10 January 2013.
[20]The Carbon Disclosure Project (CDP) is an organisation based in the United Kingdom which works with shareholders and corporations to disclose the greenhouse gas emissions of major corporations. As disclosure of emission related data as CDP's

cost of climate change could have a major impact on the Indian economy by causing a 9–13% loss in its real GDP by 2100.[21]

Developing countries like India have little ability to tolerate adverse impacts owing to high dependence climate-sensitive sectors, coupled with poor infrastructure, frail institutional mechanisms and lack of economic resources.

Hence, for India, the growth cannot be "sustainable" if it is not "inclusive",[22] which necessitates continuous investment in healthcare, education, infrastructure, poverty eradication programmes, rural connectivity, labour reforms and skills development.

## Indian Government's Initiatives and Policy Enablers

Government interventions are the key aspect for stimulating sustainable and inclusive growth. The objective of poverty eradication has a bearing on most of the developmental policies of Government of India (GoI). It has launched various initiatives to alleviate poverty such as "*Bharat Nirman*" for rural infrastructure, "*Jawahar Lal Nehru National Urban Renewal Mission*" for urban infrastructure, "*Sarva Shiksha Abhiyan*" for education, "*National Rural Employment Guarantee Act*" for rural employment and "National Rural Health Mission" for rural health, among many others. The current government launched a national level campaign, "*Swachh Bharat Abhiyan*" (Clean Indian Mission) for the cleanliness of streets and roads and infrastructure in 4041 statutory towns in the country.[23]

India has developed robust environmental governance. India's National Action Plan on Climate Change is one of the key initiatives, which focuses on solar energy, energy efficiency, sustainable habitats, water, sustaining the Himalayan eco-system, Green India, sustainable agriculture and strategic knowledge, and linking climate change and sustainable development. India pledged to decouple the emission intensity of GDP by 20–25% over 2005 levels by 2050.[24]

---

primary activity, the quality of the data reported to CDP is key. http://en.wikipedia.org/wiki/Carbon_Disclosure_Project, last accessed on 29 December 2014.

[21]Press Trust of India (2008), "India more vulnerable to climate change: UNIDO", last accessed on 10 January 2013.

[22]India Growth Strategy.

[23]Swachh Bharat Abhiyan, http://en.wikipedia.org/wiki/Swachh_Bharat_Abhiyan, accessed on 5 December 2014.

[24]Planning Commission, GoI, (2011), Low Carbon Strategies for Inclusive Growth: An Interim Report.

Responsible corporate action can be magnified by conducive Government policy interventions and incentives that encourage meaningful initiatives. GoI plays a guiding role by providing relevant policy instruments for Indian businesses and enabling them to achieve their sustainability targets. A few recent examples are:

- **The Companies Act 2013**

The Companies Act 2013, passed by the Parliament, which replaced the old Companies Act of 1956. The Act of 2013 consolidates and improves the law pertaining to corporate governance and contains a separate Clause 135 on CSR. The objective is to guide Indian Companies and help them bring sustainability into the mainstream, and contribute to inclusive growth. It requires companies with a net worth of US$10 million or more to constitute a CSR Committee, which is then given the task of formulating the company's CSR policy. It mandates such companies to spend a minimum of 2% of their average net profits of the immediately preceding 3 financial years on corporate social responsibility, and puts a binding obligation on the Boards to report on the same.

- **Department of Public Enterprises' Guidelines for CSR and Sustainable Development for Central Public Sector Enterprises (CPSEs)**

Central Public Sector Enterprises (CPSEs) are the public sector companies which have their performances reviewed by different Ministries in GoI. A Memorandum of Understanding (MoU) is signed between CPSEs and the Ministries, which specifies the deliverables and planned activities in both financial and non-financial dimensions. From 2011 onwards, as in the MoU, 5% of a CPSEs' non-financial parameter was allocated for sustainable development activities and 20% was allocated towards CSR initiatives. To facilitate adoption of CSR practices by CPSEs, the Department of Public Enterprises formulated guidelines for CSR and Sustainable Development.

- **National Voluntary Guidelines (NVGs) on Social, Environment and Economic Responsibilities of Business**

National Voluntary Guidelines (NVGs) are elucidated in nine principles, providing the basic framework for responsible business action for Indian corporations and MNCs, based on Triple Bottom Line (TBL) aspects, to make their value creating process "sustainable". After the release of NVG, the Securities and Exchange Board of India (SEBI) made it mandatory for the top 100 listed companies to report business initiatives and disclose measures taken around these nine principles. It also mandated the inclusion of a "Business Responsibility Report" in the companies' annual reports.

A few other inducing policy measures include, the Perform, Achieve and Trade (PAT) Scheme (for energy efficiency), the Green Procurement Guidelines, etc.

The sustainability challenges cannot be addressed by the Government alone. Therefore, it is imperative that the Indian corporate sector works in tandem with the government.

## Corporate Responsibility: Indian Businesses in Action

The Corporate Social Responsibility (CSR) concept is not new in India and has been followed since ancient times, albeit informally. The ethical principles of doing business were advocated by philosophers like *Kautilya* from the pre-Christian period. The idea of helping the poor and underprivileged was mentioned in primeval writings and is supported by various religions, for example the Hindu "*Dharmada*", the Sikh "*Daashaant*", and the Muslim "*Zakaat*", to name a few.[25]

Indian businesses which pioneered industrialisation followed similar notions by setting up charitable foundations, trusts, educational institutions for the development of communities, and used to give donations out of charity or philanthropy.

In recent times, there are companies which have progressed by focusing on various aspects of corporate sustainability vis-à-vis good governance, inclusivity, transformational change in process, internal and external stakeholder engagement, and innovation (Figure 8.11). This is illustrated in Table 8.3 below.

### *Sustainable Businesses: Strategies and Practices by Indian Businesses*

#### *Sustainability through leadership and good governance*

Sustainability has various facets, with leadership as the most significant, as it brings transition from business-as-usual to a sustainable business.

[25]KPMG and ASSOCHAM (2008), "Corporate Social Responsibility — Towards a Sustainable Future", A White Paper, 2008.

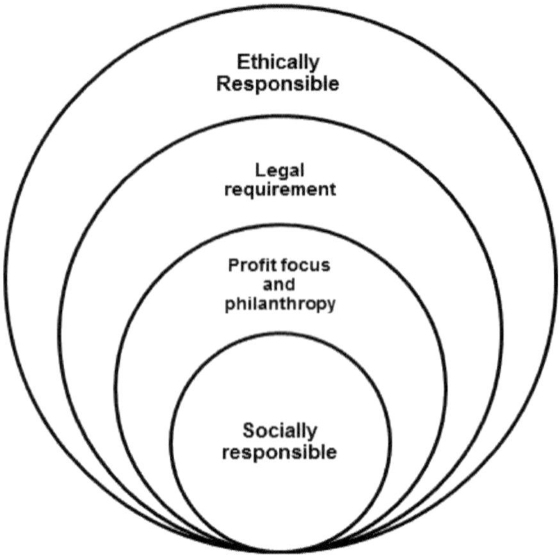


Fig. 8.11. CSR Phases in India.

Table 8.3. CSR evolutionary trends in India.

| |
|---|
| From a survey by The Energy Research Institute (TERI), named "Altered Images: the 2001 State of Corporate Responsibility in India Poll", CSR in India has evolved in four stages: |
| **The "Ethical Model" (1930s–1950s):** The key aspect of this stage is endorsement of "trusteeship", which was revitalised and reinterpreted by Mahatma Gandhi. This motivated businesses to operate in trust, in the interest of their community, and encouraged numerous family-run businesses such as the Tata group to contribute towards the well-being of society. |
| **The "Statist Model" (1950s–1970s):** This model came in the post-independence era, under the guidance of Jawahar Lal Nehru, which saw a mixed and socialist kind of economy. State ownership and legal requirements determined the responsibilities of corporations. |
| **The "Liberal Model" (1970s–1990s):** This stage was based on Milton Friedman's notion that corporate responsibility is limited to the economic bottom line. Compliance with laws, generating profits and charity towards society were entailed and deemed adequate. |
| **The "Stakeholder Model" (1990s–To date):** This came as a result of awareness of the fact that while making profits, businesses have social responsibility. It requires companies to adopt practices and processes according to the "triple bottom line", focusing on being accountable and transparent. |

Source: Adapted from KPMG and ASSOCHAM (2008), "Corporate Social Responsibility — Towards a Sustainable Future", A White Paper, 2008.

It embeds sustainability into the company's goals and core vision, thus, a strong leadership is most critical for the adoption of CSR. Corporate Governance standards are maturing and boards are increasingly recognising the need to be accountable and transparent.

In order to bring sustainability to their core business strategy, the Tata Group's Companies developed an integrated approach collectively which strengthened the link between corporate responsibility and competitiveness. The Group is always driven by its core values, namely "Integrity", "Unity", "Understanding", "Excellence", and "Responsibility".[26] The Group has devised a Tata Code of Conduct,[27] which contains 25 clauses covering almost all aspects of sustainable business; all group companies are signatory to this said Code.

Corporate Governance standards and boards are realising the significance of companies disclosing their environmental and social performance. Sustainability reporting was initially taken up by a small number of companies; their numbers have been steadily growing. Companies like ITC Limited, Tata, Wipro, Infosys and many others are now reporting annually.

### *Sustainability through new business models that deliver social and environmental benefits*

Companies offer solutions through new business models that provide answers to the sustainability challenges of our economies and respond to the sustainability indicators for change.

Yes Bank's innovative initiatives are a good example. Yes Bank[28] has adopted the "People, Planet and Profit" philosophy, making corporate leadership and creating sustainable value a part of their identity. Through its approach of "Responsible Banking", it strives to provide solutions to social and environment issues. The bank carries on research and development to interweave sustainability into its core practices. It offers financial solutions to deal with a broad range of issues like sustainable livelihoods, health, education and climate change.

---

[26]Tata Group, "Values and Purpose", http://www.tata.in/aboutus/articles/inside.aspx?artid=CKdRrD5ZDV4=, last accessed on 11 January 2013.
[27]Tata Group Website, "Tata Code of Conduct", http://www.tata.in/aboutus/articles/inside.aspx?artid=NyGNnLHkaAc=, last accessed on 11 January 2013.
[28]Yes Bank (2011–2012), Annual Report.

*Sustainability through engaging internal stakeholders*

In order to be sustainable, companies need to create stakeholder value, manage risk, achieve customer satisfaction and trust, increase engagement with employees, and develop new markets. The starting point is to identify appropriate stakeholders and appreciate their expectations and needs.

Larsen & Toubro (L&T) has a vision of "fostering a culture of caring, trust, continuous learning and meeting the expectations of employees, stakeholders and society".[29] L&T motivates and leverages its young intellectual capital by providing them with a mix of opportunities for growth. It has formulated an "Integrated Talent Management" framework that improves and fosters both employees' professional and behavioural capabilities. L&T's work culture also encourages them to have "out-ofthe-box" thinking, which improves their performance. It provides customised and innovative training programmes to cope with changing technology and business requirements.

*Sustainability through community development and societal empowerment*

Creating and maximising stakeholder value is another key factor of success for sustainable businesses. This can be achieved through interventions in partnership with stakeholders to deliver superior triple bottom line performance.

Tata Steel[30] contributes to local and regional economic and social development in numerous ways. Tata Steel Rural Development Society (TSRDS) organises most of the community development programmes. It initiated various community health programmes addressing important issues like the well-being of mother and child, and HIV/AIDS awareness. It launched "Life Express", the first hospital on rails in the world, which offers on-the-spot diagnosis and surgical treatment across rural India.

Tata Steel also provides potable water through tankers or piped water distribution systems in areas where water is scarce. TSRDS aims to provide households within the Company's operating areas with a supply of potable water. It undertook various projects to improve agricultural productivity (crop yield), which has led to an increase in household incomes and living standards.

[29]Larsen & Toubro website, last accessed on 11 January 2013.
[30]Tata Steel website, "Corporate Citizenship", http://www.tatasteel.com/corporate-citizenship/community.asp, last accessed on 12 January 2013.

**Box 8.1 Sustainability through sharing onus in national objective of helping society meet its basic needs**

According to the 2011 census, 53% of households in India do not have toilet facilities, and the figure is higher at 69.3% for rural India.[32] As per the UN report, India continues to have the largest number of people in the world defecating in the open.[33] Millions of women in India walk great distances and wait for darkness to fall to relieve themselves, which puts them at risk.

Motivated by the Clean India Mission and inspired by the Honourable Prime Minister of India, Mr Narendra Modi's speech on rural sanitation, India's apex telecom operator, Bharti Airtel Limited, in association with its development arm, decided to contribute towards improving the existing sanitation conditions in the Ludhiana District of Punjab. In this initiative, the proposal is to provide a toilet to every rural household in the District and toilet facilities for girls in every school.

On similar lines, Dubar India Limited launched a social media driven campaign '*700se7Kadam*' (700 to 7 steps) and is contributing US$ 0.02 (INR1) from the sale of every pack of Sanifresh toilet cleaner, towards toilet building, primarily in rural India.

*Sustainability through engaging the value chains*

Indian companies also work with their entire supply chains to minimise their negative impacts at each stage of the product life cycle.

Hero MotoCorp's[34] "Green Vendor Development Program" (GVDP) is an endeavour to make an environmentally conscious and responsible supply chain, in order to reduce the load on the environment and resources. This programme is based on one-to-many mentoring. To meet the increase in demand, production at facilities increase and thus increases manufacture at vendors' sites, giving rise to the environmental load.

[32]http://timesofindia.indiatimes.com/india/PM-stresses-on-cleanliness-toilets-for-girls-in-schools/articleshow/40312399.cms, last accessed on 1 December 2014.
[33]http://www.thehindu.com/news/national/india-tops-in-open-defecation/article5993478.ece, last accessed on 1 December 2014.
[34]Green Vendor Development Programme, http://www.heromotocorp.com/gvdp/green-vendor-development-program.php, last accessed on 12 January 2013.

Considering its social responsibility, Hero MotoCorp initiated an exclusive GVDP, in association with their corporate environmental group and material department, to encourage their vendors to adopt pre-specified environmental requirements and sustainable practices.

Similarly, to devise an effective engagement process with stakeholders and understand their needs and expectations, it is crucial to have feedback, which helps in continual improvement and learning. Philps' Vadodara Light Factory[35] (VFL) follows the process of collecting information from various stakeholders through surveys and other open forums. Surveys are also conducted on employees to capture their perception. The inputs are utilised for revising policies and strategies under the Quality Management System. The survey also identifies Critical Success Factors and their realisation through the PDCA approach.

### *Sustainability through skills and livelihoods creation*

Many Indian companies are working with local communities in the form of providing educational and vocational skills. This increases the communities' employability and also helps them become micro-entrepreneurs through training, funding and networking.

For example, Tata Chemicals is passionate and concerned about the communities in which it operates, and this is institutionalised in their corporate philosophy. The Company had set up the Tata Chemicals Society for Rural Development (TCSRD), which helps the community attain self-reliance on natural resource management, providing livelihood and healthcare services, education and infrastructure.

The Company implements diverse agricultural, economic and developmental programmes for its diverse locations. In Mithapur, Gujarat, it has taken the "*Okhai*" initiative, which has given a boost to local handicrafts and garments by marketing them in retails shops and on the internet. There are also various initiatives to further better pond management, which forms an integral part of the area at Haldia plant.

[35]CII-ITC Centre of Excellence for Sustainable Development (2010), Excellence in Sustainable Business: Strategies and Practices for Business Growth and Competitiveness, 2010.

Companies have also stressed on women empowerment. Bharat Heavy Electricals Ltd. (BHEL)[36] formed a Ladies Welfare Centre to provide training to children and dependents of local communities. They have been trained in secretarial skills, computers, knitting and stitching; these female students are then later employed by the company to do office-related work, receiving better opportunities than they would independently. Dependents of deceased employees are given jobs like printing, binding and other office work. The Centre also runs a school for mentally disabled children and day-care centres for the children of female employees. BHEL also supports various ancillary units with the provision of training and technical assistance.

National Thermal Power Corporation Limited (NTPC) also focuses on the vocational training of the village youth and enables them to earn a livelihood. Similarly, Jubilant Life Sciences has taken various initiatives to create sustainable livelihoods through vocational training and skill development programmes to improve agricultural productivity.

*Sustainability through innovative, inclusive business models*

In order to address the socio-economic issues and growing disparity, it becomes essential for businesses to take initiatives, making growth inclusive. By innovating new business models, Indian businesses have gone a step ahead to meet the need for inclusive growth.

Jain Irrigation[37] produces micro-irrigation systems and sells its products to small farmers, 90% of whom are smallholders with only 1 hectare of land. The company develops drip and sprinkler irrigation systems, tailored for the needs of smallholder incomes and farming situations, which reduces the water usage significantly. To address the issue of the water shortage crisis, it provides technological solutions and trains farmers on "precision farming"; how to balance the use of water, fertilisers, pesticides and energy for an increase in productivity. Considering the diversity of land type and size, and also the variety of micro-climates in India, the technology is certainly a breakthrough, incorporating sustainability into ecosystem

[36]CII-ITC Centre of Excellence for Sustainable Development (2010), Excellence in Sustainable Business: Strategies and Practices for Business Growth and Competitiveness, 2010.
[37]World Economic Forum (2011), "Redefining the Future of Growth: The New Sustainability Champions".

and water security, making positive impacts on farmer's livelihood and environment.

*Sustainability through environmental efficiency*

Sustainability has been accomplished by bringing transformational change in processes and practices, which result in a reduction of raw material use and reduces environmental impacts drastically.

Ambuja Cement Limited[38] has, since 2011, given back double the volume of water it consumes; achieving water-positive status. Its conservation efforts are built around:

- Reduction of fresh water withdrawal and other initiatives, like rain water harvesting and drip irrigation.
- Waste water recycling: It makes use of Reverse Osmosis (RO) for water recovery.
- Zero Discharge: It installed sewage treatment plants at all its units and recycles 100% of the water from its sewage treatment plants.

Various Indian businesses have integrated sustainability into their governance models and embedded it into their vision and core values. Indian businesses are recognising the drivers of transformation and overcoming sustainability challenges. Companies are proactively responding to the needs and expectation of their stakeholders and are forging ahead to develop new partnerships to engage with stakeholders to improve their sustainability performance.

**Box 8.2 ITC Limited: A Successful Transformation**

ITC Limited[39] has embedded sustainability into its business strategies and diversified itself from just a cigarette making company, having numerous social and environmental impacts, to an exemplar in sustainable business practices across the globe. It unleashed various drivers of

(*Continued*)

[38]Ambuja Cement website, https://www.ambujacement.com/index.php, accessed on 12 January 2013.

[39]ITC has a diversified presence in Cigarettes, Hotels, Agri-Business, Paperboards and Packaging, consumer goods, Information Technology, Branded Apparel, Packaged Foods & Confectionery, and Stationery. ITC's diversified status initiated from its corporate strategy to create multiple drivers of business growth by leveraging its own competencies.

**Box 8.2** (*Continued*)

growth and adopted approaches to make it inclusive and equitable, to ensure long-term business sustainability and competitiveness.

ITC has been a torch-bearer of sustainable business practices in India. Noteworthy are its initiatives towards the integration of sustainability into its business strategies, through new business models such as ***e-choupal***,[40] which is the world's largest rural digital infrastructure, which benefits over 4 million farmers in 40,000 Indian villages. It provides services on technical know-how, best practices, information on weather and market prices, and connects farmers to markets.

Its ***Social and Farm Forestry initiative***[41] was created to address the growing need for raw material by ITC's Paper and paperboard business, and their desire to acquire them in a sustainable and cost-effective way. Through this initiative, ITC linked its business requirement to the socio-economic requirements of the locals. It helped to provide sustainable livelihoods for around 6 million rural waste land owners by supporting and assisting them in transforming their wastelands into pulpwood plantations. The initiative also helped with greening over 163000 hectares of land.

ITC's Women's empowerment programme is another example of its commitment. It provided sustainable livelihoods to over 40,000 rural women by linking them with its Agarbatti Business.

The Company has been ***carbon positive*** for 9 consecutive years, ***water positive*** for 12 years in row and ***solid waste recycling positive*** for the last 7 years[41]. Its ***Watershed Development Initiative*** brings water to over 1 hundred thousand hectares of moisture-stressed areas. 33% of its total energy consumption is through renewables.

ITC has taken a leadership approach in transforming key sustainability risks into opportunities for its business growth. However, constant risks prevail when it comes to its value chains. ITC's FMCG portfolio has brought added consumer health and environmental challenges. Looking into the future, enhanced value chain responsibility can help ITC continue to be among front-runners of corporate responsibility.

[40] http://www.itcportal.com/businesses/agri-business/e-choupal.aspx, last accessed on 10 January 2013.

[41] ITC Sustainability Report, 2014, last accessed on 2 December 2014.

## Summing Up

Businesses have countless touch points in society. They can create tremendous societal value by leveraging on not only their financial resources but, more importantly, their managerial capacities and creativity to innovate. Indian businesses can bring transformational change for a better tomorrow.

As frontrunners, Indian companies deliver societal development by fostering inclusive business models, thereby acknowledging the creation of societal value by not just delivering "competitively superior value propositions" but also by innovating strategies leading to the creation of sustainable livelihoods and a positive environmental footprint. The examples demonstrate that Indian companies are aligning their strategies in accordance to their sustainability objectives and stakeholders' needs and expectations.

Sustainability has partially been infused in business strategies. However, such businesses are still limited in number, with only about a handful of companies having taken sustainability earnestly. For many, societal issues lie at the periphery and not the core; others do not know how to proceed on the path of sustainability profit and innovation. Mounting sustainability challenges, on the other hand, require intensive expansion in the number of companies looking at the integration of the triple bottom line.

Despite having one of the best corporate governance performances in the world, India has also witnessed corporate frauds, like the Satyam case of 2009 that dented the company's reputation and its stakeholders, thereby incurring doubt towards the credibility of the Indian technology sector across the world which had, on the other hand, observed other corporate fiascos like the collapse of Lehman Brothers, Enron, Worldcom, etc.

Every corporate fiasco is an indicator of existing loopholes that need to be plugged. The framework that can bridge these gaps and provide resilience towards future crises needs the integration of sustainability with governance structure and corporate objectives.

Owing to its complex nature, questions exceed the available answers when it comes to sustainability. This complexity often gives birth to challenges, and with every challenge comes an opportunity. With growing developmental challenges, it is in the best interest of Indian businesses to realise the significance of a holistic approach towards sustainability, and to address risks and convert them into opportunities.

Sustainability presents itself as an opportunity, not only in India but also across the globe. It is imperative that businesses do not just look at sustainability as an act of philanthropy, but also as a corporate responsibility. Solutions are available; however, they need to be viewed from a wider sustainability lens to create sustainable value. In this way, Indian businesses will be able to overcome challenges and emerge as champions.

Also, sustainability is not a destination. Instead, it is a long journey towards continuous learning, innovation, improvement and being responsible — and the sustainability journey of Indian businesses has just begun!

## References


Ambuja Cement website. [Online] Available at: https://www.ambujacement.com/index.php [Accessed 12 January 2013].

Brazilian Institute of Corporate Governance (IGBC), "Sustainability Guide for Companies: An Overview for Directors and Senior Executives", Corporate Governance Handbooks, 2009.

Canadian Centre for Pollution Prevention. [Online] Available at: http://www.c2p2online.com/ [Accessed 4 January 2013].

CATALYST Consortium. (2002). "What is Corporate Social Responsibility? — 8 Questions and Answers", *USAID*, Washington, July 2002.

CII-ITC Centre of Excellence for Sustainable Development. (2009). *4th Sustainability Summit: Asia 2009*, Theme Paper.

CII-ITC Centre of Excellence for Sustainable Development. (2010). "Excellence in Sustainable Business: Strategies and Practices for Business Growth and Competitiveness", *Executive Brief*.

Confederation of Indian Industry (CII) and Indian Institute of Corporate Affairs, Ministry of Corporate Affairs. (2010). *Stakeholder Engagement Handbook*, Government of India.

Green Supply Chain Management. *Green Vendor Development Programme*. [Online] Available at: http://www.heromotocorp.com/gvdp/green-vendor-development-program.php [Accessed 2 December 2014].

GRI Sustainability Reporting Guidelines.

India Today. (2012). "Economic Survey 2012–20-13: India poorest among G20 nations despite growth". [Online] Available at: http://indiatoday.intoday.in/story/economic-survey-2012-13-india-growth-poverty/1/177886.html [Accessed 10 January 2013].

India's Gender Index. (Updated 2014). [Online] Available at: http://genderindex.org/country/india [Accessed 10 January 2013].

Industrial Ecology. [Online] Available at: http://industrialecology.blogspot.com/ [Accessed 2 December 2014].

ITC Limited. *ITC Limited's official website, "e-Choupal"*. [Online] Available at: http://www.itcportal.com/businesses/agri-business/e-choupal.aspx [Accessed 10 January 2013].

ITC Limited. (2012). *Developmental Challenges for India, Reports and Accounts, 2012.*

ITC Limited. (2014). *ITC Sustainability Report.* [Online] Available at: http://itc portal.mobi/sustainability/sustainability-report-2014/index.aspx [Accessed 2 December 2014].

KPMG and ASSOCHAM. (2008). "Corporate Social Responsibility — Towards a Sustainable Future", A White Paper.

Larsen & Toubro India. *Larsen & Toubro Website.* (Updated 2015). [Online] Available at: http://www.larsentoubro.com/ [Accessed 11 January 2013].

Planning Commission, GoI. (2011). *Low Carbon Strategies for Inclusive Growth: An Interim Report.*

Press Trust of India. (2008). "India more vulnerable to climate change: UNIDO", *Thainidan News.* [Accessed 10 January 2013].

Press Trust of India. (2014). "Poverty, child, maternal deaths high in India: UN report", *Business Standard*, [Online] Available at: http://www.business-standard.com/article/current-affairs/poverty-child-maternal-deaths-high-in-india-un-report-114070800186_1.htmhttp://www.business-standard.com/india/news/a-springhope/454671/ [Accessed 13 January 2013, 11 June 2015].

Proving and Improve, "AA1000 Assurance Standard. [Online] Available at:http://www.proveandimprove.org/tools/aa1000AS.php [Accessed 7 January 2013].

Tata Group. *Values and Purpose.* [Online] Available at: http://www.tata.in/aboutus/articles/inside.aspx?artid=CKdRrD5ZDV4= [Accessed 11 January 2013].

Tata Group. (2013). *Tata Code of Conduct.* [Online] Available at: http://www.tata.in/aboutus/articles/inside.aspx?artid=NyGNnLHkaAc= [Accessed 11 January 2013].

Tata Steel. *Corporate Citizenship.* [Online] Available at: http://www.tatasteel.com/corporate-citizenship/community.asp [Accessed 12 January 2013].

TEEB. (2010). *The Economics of Ecosystems and Biodiversity: Mainstreaming the Economics of Nature: A synthesis of the approach, conclusions and recommendations of TEEB.* [Online] Available at: http://www.teebweb.org/publication/mainstreaming-the-economics-of-nature-a-synthesis-of-the-approach-conclusions-and-recommendations-of-teeb/ [Accessed 11 December 2012].

The Hindu. (2014). "India tops in open defecation". [Online] Available at: http://www.thehindu.com/news/national/india-tops-in-open-defecation/article599 3478.ece [Accessed 1 December 2014].

The Times of India. (2014). "PM stresses on cleanliness, toilets for girls in schools". [Online] Available at: http://timesofindia.indiatimes.com/india/PM-stresses-on-cleanliness-toilets-for-girls-in-schools/articleshow/40312399.cms [Accessed 1 December 2014].

UNEPA official website. (Updated 2012). *Energy Recovery.* [Online] Available at: http://www.epa.gov/epawaste/hazard/wastemin/index.htm [Accessed 2 December 2014].

Wikipedia. (Last updated on June 2015). *Swachh Bharat Abhiyan.* [Online] Available at: http://en.wikipedia.org/wiki/Swachh_Bharat_Abhiyan [Accessed 5 December 2014].

World Economic Forum. (2011). "Redefining the Future of Growth: The New Sustainability Champions", *World Economic Forum in collaboration with The Boston Consulting Group.*
Yes Bank (2011–2012), Annual Report.

**Illustrations:**


- Account Ability. (2008). *AA1000 Stakeholder Engagement Standard 2011*, Final Exposure Draft.
- Amato, A., Henderson, S., & Florence, S. (2009). *Corporate Social Responsibility and Sustainable Business A Guide to Their Leadership Tasks and Functions*, CCL Press [Imprint], Greensboro.
- Brazilian Institute of Corporate Governance. (2009). "Sustainability Guide for Companies: An Overview for Directors and Senior Executives", *Corporate Governance Handbooks.*
- Confederation of Indian Industry (2011), "The Next in Corporate Governance", *Sustainability Embedded*. (2011).
- Confederation of Indian Industry (CII) and Indian Institute of Corporate Affairs, Ministry of Corporate Affairs. (2010). *Stakeholder Engagement Handbook*, Government of India.
- Deloitte. (2012). *Current state of the Indian economy, A balancing act*, March 2012.
- Dimovska, D. (2009). *Innovative pro-poor healthcare financing and delivery models landscaping from the Rockefeller foundation-sponsored initiative on the role of the private sector in health systems in developing countries.* Results for development institute, Washington, D.C.
- Eccles, R. G., Cheng, B. & Saltzman, D. (2010). *The Landscape of Integrated Reporting Reflections and Next Steps*, Harvard Business School.
- Garner, A. Keoleian, G. A. (1995). *Industrial Ecology — An Introduction*, University of Michigan School of Natural Resources and Environment, and NPPC Research Manager.
- Global Reporting Initiative. (2000–2011). *Sustainability Reporting Guidelines,* Version 3.1. GRI Sustainability Reporting Guidelines
- Holliday, C. & Pepper, J. (2001). *Sustainability through the market: seven keys to success,* World Business Council for Sustainable Development, Conches-Geneva, Switzerland.
- Holliday, C. O. & Schmidheiny, S. (2002). *Walking the talk: the business case for sustainable development*, Greenleaf, Sheffield.

- Holme, R. & Watts, P. (2000). *Corporate social responsibility: making good business sense*, World Business Council for Sustainable Development, Conches-Geneva, Switzerland.
- International Finance Corporation. (2007). *Stakeholder Engagement: A Good Practice Handbook for Companies Doing Business in Emerging Markets.*
- International Institute for Sustainable Development. (2012). *Corporate Social Responsibility, an Implementation Guide for Business.*
- ITC. (2012). *Sustainability Report 2012.*
- ITC Limited. (2012). Reports and Accounts 2012.
- KPMG and ASSCHAM. (2008). "Corporate Social Responsibility — Towards a Sustainable Future", A White Paper.
- KPMG International. (2012). *Sustainable Insight: Water Scarcity — A dive into global reporting trends.-*
- Lynes, J. & Andrachuk, M. (2008). "Motivations for corporate social and environmental responsibility: A case study of Scandinavian Airlines", *Journal of International Management*, **14**(4), 377–390.
- Ministry of Environment & Forests, Government of India. (2004). *India's Initial National Communication to UNFCCC.*
- Ministry of Environment & Forests, Government of India. (2009). *State of Environment Report 2009.*
- Ministry of Environment & Forests, Government of India. (2011). *Sustainable Development in India, Stocktaking in the run up to Rio+20.*
- Ministry of Housing and Urban Poverty Alleviation, Government of India & UNDP. (2009). *India: Urban Poverty Report 2009 Factsheet.*
- Ministry of Statistics and Programme Implementation, Government of India. (2012). *SAARC Social Charter India Country Report.*
- OECD. (2012). *India Sustaining High and Inclusive Growth*, OECD "Better Policies" Series, October 2012.
- Peloza, J. & Shang, J. (2011). "How can corporate social responsibility activities create value for stakeholders? A systematic review", *Journal of the Academy of Marketing Science*, **39**(1), 117–135.
- Planning Commission, GoI. (2011). *Low Carbon Strategies for Inclusive Growth: An Interim Report.*
- Pop, O., Dina, G. C. & Martin, C. (2011). "Promoting the corporate social responsibility for a green economy and innovative jobs", *Procedia — Social and Behavioral Sciences*, **15**, 1020–1023.

- Porter, M. E. & Kramer, M. R. (2011). *Creating shared value how to reinvent capitalism — and unleash a wave of innovation and growth,* Harvard University, Graduate School of Business Administration, Boston, MA.
- Sabeti, H. (2011). "The For-Benefit Enterprise", *Harvard Business Review,* November 2011.
- Sachin, J. (2010). *Excellence in Sustainable Business: Strategies and Practices for Business Growth and Competitiveness*, CII-ITC Centre of Excellence for Sustainable Development.
- Tata Consultancy Services. (2012). *Corporate sustainability report 2011–12.*
- TEEB. (2010). *The Economics of Ecosystems and Biodiversity: Mainstreaming the Economics of Nature: A synthesis of the approach, conclusions and recommendations of TEEB.* [Online] Available at: http://www.teebweb.org/mainstreaming-the-economics-of-nature-a-synthesis-of-the-approach-conclusions-and-recommendations-of-teeb-launch-of-final-teeb-report/ [Accessed 11 December 2012].
- UNEP. (2011). *Towards a Green Economy: Pathways to Sustainable Development and Poverty Eradication — A Synthesis for Policy Makers.*
- UNEP Finance Initiative & World Business Council for Sustainable Development. (2012). *Translating ESG into sustainable business value*, International workshop series of the WBCSD and UNEP FI.
- United Nations Development Programme. (2012). *Annual Report 2011/2012, The Sustainable Future We Want.*
- World Business Council for Sustainable Development. (1999). *Corporate Social Responsibility: Meeting changing expectations*, Conches-Geneva, Switzerland.
- World Business Council for Sustainable Development. (2002). "The Business Case for Sustainable Development: Making a Difference towards the Earth Summit 2002 and Beyond", *Corporate Environmental Strategy*, **9**(3), 226–235.
- World Economic Forum. (2011). *Redefining the Future of Growth: The New Sustainability Champions,* Geneva, Switzerland.
- World Energy Council. (2007). *Deciding the Future: Energy Policy Scenarios to 2050*, London.
- Yes Bank. (2012). *Annual Report. (2011–2012).*

# 9

# EFFECTIVENESS OF SUSTAINABLE BUSINESS PRACTICES: CASE STUDIES OF READY-MADE GARMENT COMPANIES OF BANGLADESH

TAYEF QUADER and AUDREY CHIA

The concept of sustainable development is now firmly entrenched in current affairs. Ever since its inception at the United Nations Summit in Brazil in 1992, the term has become a significant buzzword. From this emerged sustainability and the categorisation of sustainable development into economic, social and environmental aspects. Sustainability in the business arena has equally gained much ground and is nowadays considered a base for strategic decision-making for firms. Integrating sustainable practices such as employee welfare, community involvement, sustainable procurement and ethical trading have all become imperative segments in today's globalised world for responsible business.

The textile and clothing industry of Bangladesh is no different as it is an integral source of growth for the country's rapidly developing economy. According to the Bangladesh Garments Manufacturers and Exporters Association (BGMEA), the ready-made garments (RMG) industry employs about 4 million people and is responsible for 70% of the country's total exports (BGMEA, 2013). The industry is also the second largest employment sector for the country after agriculture. As RMG factories are labour and resource intensive, their focus towards sustainability is paramount. Thus, the industry holds considerable potential to be a game changer, to transform modern day business norms by putting sustainability in the driver's seat.

## 1. Introduction

The global textile industry is a multi-billion dollar industry employing millions of people. The industry is not only responsible for providing cheap apparel to household brands, but is also a significant source of export for these countries. According to Data Monitor, a market research firm, the global apparel industry in 2009 alone generated a total revenue of US$1,031.5 billion. Figures from the World Bank show that almost 7% of global exports are in clothing and textiles, an industry significantly dominated by countries like India, China, Bangladesh, Pakistan, Vietnam and Cambodia. Furthermore, Allwood *et al.* (2006) estimated that developing countries are producing half of the world's textile exports and almost three quarters of global clothing exports. The size of the apparel industry is enormous, and with its traditional dependence on cheap labour, this sector is regarded as a high priority business sector in these economies.

As sustainability is branching to other spheres of discussion, businesses worldwide are examining the possibilities of integrating sustainability into their business models. For decades, businesses have been prominently involved in philanthropic activities. Of late, the concept of sustainability, where firms put together their economic, social and environmental issues is fast becoming present day business norms. Corporations from emerging economies have much to gain from this momentum of sustainability as they account for a large presence in the overall supply chain. The United Nations Industrial Development Organization (UNIDO) and the International Labour Organization (ILO) estimates that the apparel industry employed around 35 million people in 2011. Of the total workforce in the clothing industry, women represented almost 70% (Hernandez, 2006). Despite the impressive numbers, the clothing and textile industry in developing countries is subject to political influences, poor working conditions, human rights violations, environmental degradation and other pressing issues.

## *Sustainability in Fashion*

The United Nations Framework Convention on Climate Change (UNFCCC) in 1992 gave profound meaning to the concept of sustainable development. Although scientists and researchers argue over its true meaning, a widely accepted definition of sustainable development stands for "*development that meets the needs of the present generation without compromising the future generation's ability to meet their own needs*" (WCED Report, 1992). The crucial target of this model is to raise the quality of life by maximizing productive potential of ecosystems through the use of appropriate and innovative technologies (Gardetti, 2005).

Since the 1990s, the word "sustainability" has taken on several meanings. This approach encompasses economic, environmental and societal issues, by which the ultimate objective leads to sustainability. As argued by Frankel (1998) and Elkington (1998), sustainability means to bring a balance between economic, environmental and social challenges. The concept, in present day, has been successfully weaved into different business sectors. Particularly in the fashion and textile industry, brands are increasingly seeking to go green through cleaner production processes. As an example of sustainable fashion, a report in CNBC observed that the London Fashion Week (LFW) in 2005 had less than 5% of participating brands parading sustainable fashion products, as compared to 2013 where almost a third of designers showed eco-focused products.

Sustainability has also initiated a paradigm shift in business, raising the need to look at things from a different perspective. Similarly, in the clothing industry, sustainability not only includes the use of organic or environmentally friendly materials, but it also includes resource consumption, the supply chain, fair and ethical sourcing practices and safe working conditions for employees. This allows companies to broaden its scope from company level to employee and community level. Furthermore, companies worldwide also need to look at how their business operations affect social cohesion and community dynamics, as manufacturing firms are continuously branching out to suburban and out-of-city areas. As such, the notion of incorporating such practices into business models is gaining traction.

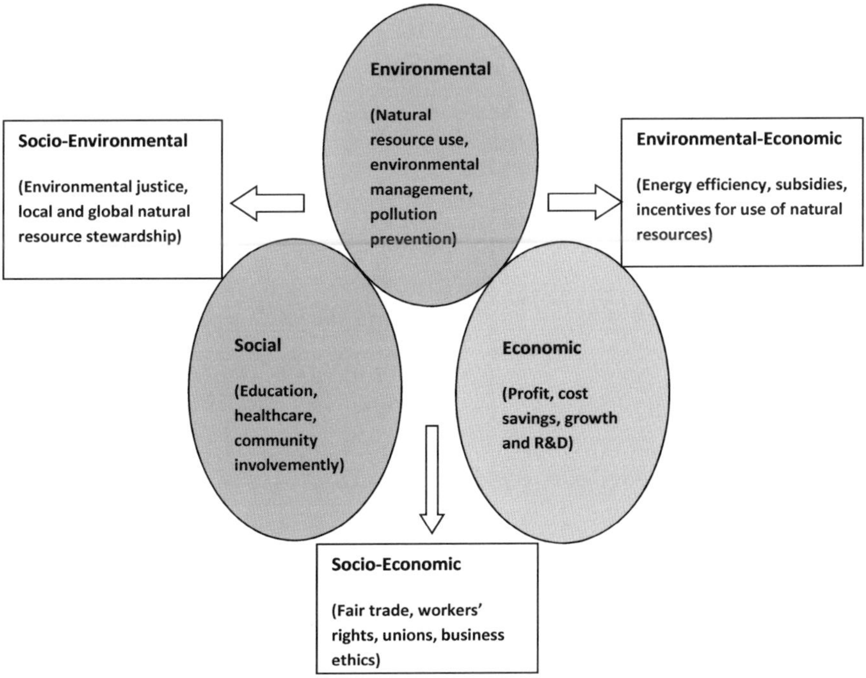


Fig. 9.1. Three spheres of sustainability (Adapted from WCED Report, 1992).

### ***Background of Bangladesh RMG Industry***

The readymade garment (RMG) industry of Bangladesh portrays a remarkable story of the country's successful transition towards an export-oriented economy. The RMG sector started its journey in the 1970s with little investment. Over the next two decades, it progressed to become one of the largest industries of the country. The contributing factors of this transformation can be largely credited to cheap labour cost, global trading agreements, government policy support and private entrepreneurship. All these factors facilitated the country to attain a handsome share of the global apparel business. From the early 1990s, the RMG sector has become the country's largest foreign exchange earning sector. It was only in recent years that remittances from expatriate workers emerged as the other booming sector. In 2010–2011, Bangladesh earned US$17.9 billion from exports of apparel products (BGMEA, 2013). McKinsey and Company, a global consultancy firm, in a report estimated that RMG exports in Bangladesh could cross the US$20 billion mark by 2015. With almost 75% of total exports, the contribution of the RMG sector towards national GDP accounts for about 13% (Ahmed and Hossain, 2006). As of financial year 2012–2013, the Export Promotion Bureau of Bangladesh reported total RMG exports of US$21.5 billion (EPB, 2013). Due to the size of the industry, it attracts significant foreign investment for the country. The sector also draws international retailers, labour rights advocates and business conglomerates to explore different areas of the RMG sector.

The industry has emerged as a carrier for advanced industrialisation. The Multi-Fibre Arrangement (MFA) of the World Trade Organization (WTO) is the regulatory structure that governed textile trading by imposing quotas on developing countries. As the MFA ended in 2005, it was feared that Bangladesh's RMG sector would suffer from increasing competition from India and China. However, the industry defied projections of a downturn and put up robust growth at a healthy rate of 20% (Kumar, 2006). Now, this industry employs roughly about 3.6 million people directly, of which 80% are women (Mahmud, 2012). The World Bank in its employment statistics report of 2006 also stated that the RMG sector of Bangladesh represents 40% of overall industrial employment. The number of garment factories in the sector too, witnessed gradual growth. With only around 50 factories in the 1980s employing a few thousand workers, the industry multiplied to around 4,490 factories all over Bangladesh to date (BGMEA, 2013).

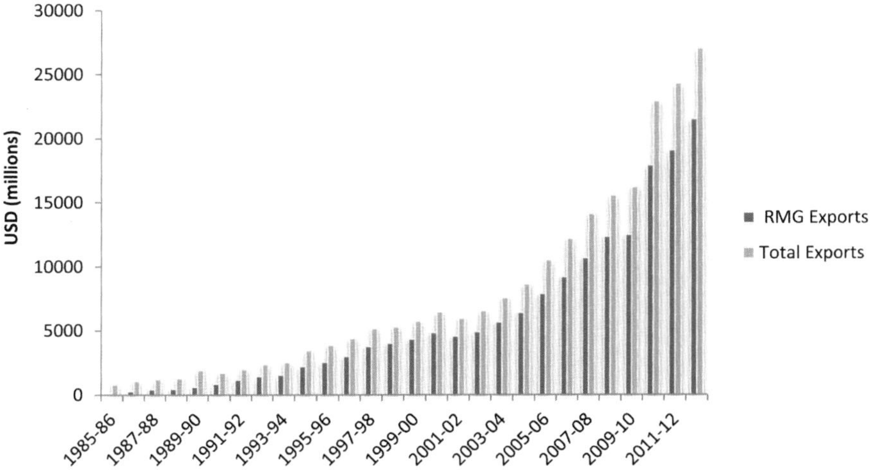


Fig. 9.2. Comparative Scenario between RMG exports and total exports of Bangladesh from 1985 to 2013. (Adapted from: Export Promotion Bureau, 2013).

In terms of exports, Bangladesh exports RMG products mainly to USA and the Europe. These two destinations account for more than 90% of the country's total earnings from RMG exports (Fig. 9.3). Although USA is the biggest importer of RMG, the industry is subject to trade and quota barriers such as the generalised system of preferences (GSP)[1]. On the other hand, the European Union imposes no restrictions on Bangladesh from the days of the MFA, which may have become the key driving force leading to the expansion of the RMG industry in Bangladesh.

Despite the impressive performance and competitiveness of the sector, several issues have recently affected the industry. There is growing concern from sourcing companies and international organisations that labour rights are often violated in the industry. The empirical evidence suggests that labour rights are not adequately addressed. Of late, fatal incidents from November 2012 through May 2013 have rocked the industry and brought world-wide attention towards Bangladesh for workplace safety violations. The fire accident at Tazreen Fashions Limited in November 2012 and the Savar Building Collapse in April 2013 have jeopardised the growth potential of the industry.

[1]GSP — Office of the United States Trade Representative. Retrieved January 5, 2013 from http://www.ustr.gov/trade-topics/trade-development/preference-programs/generalized-system-preference-gsp

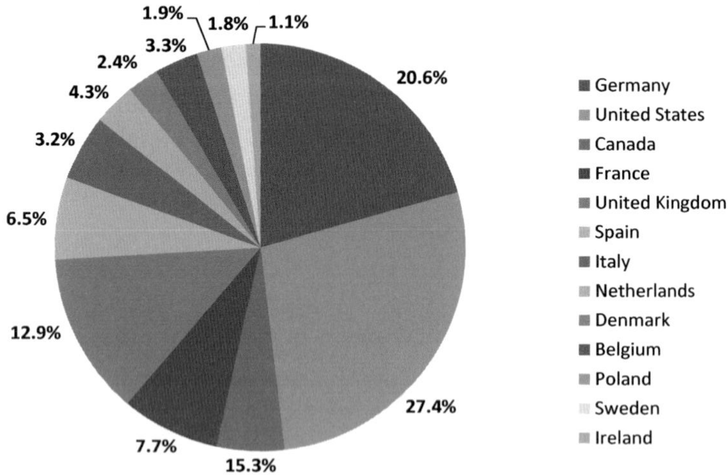


Fig. 9.3. RMG exports from Bangladesh in 2011–2012 by destination country. (Adapted from BGMEA, 2013).

Industry insiders however argue that the pressure to comply with the increasing demands of foreign buyers is what hampers the overall production process (Financial Express, 2013). This is compounded by a shortfall of space in rental units, which eventually pushes the RMG companies to sub-contract their orders to other factories with improper infrastructure and unsafe facilities. In spite of the accidents, RMG exports in the fiscal year 2011–2012 reached US$19 billion and are projected to reach US$23 billion in 2012–2013 (Nur, 2013).

## 2. Case Studies

### *Case 1: Viyellatex Group*

Viyellatex Limited (founding company of the group) was established in 1996 and is headquartered in Tongi, on the outskirts of the capital city Dhaka, Bangladesh. The core business of the group is knit garments, along with washing, dyeing and embroidery. The group is regarded as vertically integrated: from processing yarn into fibres to finished apparel products. With interests in different business sectors, the group has a diverse portfolio. They produce sleepwear, kids' wear, casual knitwear, fashionable wear and formal shirts, and have a medium degree of horizontal differentiation. It is privately-owned and employs around 14,000 workers. In 2011, the group reported net sales of around US$206 million (Table 9.1).

Table 9.1. Business characteristics of Viyellatex group.

| Viyellatex group | 2010 | 2011 | 2012 |
|---|---|---|---|
| Annual turnover | $110 million | $155 million | $206 million |
| Ownership | Privately held | | |
| Internationalisation | | | |
| # of suppliers | 09 | 14 | 19 |
| Export to # of countries | +/ − 2/3 | +/ − 3/4 | +/ − 8/9 |
| Business strategy | Product differentiation | | |
| Channels of distribution | Mostly 1st party | | |
| Customer relationship | Personal assistance | | |
| Key resources | Physical/Human labour | | |
| Key activities | Raw material importer, fabric importer, apparel & accessories manufacturer | | |
| Partnerships | NGOs/Banks/Other organisations | | |
| Cost structure | Cost-driven/Fixed | | |
| Business model | Vertical integration | | |

The group engages in-house suppliers specialising in yarn production. For woven products, the group's shirt manufacturing unit produces both formal and informal woven shirts. The group houses a printing, packaging and accessories facility located at the same site. In 2009, they diversified into tea production and logistics services. Recently, Viyellatex has become active in the electricity sector by adding Power Midland Limited to its portfolio.

The group's market reach is worldwide via its customer channels. Cotton is sourced from at least five different countries, resulting in medium internationalisation. They specialise in quality and efficient products that use a minimal amount of raw materials. To support this, they introduced an extensive IT-based SAP structure at operational level (Viyellatex, 2010).

*Business case for sustainability*

The group's vision is to contribute to the development of Bangladesh by being committed to the environment and workers' welfare. The group is active in its community through educational and healthcare projects. Viyellatex identifies the following stakeholders as key to their business growth:

- Employees
- Local community
- Buyers/retailers

- Suppliers
- Creditors/banks
- Affiliated local and international NGOs

The group reports several triple bottom line effects stemming from its sustainable supply chain management initiatives. Especially since large customers impose increasing requirements concerning sustainability, the group has chosen to integrate the triple bottom line into its business model, rather than address mere issues related to price, quality, child labour or forced labour, and other general HR issues.

*Clientele*

Viyellatex's main customers are: PUMA, PVH, Esprit, Hugo Boss, M&S, S. Oliver, Calvin Klein, DKNY and Tommy Hilfiger. They partner with this group of stakeholders as they are positioned in the mid- to high-end segment, and have relatively high quality requirements. Being strategic partners also benefits the group, as they are often prompted by large retailers to implement joint sustainable programmes. This encourages suppliers to operate more consciously and to promote social activities within their factories. An example of this would be partnerships through Plan A and Esprit (Viyellatex, 2012). Also, the group reports on customer satisfaction on product quality and product pricing annually (Viyellatex, 2011).

*Socio-economic issues*

Viyellatex acquired the 1,700 acre Ruthna Tea Estate (RTE) in Sylhet in 2009, symbolising the firm's strategic intention towards social business. Although the firm plans to start a commercial tea business by 2020, the tea estate at present functions as a plantation benefiting both local people and the firm. Other community investments include the employment of disabled individuals in its printing and packaging facilities.

Viyellatex respects the freedom of association of workers. In accordance with the law, the firm does not recruit anyone below 18 years of age. The compliance team takes responsibility for this issue by conducting audits on child labour at third-party suppliers, so as to ensure that the firm's practice is followed throughout its supply chain. This shows that by leveraging on its industry connections, the firm takes a proactive commitment to abolishing the practice of child labour (Viyellatex, 2012).

Regarding freedom of association, Viyellatex regularly meets with the Workers' Participation Committee in order to discuss any problems and negotiate on issues (Viyellatex, 2012). For health and safety purposes, workers have access to free personal protective equipment. Workers also receive training on health and safety procedures. Quality of drinking water is checked regularly. Moreover, a medical team is available to all workers and organises several free clinics throughout the year, such as the Expanded Programme on Immunisation and Blood Donation (Viyellatex, 2012).

*Environmental issues*

Viyellatex became ISO14001 certified in April 2012. The group had taken measures such as installing a cogeneration system and hot-water chillers to chill the air in their spinning mills. In a cogeneration system, condensed steam and drier exhaust is re-used, which contributes to significant reductions in energy consumption. Viyellatex also uses another cogeneration boiler as a cooling system to control temperature on the work floor — an eco-friendly alternative to air conditioning. They use waste from garments, like cotton dust and food, to produce organic compost fertiliser, which is used on the tea plantation and offered for free to local farmers. To further address the issue of climate change, Viyellatex is also building two carbon-neutral factories in Dhaka: EcoFab and Eco Couture (Viyellatex, 2012).

Concerning the issue of water, the majority of ground water consumption by the firm is taken up by garment production. A water treatment plant makes the water ready to use in other manufacturing processes. After usage, the water is discharged to the local sewage system as it is chemical-free. Water from the spinning mill is discharged into a local pond and the water used for the dyeing process is pumped into a treatment plant to be re-used for toilet flushing. This saves roughly 90 million litres of groundwater consumption per year. During the rainy season, the firm also collects rain water from the rooftop, which is filtered and stored for reuse.

To conclude, sustainability in the business process of Viyellatex has made the firm become a profitable player in the market. Integrating such practices allowed the firm to attract international buyers. While Viyellatex assumes an important position in the development of the industry, they also demonstrate a moral approach to their employees and the community in which it operates.

### *Case 2: Epyllion Group*

Epyllion group is another prominent RMG firm in Bangladesh. The company was established in 1994 as a small manufacturing and exporting unit focusing on knit products only. Gradually the company moved towards vertical integration and incorporated other aspects of textile manufacturing. Epyllion is also an RMG conglomerate with export destinations mainly in Europe and America. Besides textile manufacturing, the company recently diversified into real estate. The firm has a workforce of roughly 9,000 employees and as of 2012, a total reported revenue of US$110 million (Table 9.2).

The firm strongly believes in a three-tier philosophy of:

- Human spirit
- Ethical business practices
- Protection of the environment

Built on these pillars, the management sees human resources as the driving force of the firm. The role of human resources is vital to the business model of the firm as they believe that a combination of effort by top management and all employees have resulted in the firm's success. The firm thus wants to make its people the centre of its business development process.

Table 9.2. Business characteristics of Epyllion group.

| Epyllion group | 2010 | 2011 | 2012 |
|---|---|---|---|
| Annual turnover | $54 million | $75 million | $110 million |
| Ownership | | Privately held | |
| **Internationalisation** | | | |
| # of suppliers | 06 | 10 | 15 |
| Export to # of countries | +/ − 2/3 | +/ − 3/4 | +/ − 8/9 |
| Business strategy | Product differentiation | | |
| Channels of distribution | Mostly 1st party | | |
| Customer relationship | Personal assistance | | |
| Key resources | Physical/Human labour | | |
| Key activities | Raw material importer, fabric importer, apparel & accessories manufacturer | | |
| Partnerships | NGOs/International and local organisations | | |
| Cost structure | Cost-driven/Fixed | | |
| Business model | Vertical integration | | |

*Business case for sustainability*

For sustainability, the firm emphasises on employee and community engagement. As employees are regarded as the firm's driving force, their engagement represents the basis for a sustainable business model for Epyllion. The firm allocates a financial budget for community activities involving employee welfare and environmental awareness activities. A dedicated team under the Corporate Human Resource department is responsible for all sustainability-related activities. Furthermore, the top management and the executive board of the firm provide substantial support which enables most of the initiatives to materialise.

Ensuring workplace safety for employees is of prime importance to the firm. Besides the labour law, the firm adopts an additional Code of Conduct based on the principles of Human Rights outlined in the United Nations Global Compact (UNGC) to address labour issues. Epyllion arranges health check-ups and eye-camps facilitated towards the betterment of their employees. Apart from internal engagement programmes, Epyllion is also involved in external initiatives like providing aid and support to victims of the natural and manmade disasters. They partner with non-governmental organisations such as the Centre for the Rehabilitation of the Paralysed (CRP), the Hong Kong and Shanghai Banking Corporation (HSBC) Bank, and Grameen Communications (Epyllion, 2013).

*Clientele*

Epyllion covers 80% and 15% of exports to European and American retailers respectively. The following customer base represents Epyllion's business profile segregation:

- C&A (60%), Celio (5%), M&S (14%), G-Star (10%), H&M (10%) and Esprit (1%)

All these brands are in the mid to high brand segment with high quality requirements. The firm receives the majority of its business from these sourcing companies, and being partners, the need for sustainable initiatives becomes apparent. These large retailers persuade suppliers to adopt sustainable measures throughout the product value chain. As C&A and M&S are the biggest retailers for Epyllion in terms of export volume, issues like customer satisfaction, quality and production process take an important role when it comes to sustainability (Epyllion, 2013).

*Socio-economic issues*

Epyllion Group gives strong emphasis towards freedom of association for workers. The firm is a member of the UNGC (United Nations Global Compact) and advocates adopting the relevant principles related to labour rights. The firm is strictly against recruiting any form of child labour (below 18 years of age) and has a recruitment process that requires every worker to provide identification documents like birth certificates which are duly authenticated. A medical team conducts further tests to check authenticity of age in extreme cases. Quarterly audits on sub-contracting suppliers are also carried out with proper focus on all UNGC principles. This illustrates the firm's commitment towards the elimination of child labour by enforcing local and international standards (Epyllion, 2012).

Since majority of the employees are females, healthcare benefits are female-oriented — for example, ultra-sonogram services for pregnant females — on top of the standard medicines and medical consultation provided by the in-house medical team. All discrimination and other matters are dealt with by the HR counselling team. Regular meetings with the Worker's Participation Committee are held for any labour-related issue, and training is provided via an induction to health, safety and fire-fighting sessions. Employees' children are also provided with benefits such as immunisation programmes, blood donations and free clinics. Proper hygiene is also ensured in the workplace with regular house-keeping and quality testing of drinkings water.

*Environmental issues*

The firm is keen to adopt strategic plans based on a holistic approach of catering to economic and social needs. Hence, environmental issues are of equal importance as part of their social responsibility. Although legislation requires regular environmental monitoring by municipalities, some companies are enthusiastic about going the extra mile. A strong commitment from the management can be identified as the cause of this as all employees are instructed to strictly adhere to environmental standards. For Epyllion, they strive to work with an aim to cause 'No harm to the Environment' as the environment is considered one of the founding business pillars.

Besides these issues, the firm engages with external and internal stakeholders in the following ways:

- Eco Factory Programme with Marks & Spencer on: Renewable energy lighting, heating and insulation management, temperature management, resource and waste management and eco policy
- Energy Efficiency Program with H&M
- Sustainable Chemical Management Program (Zero Discharge) with C&A
- Cleaner Production Program with G-STAR: Assessment of current washing/dyeing/finishing operations and develop Cleaner Production (CP) action plan

*Case 3: Dulal Brothers Limited (DBL) group*

Dulal Brothers Limited (DBL) Group is one of the oldest RMG firms in Bangladesh. They are a highly diverse and vertically integrated manufacturer of knit and composite garments with a strong backward linkage in cotton processing and manufacturing. DBL was the first of the three to start its journey in the industry with basic fabric knitting in 1991. They gradually expanded into spinning, dyeing, washing, knitting and printing facilities to achieve vertical integration. Besides textile manufacturing, the firm has diversified into ceramic tile manufacturing, distribution networks and the telecommunications sector. They produce a variety of knit garments and employ about 16,000 workers. Similarly with Viyellatex and Epyllion, DBL is also privately-owned, but it exercises a family-style of business ownership as the majority of the members of the top management are from one family. As of 2012, the firm reported an annual turnover of almost US$250 million, (Table 9.3)

The firm has a total internal company base of 16 units, of which 11 are involved with textile manufacturing. In 2007, the firm started its first horizontal diversification by launching a telecommunications unit, followed by the group's ceramic manufacturing unit which started operations in early 2013 (DBL, 2013).

With medium internationalisation, DBL sources most of its cotton products from a range of four to five international destinations like USA, China and Brazil. Although cotton from Uzbekistan was in heavy use in the RMG sector up to a few years ago, with the imposition of standards by

Table 9.3. Business characteristics of DBL group.

| DBL group | 2010 | 2011 | 2012 |
|---|---|---|---|
| Annual turnover | \$211 million | \$215 million | \$243 million |
| Ownership | Privately held | | |
| **Internationalisation** | | | |
| # of suppliers | 09 | 14 | 19 |
| Export to # of countries | +/ − 2/3 | +/ − 5/6 | +/ − 9/11 |
| Business strategy | Product differentiation & Cost leadership | | |
| Channels of distribution | Mostly 1st party | | |
| Customer relationship | Personal assistance | | |
| Key resources | Physical/Human labour | | |
| Key activities | Raw material importer, fabric importer, apparel & accessories manufacturer | | |
| Partnerships | NGOs/International and local banks | | |
| Cost structure | Cost-driven/Fixed | | |
| Business model | Vertical integration | | |

American and European retailers, companies now source their cotton from nominated countries instead (DBL, 2013).

DBL's large operational base and manufacturing capacity makes the firm one of the largest textile conglomerates in Bangladesh in terms of production volume. The firm has incorporated various cost-cutting measures through Enterprise Resource Planning (ERP) software, ensuring accuracy and transparency (DBL, 2013). With a vision to be one of the top suppliers in Bangladesh, the firm has achieved recognition in terms of high product quality and environmental stewardship as they were awarded the Social and Environmental Performance Award by GIZ in 2013.

*Business case for sustainability*

DBL regards employee welfare and community engagement highly. They believe in a business model which puts employees at the centre of the firm. Thus, the firm identifies the following as the two most important pillars of business:

- Ethical trading, and
- Welfare of society

Under the branch of ethical trading, DBL places importance towards environmental stewardship. In recognition of their efforts, DBL group was one of the first textile conglomerates in Bangladesh to participate in the Cleaner Production Initiative (CPI) 2011, backed by the International Finance Corporation (IFC). The initiative targeted the upgrading of equipment of some 12 factories in the textile industry where most companies have reportedly achieved significant energy efficiency and cost savings.

### *Clientele*

DBL engages with mostly European and American retailers. According to the BGMEA, non-traditional export destinations such as Australia, South Africa and South America are gradually making inroads into the RMG supply chain, but exports to EU and USA still continue to dominate. For DBL, almost 85% of their exports are accounted for by the following European retailers: H&M, George, Esprit, PUMA, G-Star, Decathlon, C&A, MQ Sweden and Walmart Canada.

### *Socio-economic issues*

For DBL, employee and community engagement plays an important role as part of their corporate citizenship charter. Like Viyellatex and Epyllion, DBL has put in place additional compliance checks in line with international labour standards and regulations to take a stand against any form of child/bonded/forced labour. Although the local Labour Act of 2006 covers all the parameters specified by ILO, these firms willingly go the extra mile to showcase their commitment towards responsible business practices. As most employees in the RMG sector are female, DBL partnered with CARE to launch a 'Women Empowerment Programme' in 2009 to better educate women about rights, benefits and compensation. This programme equipped female employees with training for participation in management committees (DBL, 2013).

The firm puts significant emphasis on healthcare benefits and services. A range of free healthcare services are provided to employees from the on-site doctor; free medicinal support, free medical screening and testing, monthly eye and blood testing, and extended assistance for pregnant employees. Alongside, awareness programmes on healthcare issues and compensation and benefits are also provided for employees on a monthly basis. Collaborating with the Garments Training Institute, DBL also promotes training for all its employees.

Table 9.4. DBL group's year on year production and resource consumption comparison.

| Production and consumption comparison between 2010 and 2011 | | | | | |
|---|---|---|---|---|---|
| Period | Production (KG) | Gas (m3) | Electricity (KWh) | Water (m3) | Steam (KG) |
| 2010 (monthly usage) | 89,123.83 | 726,482.67 | 836.908.50 | 107.966.42 | 4911,761.07 |
| 2011 (monthly usage) | 855,123.83 | 675,740.00 | 676,777.00 | 81,677.80 | 4442,152.97 |

*Source*: DBL 2012.

*Environmental issues*

For environmental and energy resources, DBL is one of the first RMG companies in Bangladesh to participate in the 'Clean Production Initiative' (CPI) sponsored by the IFC. The CPI targets a wide range of production related activities, from wet processing plants, to yarn dyeing methods and reduction of liquor ratio. It is estimated that the initiative helped companies to save approximately US$1 million per annum through the implementation of clean and resource-efficient production methods (DBL, 2013).

Apart from the CPI project, the firm is also known for having one of the largest effluent treatment plants (ETP) in the sector. With a treatment capacity of 4000 $m^3$/hour, the firm plans to double its ETP capacity to accommodate business expansion. The production process runs in-house with a dedicated team functioning as the Energy Management Centre to monitor the overall power consumption.

The firm's Sustainable Action Programme also includes the following initiatives:

- Family Corporate Governance in collaboration with IFC
- Responsible Sourcing Initiative in partnership with the World Bank
- Energy Saving Initiative with German NGO GIZ
- Lean Manufacturing and Environmental Management System with local stakeholders
- Zero Discharge of Hazardous Chemical with Swedish retailers H&M, PUMA and G-Star

The above-mentioned initiatives have resulted in power consumption savings in dyeing machines, steam generators and boilers. An energy savings of approximately 15% has been achieved from individual heat recovery units in dyeing machines, water savings of 20% from individual flow meters located at inlet and outlet of dyeing machines was also reported, and heat recovery system in power plants resulted in about 15% energy reduction. All these initiatives factor into the firm's target of reducing carbon emissions by 15% by 2015 (DBL, 2013).

*Case study summary*

Table 9.5 highlights the similarities between the three firms in overall business sustainability initiatives. It is evident that the firms are in favour of integrating sustainability in their business operations. However, their approaches and methods differ. Although they represent one business segment, it is important to note that each firm prioritizes sustainability issues based on factors like finance, location, community and circumstances.

## 3. Analysis

Sustainability measures are practiced in varying degrees by each firm. This section discusses different aspects of sustainability from different viewpoints of (1) supply chain management, (2) product life cycle analysis and (3) the creation of shared value. Through such practices, firms express the need for businesses to work collectively in order to promote ethical and responsible business models in this industry.

### *Supply Chain Management*

In recent years, supply chain management (SCM) has developed into an essential management practice for businesses worldwide. Similarly with other industries, the SCM principles put forward more relevance to the RMG sector due to its globalised nature. As all members of the supply chain work towards achieving a common goal, SCM creates opportunities for RMG firms to be cost-competitive and sustainable. Although SCM acts on a business efficiency approach, its application in modern day business models has expanded into environment sustainability where firms look towards 'greening' the supply chain.

In Bangladesh, SCM is the dominant aspect of the industry as the majority of activities take place in this segment. For the three cases, cotton serves as the raw material for apparel manufacturing and is primarily sourced from China, USA, Uzbekistan, Turkmenistan, Kazakhstan, Brazil,

Table 9.5. Comparison of the three firms.

| | Viyellatex | Epyllion | DBL |
|---|---|---|---|
| Annual revenue range | US$ 150–200 million | US$100–150 million | US$200–250 million |
| # of employees | 14,000 | 11,000 | 16,000 |
| Male female ratio | 59:41% | 61:39% | 55:45% |
| Start of business | 1996 | 1994 | 1991 |
| # of factories | 6 | 7 | 9 |
| Internationalization | Medium | Medium | Medium |
| **Social and environmental practices** | | | |
| Environmental/Social standard | Certified ISO 14001 | Follow principles of ISO 14001 and 26000 but not certified | Adhere to principles of ISO 14001 and SA 8000 but not certified |
| Environmental reporting | GRI sustainability report, UNGC COP report, internal newsletter, press release | UNGC COP report, internal newsletter, press release | UNGC membership in process, internal newsletter and press release |
| CSR entity | Viyellatex trust | Epyllion foundation | Matin Jinnat foundation |
| CSR approach | Healthcare and education focused | Healthcare, education and social inclusion | Healthcare, education, community and local cluster development and social inclusion |
| Involvement of top management | Medium level engagement | Medium level engagement | Medium and top level engagement |
| Supply chain | Regular third-party audits prior order confirmation | Regular third-party audits prior order confirmation | Regular third-party audits prior order confirmation |
| Training & development and awareness of employees | Dedicated department with annual budget | Annual budget from CSR funding | Dedicated department with annual funding |

India, Pakistan and some African nations (BTMA, 2012). According to the BGMEA, there are companies in Bangladesh which harvest cotton but quality remains an issue.

### *Product Life Cycle Analysis*

In the globalised RMG trade, international retailers and wholesalers are now pushing suppliers in developing countries to adopt better technology and promote R&D. To improve productivity, one particular concept that picked up pace is the product life cycle analysis. Leveraging on this, international brands like Levi's Strauss, Marks & Spencer and PUMA have pioneered environmental responsibility through the use of product life-cycle analysis. According to a report by the European Environment Agency, in 2006, the concept of life-cycle thinking has proved to be significantly beneficial in evaluating environmental sustainability measures by clothing manufacturing firms (EIPRO, 2006).

According to the ISO 14040 standard, life cycle analysis (or cradle-to-grave analysis) is a technique to assess the various environmental impacts associated with all the stages of a product's life cycle. The primary objective of carrying out product life cycle analysis is to identify the environmental impacts in the different production processes in order to assess methods to reduce those impacts. Theoretically, the ISO 14040 standard serves as the most widely accepted LCA guideline (ISO, 2006).

### *Creating Shared Value*

The theme of creating shared value outlines the competitiveness of a company and the health of the communities in which it operates. They are mutually dependant, and recognising these connections will power future economic progress (Porter & Kramer, 2006). It is important to note how these three firms have leveraged on their community for their development. Shared value can be created in three levels:

- Reconceiving customer needs, products, and markets
  - *Rethinking the structure of products and markets*
- Redefining productivity in the value chain
  - *How the organization conducts its business*
- Enabling local cluster development
  - *Promoting local community development*

Fig. 9.4. Evolution of the Creating Shared Value concept (Adapted from Porter and Kramer, 2006).

### *Social and Environmental Reporting*

Environmental Reporting is a new field in the business sector of Bangladesh. With the increase in popularity of reporting guidelines and standards such as the Global Reporting Initiative (GRI) and the UNGC Communication on Progress report, organisations have stepped up their participation. Such reporting not only showcases a firm's business sustainability initiatives, but they also help stakeholders, investors and financers to get an overall picture of the company's profile, strategy and other decisive factors. As a result of this, the Bangladesh Enterprise Institute (BEI) and the CSR Centre were created; two home-grown local organisations which are promoting social and environmental reporting measures to be incorporated into a firm's activities.

Viyellatex pioneered this aspect as they became the first Bangladeshi RMG firm to enlist themselves with GRI as an Organisational Stakeholder in 2011. The firm has also produced GRI Grade C and B Sustainability Reports for 2011–2012 and 2012–2013 respectively. Epyllion has adopted the UNGC reporting guideline, and DBL is also in the process of joining both GRI and UNGC to kick-start their reporting initiatives. The role of social organisations in this process is crucial as they hold the potential to motivate RMG firms to step-up their presence for social responsibility.

### *Focus on Healthcare*

The Ministry of Labour and Employment estimates that the RMG industry employs roughly around 6 to 7 million people both from direct and indirect means. Of this, female employees represent around 70% of the entire work force. Hence, as a major part of sustainability, healthcare projects rank highly for company management. Such projects are particularly tailored

Table 9.6. Healthcare programmes of 3 RMG firms in 2012.

| Description | Viyellatex | Epyllion | DBL |
|---|---|---|---|
| Expanded Programme on Immunization (EPI) in 2012 | 12 EPI programmes conducted with treatment given to 1100 female employees | 8 EPI conducted with treatment given to over 600 female employees | 11 EPI programmes conducted with treatment given to over 1200 employees |
| Blood donation programme in 2012 | 128 management staff donated blood to various NGOs | A total of 90 management staff donated blood | Approximately 145 management staff participated in blood donation |
| Employment of physically handicapped individuals | 72 physically handicapped individuals are employed | Collaboration with CRP and other organisations in the process | Collaboration with CRP in the process |

towards female employees, their children and the local community as local government agencies are sometimes unable to provide appropriately.

Table 9.6 shows the participation of each RMG firm in different healthcare activities through 2012. They range from the Expanded Programme on Immunisation or EPI, blood donation programmes and the employment of physically handicapped individuals in the work force.

### *Sustainability as a Long-Term Business Model*

The supply chain network of an RMG firm is vast. Firms usually operate with medium internationalisation, with sourcing of raw materials from 5 to 7 international destinations. Although RMG companies rely on foreign suppliers for raw materials such as cotton, other components such as yarn and accessories are sourced from in-house suppliers. Hence, companies practise internal policing through compliance audits for both in-house business units and third-party subcontractors to ensure responsible sourcing practices. The firms also opt for voluntary sourcing certifications and social reporting mechanisms, allowing them to endorse ethical and sustainable supply chain management.

The product life cycle analysis approach is a new concept which is in the infancy stage of gaining popularity among the local RMG firms. However, many top brands such as PUMA, Marks & Spencer and Levi's have recently stepped up their efforts in order to create awareness in the industry. The

LCA approach is promoted so that firms can gradually establish adequate infrastructure to keep track of all material flow in the production process. This has the potential to assist the RMG industry to explore another fragment of sustainable practices in the future.

Lastly, corporate social responsibility has been the backbone of all three firms in terms of addressing issues related to employee and community engagement. It is important to recognize the commitment firms showcase when it comes to CSR projects that benefit the local community. Despite philanthropic components, recent initiatives undertaken by the firms have ushered in a collaborative method where partnering with local and international NGOs to address different target groups is established. Creating shared value thus falls within the periphery of CSR activities. Firms that seek to include this concept in their long-term business objectives are in good order to be a socially responsible business house.

In conclusion, the overall expanse of sustainable business practices needs further attention from local and international stakeholders than it currently receives. Apart from its existence in the RMG sector of Bangladesh, a business concept with such potential needs to be integrated into any other manufacturing industry across the country and region. With the upcoming focus on light industries in Bangladesh and neighbouring countries, attaining business sustainability through responsible practices is the impending corporate model for the future.

### *Potential for Growth of the Industry*

Within South Asia, Bangladesh is one of the few countries to have experienced robust economic growth on an average of 5–6% for the past decade. Despite previously being an agrarian economy, the country's production sector grew significantly. The export sector in particular has been the engine of industrial growth, with RMG leading the way — having grown at an average of 30% over the last 5 years and comprising more than 75% of total exports in 2011–2012 (EPB, 2013). McKinsey, a global management consultancy firm in a report in 2011 titled *"Bangladesh's ready-made garments landscape: The challenge of growth"*, highlighted the potential for growth for Bangladesh's RMG sector. The report highlighted various areas which the RMG can leverage on to attract more foreign retailers and buyers to Bangladesh. The RMG industry has been highly cost-competitive in terms of labour cost, lead times and the production and distribution network as compared to other apparel manufacturing countries such as Pakistan, Cambodia, Vietnam and India.

Furthermore, according to the Export Promotion Bureau of Bangladesh (EPB), total exports reached US$21.52 billion in the 2012–2013 financial year, and is projected to reach around US$45–50 billion by 2020. Also, the manufacturing of high quality apparel products by Bangladeshi RMG firms, which has received attention and praise from international retailers, is another area which serves as one of the strong points for many European and American retail giants. Although the RMG industry is prevalent in other Asian nations like Vietnam, Cambodia and India, the majority of their production caters to the domestic demand hence presenting Bangladeshi RMG firms an even bigger opportunity to attract more retailers to Bangladesh.

### *Challenges Facing the Industry*

The World Bank and the Asian Development Bank have projected a healthy growth for the RMG sector with exports reaching US$40 billion by 2020 at the present growth rate (World Bank, 2013). However, the following issues need careful attention from industry players, policymakers and conscious customers for the long-term development of sector:

#### *Energy supply*

An energy crisis is one area which has been the main cause for concern for all RMG factories of late. The government had endeavoured for quick rental arrangements with private companies to ensure uninterrupted power supply but, these efforts have not proved fruitful. Most firms are forced to run their own power supply from generators. In the face of multiple load-shedding in almost all regions on Bangladesh, factories incur additional costs.

#### *Price of raw materials*

The price of cotton has a high tendency of fluctuation. As mills in Bangladesh rely on importing cotton, the price in the international market plays a significant role in this trade. Although it varies due to demand and supply, many times RMG factories start with a bulk order of apparel products with one price range, but due to price volatility, the entire cost of the order shoots up drastically. Many companies in neighbouring countries such as India have started cotton production at low prices but quality remains an issue.

### *Global financial crisis*

USA and Europe are the biggest export markets for RMG products from Bangladesh. After the global downturn, exports have pulled back. Industry experts argue that the EU will take longer than estimated to stabilise, hence the RMG trade may suffer. At the same time, there is a steady growth of exports to non-traditional markets like Australia, South Africa and Brazil but matching the capacity to develop these markets will take time.

### *Competition from neighbouring countries*

Increased competition from neighbouring countries is another area that threatens the industry. Countries that offer similar wage levels like India, Pakistan, Vietnam and Cambodia are close competitors for Bangladeshi companies in this business. However, Bangladeshi RMG factories need to leverage on factors like low labour cost, quality of products and shorter lead times to deter competition.

### *Workplace safety issues*

A recent spate of accidents in 2013 regarding negligence of workplace safety has raised serious concern from local and international community. The fire at Tazreen Fashions Ltd. killed almost 112 people where fire-fighters later discovered that emergency gates were locked (Anbarasan, 2012). The Savar building collapse of Rana Plaza killed over 1,200 people when the entire building collapsed on the morning of 13 May 2013 (Butler, 2013). Both incidents attracted significant attention from foreign media, NGOs, suppliers and retailers on workplace safety issues in the industry. As such, many retailers have joined hands in a consortium to address the issue of workplace safety and suggest remedial and improvement measures.

### *Policy reforms and instability*

Following the accidents, international bodies have put pressure on the government to bring reforms to labour policies. The government thus increased the minimum wage from BDT 3,800 to BDT 5,300. However, labour advocates and NGOs argue that a comprehensive approach is required to address underlying issues related to freedom of association, workers union and compensation mechanisms. On the backdrop of the accidents, western trading bodies of EU and USA have withdrawn the generalised system of preferences (GSP) facilities for the time being on apparel products from Bangladesh, to encourage faster changes to the system.

Bangladesh's political scenario also doesn't help the industry's cause. In spite of the incentives provided by the government through Export Processing Zones (EPZ) and tax breaks to lure investment, local firms cite political violence to be the prime cause for loss of business. Strikes and road blockades hamper movement of shipment around the country causing severe delays. Many owners of RMG firms have expressed their grief and disappointment as the industry estimates to make a loss of around 3 to 4 billion USD in just around three months due to the political situation of the country.

### *Recommendations*

It is noteworthy to express that all three firms have integrated different aspects of sustainability as a part of their business strategy. However, there remains a huge scope for initiating a comprehensive sustainability measure regarding SCM and product life cycle — far more can be done to improve efficiency in this industry. It is also important to recognise that for big firms, it is easier to implement CSR initiatives as compared to small and medium firms due to resource constraints. Conversely, addressing the following issues may allow RMG firms of any capacity to integrate sustainability into their business processes:

- Addressing the subcontractor issue in the supply chain network. It is an area where firms of any capacity can showcase their sustainability potential. Concepts such as greening the supply network and inclusion of third-party suppliers can be within the firm's business strategy.
- International retailers and buyers need to understand that integrating sustainable practices by RMG firms will entail high operating costs in the short run. However, as they will prove profitable in the long run, business decisions by retailers should reflect both cost efficiency along with sustainability.
- Increased support from retailers regarding product life cycle. The supply chain network being highly globalised, it is another area for major savings in environmental footprint for suppliers and retailers.

- Increase in awareness from retailers, buyers and trade bodies towards promoting resource consumption on factory level. More educational and awareness programmes should be conducted for all employees of RMG factory to have more inclusion in the entire process of sustainability.
- Creating shared value may be the way forward for businesses in any industry to fully integrate sustainability into business strategies, models, decision-making process and objectives — support from local, governmental and international agencies can be enhanced in this area which can include a brand of CSR activities and measures
- Reform of Labour Policies — although many regard this area as a highly neglected aspect of this industry, retailers should step up their efforts towards helping both local government and firms to establish socially acceptable and ethically responsible labour policies that cater to the interests of both employers and employees.

To conclude, there is indeed a broader scope of potential where further issues like supply chain and community engagement can be explored. The clothing sector is an industry which will only expand as populations across countries continue to grow. A host of opportunities certainly lies ahead where sustainability in business can be transformed into business ideologies.

## References

About the Group — DBL Group, (2013). Retrieved January 2, 2013 from http://www.dbl-group.com/index.php/about/dbl-group

Ahmed, J.U., & Hossain, T. (2009). "Industrial Safety in the Readymade Garment sector — A developing country perspective", *Sri Lankan Journal of Management*, **14**(1).

Alam, M.J., Mamun, M.Z., & Islam, N. (2004). "Workplace Security of Female Garments Workers in Bangladesh", *Social Science Review*, **21**(2), 191–200.

Allwood, J.M., Laursen, S.E., Malvido De Rodriguez, C., & Bockem, N.M.P. (2006). "Well Dressed? The Present and Future Sustainability of Clothing and Textiles in the United Kingdom". University of Cambridge, Institute of Manufacturing. Retrieved November 31, 2013 from http://www.ifm.eng.cam.ac.uk/uploads/Resources/Other_Reports/UK_textiles.pdf

Andersen, M., & Skjoett-Larsen, T. (2009). "Corporate Social Responsibility in Global Supply Chains", *Supply Chain Management: An International Journal*, **14**(2), 75–86.

Awaysheh, A., & Klassen, R.D. (2010). "The impact of supply chain structure on the use of supplier socially responsible practices", *International Journal of Operations & Production Management*, **30**(12), 1246–1268.

Bangladesh Textile Mills Association. Retrieved December 1, 2013 from http://www.btmadhaka.com

Bangladesh: RMG Exports to Non-Traditional Market Grow Again. (2013). Retrieved September 4, 2013 from http://textilesupdate.com/bangladesh-rmg-exports-to-non-traditional-market-grow-again

Belal, A. (2001). "A Study of Corporate Social Disclosures in Bangladesh", *Managerial Auditing Journal*, **16**(5). 274–289.

Belal, A., & Owen, D. (2007). "The Views of Corporate Managers on the Current State of and Future Prospects for Social Reporting in Bangladesh: An Engagement-based Study", *Accounting, Auditing and Accountability Journal*, **20**(3), 472–494.

Berg, A., Hedrich, S. & Tochtermann, T. (2012). *Bangladesh: The next hotsport in apparel sourcing?* Retrieved August 31, 2013 from http://www.mckinsey.com/insights/consumer_and_retail/bangladesh_the_next_hot_spot_in_apparel_sourcing

Carroll, A.B. (1991). "The Pyramid of Corporate Social Responsibility: Toward the moral management of organizational stakeholders", *Business Horizons*, July–August 1991.

Centre for Policy Dialogue, (2002). "Corporate responsibility practices in Bangladesh: Results from a Benchmark study", *Occasional Paper Series*, Paper-18, p. 12.

Certification logos, (2013). Retrieved from http://www.cotton-made-in-africa.com/en/, http://www.cottonusa.org/, http://www.global-standard.org/ and http://textileexchange.org/content/oe-standards

Chowdhury, R. (2011, November 17). "Viyellatex vows to be fully carbon-neutral by 2016", *The Financial Express*. Retrieved September 05, 2013 from http://www.thefinancialexpress-bd.com/

Clothing: Global Industry Overview. (2010). Retrived October 23, 2013, from http://www.marketresearch.com/Datamonitor-v72/Clothing-Global-Overview-2819866/

Cole, V.J. (2013). *Life Cycle Assessment*. The Pennsylvania State University. Retrieved December 10, 2013 from https://www.e-education.psu.edu/egee401/content/p10_p5.html

Country Profile Bangladesh — World Bank Data. Retrieved October 31, 2013 from http://data.worldbank.org/country/bangladesh

Dasgupta, S. (2002). "Attitudes towards Trade Unions in Bangladesh, Brazil, Hungary and Tanzania", *International Labour Review*, **14**(1), 2002.

DBL Group, (2013). Academic thesis research folllow-up questions.

Department of Labour of the Ministry of Labour and Employment. Retrieved December 5, 2013 from http://dol.gov.bd/

Elkington, J. (1998). *Cannibals with Forks*. Gabriola Island, BC: New Society Publishers.

Elance. (2009). *Elance Company Profile: Viyellatex Group Ltd.* Retrieved October 6, 2013, from http://www.elance.com

Epyllion, (2013). *Epyllion Group UN Global Compact Communication on Progress Report 2013*. Retrieved November 31, 2013 from www.epyllion group.com

Epyllion, (2013). Academic thesis research follow-up questions.

European Environment Agency, (2006). *Environmental Impact of Products (EIPRO)*. Retrieved from http://ec.europa.eu/environment/ipp/pdf/eipro_report.pdf

Frankel, C. (1998). *In Earth's Company*. Gabriola Island, BC: New Society Publishers.

Factory growth in Bangladesh — Bangladesh Garments Manufacturers and Exporters Association, (2013). Retrieved January 5, 2014 from http://www.bgmea.com.bd/chart_test/factory_growth_in_bangladesh

Gardetti, M.A. (2005) DesarrolloSustentable, Sustentabilidady Sustentabilidad-Corporativa in M.A. Gardetti (ed.), *Textos en SustentabilidadEmpresarial: Integrandolasconsideracionessociales, ambientales y económicas con el corto y largo plazo*, Buenos Aires: LA-BELL.

Harland, C.M. (1996). "Supply Chain Management: Relationships, Chains and Networks", *British Journal of Management*, **7**(1), S63–S80.

Hernandez, M.P. (2006). *The Outcome of Hong Kong: Reflections from a Gender Perspective*. Geneva: IGTN International Gender and Trade Network and Center of Concern.

ILO, (2006). *Global Employment Trends Brief*. Retrieved October 28, 2013 from www.ilo.org/wcmsp5/groups/public/—ed_emp/—emp_elm/—trends/documents/publication/wcms_114294.pdf

Industrial Statistics Database — UNIDO. (2013). Retrieved November 13, 2013 from http://www.unido.org/en/resources/statistics/statistical-databases/indstat4-2013-edition.html

ISO 14040:2006. Environmental Management — Life Cycle Assessment — Principles and Framework. Retrieved December 5, 2013 from http://www.iso.org/iso/catalogue_detail.htm?csnumber=37456

Karim, S. (2011, October 17). *IFC Cleaner-Production Initiative Yields Significant Savings for Bangladeshi Textile Factories*. Retrieved November 1, 2013, from http://ifcext.ifc.org/IFCExt/Pressroom/IFCPressRoom.nsf/0/3F39F8701FDB88378525792C004BBC1E

Kharpal, A. (2013, September 13). *London Fashion Week: Is sustainability on trend?* Retrieved from http://www.cnbc.com/id/101032638

Kumar, A. (2006). *Bangladesh: Industrial Chaos Worsens Political Instability*. South Asia Analysis Group, Paper No. 1852. Retrieved from http://www.southasiaanalysis.org

Lam, J.K.C., & Postle, R. (2008). "Textile and apparel supply chain management in Hong Kong", *International Journal of Clothing Science and Technology*, **18**(4), 265–277.

Mahmud, R.B. (2012). "Skills development in Bangladesh RMG sector", *The News Today*. Retrieved November 01, 2013 from http://www.newstoday.com.bd

Majumder, P.P. (1998). *Health status of the Garment workers in Bangladesh: Findings from a survey of employer and employees*. Bangladesh Institute of Development Studies (BIDS), Dhaka, Bangladesh.

Mohammed, A. (2012, March 14). "Viyellatex licks its lips to get increased orders", *The Financial Express*. Retrieved September 29, 2013 from http://www.thefinancialexpress-bd.com/

Newbold, A. (2012, November 30). *Zara bans toxic fashion.* Retrieved December 5, 2013 from http://fashion.telegraph.co.uk/news-features/TMG9713593/Zara-bans-toxic-fashion.html

Nur, S.A. (2013, May 1). "Space shortages at rented houses make RMG units non-compliant", *The Financial Express.* Retrieved from http://www.thefinancialexpress-bd.com/old/index.php?ref=MjBfMDVfMDFfMTNfMV8xXzE2ODExNg

Parvez, A. (2007, May 18). "The Project is in Difficulties", *The Daily New Age.* Retrieved October 23, 2013 from http://www.newagebd.com/

Pakistan Textile Journal. (2011). "Textile Briefs International", *Pakistan Textile Journal*, **60**(10), 10.

Perry, G. (2010, January 29). *The essential CSR best practices.* Retrived October 25, 2013 from http://www.greenbiz.com/blog/2010/01/29/essential-csr-best-practices

Perry, P. (2010). "Garments without guilt?", *Sustainability in Fashion and Textiles.* Sheffield: Greenleaf Publishing.

Porter, M.E., & Kramer, M.R. (2006). "Strategy and Society: The link between competitive advantage and corporate social responsibility", *Harvard Business Review*, 78–94. Retrieved from http://www.hbr.org

Quddus, M., & Rashid, S. (2000). "Entrepreneurs and Economic Development: the remarkable story of garment exports from Bangladesh", *Journal of Bangladesh Studies*, **2**(1), 65–66.

Report of the World Commission on Environment and Development: Our Common Future. (1992). Retrieved September 12, 2013 from http://www.un-documents.net/wced-ocf.htm

Seuring, S., & Müller, M. (2008). "From a Literature Review to a Conceptual Framework for Sustainable Supply Chain Management", *Journal of Cleaner Production.* **16**(15), 1699–1710.

Sobhan, F. (2006, September 16). "CSR: Its Importance for Bangladesh", *The Financial Express.* Retrieved from http://www.thefinancialexpress-bd.com/

The Financial Express. (2009, February 20). "Rahimafrooz, Viyellatex receive StanChart-FE CSR award '08", *The Financial Express (Bangladesh).* Retrieved from http://www.thefinancialexpress-bd.com/

Trade Information. (2013). BGMEA. Retrived from http://bgmea.com.bd/home/pages/TradeInformation#.UtOLyuInnWg

United Nations Framework Convention on Climate Change (UNFCCC). (1992). Retrieved October 25, 2013 from http://unfccc.int/2860.php

United Nations Global Compact (UNGC). (2007). Retrieved November 21, 2013 from http:www.unglobalcompact.org

Viyellatex, (2012a). *ViyellaTalk Corporate Newsletter.* Retrieved August 4, 2013, from www.viyellatexgroup.com

Viyellatex. (2012). *Viyellatex Group UN Global Compact Communication on Progress Report 2012.* Retrieved August 6, 2013, from www.unglobalcompact.org

Viyellatex, (2010). *Viyellatex Sustainability Report 2010.* Retrieved August 6, 2013, from www.viyellatex.com

Viyellatex, (2011). *Viyellatex Sustainability Report 2011*. Retrieved August 6, 2013, from www.viyellatex.com

Viyellatex, (2013). Academic thesis research follow-up questions.

WTO Agreement on Textiles and Clothing (ATC). (1995). Retrieved October 28, 2013 from http://web.ita.doc.gov/tacgi/eamain.nsf/6e1600e39721316c852570ab0056f719/28f761701228564985257 3a000641ff1?OpenDocument

World Bank Report (2013, April 13). *Bangladesh Development Update: Slower yet healthy growth with remarkable development progress*. Retrieved November 25, 2013 from http://www.worldbank.org/en/news/feature/2013/04/13/bangladesh-development-update-slower-yet-healthy-growth-with-remarkable-development-progress

# 10

# INTEGRATED POLICY DESIGN THROUGH LIFE CYCLE SUSTAINABILITY ASSESSMENT: A CASE STUDY OF COTTON GARMENTS

THIDA TUN and KUA HARN-WEI

Due to consistent growth in our demand for, and supply of, goods and services, we are depleting natural resources, producing large amounts of pollutants and wastes, and contributing to global warming at an accelerated pace. Based on the theory of sustainable development, life cycle sustainability assessment (LCSA) is a useful tool to be applied to the whole supply chain of products and services. Recently, a wide range of environmental and social issues have been raised on the supply chain of the textile and clothing industry. The cotton T-shirt was selected as the subject for this sustainability assessment. The data was collected from literature review, informal interviews and surveys. Existing life cycle assessment techniques are discussed, followed by a discussion on the relatively new LCSA methodology. From the results of the study, integrated policies are proposed to address any cross-over effects and ensure that, as much as possible, solutions sought for one set of indicators would not end up creating new problems for other indicators or/and sustainability domains.

## Introduction

The prices of mass market clothing have been dropping significantly in the last decade and this, as a result, has escalated the sales volume of cotton clothes around the world. These sales have (in turn) resulted in huge quantities of waste being generated. Specifically, about 13.1 million tons of textiles were generated in 2010 and these accounted for about 5.3% of the total municipal solid wastes in the United States (USEPA, 2012). Henceforth, increasing attention has been paid to assessing the environmental impacts of the fashion industry in recent years.

However, in order to comprehensively assess the sustainability impacts of cotton clothes, one needs to look into how the triple bottom lines of sustainable development can be assessed throughout the entire life cycle of the products. Many efforts have been undertaken to extend the methodology of traditional life cycle assessment (LCA) to what is now widely known as life cycle sustainability assessment (LCSA). The main aim of this study is to develop an integrated policy framework to improve the sustainability of the cotton garment industry by applying LCSA to assess the total life cycle sustainability of cotton garment.

### *Impacts of Cotton Garment Industry on Sustainability*

Technological innovation decreases the price of production in the clothing industry. According to the United Nations Environmental Programme, the clothing sector is the second largest global economic activity in terms of international trade. Demand for fibre has nearly doubled in the last 15 years (Technical Textile Markets, 2007), and has been responded to by an increase in productivity of the industry. For example, from 1945 to 2007, the cultivated land area increased by 35%; whereas the output per hectare quadrupled from 0.2 to 0.8 tons per hectare (International Cotton Advisory Committee, 2009).

However, there are sustainability concerns on the development of the industry. For example, the nitrogen and phosphorus contents in fertilisers used in cotton fields contribute to the problem of eutrophication (Horrigan *et al.*, 2002). An increase in nutrient levels in water also allows new species of organisms to invade and compete with incumbent species; an example of this is an algal bloom that poses a threat to the local food chain (Anderson, 1994). The agricultural sector as a whole is both a major consumer of water and contributor to water pollution. According to UNIDO worldwide statistics (1996), the agricultural sector is responsible for 69% of total water withdrawal. Moreover, textile industries discharge millions of gallons of highly polluted water filled with chemicals (such as formaldehyde and chlorine) and a sizable quantity of these are channelled into various waterways. In fact, the World Bank estimated that 17% to 20% of industrial water pollution comes from textile dyeing and treatment. In addition, carbon dioxide emissions from field clearance, electricity generation and transportation, and emissions of nitrous oxide from the use of fertilisers in plantations, contribute to emissions of greenhouse gases (GHGs).

While a few LCA studies have been conducted on cotton garment, social life cycle assessment and life cycle costing of cotton garments have not been extensively done so far. Nielson *et al.* (2009) compared the life cycle assessment of two different types of garments produced with biotechnology and conventional technologies, and showed that shifting from conventional to new technology saves energy and water, hence reducing overall environmental impacts. An LCA for shirts made with polyvinyl chloride (PVC) fibres was conducted by Alcover (2010) and the highest environmental impacts were found to be photochemical oxidation and water consumption. Chapman (2010) pointed out that natural fibres used less energy in their production but required more of other resources, such as water, and also had higher eco-toxicity levels than synthetic fibres. Furthermore, during the cloth-making process, sewing had the highest environmental impact due to the high electricity demand of the machines involved. Our study differs from these in that we also looked into the impacts due to distribution/retail and consumers' use phases.

The economic and social impacts of the cotton garment industry must be better understood. The cotton industry makes significant economic contributions to the agricultural sector in developing countries. China is the key global cotton producer and a major consumer of cotton. From 2010–2011, China ranked as the top producer (about 33 million 480 lb bales), importer (16 million 180 lb bales) and consumer (48 million 480 lb bales) of cotton in the world (NCCA, 2012). Substantial growths in apparel and textile exports were also seen in Bangladesh, Pakistan and Turkey. Direct employment in the apparel and textiles industry in China accounted for about 9% of all jobs in China's manufacturing sector (about 23 million people) (VICS, 2009).

In China, most of the labourers in the garment factories are migrants from the rural areas with a limited level of education. Uneven economic development policies in many developing countries have led to underdevelopment and poverty in rural areas. For example, there were over 120 million internal migrant workers in China (Chan, 2005). Workers' rights have also been breached at the workplace in many developing countries, in forms including having long working hours and low wages (Chaw, 2003). The garment workers in Bangladesh, the second largest garment exporters after China, earn only about 14.60 Taka per hour (which is equivalent to US$0.21); the hourly wage rates in China and India are US$0.93 and US$0.55 respectively (Institute for Global Labour and Human Rights, 2010). Health and safety of workers are other important factors in garment factories. These include the stress arisen from shift work schedules, workplace accidents, respiratory diseases from exposure to fibre dust and other health effects.

By understanding the different effects of the cotton garment industry in all the three domains of sustainable development, one can gain a fuller understanding on the multi-dimensional challenges faced by this industry.

### *Life Cycle Sustainability Assessment (LCSA)*

The concept of LCSA is developed to examine the sustainability of a product in the environmental, social and economic domains (the so-called triple bottom lines of sustainability). LCSA can be referred to as the evaluation of all environmental, social and economic impacts in decision making processes towards more sustainable products throughout their life cycle (UNEP/SETAC, 2011). There are many benefits of analysing the sustainability of the product and service through the life cycle concept. First, it allows for improvement of corporate social responsibility by taking into consideration social related factors in the supply chain. Further, LCSA helps to clarify the trade-offs among the social, economic and environment factors in a product's life cycle. This helps decision-makers choose the most appropriate products and services that have the best overall sustainability performance. On the other hand, difficulties in balancing the relative weights of individual indicators exist. Additionally, the result is difficult to understand for non-technical audiences and therefore comprehensive presentation of results is essential.

Limited research has been done on LCSA; and, only economic and environmental sustainability indicators are comprehensively assessed in most of the studies. In an LCSA of fuel, Zhou *et al.* (2007) examined the sustainability impacts with respect to four different categories of indicators — namely economy, environment, renewability and energy. The assessment set included life cycle costs, global warming potential, net energy yield, and non-renewable resource depletion. Capitano *et al.* (2011) conducted an LCSA to compare two production processes for marble products in the Sicilian region by using the multi-criteria analysis approach. The assessment indicators comprised of human toxicity potential, acidification potential, eutrophication potential, photochemical oxidation, global warming potential, cost of extraction and production, fuel costs, waste disposal cost, electricity cost, total number of employees, women in administration and immigrants, health insurance, annual health check and monthly salary.

Our study differs from these studies in that a set of more comprehensive life cycle costing criteria at each stage of the life cycle, such as cost of energy and water, wastewater treatment, waste disposal and rental fee for distribution, were taken into account. In addition, we considered the employment of workers with disabilities as one of the social indicators, as this is becoming a factor of concern around the world.

## Research Methodology

The LCSA applied here is composed of three components: (1) LCA — for environmental impact assessment, (2) social life cycle assessment (SLCA) — applied to every life cycle stage that is defined and considered in the LCA, and (3) life cycle costing assessment (LCCA) of the considered lifecycle stages.

### *Lifecycle Assessment (LCA)*

The objective of the LCA was to examine the environmental impacts of cotton shirt with respect to the following impact categories: (1) acidification, (2) eutrophication, (3) global warming potential (GWP), (4) water consumption, and (5) ozone layer depletion. The two main types of input considered for each life cycle stage were energy (including electricity and fuel) and water consumptions. Primary data was collected from onsite observations and interviews. Specifically, observations were made to compile qualitative information of the working conditions of the staff members in the studied factories or facilities. Formal and informal interviews were conducted with retailers' staff, and consumers. Secondary data were collected from published documents from international organisations, and company reports on corporate social responsibilities.

### *Functional Unit and Data Sources*

The cotton shirt assessed in this study is produced in China. The cotton plantation is a countryside cotton field in Suzhou, which is situated on the lower part of the Yangtze River and a part of the delta region. After the cotton textiles are manufactured in the Suzhou textile manufacturing factory, they are transported to the garment factory in Jintan City, Jiangsu Province. The finished products are then transported to Singapore on container ships. The distribution and use is in Singapore. Used shirts are reused as rags several times and finally disposed together with domestic waste through incineration.

The functional unit considered was 250 g of plain cotton shirt. This quantity was chosen because 250 g is roughly the mass of a short-sleeved cotton shirt. However, since about 15% of the fabric is wasted during the shirt making process, the functional unit was taken as 295 g of cotton shirt instead. It was assumed that these shirts were worn in a tropical area, all year round. The lifespan of a shirt was taken to be 1 year.

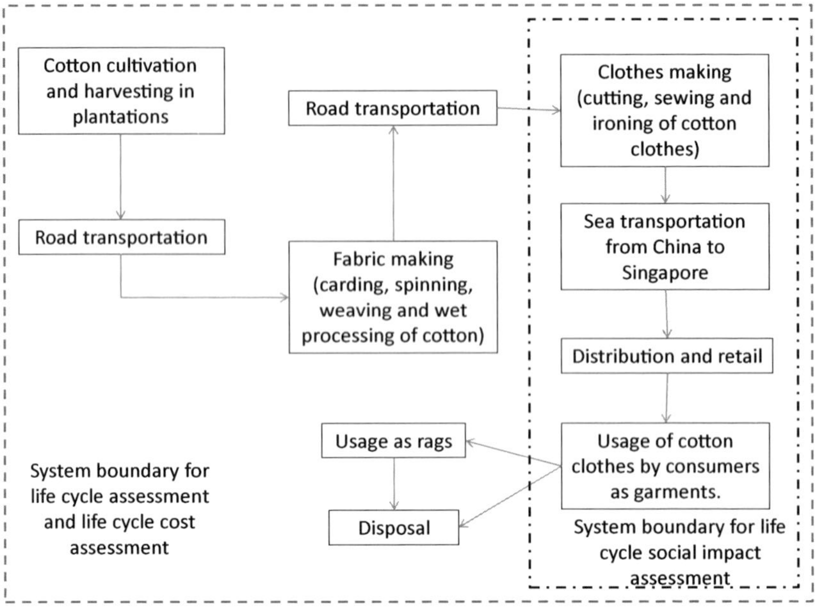


Fig. 10.1. The different system boundaries defined for lifecycle assessment, lifecycle social impact assessment and lifecycle costing.

Other than the stated sources of primary and secondary data, Ecoinvent databases were used to estimate the environmental impacts. Data collected was confined to the period between 2000 and 2012.

### *System Boundary and Cut-off Criteria*

The lifecycle stages considered in this study are shown in Fig. 10.1.

The data sources and uncertainties associated with each of these life cycle stages are summarised in Table 10.1.

Due to either the unavailability of data or the fact that there is no direct relation with the life cycle stages considered, the following activities are ignored in this study:

- Construction of the factory buildings,
- Land used by the industry and plantation,
- Fabrication and transport of the machines,
- Transportation of workers to factories or facilities,
- Materials that compose less than 5% of finish process (for example, stitch and label), and

Table 10.1. Data sources and uncertainties for every lifecycle stage considered for the lifecycle assessment.

| Lifecycle stages | Descriptions | Data source | Uncertainties in data (source, magnitude etc.) |
|---|---|---|---|
| Cotton cultivation and harvesting in plantations | Includes land conversion, consumption of electricity, water, fuel energy and fertilizer. | Dagistan *et al.*, 2009; Defra, 2007; Bluesign Technologies, 2008; Chapagain *et al.*, 2005; USDA, 2011. | The plantation practices and energy usage may differ slightly from the sources of data. |
| Carding, spinning, weaving and wet processing of cotton fibres | Includes water and electricity consumption. | Bluesign Technologies, 2008; Palamutcu, 2010. | Different technologies from those in the sources of data may be used. |
| Cutting, sewing and ironing of cotton cloth | Includes electricity consumption. | Altun, 2012 | Different technologies from those in the sources of data may be used. |
| Road transportation within China | Refer to annex 1 | Data was calculated based on average types of trucks used. | Some trucks used may differ from the truck assumed in calculation. |
| Sea transportation from China to Singapore | Refer to annex 2 | Data was calculated based on average types of trans-oceanic ships used. | Some ships used may differ from the ships assumed in calculation. |
| Distribution | Includes electricity usage. | Data was calculated based on average type of LED light bulb. | Total number and types of light bulbs used may differ in different stores. |
| Usage of shirt as garment; reuse as rags | Includes electricity and water consumption. Assumes that washing and ironing are done once per week per shirt, using 6.5 kg top loaded machine with normal water level mode, line drying and 1000 W iron. | TFTC, 2012; Pakula, 2010. Data was calculated based on 700 W, 6.5 kg work load washing machine and 1000 W iron. | The washing machines and iron used in the households may differ from the ones we assumed in our calculations. |
| Disposal | Include energy consumption. | Khoo *et al.*, 2009 | Technology used may differ. |

- Energy consumption of the factories apart from the manufacturing of products.

We assumed that road transportation between these stages within China and Singapore is by 5-tonne trucks, running at full load at any one delivery. Using the methodology detailed in annexes 1 and 2, it could be deduced that land and ship transportation of 1 unit of shirt required about 2.09 g of diesel and 1.28 g of heavy fuel oil respectively. The environmental impacts due to transportation were then calculated based on these resource consumptions.

### *Social Life Cycle Assessment (SLCA)*

The second component of the LCSA is SLCA, in which we evaluate the impact of cotton shirt on two main stakeholder groups — employees and the community within which the factories or facilities operate. Each lifecycle stage was accessed separately. However, due to a lack of rigorous data, these assessments were done only for the shirt-manufacturing (which includes the cutting, sewing and ironing of shirts) and retail stages. The sea transportation stage was not included because this service extends to a very wide range of other types of goods (such as food products) apart from cotton shirts. After applying mass allocation to these various transported goods and proportionate the impacts to cotton shirts, these impacts were found to be negligible.

In the present SLCA, focus was placed on the social impacts on employees. Working hours is one of the important factors to be considered to evaluate the effect on worker's health, safety and work-life balance. Many countries regulate the maximum number of working hours permitted by law. The effects of long working hours can be classified into short term and long term effects. Short term effects include fatigue, sleeping problems, blood pressure and acute injuries. Cardiovascular disease, depression, tension, anger and stress can occur as long term effects. In addition, the lack of work–life balance can cause marriage dysfunction, and problems with and in childcare and child development. In addition to this, routine shift work can be related to heart disease because of the psycho-social, behavioural and physiological stress caused by working non-standard work hours (Institute for Work and Health, 2010).

Recently, China put in place important laws to promote the employment of disabled people. For example, the Employment Promotion Law (2007) contains an anti-discrimination provision relating to persons with

Table 10.2. Data sources and uncertainties for every lifecycle stage considered for the lifecycle social impact assessment.

| Lifecycle stages | Descriptions | Data source | Uncertainties in data (source, magnitude etc.) |
|---|---|---|---|
| Cutting, sewing and ironing of cotton cloth | Include number of working hours, wages and workers. | Company's corporate social responsibility (CSR) book | Data is expected to be accurate. |
| Distribution and retail | Includes number of working hours, workers, wages and number of disabilities. | Company CSR books; informal interview with staffs | Data is expected to be accurate, especially at the time of interviews. |

disabilities and the 12th Five Year National Programme on Disability (2011–2015) has, as one of its key objectives, the creation of 1 million jobs for people with disabilities by 2015 (ILO, 2008). In Singapore, the Tripartite Guidelines on Fair Employment Practices sets out guidelines that prevent discrimination of any form, including against those with disabilities. However, there are no laws to safeguard the employment of disabled people. The number of disabled people employed in the life cycle of cotton shirt was considered as a factor to assess the social impact of the industry.

In summary, while there can be many other types of social related indicators, we focused on three of the most important and yet common indicators of impact on employees — working hours, wage and number of disabled people employed. Table 10.2 summarises the descriptions, data sources and uncertainties in this SLCA.

### *Life Cycle Cost Assessment (LCCA)*

In this third component of LCSA, the cost incurred at each lifecycle stage of the cotton shirt by different stakeholders was studied. It is worth noting that certain costs of externalities were also included, such as the cost of wastewater treatment and disposal. We defined the total life cycle cost as:

$$\mathrm{LCC} = \mathrm{C_{cp}} + \mathrm{C_{fm}} + \mathrm{C_{sm}} + \mathrm{C_{t}} + \mathrm{C_{dt}} + \mathrm{C_{cu}} + \mathrm{C_{di}} \qquad (1)$$

where LCC is the life cycle cost of the cotton shirt, $\mathrm{C_{cp}}$ is the cost of cotton production, which includes the costs of diesel, water, electricity, fertilisers, pesticide, ginning and harvesting; $\mathrm{C_{fm}}$ is the cost of fabric manufacturing, which includes costs of energy, water and wastewater treatment; $C_{sm}$ is the cost of shirt making, which includes costs of manpower, machineries, energy, water and materials; $C_t$ is the cost of transportation, which includes costs of service and fuel; $C_{dt}$ is the cost of distribution, which includes warehouse and shop rental, electricity and wages; $C_{cu}$ is the cost of consumer usage, which includes the cost of purchase of the shirt, and utility costs for water and electricity usage to maintain the shirt by washing and ironing; and $C_{di}$ is the cost of disposal, that is the cost borne by the customer for waste discharge. Table 10.3 summarises the descriptions, data sources and uncertainties in this LCCA.

It is worth noting that externalities have also been considered in this model. In particular, the cost of wastewater treatment, the water conservation tax, and the waterborne fee in the cost of water consumption were considered. The wastewater treatment cost is incurred at the fabric manufacturing stage to reduce pollutant levels in the water before it is discharged. This cost is charged by the city municipality and the cost for water treatment at the factory itself is not included in this calculation. The waterborne fee is taken into account at the consumer use stage and is levied to offset the cost of used water treatment and for operating and maintain used water network in Singapore.

## Results

### *Environmental Impacts*

Air emissions of gaseous sulphur dioxide, oxides of nitrogen, hydrogen chloride, hydrogen fluoride and ammonia contribute to acidification. The main sources of acidification are agricultural activities and fossil fuel combustion for electricity generation. Figure 10.2 shows that while distribution and disposal have minimum contributions to acidification, the production of cotton for shirts produces the highest amount of acidification (0.017 kg $SO_2$-eq, which accounts for 37.8% of the total), followed by the fabric manufacturing (0.014 kg $SO_2$-eq, which accounts for 31.1% of the total), and shirt-making (0.011 kg $SO_2$-eq, which accounts for 24.4% of the total) stages. The main causes for acidification are the nitrogen and phosphorous from fertilisers used for cotton-farming.

Figure 10.3 illustrates that the cotton production stage has the largest contribution to eutrophication — $4.1 \times 10^{-3}$ kg $PO_4$-eq (62.5% of the total eutrophication); this again is due to the usage of fertiliser for plant growth. Surprisingly, the second largest impact comes from the use stage, which

Table 10.3. Relevant lifecycle stages, data sources and uncertainties for every cost component considered for the lifecycle cost assessment.

| Cost component | Lifecycle stages relevant to cost component | Descriptions | Data source | Uncertainties in data (source, magnitude etc.) |
|---|---|---|---|---|
| Cost of cotton production, $C_{cp}$ | Cotton cultivation and harvesting in plantations | Includes costs of energy, fertilisers, pesticide, ginning and harvesting. | Zhou *et al.*, 2000; Chaudhry, 2008; JSChina, 2012. | The cost assumed may differ with current market price and exchange rate, due to price fluctuations. |
| Cost of fabric manufacturing, $C_{fm}$ | Carding, spinning, weaving and wet processing of cotton fibres | Includes costs of machineries, energy, water and materials. | DBS, 2011; Zhou *et al.*, 2005; Fu *et al.*, 2008. | The cost assumed may differ with current market price and exchange rate. |
| Cost of shirt making, $C_{sm}$ | Cutting, sewing and ironing of cotton cloth | Includes costs of manpower, machineries, energy, water and materials. | CFLA, 2011; DBS, 2011 | The cost assumed may differ with current market price and exchange rate, due to price fluctuations. |
| Cost of transportation, $C_t$ | Road transportation within China<br>Sea transportation from China to Singapore | Includes costs of service and fuel. | Calculated based on average shipping and fuel price. | The cost assumed may differ with current market price due to fluctuations in fuel prices. |
| Cost of distribution, $C_{dt}$ | Storage in warehouses | Includes rental cost of warehouses. | Calculated based on average warehouse rental cost in specific area. | Cost is expected to be accurate but prices will differ with locations of warehouses. |

(*Continued*)

Table 10.3. (*Continued*)

| Cost component | Lifecycle stages relevant to cost component | Descriptions | Data source | Uncertainties in data (source, magnitude etc.) |
|---|---|---|---|---|
| | Distribution and retail | Includes cost of manpower, electricity and rental. | Calculated based on average rental cost in the area, company recruitment advertisement and informal interview with staffs. | Cost is expected to be accurate. |
| Cost of consumer usage, $C_{cu}$ | Usage of shirt as garment; reuse as rags | Includes the utility costs for water and electricity usage to maintain the shirt by washing and ironing. | SPService, 2013. | Cost may differ based on type of machines used and usage pattern of consumers. |
| Cost of disposal, $C_{di}$ | Disposal | Cost of disposal incurred by customer. | NEA, 2013. | Cost is expected to be accurate. |

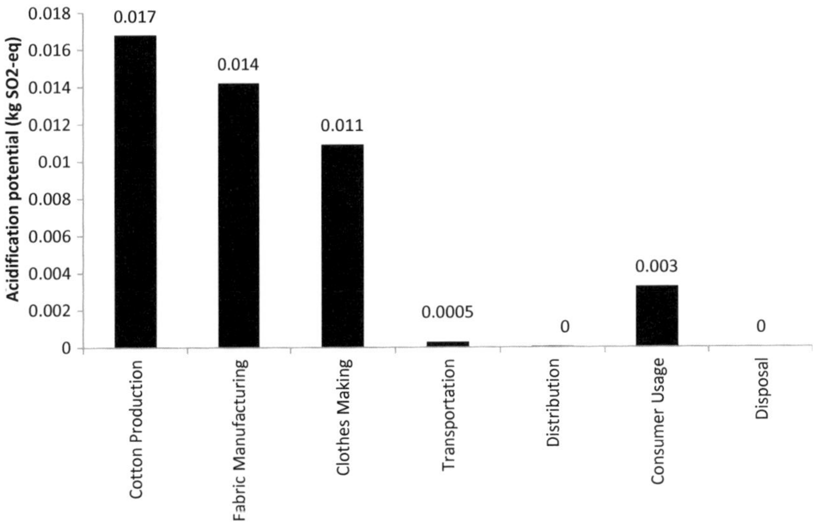


Fig. 10.2. Acidification potentials of the different lifecycle stages.

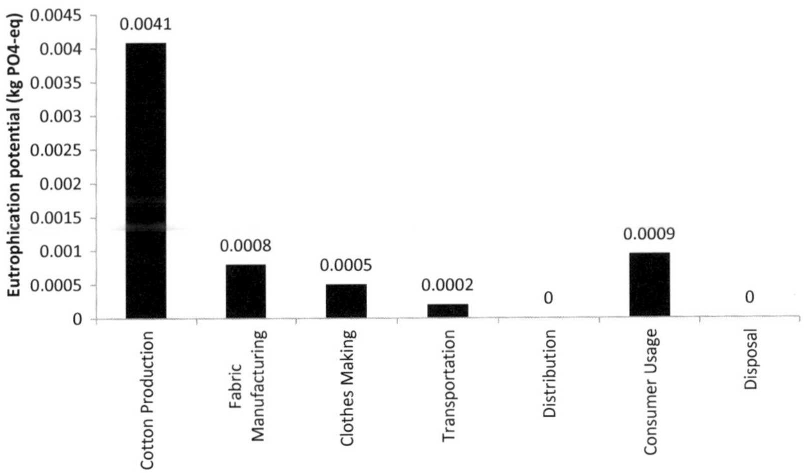


Fig. 10.3. Eutrophication potentials of the different lifecycle stages.

contributes $9.46 \times 10^{-4}$ kg $PO_4$-eq (14.1% of total) and this is due to the high consumption of electricity and water required in the washing and ironing of clothes.

This result for the use stage is also supported by data for GWP; Fig. 10.4 shows that the use stage has the highest GWP, accounting

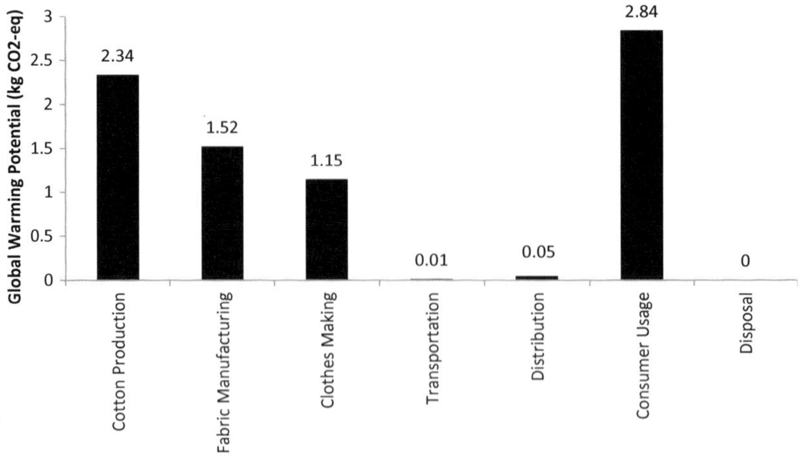


Fig. 10.4. Global warming potentials of the different lifecycle stages.

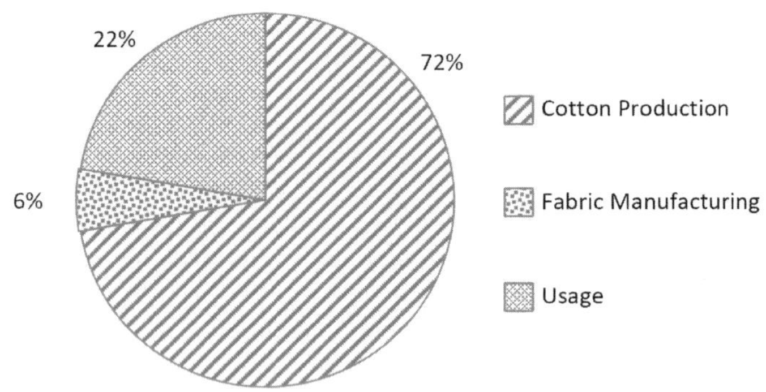


Fig. 10.5. Water consumptions in the different lifecycle stages.

for about 35.9% (2.84 kg $CO_2$-eq) of the total life cycle GWP. Cotton production is the second largest contributor of greenhouse gases, which constitutes approximately 29.6% (2.34 kg $CO_2$-eq) of the total emissions.

Transportation and disposal stages account for the least. In fact, these results are also caused by the use stage consuming the highest amounts of water, as shown in Fig. 10.5.

The potential of chlorofluorohydrocarbons (CFC) and chlorinated hydrocarbons (HCs) emissions for depleting the ozone layer is called the ozone depletion potential. Five out of seven life cycle stages have insignificant impact on ozone depletion. The major impacts are from the use stage ($3.87 \times 10^{-7}$ kg CFC-11-eq (70%)) and cotton production stages ($1.55 \times 10^{-7}$ kg CFC-11-eq (28%)).

### *Social and Economic Impacts*

Workers in the cotton garment factory work on a 9.5-hour shift, which includes a 90-minute lunch break. On average, workers need about 4.8 minutes to produce a finish product. On the other hand, at the distribution stage, workers need to spend about 3 minutes to complete a series of work, including unpacking, categorising into various sizing, folding and, if necessary, re-packing into custom packages. Staff members typically have to work 10.5-hour (including 90 minutes of lunch break) or/and 3.5-hour shifts. The first shift is from 8:00 am to 6:30 pm and the second shift finishes at 11:00 pm. According to the Labour Law of China, Article 36, workers must not work more than 8 hours a day and 40 hours a week. The Ministry of Manpower in Singapore regulates the maximum number of working hours as 12 hours per day. Hence, the working hours in both stages are in compliance with the laws of the respective nations, although those in China typically meet the maximum duration allowable by legislation.

The minimum monthly wage in Jiangsu province, China, is set at 1,140 RMB (USD 185.72), whereas the garment factory worker on average earns 1,230 RMB (USD 200.35) (China Labour Statistics Yearbook, 2009). In Singapore, there is no minimum wage restriction, but workers who receive less than S$2,000 (USD 1,582) per month are provided additional benefits, such as rest days, annual leave and sick leave, under Part IV of the Employment Act (Ministry of Manpower, 2008).

Finally, while no physically disabled staff member was found in the garment factory, it was observed that a physically disabled worker was employed by the retailer in Singapore as a "back office" staff. This employment complies with the company's internal corporate policy of employing at least one disable staff per shop. No disabled staff were employed by the garment factory.

Figure 10.6 compares the different cost components that make up the life cycle cost. It is clear that $C_{cu}$ outweighs the other costs, accounting for 61.8% of the total cost of the shirt. Distribution ($C_{dt}$) contributes to 28% of the total cost. It was found that approximately one quarter of $C_{cu}$ is incurred on the electricity and water required for maintenance of the shirts by consumers.

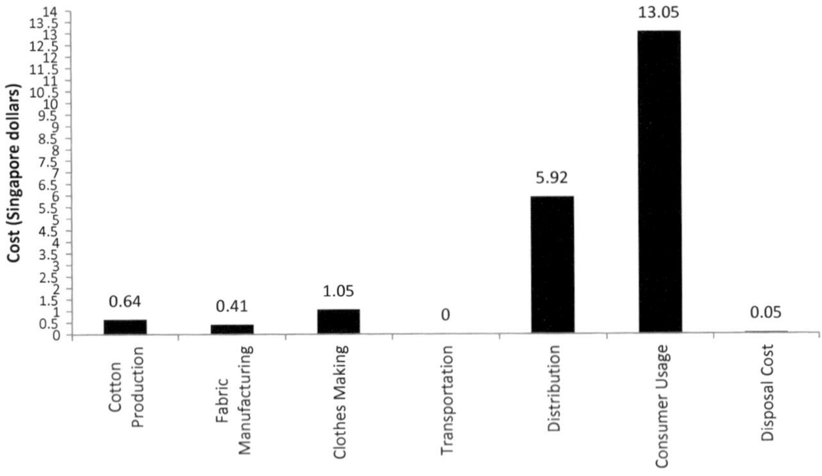


Fig. 10.6. Breakdown on life cycle cost into the cost of different stages. Here, S$1.26 is equivalent to USD 1.

## Sensitivity Analysis

The sensitivity analysis of our results was conducted by changing the parameters of the assumption in this study. Since the use stage has the highest impact in some of the assessment indicators, such as ozone depletion potential and GWP, the assumptions for this stage were modified and the results assessed. In this base case scenario (as described in Table 10.1), it was assumed that the washing machine can take a maximum load of 6.5 kg, is top loaded, and has a normal water level mode with medium efficiency for energy and water. Further, washed clothes are line dried and ironing is done with a 1000 W electric iron, and washing and ironing are done 52 times per year (that is, once per week).

Since consumer behaviour is largely varied, and it is possible to underestimate or overestimate the value in LCA, three alternative scenarios were considered. In scenario 1, we assumed that the washing machine and electric iron used are all energy and/or water efficient. Similar to the base scenario, washing and ironing are done once per week and washed clothes are line dried. In scenario 2, we assumed that all conditions are similar to the base scenario, except that there is no ironing required. Finally, conditions in scenario 3 are similar to those in the base scenario, except that the washing machine has a higher capacity and is operating at full load every time it is used.

Table 10.4. Percentage reductions in environmental impacts during the use stage, due to different alternate scenarios of consumer usage.

| | acidification | Global warming potential | Eutrophication | Ozone depletion |
|---|---|---|---|---|
| Scenario 1: using of energy and water efficient washing machine, and energy efficient iron. Washing and ironing done once per week. Washed clothes are line dried. | 32.24% | 22.06% | 49.18% | 19.70% |
| Scenario 2: similar conditions as base scenario but the only difference is that no ironing required. | 49.03% | 64.59% | 23.11% | 68.21% |
| Scenario 3: similar conditions as base scenario but the only difference is that a large capacity (8 kg) washing machine with full load. | 7.09% | 9.34% | 3.34% | 9.86% |

As shown in Table 10.4, it was found that using more energy and water efficient electronic appliances can reduce GWP and ozone depletion potential by about 22% and 20% respectively. When ironing is omitted entirely, reductions in impacts can run up to about 65% and 68% respectively. The cost incurred for energy and water consumptions will correspondingly be reduced. On the other hand, it was found that no significant decrease is found in washing with a higher capacity machine operating at full load. These results show the sensitivity of our results and hint at how changes in consumer behaviour, and policies that are successful in causing these changes, may have significant sustainability effects.

## Integrated Policies to Improve Overall Sustainability of Cotton Garment Industry

Integrated sustainability policies were defined by Kua (2007) as policies that fulfil any of the four characteristics in sustainability-oriented policy design and implementation. In short, such policies aim to address all the three domains of sustainability, including examining any possible conflicts among the indicators in the three domains. Furthermore, a variety of policy instruments and stakeholders are employed and engaged in the design and implementation of these policies. Integrated sustainability policies were shown to have higher likelihood of success in addressing complex sustainability problems. Henceforth, the objective of this study is to reduce the impacts identified by the applied LCSA using integrated policies, instead of proposing using improvements in one domain (or indicator) to make up for the negative impacts in another domain (or indicator).

Based on the results of the LCSA, the sustainability impact matrix (Table 10.5) was created to summarise and compare the sustainability performances of the different life cycle stages considered. If based on only LCA results, the distribution stage has much less impact than the shirt-making stage; however, LCCA revealed that the distribution stage contributes much more to the net life cycle cost. This exemplifies the usefulness of adopting a more holistic approach in evaluating the life cycle stages afforded by an LCSA approach, as opposed to using the conventional LCA approach. As mentioned, the net lifecycle cost of the use stage is the highest. The cost of the shirt is determined by market conditions, including consumers' valuation of the shirts they purchase from retailers. It is worth noting that the use stage also consumes the highest amount of energy and hence incur the highest energy costs. The main reason behind the low environmental impacts of transportation, distribution and disposal is that large quantities of shirts are involved in these stages and as a result of mass allocation, the impacts attributed to one unit of shirt is relatively low.

To address the above mentioned problems, what type of integrated policies should be implemented? First, working hours in the shirt-making stage in China can be improved by exploring a certain extent of automation in the process at the factory. However, while this can increase overall productivity and quality of the shirts, it may result in the retrenchment of workers. Besides, automation may result in an increase in energy consumption and its associated costs, leading to higher GWP. This highlights a possible cross-over effect, or conflict, among the three sustainability domains. A possible rectifying measure is to conduct a complete internal energy audit of the factory and reduce as much energy wastage in other areas of the factories' daily operations as possible. Non-governmental organisations, workers' unions and local governmental agencies may also identify areas or positions in which physically-disabled persons can be employed.

To reduce the environmental and economic impacts of the use stage, energy consumption for washing and ironing should be reduced. The conventional approach, which is especially supported by different green

Table 10.5. Sustainability impact matrix summarising the environmental, economic and social impacts from the life cycle sustainability analysis. Values shown are the percentage contributions by the life cycle stages. The shaded values are the highest contribution by the respective lifecycle stage in the associated impact.

| | Cotton production | Fabric manufacturing | Shirt making | Transportation | Distribution | Use | Disposal |
|---|---|---|---|---|---|---|---|
| Acidification | 37.8 | 31.1 | 24.4 | 0.0 | 0.0 | 6.7 | 0.0 |
| Global Warming Potential | 29.6 | 19.2 | 14.5 | 0.1 | 0.6 | 35.9 | 0.0 |
| Eutrophication | 62.5 | 12.5 | 7.8 | 3.1 | 0.0 | 14.1 | 0.0 |
| Ozone depletion | 28.0 | 2.0 | 0.0 | 0.0 | 0.0 | 70.0 | 0.0 |
| Life cycle cost | 3.0 | 1.9 | 5.0 | 0.0 | 28.0 | 61.8 | 0.2 |
| Working hours | | | Meet maximum allowable number of hours | | Below maximum allowable number of hours | | |
| Income | | | In the $0^{th}$–$9^{th}$ percentile of per capita income (2010) | | In about the $3^{rd}$ percentile of per capita income (2012) | | |

building standards, including the US Leadership in Energy and Environmental Design (LEED) and Green Mark Scheme of Singapore, is to promote the use of energy-efficient washing machines. However, as shown in our sensitivity analysis (Table 5), it is also necessary to eliminate or, at least reduce, the need for ironing, which can be realised by mixing cotton with a certain proportion of wool, especially for the front panel of the shirt.

Finally, to reduce the eutrophication of the cotton production stage, organic farming should be promoted; however, this may increase the production cost (hence the life cycle cost) of a cotton shirt. This increase is likely caused by greater labour inputs per unit of output, and a lack in the diversity of enterprises offering organic cotton implies that economies of scale cannot be achieved. To moderate price increase, subsidies for organic cotton are as important as increasing the consumer demand for organic cotton (so as to increase the volume of production, and subsequently spur economies of scale). The latter may be realised by highlighting the environmental benefits of organic cotton farming.

In conclusion, the LCSA proposed was able to not only assess the life cycle's environmental, social and economic impacts separately, but also underline certain cross-over effects or conflicts. Integrated policies were proposed to address these cross-over effects to ensure that, as much as possible, solutions sought for one set of indicators will not end up creating new problems for other indicators or/and domains.

## References

A Legal Aid and Human Rights Organisation, 2012. *Investigation Report of Garment Factory Fire.* [Online] Available at http://www.askbd.org/web/ [Retrieved on 14 Jan 2013].

AECOM and URS (2012). *North Carolina Maritime Strategy Vessel Size vs. Cost*, North Carolina Department of Transportation.

Alcover, F.C. (2010). *Life Cycle Analysis of a T-shirt*, ENSISA, Universitat Politècnica de Catalunya. Escola Tècnica Superior d'Enginyeria Industrial de Barcelona. Escola Tècnica Superior d'Enginyeria Industrial de Barcelona. Mobilitat. [Online]. Available at http://upcommons.upc.edu/pfc/bitstream/2099.1/14307/1/Mem%C3%B2ria.pdf [Accessed 15 Feb 2013].

Alesina, A. and Rodrick, D.D. (1994). "Distributive Politics and Economic Growth", *The Quarterly Journal of Economics*, **109**(2), 465–490.

Altun, S. (2012). "Life Cycle Assessment of Clothing Process", *Research Journal of Chemical Sciences*, **2**, 87–89.

Anderson, D.M. (1994). "Red tides", *Scientific American*, ***271***, 62–68.

Baker, D. (2007). *Health Effects of Long Working Hours, Workshop on Occupational Health and Mental Health*, Peking University, 24–28 September. [Online]. Available at http://www.workhealth.org/Adobe%20Acrobat%20 files/China%20Baker%201.pdf [Accessed on 10 Jan 2013].

Barnes, J.R. (2009). *Life Cycle Inventory for Cotton, Cotton Incorporated.* [Online]. Available at http://cottontoday.cottoninc.com/Sustainability-About/Life-Cycle-Inventory-Data-For-Cotton/Life-Cycle-Inventory-Data-For-Cotton.pdf [Accessed on 20 Feb 2013].

Baugh, G. (2008). *Polyester vs. Cotton — Which Is Better for the Environment?, Apparel Design & Merchandising*, San Francisco State University. [Online]. Available at http://www.udel.edu/fibre/issue2/responsibility/ [Accessed on 13 Jan 2013].

Beijing Orient Agribusiness Consultant (2012). *Biweekly Market Report*, Issue 20120110. [Online]. Available at: http://en.cnagri.com/report/woodcotton/ 20120201/252838.html# [Accessed on 15 Jan 2013].

Berniell, M. and Sanchez-Paramo, C. (2012). "Overview of Time Use Data Used for the Analysis of Gender Differences in Time Use Patterns", *Background paper for the WDR.*

Bluesign Technologies (2008). "Water Management in China's Apparel and Textile Factories", *AFIRM RSL Seminar presentation*, 27 Sep 2007. Reprinted in BSR.

Business for Social Responsibility (2009). *Apparel Industry Life Cycle Carbon Mapping.* [Online]. Available at: www.bsr.org.

Caniato, F., Caridi, M., Crippa, L., and Moretto, A. (2012). "Environmental Sustainability in Fashion Supply Chains: An exploratory case based research", *International Journal of Production Economics*, **135**, 659–670.

Capitano, C., Traverso, M., Rizzo, G., and Finkbeiner, M. (2011). "Life Cycle Sustainability Assessment: An Implementation to Marble Products", *Proceedings of the LCM 2011 Conference.* Berlin, 29–31 August.

Center for Research and Policy Making (2005). *World Trends in Clothing Manufacturing*, Occasional Paper No. 9, Macedonia, 31 Nov.

Chan, W.L. (2005). "Gender and Global Labor Organising: Migrant Women Workers of Garment Industry in South China", *Conference by Sweatshop Watch*, May 8–9.

Chand, A. (2006). "Physical and Psychological Health Problems of Garment Workers in the Fiji", *Pacific Public Health 3*, **13**(2), 65–70.

Chapagain, A.K., Hoekstra, A.Y., Savenije, H.H.G., and Gautam, R. (2005). "The water footprint of cotton consumption", *Research Report Series 18*, UNESCO-IHE.

Chapagain, A.K., Hoekstra, A.Y., Savenije, H.H.G., and Gautam, R. (2006). "The Water Footprint of Cotton Consumption: An assessment of the Impact of Worldwide Consumption of Cotton Products on the Water Resources in the Cotton Producing Countries", *Ecological Economics*, **60**, 186–203.

Chapman, A. (2010). "Mistra Future Fashion — Review of Life Cycle Assessments of Clothing", *Oakdene Hollins Research & Consulting*, Sweden.

Chaudhry, M. (2008). "World Cotton Outlook and production cost", *Presented at the 9th Meeting of the Southern and Eastern African Cotton Forum (SEACF) Dar es Salaam*, Tanzania, 7–8 April.

Chaw, C. (2003). "Rural Women Migrating to Urban Garment Factories in Myanmar", *Social Challenges for the Mekong Region*, Chiang Mai.

Cherrett, N., Barrett, J., Clemett, A., Chadwick, M., and Chadwick, M.J. (2005). *Ecological Foot Print and Water Analysis of Cotton, Hemp and Polyster*, Stockholm Environment Institute.

*China Environmental Statistical Yearbook 2006, 2007, 2008, 2009 and 2010.* [Online]. Available at http://english.mep.gov.cn/standards_reports/EnvironmentalStatistics/yearbook2006/ [Retrieved on 20 Jan 2013].

*China Labour Statistical Yearbook, 2009.* [Online]. Available at http://chinadataonline.org

China Fair Labour Association (2011). *Independent Verification External Report: Chengfeng Group.* [Online]. Available at http://portal.fairlabor.org/fla/go.asp?u=/pub/zTr5&pm=0&hfn=3&hfs=28

Chougule, M.B., and Sonaje, N.P. (2012). "Novel Techniques of Water Recycling in Textile Wet Processing through Best Management Practices", *International Journal of Applied Science and Advance Technology*, **1**, 29–33.

Ciroth, A., and Franze, J. (2011). "Social Aspects of Products over the Whole Life Cycle", *Conducting Social LCA Workshop*, Berlin, 30 May.

*Company CSR Report 2010, 2011.*

Curran, M.A. (2006). "Life Cycle Assessment: Principles and Practice", *Scientific Applications International Corporation, Work Assignment 3–15.*

Dagistan, E., Akcaoz, H., Demirtas, B.R., and Yilmaz, Y. (2009). "Energy usage and benefit-cost analysis of cotton production in Turkey", *African Journal of Agricultural Research*, **4**(7), 599–604.

DBS Group Research (2011). *China: A Case Study of Price Control on Electricity.* [Online]. Available at https://www.dbsvresearch.com/...nsf/(vwAllDocs)/.../CN_ 110520.pdf [Retrieved on 15 Feb 2013].

Department of Environment and Rural Affairs, London (2007). *Mapping of Evidence on Sustainable Development Impacts that Occur in the Lifecycles of Clothing.*

Department of Environment and Rural Affairs, London (2009). *Reducing the environmental Impact of Clothes Cleaning.* [Online]. Available at: www.defra.gov.uk

Department of Environment and Rural Affairs, London (2011), *Sustainable Clothing Roadmap.* [Online]. Available at https://www.gov.uk/government/uploads/system/uploads/attachment_data/file/69299/pb13461-clothing-actionplan-110518.pdf

Dreyer, L.C., Hauschild, M.Z., and Schierbeck, J. (2006). "A Framework for Social Life Cycle Impact Assessment", *International Journal of Life Cycle Assessment*, **11**, 88–97.

Hamdy, E.M., Barghouti, S., Gillham, F., and Al-Saffy, M.T. (1994). "Cotton Production Prospects for the Decade to 2005: A Global Overview", *World Bank Technical Paper No. 231.*

Environmental Justice Foundation (n.d.). *Cotton and Water*. [Online]. Available at: http://ejfoundation.org/cotton/cotton-and-water [Retrieved on 10 Jan 2013].

Finkbeiner, M., Schau, E.M., Lehmann, A., and Traverso, M. (2010). "Towards Life Cycle Sustainability Assessment", *Sustainability*, **2**, 3309–3322.

Food and Agriculture Organisation of the United Nations (2012). *Fertiliser consumption by country*, FAOSTAT.

Fu, T., Chang, M., and Zhong, L. (2008). *Reform of China's Urban Water Sector*, 1st Edition, IWA publishing.

Glucha, P.. and Baumann, H. (2004). "The Life cycle Costing (LCC) approach: A Conceptual Discussion of its Usefulness for Environmental Decision Making", *Building and Environment*, **39**, 571–580.

Guinee, J.B., Heijungs, R., and Huppes, G. (2011). "Life Cycle Assessment: Past, Present and Future", *Environmental Science & Technology*, **45**, 90–96.

Guinee, J.B., Heijungs, R., and Huppes, G. (2009). *Research Strategy, Programmes and Exemplary projects on LCSA*, CALCAS project No. 037075.

Heffer, P. (2009). *Assessment of Fertiliser Use by Crop at the Global Level 2006/07-2007/08*, International Fertiliser Industry Association.

Horrigan, L., Lawrence, R.S., and Walker, P. (2002). "How sustainable agriculture can address the environmental and human health harms of industrial agriculture", *Environmental health perspectives*, **110**(5), 445–456.

Hunkeler, D., Lichtenvort, K., and Rebitzer, G. (2008). *Environmental Life Cycle Costing*, SETAC Press, Florida.

IFC (2007). *Environmental, Health and Safety Guidelines for Textile Manufacturing*, The World Bank Group.

Institute for Global Labour and Human Rights (2010). *Comparative Garment Worker Wages*. [Online]. Available at: http://www.globallabourrights.org/alerts?id=0297 [Retrieved on 14 Jan 2013].

Institute for Work and Health (2010). "The Health Effects of Shift Work: Summary Report", *Scientific Symposium*, Toronto, 12 April. [Online]. Available at: http://www.iwh.on.ca/system/files/documents/shift_work_2010_summary_report.pdf

Intergovernmental Panel on Climate Change (2000). "Special Report on Emission Scenarios, *A Special Report of IPCC Working Group III.*

Intergovernmental Panel on Climate Change (2006). "Mobile Combustion", *Volume 2: Energy*, IPCC Guidelines for National Greenhouse Gas Inventories, 3.47–3.50.

International Cotton Advisory Committee (2009). *Economics and Statistics*. [Online. Available at: http://icac.org/economics-and-statistics/background [Retrieved on 13 Feb 2013].

International Labour Organisation (2013). *Working Time: Importance and Applications*. [Online]. Available at: http://www.ilo.org/global/statistics-and-databases/statistics-overview-and-topics/working-time/importance-and-applications/lang–en/index.htm [Retrieved on 13 Jan 2012].

International Labour Organisation (2008). *Decent Work Country Programme: The People's Republic of China.* [Online]. Available at: http://www.ilo.org/public/english/bureau/program/dwcp/download/china.pdf

ISO 14040 (2006). *Environmental management — Life cycle assessment — Principles and framework.*

ISO 14044 (2006). *Environmental management — Life cycle assessment — Requirements and guidelines.*

Istvánovics, V. (2009). "Eutrophication of Lakes and Reservoirs", *Encyclopaedia of Inland Waters*, pp. 157–165.

*JSChina* (2012). "Jiangsu to adopt new electricity pricing system in June, 2012", Apr 10, *JSChina.* [Online]. Available at: http://english.jschina.com.cn/TodayJiangsu/201204/t983035.shtml

Kalliala, E.M., and Nousiainen, P. (1999). "Life cycle assessment environmental profile of cotton and polyester cotton fabrics", *AUTEX Research Journal*, **1**(1), 8–20.

Khoo, H.H., Lim, T.Z., and Tan, B.H. (2010). "Food Waste Conversion Options in Singapore: Environmental Impacts based on an LCA Perspective", *The Science of the Total Environment*, **408**(6), 1367–1373.

Kua, H.W. (2007). "Information Flow and Its Significance in Coherently Integrated Policymaking for Promoting Energy Efficiency", *Environmental Science and Technology*, **41**(9), 3047–3054.

Metz, B., Davidson, O.R., Bosch, P.R., Dave, R., and Meyer, L.A. (2007). *Contribution of Working Group III to the Fourth Assessment Report of the Intergovernmental Panel on Climate Change*, Cambridge University Press, Cambridge, United Kingdom and New York.

Miller, G. (2004). *Environmental Science: Working with the earth*, Tenth edition, Brooke/Cole.

Ministry of Environment and Water Resources (n.d.), *Solid Waste Management.* [Online]. Available at: http://app.mewr.gov.sg/web/Contents/contents.aspx?ContId=680 [Accessed on 07 Feb 2013].

Ministry of Manpower (2008). *Changes to Employment Act.* [Online]. Available at: www.mom.gov.sg/Documents/employment-practices/changes-to-the-ea.pdf

Morgan, A. (2011). "The Environmental impacts of Cotton: Moving toward Solution", *Cotton and water management roundtable*, UC Berkeley, Haas Business School, 02 Mar. [Online]. Available at: http://responsiblebusiness.haas.berkeley.edu/documents/2-Alexis.pdf [Accessed on 13 Jan 2013].

National Aeronautics and Space Administration {n.d.), *Key Indicators of Global Climate Change*,=. [Online]. Available at: http://climate.nasa.gov/key_indicators/ [Retrieved on 22 Jan 2013].

National Cotton Council of America (2012). *Rankings.* [Online]. Available at: http://www.cotton.org/econ/cropinfo/cropdata/rankings.cfm [Accessed on 13 Jan 2013].

National Environmental Agency Singapore (n.d.). *Waste Collection Fee.* [Online]. Available at: http://cms.nea.gov.sg/PublicWasteCollection.aspx [Accessed on 19 Jul 2013].

National Environment Agency Singapore (2011). *Waste Statistics and Recycling Rate.* [Online]. Available at: http://app2.nea.gov.sg/topics_wastestats.aspx [Accessed on 10 Jan 2013].

Natural Resources Defense Council (2012). *From Field to Store: Your T-shirt's Life Story.* [Online]. Available at: http://www.nrdc.org/living/stuff/t-shirt-life-story.asp [Accessed on 20 Jan 2013].

Nielsen, A., and Nielsen, P. (2009). *Comparative life cycle assessment of the Elemental T-shirt produced with biotechnology and a Conventional T-shirt produced with conventional technology.* [Online]. Available at: http://www.novozymes.com/en/sustainability/Published-LCA-studies/Documents/Comparative%20LCA%20of%20a%20t-shirt%20produced%20with%20biotechnology%20and%20conventional%20technologies.pdf [Accessed on 20 Jan 2013].

OECD (n.d.). *Water Use in Agriculture.* [Online]. Available at: http://www.oecd.org/environment/wateruseinagriculture.htm [Accessed on 11 Jan 2013].

Organisation for Economic Co-operation and Development, Paris (1982). *Eutrophication of Water, Monitoring, Assessment and Control.* [Online]. Available at: http://www.chebucto.ns.ca/ccn/info/Science/SWCS/TPMODELS/OECD/oecd.html

Pakula, C., and Stamminger, R. (2010). "Electricity and Water Consumption for Laundry Washing, by Washing Machine Worldwide", *Energy Efficiency,* **3**, 365–382.

Palamutcu, S. (2010). "Electric energy consumption in the cotton textile processing stages", *Energy,* **35**(7), 2945–2952.

Paul, W., and George, W. (2007). "Fertilisers and other farm chemicals Review Article", *Handbook of Agricultural Economics,* **3**, 2741–2777.

Rebitzer, G., and Hunkeler, D. (2003). "Life cycle costing in LCM: Ambitions, opportunities, and limitations — Discussing a framework", *International Journal of Life Cycle Assessment,* **8**(5), 253–256.

Roth, G. (2010). *Economic, Environmental and Social Sustainability indicators of the Australian Cotton Industry.*

Saicheua, V., Cooper, T., and Knox, A. (2011). "Public Understanding Towards Sustainable Clothing and the Supply Chian", *Fashion Colloquia,* London College of Fashion.

Schwab, K. (2012). "The Global Competitiveness Report 2012–2013", *World Economic Forum,* Geneva.

SETAC (1992). "A Conceptual Framework for Life Cycle Impact Assessment", *SETAC and SETAC Foundation for Environmental Education, Inc Workshop Report,* 1–7 Feb, Florida, USA.

Singapore Department of Statistics (2008). *Report on the household expenditure survey Singapore Yearbook of Manpower Statistics,* 2012. [Online]. Available at: http://www.mom.gov.sg/documents/statistics-publications/yearbook12/mrsd_2012yearbook.pdf

SP Service (2013). *Electricity Tariff.* [Online]. Available at: http://www.singaporepower.com.sg/irj/go/km/docs/wpccontent/Sites/SP%20Services/Site%20Content/Tariffs/documents/latest_press_release.pdf [Accessed on 14 Jan 2013].

Technical Textile Markets (2007). *Global Trends in Fibre Prices, Production and Consumption.*

The Federal Trade Commission (2012). *Appliance Energy Data: Standard Clothes Washer Energy Data.* [Online]. Available at: http://www.ftc.gov/bcp/ conline/edcams/eande/appliances/data/2012/clwasher/standard_brand.pdf [Accessed on 22 Jan 2013].

The National Institute of Environmental Health Sciences (NIEHS) (n.d.). *Pesticides.* [Online]. Available at: http://www.niehs.nih.gov/health/topics/agents/pesticides/index.cfm [Accessed on 14 Jan 2013].

Thomas, E.G. (1998). *Streamlined Life-Cycle Assessment*, Prentice Hall, New Jersey.

U.S Bureau of Labour Statistics (2012). "How do U.S. expenditures compare with those of other countries?", *Focus on Price and Spending*, **2**(16). [Online]. Available at: http://www.bls.gov/opub/focus/volume2_number16/cex_2_16.htm [Accessed on 15 Feb 2013].

U.S. Energy Information Administration (2012). *Country Analysis Brief Overview: China.* [Online]. Available at: http://www.eia.gov/countries/country-data.cfm?fips=CH [Accessed on 13 Jan 2013].

U.S Environmental Protection Agency (n.d.). *Wastes-Resource Conservation-Common Wastes & Materials.* [Online]. Available at: http://www.epa.gov/osw/conserve/materials/textiles.htm

UNEP and SETAC (2009). *Guidelines for Social Life Cycle Assessment of Products*, UNEP/SETAC Life Cycle Initiative.

UNEP and SETAC (2009). *Life Cycle Management — How business uses it to decrease footprint, create opportunities and make value chains more sustainable*, UNEP/SETAC Publications.

UNEP and SETAC (2011). *Towards A Life Cycle Sustainability Assessment: Making Informed Choice on Products*, UNEP/SETAC Life Cycle Initiative.

UNEP (2002). Integrated Assessment of Trade Liberalization and Trade-Related Policies: A Country Study on the Cotton Sector in China, *Synthesis Report of UNEP Country Projects-Round II.* [Online]. Available at: www.unep.ch/etu/publications/CSII_China.pdf [Accessed on 15 Feb 2013].

United Nations Industrial Development Organisation (1992). *Output of a Seminar on Energy Conservation in Textile Industry.* [Online]. Available at: http://www.unido.org/fileadmin/import/userfiles/puffk/textile.pdf [Accessed on 15 Feb 2013].

USDA Foreign Agricultural Service (2011). *China-Peoples Republic of Cottons and Products Annual*, GAIN Report No. CH 11018.

Voluntary Inter-industry Commerce Solutions (2009). *Chinese Apparel & Textile Industry Revitalization Underway.* [Online]. Available at: http://www.vics.org/docs/committees/Voice_of_VICS_Aug_2009_CLIP.pdf [Accessed on 15 Feb 2013].

World Bank (2013). *Employment data of China.* [Online]. Available at: http://data.worldbank.org/indicator/SL.UEM.TOTL.ZS [Retrieved on 13 Mar 2013].

World Economic Forum (2010). *The Global Gender Report.* [Online]. Available at: http://www.weforum.org/pdf/gendergap/report2010.pdf [Accessed on 22 Jan 2013].
World Energy Council (2011). *Global Transport Scenarios 2050.* [Online]. Available at: http://www.wec-france.org/DocumentsPDF/Etudes%20CME/wec_transport_scenarios_2050.pdf [Accessed on 13 Jan 2013].
WWF (2007). *Cleaner, Greener Cotton: Impacts and Better Management Practices.*
Zamagni, A., Buttol, P., Amerighi, O., Felici, B., Roberto, R., and Masoni, P. (2011). "Definition of Social Indicators: A case study of an innovative technology", *SETAC 17th LCA Case Study Symposium Sustainable Lifestyles*, Budapest, Hungary, 28 Feb–01 Mar.
Zevenhoven, R., and Kilpinen, P. (2001). *Control of Pollutants in Flue Gases and Fuel Gases.* [Online]. Available at: http://users.abo.fi/rzevenho/greenhou.PDF [Accessed on 19 July 2013].
Zhang, T. (2011). *Country Sector Overview: The Cotton Sector in China*, The Sustainable Trade Initiative.
Zhou, Z., Jiang. H., and Qin, L. (2007). "Life Cycle Sustainability Assessment of Fuels", *Fuel*, **86**(1–2), 256–263.
Zhou, Y., and Tol, S.J. (2005). *Water Use in China's Domestic, Industrial and Agricultural Sectors: An Empirical Analysis, Working Paper FNU-67.* [Online]. Available at: http://www.fnu.zmaw.de/fileadmin/fnu-files/publication/working-papers/WD_ZhouFNU67.pdf [Accessed on 23 Jan 2013].
Zhou, Y.Z., and Wei, B. (2000). "Pricing Irrigation Water in China", *Pricing Irrigation Water*, Washington D.C., Resources for the Future.

## Annex 1: Land Transportation within China and Singapore

The distance from the Suzhou cotton field to the fabric manufacturing factory under study is about 9.6 km. The distance between Suzhou to the garment factory in Jintan city is about 131 km. It is very common for companies to utilise 5-tonne trucks for the transportation of goods. At full load, such a truck can carry 20 tonnes of cargo. From our interview with transportation companies in Singapore,[1] a truck can typically cover an average of about 3 km on 1 litre of diesel at full load. Assuming that a truck is always at full load, it will consume:

$$[140.6\,\text{km}/3\,\text{km}] \times 1 \text{ litre} = 46.87 \text{ litre of diesel} \quad \text{(A.1)}$$

46.87 litres of diesel weighs about 39 kg. So, one unit of shirts needs about 1.95 g of diesel.

[1]We interviewed a total of around 20 companies in Singapore, including PL Scientific Services Pte Ltd, for this estimates.

The distance between the seaport and retail store in Singapore is averaged as 10 km. Using the above assumptions, it can be easily deduced that one unit of shirts expends about 0.14 g of diesel. Therefore, by mass allocation and direct proportion, one unit of shirt requires 2.09 g of diesel for land transportation.

### Annex 2: Sea Transportation within China and from China to Singapore

Shirts are transported from the garment factory in Jintan city to Changzhou port by ship or boat over about 40 km, which requires about an hour of travel. Shipping from Changzhou to Singapore covers a distance of about 2,242 nautical miles, at a speed of about 20 knots. This requires about 4.6 days in duration. Hence, the total duration of ship transportation is about 4.64 days.

Normally, the capacities of these carriers varies from 30,000 to 35,000 tonnes (an average of 32,500 tonnes was considered for this study) and they are powered by heavy fuel oil. The guidelines provided by the IPCC (2006) estimates that the *daily* fuel consumption of these trans-ocean carriers going at full power can be computed from the formula:

$$20.186 + (0.00049 \times \text{GRT}), \tag{A.2}$$

where GRT stands for the gross tonnage of the carrier. Since the GRT is taken to be 32,500 tonnes, the daily consumption of heavy fuel oil by the carrier would be about 36 tons. Over a period of 4.64 days, the bulk carrier would require about 167 tonnes of heavy fuel oil. By mass allocation and direct proportion, a unit of shirt requires about 1.28 g of heavy fuel oil.

# 11

# SUSTAINABILITY ISSUES AND STRATEGIES OF BIOFUEL DEVELOPMENT IN SOUTHEAST ASIA*

MALLIKA NAGURAN

Biofuels remain the crux of contention as ASEAN countries continue to entrench these agro-based fuels within the energy security agenda, at the cost of forest-dependent communities. ASEAN as a regional bloc is yet to drive sustainability in biofuels, but could do so if it pays heed to the resource nexus of land, food, energy and water within the contexts of governance and market forces. Governance in biofuels is important, and is no longer dependent on conventional actors such as governmental entities, but on non-governmental organisations, financial institutions, and civic groups. Resource trade cycle and life cycle analyses are two models that need to be included in the evaluation of biofuel influences and impacts. Recommendations are made towards the sustainable development of biofuels that include making policy changes, regulating capital and market flows, as well as instituting standards and best practices. A new definition of sustainable development is offered in this light.

## Biofuel Development Overview

In light of biofuels being presented as an alternative energy source to carbon intensive oil, gas and coal, the question of whether biofuels are indeed clean alternatives to fossil fuels must be posed, given the huge environmental degradation and suffering experienced by the affected communities in the course of biofuel development.

The world's biofuel production grew by about 13.8 percent, from 52 million tonnes of oil equivalent (Mtoe) in 2009 to 59 Mtoe (or slightly over 100 billion litres) in 2010, led by market leaders in the European Union, USA and Brazil (Table 11.1 and Table 11.2).

Biofuels typically refer to ethanol and diesel derived from biomass (plant matter). Bio-ethanol is made from sugar-based feedstock such as sugar cane, cassava and sorghum. Biodiesel (methyl esters) comes from oil seed crops such as palm oil, jatropha and coconut.

*This paper was supervised by Associate Professor Kua Harn Wei, Department of Building, School of Design and Environment, National University of Singapore.

Table 11.1. World fuel ethanol outlook (billion litres).

| Supply | 2010 | 2020 |
|---|---|---|
| United States | 50.09 (of which cellulosic ethanol credits 0.00) | 68.00 (of which cellulosic ethanol credits 10.00) |
| Brazil | 25.53 | 60.00 |
| European Union | 4.26 (of which cellulosic ethanol credits 0.00) | 8.10 (of which cellulosic ethanol credits 10.00) |
| China | 2.05 | 2.58 |
| Canada | 1.31 | 2.30 |
| Australia | 0.27 | 0.65 |
| India | 0.24 | 1.69 |
| Thailand | 0.43 | 1.16 |
| Others | 0.95 | 5.59 |
| **Total** | **85.12** | **150.07** |

Source: Informa Agra & Commercial, 2011.

Table 11.2. World biodiesel outlook (1,000 tonnes).

| Supply | 2010 | 2020 |
|---|---|---|
| European Union | 8,949 | 11,400 |
| Brazil | 2,110 | 3,700 |
| Argentina | 1,815 | 3,500 |
| United States | 1,080 | 3,450 |
| Indonesia | 650 | 1,900 |
| Thailand | 560 | 1,400 |
| Canada | 190 | 1,100 |
| Malaysia | 117 | 1,200 |
| Others | 1,197 | 2,400 |
| **Total** | **16,668 (18.94 billion litres)*** | **30,090 (34.2 billion litres)*** |

*Author's calculation (conversion of tonnes to litres).
Source: Author's creation based on Informa Agra & Commercial (2011).

Biofuel production in their various forms has been pitched as being less polluting than coal or oil production (Figure 11.1). However, environmental activists (e.g. Greenpeace and ActionAid) have brought to attention specific practices that paint a contradictory picture indicating their multi-dimensional destructive impact on the environment, including carbon dioxide ($CO_2$) emissions (Greenpeace, 2010; ActionAid, 2012). $CO_2$ is a colourless, odourless and non-poisonous gas that is formed by the

Table 11.3. An ASEAN overview of population, territory and economy.

| Country | Total land area (sq km) | Total population (thousand) | Gross domestic product: at current prices (US$ Mn) | Gross domestic product: at current prices (PPP$ Mn)[2] | Gross domestic product: per capita (US$) | Gross domestic product: per capita (PPP$) |
|---|---|---|---|---|---|---|
| Brunei Darussalam | 5,765 | 415 | 12,402 | 19,406 | 29,915 | 46,811 |
| Cambodia | 181,035 | 15,269 | 11,168 | 28,985 | 731 | 1,898 |
| Indonesia | 1,860,360 | 234,181 | 708,032 | 1,030,998 | 3,023 | 4,403 |
| Lao PDR | 236,800 | 6,230 | 6,508 | 16,105 | 1,045 | 2,585 |
| Malaysia | 330,252 | 28,909 | 238,849 | 415,157 | 8,262 | 14,361 |
| Myanmar[1] | 676,577 | 60,163 | 43,025 | 76,601 | 715 | 1,273 |
| Philippines | 300,000 | 94,013 | 189,326 | 351,686 | 2,014 | 3,741 |
| Singapore | 710 | 5,077 | 223,015 | 291,934 | 43,929 | 57,505 |
| Thailand | 513,120 | 67,312 | 318,709 | 585,698 | 4,735 | 8,701 |
| Viet Nam | 331,051 | 86,930 | 107,650 | 291,260 | 1,238 | 3,351 |
| ASEAN | 4,435,670 | 598,498 | 1,858,683 | 3,107,829 | 3,106 | 5,193 |
| CLMV[3] | 1,425,463 | 168,592 | 168,351 | 412,951 | 999 | 2,449 |
| ASEAN6[4] | 3,010,207 | 429,907 | 1,690,332 | 2,694,878 | 3,932 | 6,269 |

[1] Myanmar: USS-Kyat exchange rate is based on the parallel rate as used in IMF-WEO April 2011.
[2] GDP per capita in PPPS is GDP converted to international dollars using purchasing power parity (PPP) rates; hence PPP $1 in a country, say Cambodia has the same purchasing power as PPP $1 in all other countries in the world. PPP is an index of purchasing power, showing the purchasing power of US$1 in a country compared to US$1 in the benchmark country (USA).
[3] CLMV includes Cambodia, Lao PDR, Myanmar and Viet Nam
[4] ASEAN6 includes Brunei Darussaiam, Indonesia, Malaysia, Philippines, Singapore and Thailand
Source: ASEAN, 2011b.

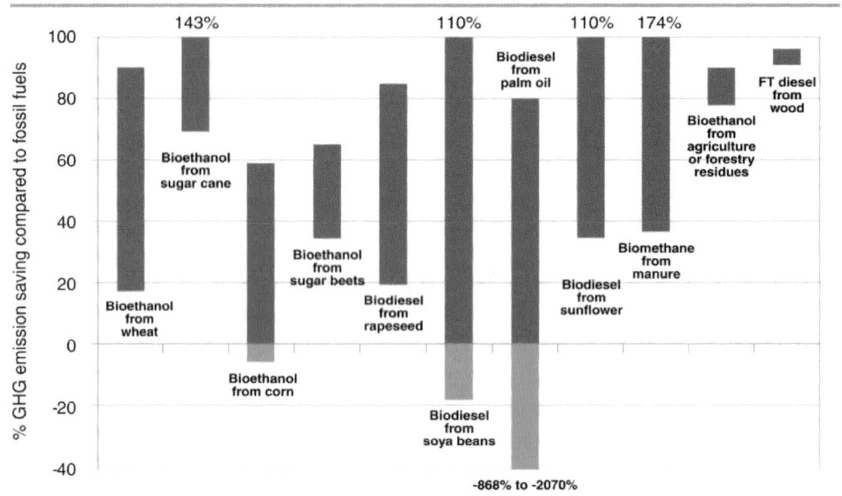


Fig. 11.1. Biofuel varieties and greenhouse gas emission savings (Source: UNEP, 2009).

combustion of carbon and living organism respiration. It is considered a greenhouse gas, or GHG (United Nations, 1992). $CO_2$ emitting activities include slash-and-burn crop cultivation,[1] deforestation, water resource depletion, agrochemical pollution (due to fertilisers, pesticides), among others. This explains the negative figure of GHG savings (up to minus 2,070 percent) of bio-diesel from palm oil in Figure 11.1 (UNEP, 2009).

Bio-ethanol's ability to replace liquid fossil energy as a renewable fuel is the foremost reason for its increased demand and consumption in many parts of the world, particularly in Brazil and the USA, reducing its dependence on fossil energy to fulfil energy security requirements.

Blending targets or mandates for motorised vehicles in more than 50 countries have led to an explosion in biofuel production quantity. Figure 11.2 illustrates this dramatic increase, led by the European Union (IEA, 2011). Global biofuel production increased from 16 billion litres in year 2000 to 100 billion litres within a decade.

[1]The slash-and-burn agriculture is a method of cultivation whereby areas of the forest are burnt and cleared for planting. It is a conventional practice as well. When soil fertility declines, cultivation shifts to a new plot. Farmers often resort to slash-and-burn technique on forests to clear the land for biofuel crops. The slash and burn agriculture definition was taken from United Nations (1992).

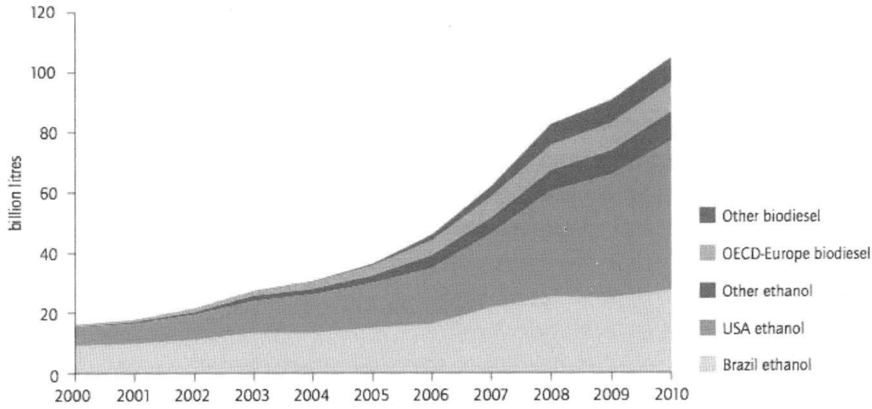


Fig. 11.2. Growth in biofuel production worldwide (Source: IEA, 2011).

Growing such biomass, however, requires land. Often, available land for biofuel production is scarce, leading to land exploitation in the form of direct land conversion (LC) and indirect land use change (ILUC). Furthermore, there have been land grabs by big and foreign owned corporations, which contribute to land ownership issues (peasants' loss of land) and the eradication of subsistence farming. This, if left unchecked, will contribute to even greater and widespread poverty in biofuel-producing developing countries.

Recognising that biofuels may not be a holy grail for those seeking a clean alternative to fossil-based energy sources, the International Energy Agency (IEA) repeatedly stressed on the need for sustainable production and more technologically advanced feedstock (e.g. algae) for biofuels. They are used to substitute for a portion of transportation fuel, in order to meet greenhouse gas reduction targets (to be further discussed later).

In a report requested by G8 energy ministers in 2011, the IEA stated that biofuels have the potential to meet more than a quarter of the world demand for transportation fuels by 2050 (IEA, 2011). The report set a target of a 50 percent reduction in energy-related carbon dioxide emissions by 2050 (from 2005 levels). This is based on the assumption that conventional technologies became more efficient at converting crops, algae and other organic material into energy (IEA, 2011).

## An ASEAN Overview

Fast-growing ASEAN has a green vision. Each ASEAN member country (AMC)[2] has plans to adopt greener options to conventional energy sources (oil, gas and coal) to sustain its high economic growth. These plans serve primarily to meet energy security objectives by creating diversification of energy sources, as well as reduce reliance on imported fuels by exploring and developing their own domestic fuel sources. A secondary objective is to clean up the air and mitigate climate change through biofuels by reducing an estimated 25 million tonnes of greenhouse gas emissions (GHG) in 2015 (*Ibid.*). The group envisions biofuels stepping up to these tasks by contributing to at least five percent of the fuel needed for the regional transport market by 2015 (*Ibid.*; ASEAN, 2011). A third objective is tied to job creation, income provision and economic growth (Zhou and Thomson, 2009).

Malaysia, Indonesia, Thailand and the Philippines have shown tremendous growth in biofuel production within the last decade. These countries have concrete plans or ongoing mandates on biofuel blends for transportation fuel (Table 11.4). The proportion of biofuel that is blended is denoted by the percentage; for instance, E3 would refer to three percent of agro-based ethanol blended with fossil-based ethanol while B2.5 would refer to 2.5 percent of agro-based diesel that has been blended with fossil based diesel. Table 11.5 outlines the feedstock used in biofuels.

The dramatic growth in biofuel production in Malaysia, Thailand and Indonesia specifically has been incentivised by comprehensive financial instruments such as price subsidies and tax rebates. For countries such as Vietnam, biofuels is a dawning agro-business as it prepares to stimulate its sluggish growth with production targets. In resource rich but poverty-stricken Cambodia and Laos, wealth via biofuels is tomorrow's dream.

In 2010, ASEAN as a regional entity supplied 1,327 billion litres of biodiesel and 486,000 cubic metres of bio-ethanol to buyers (Tables 11.6 and 11.7). Palm oil and its fraction (sub and by-products of palm oil) ranked seventh highest (2.4 percent) in terms of its share of ASEAN's top 20 export commodities (43.4 percent of total exports).

[2]ASEAN 10-member countries are Singapore, Malaysia, Indonesia, Brunei Darussalam, Thailand, The Philippines, Lao PDR, Vietnam, Cambodia and Myanmar.

Table 11.4. ASEAN biomass and biofuel targets and mandates.

| Country | Current Mandate/Target | Future Mandate/Target |
|---|---|---|
| Indonesia | Mandate: E3, B2.5 | Mandate: E5, B5 (2015); E15, B20 (2025)<br>By 2025, biofuels component of energy mix – 5%; biomass – 0.74% |
| Malaysia | Mandate: B5 | Installed renewable energy capacity by 2030: biomass 1340 MW; biogas 410 MW<br>Biofuels to displace 5% of diesel in road transport |
| Thailand | Mandate: B3<br>Biofuels to displace 12.2% of transport energy demand | 3 Ml/d ethanol, B5 (2011); 9 Ml/d ethanol (2017) |
| The Philippines | Mandate: E5, B1 (by 2007), B2 (by 2009) | E10 (February 2012), B2<br>By 2030, 102 MW of installed new biomass capacity. Displace 15% of diesel and 20% of gasoline with biofuels |
| Vietnam | No Biofuels Target | Target: 50 Ml biodiesel; 500 Ml ethanol (2020); E5 blend in 2013<br>1.8 million tonnes of biofuels from 2015 to meet 5 percent of domestic petrol and diesel demand in the next 15 years<br>By 2030, 400 MW installed biomass capacity |
| Laos | No Biofuels Target | No Biofuels Target |
| Cambodia | No Biofuels Target<br>Biomass gasification 87 kW | No Biofuels Target |
| Myanmar | No Biofuels Target | Displace 8% of transport fuel with biofuels |
| Brunei | No Biofuels Target | No Biofuels Target |
| Singapore | No Biofuels Target | No Biofuels Target |

Source: Author's creation based on the following: IEA (2011), ASEAN (2011c), Platts (2011), Viet Nam News (2012).

In **biodiesel supply**, Indonesia led ASEAN by churning out 650 billion litres of the renewable source of energy in 2010. When it comes to **bio-ethanol** supply, Thailand topped the list with a production volume of 426,000 cubic metres in 2010, made from sugarcane.

Table 11.5. Type of feedstock used in liquid biofuel producing countries.

| Country | Biodiesel feedstock | Bio-ethanol feedstock |
|---|---|---|
| Indonesia | Palm oil, Jatropha | Sugarcane, Cassava |
| Malaysia | Palm oil | Nil |
| Thailand | Palm oil, Jatropha | Sugarcane, Cassava |
| The Philippines | Coconut methyl ester (CME), Palm oil | Sugarcane, Sweet Sorghum, Cassava |
| Vietnam | Palm oil | Cassava |
| Laos | Nil | Nil |
| Cambodia | Nil | Nil |
| Myanmar | Nil | Nil |
| Brunei | Nil | Nil |
| Singapore | Palm oil (imported), Recycled Cooking Oil | Nil |

Source: Author's creation based on APEC (2012).

Table 11.6. Fuel ethanol world production from 2006 to 2010 (1,000 cubic metres).[3]

| Asian Country | 2006 | 2010 | % Difference |
|---|---|---|---|
| Thailand | 135 | 426 | +215 |
| Indonesia | 0 | 30 | — |
| Philippines | 0 | 30 | — |
| China | 1,685 | 2,050 | +22 |
| India | 105 | 238 | +127 |
| Japan | 0 | 10 | — |
| S.Korea | 0 | 8 | — |
| Pakistan | 0 | 10 | — |
| Taiwan | 0 | 5 | — |
| Other Asia | 0 | 100 | — |
| **Asia** | **1,925** | **2,907** | **+51** |
| USA | 18,381 | 50,088 | +172 |
| Brazil | 16,700 | 25,530 | +53 |
| World | 39,192 | 85,150 | +117 |

Source: Informa Agra & Commercial, 2011.

The European Union (EU) is an important market for the biofuel-exporting AMCs. The new EU energy directive (Directive 2009/28/EC) adopted on 23 April 2009 requires the delivery of 20 percent of the EU's total energy to originate from renewable sources by 2020 (Chiavari, 2013).

[3]Asia as a percentage of world in fuel ethanol production in 2006–4.9%; 2010–3.4% (Informa Agra & Commercial, 2011).

Table 11.7. Southeast Asia fuel biodiesel outlook (billion litres).

| Country | 2010 | 2020 |
|---|---|---|
| **Demand** | | |
| Indonesia | 196 | 1,000 |
| Malaysia | 6 | 600 |
| Thailand | 554 | 1,400 |
| **Supply** | | |
| Indonesia | 650 | 1,900 |
| Malaysia | 117 | 1,200 |
| Thailand | 560 | 1,400 |

Source: Author's creation based on data from Informa Agra & Commercial (2011).[4]

A ten percent share of renewable energy is also set for consumption by the EU transport sector, which generates biofuel demand (*Ibid.*).

The environmental sustainability criteria in the new directive have a direct impact on biofuel-exporting countries, including the ASEAN ones. Thus, AMCs have to fulfil strict standards within the associated certification criteria to maintain its export volume, which remains a challenge due to social and environmental problems in AMCs, particularly Indonesia and Malaysia, to be discussed in this report.

## *ASEAN Biofuel Country Highlights*

### a) *Indonesia*

Indonesia is the world's largest palm oil producer. The huge development in biofuel production is supported by national policies, investment guarantees and protections that specifically encourage national and foreign investments in the country (Biofuel Indonesia, n.d.). Export growth has been driven by subsidised sales volume by state-owned Pertamina, topped by an advantageous export tax structure for biodiesel (Informa Agra & Commercial, 2011).

Indonesia sees biofuels as one of the key instruments to accelerating economic growth, alleviating poverty, and creating employment opportunities while also, under the Kyoto Protocol, mitigating greenhouse gas emissions.

[4]Source: Author's creation based on data from Informa Agra & Commercial. (2011). *World Ethanol and Biodiesel Markets — A Global Industry Outlook* (United Kingdom: Informa Agra & Commercial).

Table 11.8. Malaysia's renewable energy targets.

| Year | Against total energy mix |
|---|---|
| By 2015 | 5% |
| By 2020 | 9% |
| By 2050 | 13% |

Source: Author's creation based on KeTTHA (2008).

The Indonesian national energy policy has stated goals of achieving two percent biofuels in the energy mix by 2010 (5.29 million kilolitres), growing to three percent by 2015 (9.84 million kilolitres) and five percent by 2025 (22.26 million kilolitres) (Legowo *et al.*, 2007).

b) *Malaysia*

In 1999, the Malaysian Prime Minister declared renewable energy as the "fifth fuel" (Ibrahim, 2002). A National Biofuel Policy in 2005 provided a framework to develop the biofuels industry while cultivating biodiesel as one of the five main energy sources (Lopez and Laan, 2008). Malaysia is the second largest producer and the largest exporter of crude palm oil (*Ibid.*). Malaysia is now targeting for renewable energy to contribute to at least five percent of the total energy mix by 2015 (Seda, 2008) (Table 11.8).

Malaysia's 2013 budget continues to bolster the palm oil industry — the biggest agriculture sector there — as a pinnacle of its Economic Transformation Programme (Pemandu, 2012). The replanting of oil palm alone gets a funding of RM432 million (*Ibid.*). Malaysia's Renewable Energy Policy and Action Plan, effective since June 2010, has set a target of 4,000 MW of installed renewable energy capacity by 2030, of which biogas and biomass will account for 410 MW and 1,340 MW of installed capacity by 2030 respectively (*Ibid.*). Considerable investments in infrastructure have to be made to realise these new targets.

c) *Thailand*

Thailand seeks to strengthen domestic energy security and hopes to leverage on the biofuel sector to promote the use of integrated green energy in communities and develop greater alternative energy industries. The Thai cabinet's 10-year Alternative Energy Development Plan (2012–2021) mapped an increase in the share of renewable and alternative energy, from the existing 9.4 percent of the country's total energy consumption to 25 percent by 2021 (against 20 percent in the former plan) (USDA, 2012). It also moderated production and consumption targets of biofuels (sugarcane feedstock supplemented by cassava) to address shortfalls in previous years' performances.

The Thai government uses various financial instruments to steer its course.[5] The manufacture of Eco-car (running on E20) and flex-fuel vehicles (FFV) are being encouraged through excise tax reductions as well (*Ibid.*). A mandatory B5 (or five percent biodiesel blend with fossil diesel) took effect on 1 January 2012, which should drive a steady growth of this fuel in the short term.

d) *The Philippines*

Following the Philippines' Renewable Energy Law, the enactment of the Biofuels Act of 2006 (Republic Act 9367) opened up new markets for biofuel. Biodiesel (mostly from coconut palm feedstock) and bio-ethanol (mainly from sugarcane feedstock) contribute towards a greater supply of locally-produced and imported biofuel to serve an ever increasing demand for cleaner transportation fuel.

Biofuel is seen as a way forward for the Philippines, a grouping of more than 7,000 islands, to become more energy self-sufficient by reducing the country's dependence on foreign fossil fuels, and to mitigate greenhouse gas emissions while creating more rural employment (Philippines Board of Investment, 2011). By 2030, it hopes to be a "leading net exporter" of biofuels (USDA, 2012).

The government introduced the B-1 mandate (one percent biodiesel component added to 99 percent of fossil diesel) in 2007. Two years later, on 6 February 2009, the minimum blend was doubled. A blending requirement for biodiesel at five percent is on the cards, but the market may not be ready for such a change. Bio-ethanol production within the country is low, however, and as such the fuel is being imported from elsewhere, such as Thailand, Vietnam and Brazil (Talkvietnam.com, 2012; Aguiba, 2012; USDA, 2012).

---

[5]This include an E20 gasohol user subsidy of 3.0 baht/litre (36 US cents/gallon) less than the retail price of Octane 95 gasohol (a regular ethanol-gasoline blend sold at pump stations), and by giving gasohol pump stations 0.5 baht/litre (6 US cents/gallon) more than the regular gasoline sales (USDA, 2012).

Fig. 11.3. Neste Oil biofuel refinery in Tuas, Singapore (Photo credit and Source: Neste Oil Singapore Pte Ltd).

e) *Singapore*

Singapore as the region's financial centre has boosted palm oil expansion. A number of Singapore-registered companies invest in biofuel production of neighbouring countries, and these include Bunge Agribusiness Singapore Pte Ltd (subsidiary of Bunge Ltd) and Wilmar, which, through a series of mergers and acquisitions, owns one of the largest plantation areas in Southeast Asia at around 573,405 ha (Bunge, 2012; Jiwan, 2013).

In addition, Finnish-owned Neste Oil has an 800,000 tonnes per annum biofuel capacity plant in Tuas, at the west end of Singapore (Figure 11.3) — a strategic location between two major biofuel producer countries — Malaysia and Indonesia. Neste Oil imports crude palm oil (CPO) and side stream products from producers in Malaysia and Indonesia who comply with the International Sustainability and Carbon Certification, and then processes the liquid to produce its proprietary NExBTL (Neste Oil, 2011).[6]

## Biofuel Issues and Challenges

In the world of biofuels, various actors, such as industry associations and non-governmental organisations (NGOs) have introduced sustainability criteria.

### *European Union Sustainability Criteria*

Debates over sustainability criteria by European Union member countries, pressured by international non-governmental organisations (INGOs), have

[6]This is a high-octane biodiesel suitable for powering motorised vehicles as well as aircrafts (Neste Oil, 2011).

led to the establishment of Directive 2009/28/EC. This renewable energy directive (RED) is a primer for EU's responsible biofuel trade with climate change mitigation imperatives, and has the following requirements:

i. Minimum GHG savings
ii. Ban on using biomass from primary forest, protected natural areas or highly bio-diverse grassland
iii. Ban on using biomass from land converted from wetland, forest or undrained peatland
iv. Reporting on certain environmental and social impacts

While these requirements under the binding sustainability criteria are mandatory by law, they are inadequate in a number of areas (Chivari, 2013). Social impacts were not required in the EU RED's impact assessment due to difficulties in administering this aspect and complications with regards to international law. This is most unfortunate as the exclusion of social sustainability prevents opportunities in rural development through these possible measures: participation of small-scale farmers, encouraging rural labour, preventing forced displacement, and ensuring access to key resources (fuel wood, plants for medicinal purposes, etc.) by rural communities (*Ibid.*).

While addressing the direct impacts of biofuel demand on primary resources, the EU failed to address consequential impacts such as indirect land use change. Under intense pressure by NGOs over the years to close this gap, three years later, the European Commission issued a proposal to revise its directive on 17 October 2012 with the added consideration of the impacts of ILUC in assessing the GHG performance of biofuels. It now also seeks to limit the use of food-based biofuels to five percent, down from 10 percent, to meet its renewable energy target in the RED. The remaining five percent will have to come from second-generation biofuels from non-feedstock such as waste and switchgrass, which do not compete with food or pose land-related issues (European Commission, 2012).[7] This recent development remains a proposal and has yet to be put into motion.

### *Land-related Changes and Issues*

Land use change is perhaps the greatest cause for concern in the cultivation of biofuel crops in Southeast Asia. Issues concerning economic, social,

[7]In its new proposal, the EC is also proposing to consider investment with entities that produce biofuels that lower GHG by 60% (in 2020), in addition to its previous 35% GHG savings for existing plants or 50% for new ones in 2017 (European Commission, 2012).

cultural and environmental dimensions are due to deforestation, land grabbing, pollution of land, sea and air, as well as home and livelihood displacement for local communities.

Food insecurity has led to acts of sourcing and seeking control over large tracts of land overseas to secure food supply, primarily in Africa. It has been claimed that 80 percent of all land grabs took place in Africa (Borras, 2013).

But recent studies have pointed at land grabbing and land rush (peasants expelled from their land, corruption in land deals, etc.) taking place in the former Soviet Eurasia (Vissoor and Spoor, 2011) as well as Southeast Asia (Borras and Franco, 2011). The context and conditions of land-related issues may differ in these areas. This report investigates relevant issues in top biofuel producing countries in Southeast Asia.

a) *Definition of Land Grabbing*

There are broad and narrow interpretations in defining what land grabbing means. Early studies look at "scale and character of land acquisitions", which often involves land takeover that exceeds the 1,000 ha benchmark. Recent studies consider the "scale and character of capital" involved in land acquisition as this broadens the understanding of the operation of capital. For instance, the nature of land control is better understood if capital, not land, becomes the unit of analysis, i.e., purchase, lease or forest conservation of the land concerned (Borras, 2013).

There is also a need to embed land grabbing within the context of global capitalist development, which brings with it the convergence of multiple elements with their related challenges such as food, energy, water, climate change and finance capital. In addition, there is the need for greater land-based resources by "newer hubs of global capital" such as the BRICS.[8]

Saturnino M. Borras offers a comprehensive definition of land grab (*Ibid.*). Contemporary land grabbing, he says, is:

> "the capturing of control of relatively vast tracts of land and other natural resources through a variety of mechanisms and forms that involve large-scale capital that often shifts resource-use orientation into an extractive character, whether for international or domestic purposes, as capital's response to the convergence of food, energy and financial crises, climate change mitigation imperatives, and demands for resources from newer hubs of global capital."

[8]BRICS is short for Brazil, Russia, India, China and South Africa.

b) *Definition of Green Grabbing*

Borras' definition of land grabbing fits well within the oil palm context as its contribution towards climate change mitigation has begun to appear in mainstream discourse, particularly by governments in seeking to fulfil global environmental requirements towards greenhouse gas emission reduction. Such a reason can be considered under "green grabbing" discourse, where "green credentials are called upon to justify appropriations of land for food or fuel" (Borras and Franco, 2011).

Oil palm provision for biofuel supply has legitimised a commercial deal towards a green agenda. The idea of green grabbing does not only refer to total land takeover or alienation from existing or rightful owners. It also involves "the restructuring of rules and authority over the access, use and management of resources", which in turn affects existing socio-economic norms, as in the case of the indigenous peoples of Borneo, in the next section (*Ibid.*).

There are even greater differentiated impacts of green grabbing on local communities than what is usually represented in discourse. Such discourse, predominantly on climate change mitigation, is particularly perpetuated through government propaganda. As such the role of activists is important as they provide an alternative viewpoint in the debate on green grabbing.

### *Transboundary Haze Pollution*

Land cultivation for oil palm plantations, especially with the slash-and-burn method of agricultural clearance, is the reason for the acute loss of primary forests and its wildlife, chiefly in Malaysia and Indonesia (Chin, 2011). There is evidence that the 1997–1998 forest fires in Indonesia that contributed to the regional haze took place mostly in oil palm plantations (Jiwan, 2013). The raging fires were mostly due to the draining and clearing of peatlands by fire, thereby releasing toxic gases (*Ibid.*).

While Indonesia regards most of these issues as domestic problems, the impact of smoke emission with the burning of agro-crops for land conversion in Indonesia has resulted in a transnational smog or haze problem. The combination of carbon, soot and particulate matter in the air was severe enough (especially for Singapore and Malaysia — the main victims) to see ASEAN leaders colluding to find ways to resolve the issue since the 1990s, culminating in the enactment of the ASEAN Agreement on Transboundary Haze Pollution in 2002. Indonesia was the only AMC that had not ratified the agreement, until it agreed to do so in September 2014 (MEWR, 2012 and

Sentana, 2014). Urgent action to control the spread of fire in the peatlands also led to the formation of the ASEAN Peatland Management Initiative (APMI) in 2003.

Indonesian farmers are still burning their harvested agrofields to make way for a new season of planting — a method that has been used for generations — rebuffing environmental protection laws. While penalties exist for forest clearance by fire in Indonesia (punishable by up to 10 years of imprisonment and a fine up of up 10 billion rupiah), corruption is also cited as the other weak link in regulating and monitoring such an offence.

A recent incident in 2012, in the Aceh province, concerned the persistent burning of the Tripa peat swamp to make way for oil palm plantations (Lang, 2012). Not only did it flout Indonesia's environmental protection law, it also received the blessings of the governor of Aceh, who altered the Indicative Moratorium Map (IMM) drawn for forest conservation under the US$1 billion Reduced Emissions through Deforestation and Forest Degradation (REDD) agreement with Norway. In response, environmental NGO Friends of the Earth Indonesia (WALHI) sought legal recourse by suing the governor as well as the company concerned, which is PT Kalista Alam. While the Banda Aceh State Administrative Court rejected the case in April 2012, on further appeal, the Medan State Administrative Court ruled in favour of the plaintiffs.

While citizens suffer, ASEAN leaders continue to take a diplomatic approach in dealing with the yearly haze problem; however, without much result. In 2012, the haze problem peaked, with a higher cumulative number of hotspots recorded in Sumatra, Indonesia than in previous years (Figure 11.4). Kalimantan, Indonesia also recorded high hotspot counts in 2011 and 2012 (Figure 11.5). This prompted AMCs, especially Singapore, to pressure Indonesia to ratify the ASEAN Agreement on Transboundary Haze Pollution. Singapore also passed a bill that would enable action to be taken against errant polluters located in Singapore or overseas (Shen, 2014).

In ten years, the haze pollution has caused significant problems to health, the air, land and water environments, and the tourism industry of affected countries in the region. Taking this as evidence, it is fair to comment that the ASEAN institution falls short of its ASEAN Security Community vision formulated at the Bali Concord II in 2003 (Severino, 2007).[9] This calls for a

[9]The second Declaration of the ASEAN Concord (or Bali Concord) formalised three components of the ASEAN Community: the ASEAN Security Community, the ASEAN Economic Community and the ASEAN Socio-Cultural Community (Severino, 2007).

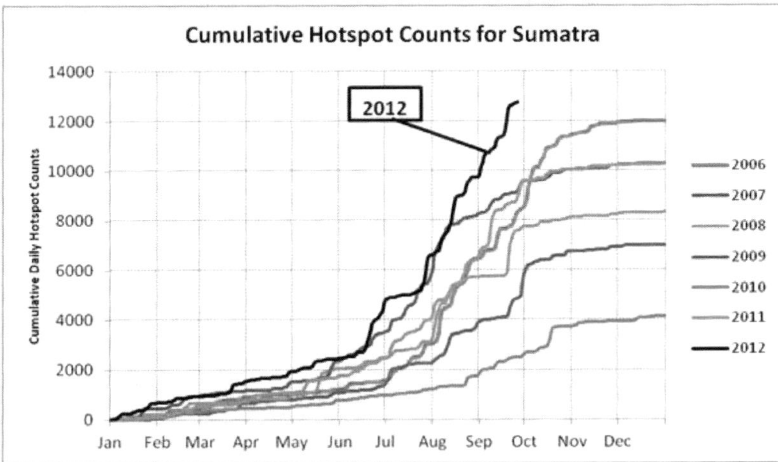


**SUMATRA**
**( West Sumatra, Riau, Jambi and South Sumatra)**

| 2006 | 2007 | 2008 | 2009 | 2010 | 2011 | 2012 |
|---|---|---|---|---|---|---|
| 12014 | 7017 | 8349 | 10297 | 4147 | 10320 | 12750 |

Fig. 11.4. (Colour online) Cumulative hotspots in Sumatra 2006–2012: In 2012, hotspots numbering 12,750 were recorded in Sumatra, the highest reading since 2006 (Note: Readings for 2012 were taken up until 30 September; Source: Meteorological Service Singapore)

unified community that adheres to a comprehensive concept of security, which includes environmental pollution.

## *Top Producer Countries' Issues and Conflicts*

### a) *Malaysia*

Malaysia in 2010 had 4.85 million hectares (mil ha) of oil palm plantations, covering 14 percent of the total land area in both Peninsular and Borneo Malaysia. This is a huge transformation from the 1970s where oil palm plantation was just a jot on the country's mass, occupying less than 1 mil ha. Land conversion for oil palm also takes place on peatlands, contributing to the threat of prolonged burning beneath the ground, and thus causing higher levels of carbon emissions. A rise in the conversion of peat soil has been tracked in Sarawak. In Figure 11.6, a record of oil palm plantation on peatlands is shown from 2003–2009, while an increase is predicted for 2010.

Despite environmental concerns and activists' objections on the negative impact of palm oil plantations on nature, biodiversity and wildlife population decline, Malaysia is ramping up its palm oil for biodiesel

**KALIMANTAN**

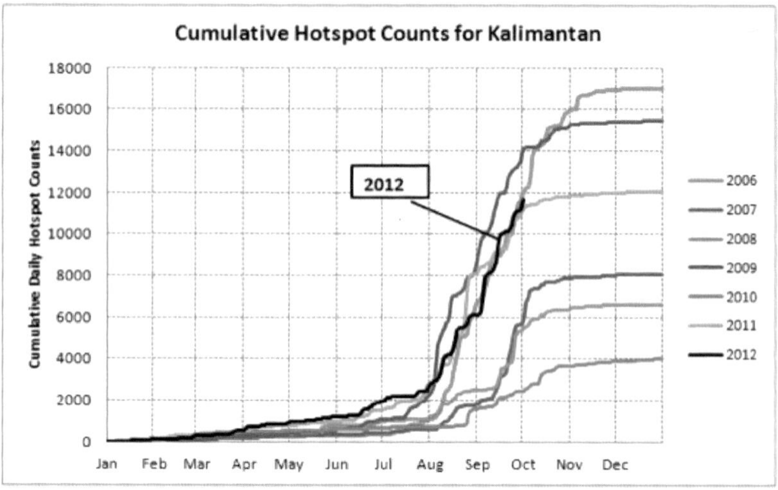


| KALIMANTAN (West, Central, East and South Kalimantan) | | | | | | |
|---|---|---|---|---|---|---|
| 2006 | 2007 | 2008 | 2009 | 2010 | 2011 | 2012 |
| 17045 | 8124 | 6633 | 15495 | 4033 | 12066 | 11679 |

Fig. 11.5. (Colour online) Cumulative hotspots in Kalimantan 2006–2013. The air pollution in Kalimantan continues to cause concern. In September 2012, it recorded a high of 11,679 hotspots compared to 12,066 in December 2011.[10]

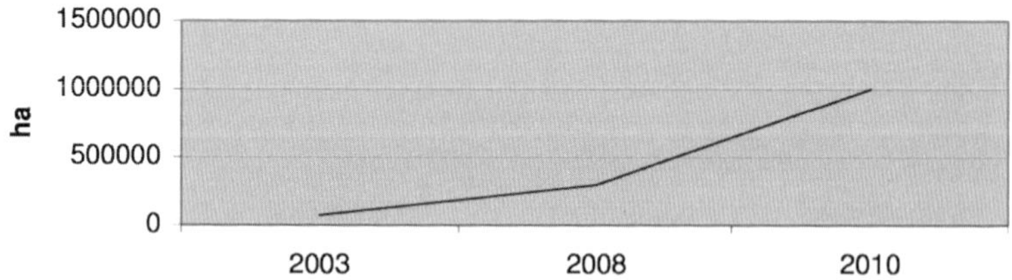


Fig. 11.6. (Colour online) Oil palm plantation on peat soil in Sarawak (Source: Tropical Peat Research Institute, 2009).

programmes. Contributing to RM62.86 billion of gross national income in 2010, palm oil is seen as a "key to Malaysia achieving high-income status within the next 10 years", according to Commodities Minister Datuk Hamzah Zainudin, who projected that the feedstock will generate RM178 billion in returns by 2020 (The Star, 2011).

[10]Source: Meteorological Service Singapore. Retrieved on October 16, 2012 from http://www.weather.gov.sg/wip/c/portal/layout?p_l_id=PUB.1003.613.

The Minister dismissed the environmental impact of oil palm plantations as "a non-issue", stating that presently, about 56.4 percent of the country still has natural forests (*Ibid.*). According to recent reports, the pervasiveness of oil palm plantations can be pegged at 15 percent of Malaysia, less than a third of the extent of remaining natural forests (Bayar, 2011).

In Malaysia, current laws do not appear to favour indigenous peoples' rights. In Sarawak, the government limits the exercise of 'Native Customary Rights' (NCR) while in Sabah, few NCR areas are clearly demarcated. United Nations experts have called on Malaysia regularly to review its current policies, towards greater respect of human rights, as recognised in the UN Declaration on the Rights of Indigenous Peoples (UN News Centre, 2013). Such evidence can be found, not just in agroplantation sectors, but also in other industries, such as energy production. Free, prior and informed consent (FRIC) of Penan communities was found to be absent in the deployment of the Murum Hydroelectric Dam, which led to the forced resettlement of those indigenous peoples (Bulan, 2010). This is just one of many incidences that can be found in literature. Renewed concerns have been raised on the status of native customary land rights, following plans by the Rural and Regional Development Ministry to amend the Aboriginal People's Act 1954. The likely impact is the loss of up to 645,000 ha of community land (Hamid, 2013).

In addition, independent farm land smallholders are increasingly becoming a minority as the government favours large estate development by the private sector. Greater conflict will be experienced due to ineffective and unsatisfactory relocation efforts of affected communities coupled by inadequate compensation (Chin, 2011).

b) *Indonesia*

Indonesia is the top producer of palm oil as well as a top emitter of GHG from land conversion, resulting in deforestation and carbon emissions (Carlson *et al.*, 2012). Sumatra and Kalimantan alone produce half of palm oil supplies worldwide, and the Government intends to double palm oil production by expanding landholdings in Kalimantan and Papua (*Ibid.*). Around 18 mil ha of tropical rainforests have been cleared (Colchester *et al.*, 2006). In 2006, Indonesia planned for another 20 mil ha of land for oil palm expansion (*Ibid.*).

A *Nature Climate Change* study revealed that over ten years, beginning in 1990, 90 percent of oil palm plantations in Kalimantan alone were established on forested land (47 percent intact, 22 percent logged and 21 percent agroforests) (*Ibid.*). In September 2012, there were 868 open fires, due to slash-and-burn agricultural practice, or "hotspots", in Central Kalimantan, the region that is the beneficiary of the Indonesia–Norway US$1 billion REDD agreement. The continuing practice of burning forests can be attributed to the fact that the required moratorium on forest concessions for agriculture conversion applies to new concessions (which have yet to be executed), not existing ones. The study also reported that 79 percent of allocated leases for farm conversion have yet to be developed, in spite of an oil palm expansion of 278 percent from 2000 to 2010 (*Ibid.*). The effectiveness of REDD in preserving forests is yet to be seen given the challenging socio-political conditions described earlier.

Since independence, Indonesia has been undermining the rights of indigenous peoples by dismantling customary institutions and forcing them into the national mainstream through resettlement (Colchester, 2010). Policies originating from the colonial era still apply in the forestry and agriculture sectors, which maintain that the state has the ultimate power of land rights. In fact, following independence, current rulers continue the colonial practice of appropriating peoples' lands and resources by perpetuating the principal of "eminent domain" of the state through the Mining Act, the Forestry Act (1997) and the Plantation Act (2004). Specifications such as the Forestry Law 41/1999 and the Plantation Act 18/2004 function to "limit or effectively extinguish rights of local communities and indigenous peoples" (*Ibid.*).

Customary rights are regarded as weak usufructuary rights on State lands that are better off being developed for logging and plantation. Due to unclear regulations, only less than 0.2 percent of the 70 percent of 'forest' territory has been allocated to rural communities (*Ibid.*).

Social disruptions over lands and livelihoods in Indonesia due to oil palm expansion have also been widely reported (Jiwan, 2013). These affect not just the indigenous people but local communities, smallholders and farmers (due to exploitative industrial partnership schemes), workers, women, and children as well. Human rights abuse rank high in the 3,500 land conflicts recorded (*Ibid.*).

Dissent and unhappiness from not getting justice have escalated into acts of protests such as public demonstrations, petitions, blockades of company roads and attacks on company properties (Colchester, 2010). These protests are sometimes met by repression from company security forces or state agencies, leading to conflicts that result in violence by

Fig. 11.7. **Land rights protest in Jakarta, Indonesia:** People take to the streets in protest against land rights issues due to oil palm expansion on 21 December 2011 (Source: Earth First!, 2011; Photograph by Jonathan McIntosh[11]).

state actors such as the police or military forces (Jiwan, 2013). Beatings, shootings and murders have been documented that are believed to be linked to land-related issue, one involving a farmer and mother in a public demonstration by smallholders on 8 June 2010 in Riau. The acts of murder and human rights violation led to a demonstration by the public, civic groups and NGOs outside the national police headquarters in Jakarta (Figure 11.7). Other forms of demonstration involved self-inflicted pain and starvation to send a strong signal, such as Riau peasants who sewed their mouths shut in protest (Figure 11.8) (*Ibid.*).

Sawit Watch, an environmental NGO, recorded 664 unresolved land disputes involving palm oil companies in 2011. Sawit Watch and Jatam (a mining advocacy network) mapped the areas of civil conflict as a result of oil palm plantation and mining activities (Figure 11.9).

[11]Source: Earth First! (December 29, 2011). Newswire. Indonesian protesters sew their mouths shut over deforestation/palm oil. Retrieved on October 16, 2012 from http://earthfirstnews.wordpress.com/2011/12/29/indonesian-protesters-sew-their-mouths-shut-over-deforestationpalm-oil/.

Fig. 11.8. **Silent but painful protest over deforestation/palm oil:** Riau protesters sewed their mouths outside Jakarta's House of Parliament and were hospitalised after starving for five days (Source: Earth First!, 2011; Photograph by Yudhi Sukma Wijay).

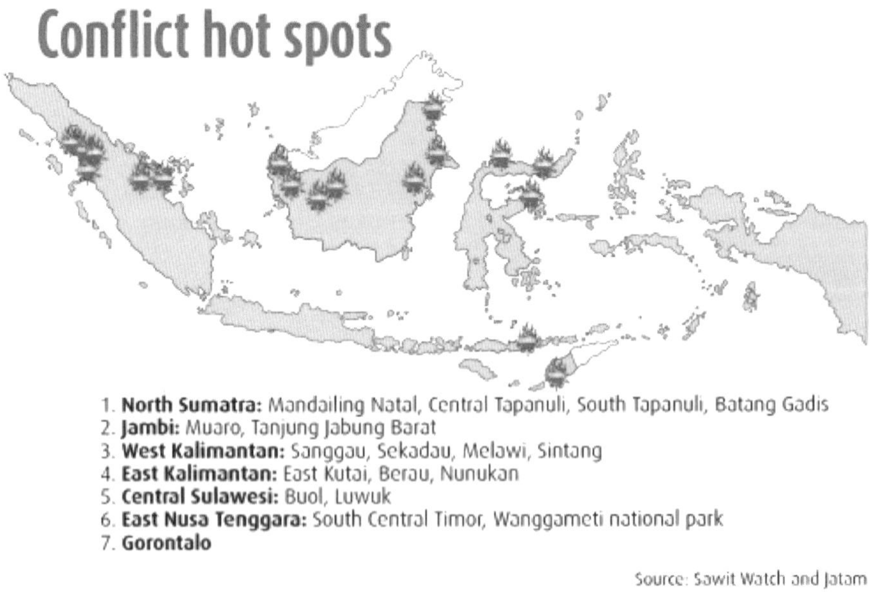


Fig. 11.9. **Conflict areas identified in Indonesia:** Several regions across Indonesia are conflict prone due to the oil palm and mining industries, according to NGO actors Sawit Watch and Jatam (Source: *The Jakarta Post*, "Mining, plantation disputes to intensity", 29 December 2011).

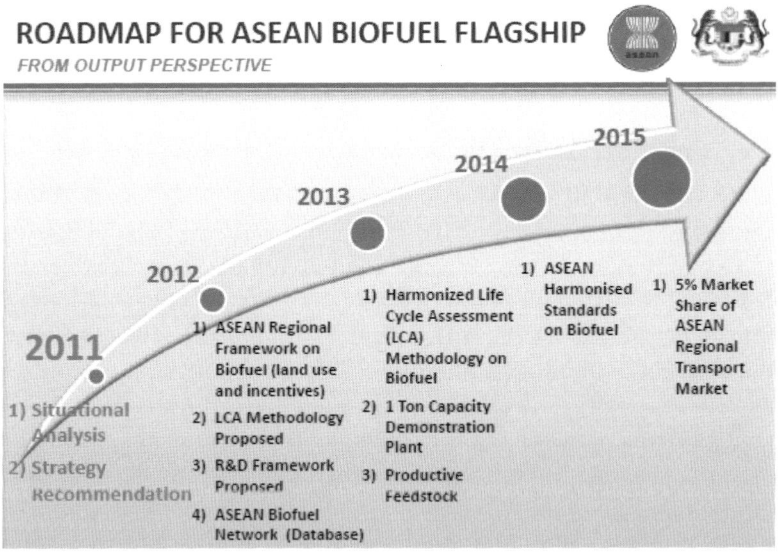


Fig. 11.10. **ASEAN biofuel flagship roadmap** (Source: Malaysia, 2011[12])

## Influences of Key Actors

It is important to identify who the key actors are in biofuel development and promotion, so as to pinpoint domains and links where concepts of responsible operations and ethics can be perpetuated. The role of central states, transnational corporations, domestic capital and global commodity chains figure prominently, and have to be considered when analysing the trajectories of biofuel development (Jiwan, 2013). A sample of typical actors in the palm oil industry for biofuels can be seen in the resource trade cycle analysis (RETRAC) model (Figure 11.10).

### a) *Corporations*

Corporations figure prominently in shaping biofuel sustainability, or the lack thereof. As Malaysia experienced high growth in oil palm production, the industry went from a national to transnational expansion phase with upstream investments through new plantations, especially in Indonesia (due

[12]Malaysia, the lead country of the biofuel flagship programme within ASEAN COST+3, presented these several initiatives that could be undertaken by ASEAN over a five-year timeframe. (Malaysia's presentation at the 6th Meeting of the ASEAN COST+3 at Jeju, Republic of Korea, 2011, 6 December, p. 10)

to limited land in Malaysia), and downstream investments in Europe, China and India.[13] An estimate of 50 Malaysian companies control a considerable share of the oil palm plantations in Indonesia (Teoh, 2013).

### b) *Transnational Activists*

Not all NGOs are alike in the oil palm sectors — each reacts differently and produces different results. Transnational activists can be categorised into three different types of transnational organisations and they are: International Non-Government Organisation (INGOs); Transnational social movement organisations (TSMOs); and transnational advocacy networks (TANs) (Oliver Pye, 2013). There are differences to the structure, membership, and activities of these groups, as described in Table 11.9. These distinct NGOs act differently to a common set of problems, giving rise to different campaigns and outcomes, as illustrated by the following example.

In the case of biofuel advocacy, TSMOs (e.g., Friends of Earth, WALHI Indonesia, Sahabat Alam Malaysia) and TANs (WRM, Agrofuel Network), together with think tanks such as Biofuelwatch, UK, and Sawit Watch Indonesia, plus local groups in Europe and Southeast Asia, were part of the campaign coalition for a moratorium on agrofuel targets in the EU (*Ibid.*). This has the objective of getting the EU to review its mandate of substituting 10 percent of its transportation fuel with solely agrobased biofuels. The RSPO-certification system had become the default system for palm oil to qualify for the sustainability criteria, as spelt out in Article 17 of the Renewable Energy Directive (EU 2009). By politicising biofuels, and by creating a new transnational campaign coalition, the TSMOs and the TANs contributed to a de-legitimisation of the RSPO (*Ibid.*).

### c) *The ASEAN Institution*[14]

ASEAN went from being a net oil exporter to net oil importer since 1995. In 2005, it imported 30 percent of its oil requirements (ASEAN, 2010). To reduce their dependence on imported oil, and in line with national and

---

[13]Upstream activities involve oil palm cultivation, production of fresh fruit bunches (FFB) and processing them into crude palm oil (COP) and palm kernel oil (PKO). Downstream typically refers to palm oil refiners, palm kernel crushers, manufacturers of palm-based edible products and specialty oils and fats (Malaysian Rating Corporation Berhad).

[14]The author gathered much of the information pertaining to ASEAN's endeavours in biofuel development through email interviews and minutes of meetings.

Table 11.9. Three types of transnational organisations (Pye, 2013).

| | Structure | Membership | Activities | Examples |
|---|---|---|---|---|
| International Non-governmental Organisations (INGOs) | Formalised, centralised and hierarchical | Full-time staff main agents, largely passive supporters and donations | Project management, lobbying, media work | WWF<br>Oxfam<br>Greenpeace |
| Transnational Social Movement Organisations (TSMOs) | Formalised with a large degree of autonomy for national federations | Some full-time staff but grassroots organising of voluntary activists is important or central | Lobbying, media work and grassroots organising | Friends of Earth<br>Via Campesina |
| Transnational Advocacy Networks (TANs) | Informal networks of autonomous organisations, groups, and individuals | INGOs and TSMOs can be part of network and their staff can be central, but voluntary activism is often key | Lobbying, public education, civil disobedience | WRM<br>CJN!<br>Agrofuel Network |

regional energy security goals, certain AMCs have invested in bio-energy through biomass and biofuels, in addition to other alternative sources of power such as hydro, wind and solar.

ASEAN saw merit in biofuel as a supplementary solution to conventional energy sources, while developing rural areas and industry for additional income streams. A string of ASEAN initiatives were announced over time on harnessing energy security independently of food security (ASEAN, n.d.).[15] This changed in September 2011 with the ASEAN agriculture ministers issuing a statement to endorse "in principle" the ASEAN Plus Three Comprehensive Strategy on Food Security and Bioenergy Development (APTCS-FSBD) (ASEAN, n.d.). ASEAN leaders have begun to take sustainability concerns for biofuels seriously, both from the developmental and policy perspectives.

### *ASEAN Biofuel Flagship Programme*

Three months on from the ASEAN leaders' endorsement of the APTCS-FSBD, Malaysia's Renewable Energy Research Centre (the lead agency of the ASEAN COST Biofuel Flagship programme) presented on the proposed framework and strategies of the ASEAN Biofuel Flagship Programme on 6 December 2011 at the 6$^{th}$ Meeting of the ASEAN COST Plus Three in Jeju, Republic of Korea (ASEAN, 2011).

Under the proposed ASEAN Biofuel Flagship Programme, ASEAN envisions that by 2015 it will "significantly use biofuels to reduce dependence on fossil fuel" (*Ibid.*, p. 3). By 2015, at least five percent share of the ASEAN regional transport market will be captured through biofuel (*Ibid.*, p. 4). Recommended strategies include aspects of sustainability, best practices based on biofuel life cycle assessment (LCA), harmonisation of standards to meet international transportation needs, and R&D among member states. A recommended R&D framework includes a one tonne capacity demonstration plant, diversifying from non-food feedstock to 2$^{nd}$ and 3$^{rd}$ generation biofuels technology, along with the capture of biogas for energy (*Ibid.*, p. 7). Of interest is the emphasis on having a

[15]The ASEAN Integrated Food Security (AIFS) Framework, ASEAN Plus Three Emergency Rice Reserve (APTERR), the ASEAN Food Security Information System (AFSIS), the ASEAN Plan of Action on Energy Cooperation (APAEC), Asia Biomass Energy Principles, and ASEAN Plan of Action on Science and Technology (APAST) Flagship Programme on Biofuels. (ASEAN Food Security and Bioenergy Development, ND).

harmonised LCA methodology, possibly owing to the wide acceptance of this methodology in addressing environmental impacts with its structured approach in accounting for impacts along the entire value chain, from the extraction of resources to delivery of product. Figure 11.10 shows a proposed plan in achieving a regional collaboration on biofuel development.

Although Malaysia had pitched this Roadmap for ASEAN Biofuel Flagship covering a five-year period from 2011, it is unclear how this programme will co-exist with the APTCS-FSBD, as there are overlaps in the required deliverables.

An ASEAN-led LCA study needs to account for land use change and consider resource trade cycle analysis for a more accurate pinpointing of responsibility nodes within connectivity chains.

## Resource Trade Cycle and Life Cycle Assessments

### *RETRAC Model*

A resource trade cycle analysis (RETRAC) model developed by Foundation Aidenvironment is an analytic framework designed to study linkages between the consumption and production of natural resource-based products and commodities, such as timber, soy bean and palm oil. Figure 11.11 depicts such linkages in a palm oil industry, based on the author's interpretation of the model (Wakker, 2013). RETRAC demonstrates how trade and investment cause impacts at the ground level, and more importantly, address misleading consumer information.

The RETRAC model differs from the general life cycle assessment (LCA) frameworks, where the latter calls for the accounting of the carbon and/or environmental footprints of specific production or distribution acts along the entire life cycle of biofuels. An LCA can be onerous as it may seek to include nearly every item of cradle to grave elements into the accounting process, hence making it a long, slow and tedious process. An LCA typically does not consider market forces or capital flows, or the behavioural roles of key actors and their interaction with one another.

There are more recent theories on the *consequential* LCA of biofuel production by taking into account, primarily, the impact caused by indirect land use change (Lapola *et al.*, 2010). Because land functions as a carbon sink, which sequesters a fifth of man-made emissions each year, any degradation or destruction of land by cutting, burning or ploughing soil would result in the release of greenhouse gas (GHG) emissions. This refers to land use change, or LUC (Smith, 2010). There are many incidences where

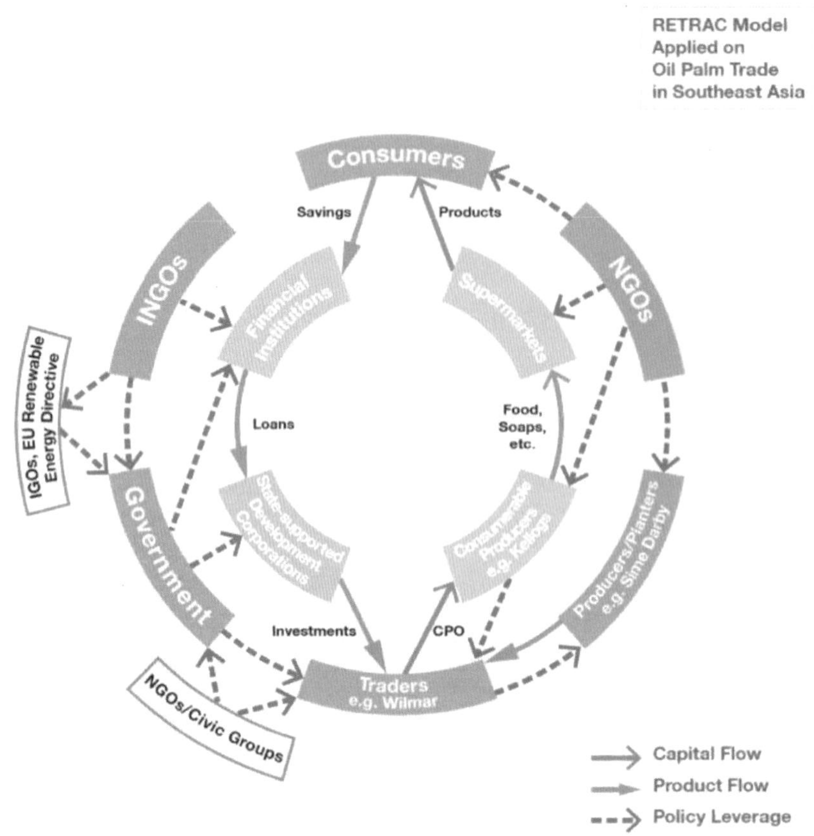


Fig. 11.11. Resource trade cycle analysis model applied to a palm oil industry; the author's creation based on Wakker (2013).[16]

biofuel feedstock is planted on land that was intended for another purpose, and as such, an additional demand for land is created somewhere else. This concerns indirect land use change or ILUC (*Ibid.*, p. 53).

## Implications for Southeast Asia

ILUC factors have recently been accounted for by the US and EU in their drives towards sustainable bio-energy production from agrocrops, through the US EPA's RFS2 standard and European Commission's Renewable

[16]Creation of a RETRAC model based on a depiction of stakeholder and player interdependent roles that are typical to a palm oil industry in Southeast Asia.

Energy and Fuel Quality Directives respectively.[17] On a regional basis, ILUC is *not* being factored in the LCA models currently under consideration by ASEAN. This should be cause for major concern.[18] It must be noted that while international sustainability policies consider the use of "marginal" or "idle" land for biofuel cultivation as a positive act, this has not been the case in Southeast Asia.

In Malaysia and Indonesia, federal or state governments interpret the concepts of "marginal" or "idle" land loosely. Such lands are often occupied by indigenous communities (apart from forests) and are considered as communal land by locals and safety nets to the rural poor, but have been taken over and converted for agrocrop cultivation. Indigenous peoples' lack of title deeds leads state governments to justify that such land belongs to the state instead, which then issues long-term concession rights to business operators (up to 100 ha in Indonesia).

The conversion of land use to oil palm "generally results in a weakening or a termination of customary tenure systems" (Jiwan, 2013). This becomes the bone of contention between marginalised communities in Borneo Malaysia and Indonesia and the ruling powers, leading to conflicts and demonstrations organised by INGOs as described earlier in this report. Impacts of direct land use change in such instances include a range of social and environmental factors such as loss of communal land and reduction of ecosystem goods and services. Forest fires present greater negative externalities such as flooding and loss of access to clean water (*Ibid.*). Livelihood diversification of indigenous peoples is a consequence of ILUC, which in turn generates negative impacts on their health and well-being in addition to changes to social, cultural and economic statuses.

## Proposed Sustainability Strategies

### a) *Policy Recommendations*

ASEAN should address its shortcomings with respect to the nature of its institution in administering biofuel development. ASEAN leaders need to agree on stricter enforcement for the non-compliance of the transboundary

[17]The European Commission's Directive 2009/28/EC applies to the promotion and use of energy from renewable sources while its Directive 98/70EC relates to the quality of petrol and diesel fuels (EC).

[18]Observations made by the author based on interviews with ASEAN biofuel representatives.

haze agreement, which could include penalties such as fines (that only Singapore has done) and/or trade embargoes on the agricultural produce arising from the errant country. Strong signals have to be sent to Indonesia to introduce effective laws and policies at federal and state levels, and stamp out corruption.

Second, forest preservation programmes through REDD+ are seen as a way forward in reducing carbon emissions while saving precious nature reserves and biodiversity (*Ibid.*, Redd-Monitor). This, as described earlier in this report, benefits Indonesia partially. Significant problems exist at the fundamental level in Indonesia, which threaten to derail the goodwill behind the concept. These include: corruption at the level of forest conservation enforcers; general corruption, empathy and lack of resolve at the authority level; oil palm plantation traction with the authorities; palm oil agricultural importance in national economic growth, hence, taking priority over environmental and food security; and State actors wearing multiple hats including having stakes in palm oil companies (Lang, 2012 and Jiwan, 2013).

REDD+ administrators, typically countries in the EU, should ensure that Indonesia alters its present policy and operational frameworks towards better social and environmental sustainability as well as adherence to international human rights laws. Furthermore, without internal convictions to preserve forests as natural assets (and not just view them as nature capital), motivators such as greed and power coupled with external factors like economic development, income generation and market demand will rule at unsustainable levels.

Third, palm oil producing countries should take active steps to align national laws and policies with international commitments on human rights. These include the United Nations Declaration on the Rights of Indigenous Peoples, International Convention on Civil and Political Rights, the International Convention on Economic, Social and Cultural Rights, the International Convention on the Elimination of Racial Discrimination, the Convention of the Elimination of Discrimination Against Women. State repression resulting in palm oil conflicts should be addressed by ASEAN leaders as well when evaluating the progress of transnational biofuel programmes. In short, ASEAN as a group should be held responsible for its failure to act (as a unified ASEAN community) against human rights violation of member countries.

Indonesia should recognise the rights of indigenous communities and respect their livelihoods, which are dependent on forests. A human rights approach to development has been called for (Jiwan, 2013).

Fourth, ASEAN should factor in "land" and "water" in the present energy-food nexus to bring about harmonised energy-land-food-water policies, standards and guidelines. Indirect impacts caused by land use change have to be factored in as well in LCA studies, bearing in mind that the European Commission has in October 2012 proposed to make changes to its renewable energy directive to incorporate ILUC impacts.

### b) *New Perspective on Sustainable Development*

There are inadequate definitions of sustainable biofuel development, even as the EU outlines areas of environmental importance in its renewable energy directive (RED) on 23 April 2009 (Directive 2009/28/EC). This author offers an alternative definition on sustainable development that could apply to any entity, including biofuels:

> "Sustainable development is the process of acquiring social and economic gains from natural resources while contributing to their continued enlargement and enrichment by way of conservation and/or restoration for the betterment of living things and non-living things of heritage value."

Sustainable development itself, when considering the phrase, is an oxymoron. Moreover, environmental sustainability and sustainable development is not a zero sum game.[19] Sustainable development by corporations should instead seek to augment natural resources and improve the lives of people that are dependent on them while reaping profits.

### c) *Regulating Markets and Capital Flows*

Financial institutions and banks have important roles to play in influencing sustainable developments and preventing environmental degradation (Figure 11.10). This can be assisted by having rigorous checklists based on INGO guidelines on environmental-friendly investments when issuing loans to oil palm corporations for land acquisition or production facilities.

---

[19]Environmental sustainability includes people and hence, diverse communities within societies.

Supply chains supporting agrocrops for oil, whether for food or fuel, should be subject to such guidelines as well to ensure that as many market players contribute towards the common goal of sustainable development.

Governments should heed studies on the value of nature and ecosystems, and consider the benefits they bring to the economy in the long run. In 2006, a cost analysis study of the conversion of mixed-use forest and agroforest lands to oil palm plantations in West Kalimantan revealed a total loss of 272.26 billion rupiah (around US$28 million). The land use change resulted in a cumulative loss of income from the agriculture sector, and valuations from forest products, biodiversity and ecological services.[20]

### d) *Standards and Best Practices*

Studies done on policy frameworks, land conversion methods, sound agricultural practices and such should be shared at the relevant ASEAN tracks with the biofuel industries.

As land ownership has been the cause of much grievance and conflict among farmers, smallholders and particularly the indigenous peoples, solutions should be sought as to maintain an equilibrium between fulfilling economic needs as well as social harmony. A good case study to refer to is the pro-human rights approach in Sibak village of Bengkulu Province, Indonesia. The customary land reserve is respected and maintained while communities earn an income by growing palm oil crops on their land and delivering the fruit bunches to oil palm producers. This is also a strategic approach in ensuring that high value conservation forests are maintained in the midst of surrounding agrocrop lands.

## Conclusion

Even with Indonesia joining the other AMCs in ratifying the Agreement on Transboundary Haze Pollution, it is unlikely that the region will see dramatically clearer skies and clean air during the harvest seasons. As long as ASEAN avoids having liability clauses within multi-lateral agreements or formal punitive instruments targeted at polluting countries, the region's air, health, tourism, biodiversity and economies will continue to be severely impacted.

[20]This study was conducted by NGO WALHI/Friends of the Earth Indonesia based on monographs from surveys with villagers in Manis Mata Sub-District, Ketapang District, West Kalimantan.

As such, strict environmental guidelines by the UN and sustainable production directives set by the EU are more likely to be the drivers for change in sustainable practices in Asian biofuel producing companies, rather than ASEAN itself.

The role of governance in the biofuel sector is dependent on social institutions, various types of NGOs especially at local levels, and citizens. Financial institutions are particularly powerful and are in an inarguable position to tap their sphere of influence to generate greater social and environmental responsibility, and ethics, in agrocrop investments. A RETRAC model (Figure 11.10) identifies key actors, their commitments towards sustainability practices and the flow of command chains, including markets and capitalisation.

ASEAN should also urgently address the energy-land-food-water nexus to prevent land-use conflicts while providing for alternative energy sources without impacting on food and water security. ASEAN with more bite than bark will lay a firm path towards sustainable biofuel production, fulfilling its goals of energy security.

The interaction between national-level RETRAC models and ASEAN-level RETRAC models is also an excellent subject to explore in future dissertations or theses. Market forces, driven by the latest revisions in the EC RED, coupled with targeted and distributed governance by multiple biofuel actors, as represented in this author's RETRAC model, will enable such a change to take place towards greater sustainability.

## References

ActionAid. (2012). *Fuel for Thought: Addressing the social impacts of EU biofuels policies.* [Online] Available at: http://www.actionaid.org.uk/103219/news.html. [Accessed 7 March 2012].

Aguiba, M. (2012). "Gov't May Lose P48-B Investments", *Manila Bulletin Publishing Corp.*, 3 September. [Online] Available at: http://mb.com.ph/node/347432/govt-may-lo. [Accessed 10 August 2012].

Andrews-Speed, P., Bleischwitz, R., Boersma, T., Johnson, C., Kemp, G., and VanDeveer, S. (2012). *The Global Resource Nexus: The struggles for land, energy, food, water and minerals*, Transatlantic Academy, Washington DC.

APEC. (2012). *APEC Biofuels.* [Online] Available at: http://www.biofuels.apec.org/me_philippines.html. [Accessed 15 May 2012].

ASEAN. (2010). *ASEAN Plan of Action for Energy Cooperation 2010–2015*, The ASEAN Secretariat, Jakarta.

ASEAN. (2011a). *Report of the 6$^{th}$ Meeting of the ASEAN Committee on Science and Technology Plus Three* in Jeju, Republic of Korea, p. 8.

ASEAN. (2011b). *Community in Figures*, ASEAN Secretariat, Jakarta, p. 1. [Online] Available at: http://www.aseansec.org/documents/ASEAN%20community%20in%20figures.pdf. [Accessed 15 April 2012].

ASEAN. (2011c). *The 3rd ASEAN Energy Outlook*, ASEAN Center for Energy, Jakarta.

ASEAN. (2012). *ASEAN Economic Community Blueprint*, The ASEAN Secretariat, Jakarta.

Bayar, T. (2011). "Malaysia explores its renewables option", *Renewable Energy World*, 16 September. [Online]. Available at: http://www.renewableenergyworld.com/rea/news/article/2011/09/malaysia-explores-its-renewables-options. [Accessed 18 June 2012].

Biofuel Indonesia. (2007). *Government Initiatives.* [Online] Available at: http://www.biofuelindonesia.com/government.html. [Accessed 23 July 2012].

Biofuelsdigest.com. (2011). "Biofuel Mandates Around the World", *Biofuels Digest*, 21 July. [Online] Available at: http://www.biofuelsdigest.com/bdigest/2011/07/21/biofuels-mandates-around-the-world/. [Accessed 22 May 2012].

Borras, S. Jr. (2013). "Foreword", *The Palm Oil Controversy in Southeast Asia — A Transnational Perspective*, ISEAS, Singapore.

Borras, S. Jr. and Franco, J. C. (2011). *Political dynamics of land grabbing in Southeast Asia: Understanding Europe's role*, Discussion Paper. Transnational Institute, Amsterdam.

BP. (2011). *BP Statistical Review of World Energy June 2011*, BP, London.

Bulan, R. (2010). *Indigenous Peoples and the Right to Participate in Decision Making in Malaysia. Centre for Malaysian Indigenous Studies*, University of Malaya, Kuala Lumpur, Malaysia. [Online] Available at: www2.ohchr.org/english/issues/indigenous/.../UniversityMalaya.doc [Accessed 11 April 2015].

Bunge website. (ND). [Online] Available at: http://www.bunge.com/Asia-Pacific.Bloomberg [Accessed 27 September 2012].

Business Time Malaysia. (ND). "Biofuel a promising alternative", *New Straits Times Online.* [Online] Available at: http://www.btimes.com.my/Current_News/BTIMES/articles/20121016155557/Article/index_html. [Accessed 25 October 2012].

Carlson, K., Curran, L., Asner, G., Pittman, A., Trigg, S., and Adeney, J. (2012). "Carbon emissions from forest conversion by Kalimantan oil palm plantations", *Nature Climate Change.* [Online] Available at: http://www.nature.com/nclimate/journal/vaop/ncurrent/full/nclimate1702.html. [Accessed 18 October 2012].

Chiavari, J. (2013). "Implications of EU Biofuel Policies for Southeast Asia", *The Palm Oil Controversy in Southeast Asia — A Transnational Perspective,* ISEAS, Singapore, p. 213.

Chin, M. (2011). "Biofuels in Malaysia: An Analysis of the legal and institutional framework", *Center for International Forestry Research.* Working Paper 64.

Clean Fuels Development Coalition. (2007). *The Ethanol Fact Book*, Clean Fuels Development Coalition, Washington, D.C., p. 47.

Colchester, M. (2011). "Palm oil and indigenous peoples of South East Asia", *Forest Peoples Programme,* International Land Coalition, Rome.

Colchester, M., Jiwan, N., Andiko, Sirait, M., Firdaus, A.Y.F., Surambo, A., and Pane, H. (2006). "Promised Land: Palm Oil and Land Acquisition in Indonesia — Implications for Local Communities and Indigenous Peoples", *Forest Peoples Programme*, Sawit Watch, HuMA and ICRAF, Bogor. [Online]. Available at: http://www.forestpeoples.org/sites/fpp/files/publication/2010/08/promisedlandeng.pdf.

Dewi, O. (2013). "Reconciling development, conservation, and social justice in West Kalimantan." in *The Palm Oil Controversy in Southeast Asia: A Transnational Perspective*, ISEAS, Singapore, pp. 174–176.

European Commission. (2012). "New Commission proposal to minimise the climate impacts of biofuel production", *press release,* 17 October 2012. [Online] Available at: http://europa.eu/rapid/press-release_IP 12 1112_en.htm#footnote-2). [Accessed 13 January 2013].

Ewing, J.J. and McRae, E. (2012). Transboundary haze in Southeast Asia: Challenges and pathways forward. *NTS Alert*, RSIS Centre for Non-Traditional Security (NTS) Studies for NTS-Asia, Singapore.

FAO. (2010). *Oilseeds Oils & Meals Monthly Price and Policy Update.* No. 16. Food and Agriculture Organization of the United Nations, Rome.

Greenpeace. (2010). *Driving to destruction: The Impact of Europe's biofuel plans.* [Online] Available at: http://www.greenpeace.org/eu-unit/Global/eu-unit/reports-briefings/2011percent20pubs/1/driving-to-destruction-08-11-10.pdf. [Accessed 7 March 2012].

Hamid, A.R. (2013). "Organisations protest against decision to amend Aboriginal People's Act", *The Star*, 13 July. [Online] Available at: http://www.thestar.com.my/News/Nation/2013/07/13/Orang-asli-fear-loss-of-land-Organisations-protest-against-decision-to-amend-Aboriginal-Peoples-Act/ [Accessed 11 April 2015].

Ibrahim, H. (2002). "Small renewable energy programme for the promotion of renewable energy power generation", Presentation at the *ASEM Green IPPs Network*, 27 October, Bangkok.

Informa Agra & Commercial. (2011). *World Ethanol and Biodiesel Markets — A Global Industry Outlook 2011*, Informa Agra & Commercial, London.

Ing., Legowo, E.H., Kuusuryani, Y., Reksowardojo, I.K. (2007). "Biofuel Development", Presentation at the *USDA Global Conference*, 20–22 July, Indonesia Ministry of Energy and Mineral Resources, Jakarta.

International Energy Agency. (2010). "Renewables for transport," *World Energy Outlook 2010.* International Energy Agency, Paris, p. 359.

International Energy Agency. (2011). Technology Roadmap: Biofuels for Transport, International Energy Agency Renewable Energy Division, Paris.

International Energy Agency. (2012). *Combustible Renewables and Waste Definitions.* [Online] Available at: https://www.iea.org/statistics/resources/balancedefinitions/#biofuelsandwaste [Accessed 20 October 2012].

Jiwan, N. (2013). "The political ecology of the Indonesian palm oil industry" in Pye, O., Bhattacharya, J. (Eds.), *The Palm Oil Controversy in Southeast Asia — A Transnational Perspective,* ISEAS, Singapore, pp. 42–73.

Katsumata, H. and Tan, S. (2007). "An Ideal ASEAN for People and Governments", *People's ASEAN and Governments' ASEAN*, S. Rajaratnam School of International Studies, Singapore, p. 15.

Kementerian Tenaga Teknologi Hijau dan Air. (2008). *National Renewable Energy Policy & Action Plan*, KeTTHA, Kuala Lumpur.

Kirschke, J. (2011). "Newswire, Indonesian protesters sew their mouths shut over deforestation/palm oil", *Earth First!* [Online] Available at: http://earthfirstnews.wordpress.com/2011/12/29/indonesian-protesters-sew-their-mouths-shut-over-deforestationpalm-oil/. [Accessed 29 October 2012].

Krishnamurti, B. (2008). "Government Strategic Policies on Sustainable Palm Oil Development", paper presented at the *Indonesian Palm Oil Conference and Price Outlook 2009*, 3–4 September, Bali International Convention Center — The Westin Resort, Denpasar, Bali.

Lang, C. (2012). "Indonesian court revokes oil palm concession in Tripa peat swamp", *REDD-Monitor.org*, 7 September 2012, [Online]. Available at: http://www.redd-monitor.org/2012/09/07/indonesian-court-revokes-oil-palm-concession-in-tripa-peat-swamp/. [Accessed on 16 October 2012].

Lapola, D. Schaladach, R., Alcamo, J., Bondeau, A., Koch, J., Koelking, C., and Priess, J.A. (2010). "Indirect land-use changes can overcome carbon savings from biofuels in Brazil", *PNAS Direct Submission, Proceedings of the National Academy of Sciences and of the United States of America*, Washington. D.C.

Lopez, G.P. and Laan, T. (2008). "Biofuels — at what cost? Government support for biodiesel in Malaysia", *The Global Subsidies Initiative (GS) and The International Institute for Sustainable Development (IISD)*, p. 24. [Online]. Available at: http://www.iisd.org/gsi/sites/default/files/final_malaysia_2.pdf. [Accessed 3 June 2012].

Malaysia Government. (ND). *National Biomass Strategy 2020: New wealth creation for Malaysia's palm oil industry*,

Menguita-Feranil, M.L. (2013). "Contradictions of palm oil promotion in the Philippines" in Pye, O., Bhattacharya, J. (Eds.), *The Palm Oil Controversy in Southeast Asia — A Transnational Perspective*, ISEAS, Singapore, pp. 97–117.

Meteorological Service Singapore. (ND). [Online] Available at: http://www.weather.gov.sg/wip/c/portal/layout?p_l_id=PUB.1003.613 [Accessed 16 October 2012].

Ministry of the Environment and Water Resources. (2012). *ASEAN Ministerial Meeting on the Environment* in Bangkok, Thailand, 26 September.

[Online] Available at: http://app.mewr.gov.sg/web/contents/contents.aspx?contid=1730. [Accessed 16 August 2012].

Murphy, R., Woods, J., Black, M., and McManus, M. (2011). "Global developments in the competition for land from biofuels", *Food Policy*, **36**(36), S52–S61.

Neste Oil website. (2011). *Neste Oil celebrates the grand opening of its ISCC-certified renewable diesel plant in Singapore*. [Online] Available at: http://www.nesteoil.com/default.asp?path=1,41,540,1259,1260,16746,17030. [Accessed 14 August 2012].

Pemandu. (ND). *Overview of the National Key Economic Areas*. [Online] Available at: http://etp.pemandu.gov.my. [Accessed 27 July 2012].

Philippines Board of Investment. (2011). *Philippine Biofuels*. [Online] Available at: http://www.boi.gov.ph/pdf/industryprofiles/Renewable%20Energy/Biofuels.pdf. [Accessed 27 July 2012].

Ho, M. (2011). "PetroVietnam to export 85% of fuel ethanol production over 2011–2013", *Platts*, 24 May. [Online] Available at: (http://www.platts.com/RSSFeedDetailedNews/RSSFeed/Petrochemicals/7677340. [Accessed 2 May 2011].

Pye, O. (2013). "An analysis of transnational environmental campaigning around palm oil" in Pye, O., Bhattacharya, J. (Eds.), *The Palm Oil Controversy in Southeast Asia — A Transnational Perspective*, ISEAS, Singapore, pp. 179–196.

Pye, O., Bhattacharya, J. (Eds.). (2013). *The Palm Oil Controversy in Southeast Asia — A Transnational Perspective*, ISEAS, Singapore.

Remo, A. (2011). "Oil firms allowed to bring in ethanol", *Business.Inquirer.Net*, 24 July. [Online] Available at: http://business.inquirer.net/8653. [Accessed 10 September 2012].

RSIS. (2007). *People's ASEAN and Governments' ASEAN*, S. Rajaratnam School of International Studies, Singapore.

SEDA (Kementerian Tenaga Teknologi Hijau dan Air). (2008). *National Renewable Energy Policy & Action Plan*, p.v Table (ii), KeTTHA, Kuala Lumpur. [Online]. Available at: http://seda.gov.my/go-home.php. [Accessed 27 July 2012)].

Shen, R. (2014). "Singapore approves bill to fine domestic, overseas polluters", *Reuters*, 5 August. [Online]. Available at: http://www.reuters.com/article/2014/08/05/us-singapore-haze-lawmaking-idUSKBN0G50WB20140805. [Accessed 19 November 2014].

Smith, J. (2010). *Biofuels and the Globalization of Risk*, Zed Books Ltd, London.

Sentana, M. and Hariyanto, J. (2014). "Indonesia ratifies regional pollution treaty as haze spreads to neighbors", *Wall Street Journal*, 16 September. [Online]. Available at: http://online.wsj.com/articles/indonesia-ratifies-regional-pollution-treaty-as-haze-spreads-to-neighbors-1410869959. [Accessed 10 April 2015].

Severino, R. (2007). "A Sense of Community for Southeast Asia" in RSIS, *People's ASEAN and Governments' ASEAN*, S. Rajaratnam School of International Studies, Singapore.

Sovacool, B. (2009). "Energy Policy & Cooperation in SEA: The history, challenges and implications of the trans-ASEAN gas pipeline (TAGP) network", *Energy Policy*, **37** (2009), 2356–2367.

Talkvietnam.com. (2012). *Dung Quat plant exports biofuel to Philippines.* [Online]. Available at: http://talkvietnam.com/2012/06/dung-quat-plant-exports-biofuel-to-philippines/?utm_source=feedburner&utm_medium=twitter&utm_campaign=Feed%3A+TalkVietnam+%28Talk+Vietnam%29. [Accessed 10 September 2012].

Teoh, C.H. (2013). "Malaysian corporations as strategic players in Southeast Asia's palm oil industry" in RSIS, *The Palm Oil Controversy in Southeast Asia: A Transnational Perspective*, ISEAS, Singapore.

The Jakarta Post (2011). "Mining, plantation disputes to intensify", *The Jakarta Post*, 29 December. [Online]. Available at: http://www.thejakartapost.com/news/2011/12/29/mining-plantation-disputes-intensify.html. [Accessed 29 October 2012].

The Star (2011). "Commodities not affected by economic uncertainties", *The Star*, 27 September. [Online]. Available at: http://thestar.com.my/news/story.asp?file=/2011/9/27/nation/9576954&sec=nation. [Accessed 16 October 2012].

Tropical Peat Research Institute. (2009). Extracted by Ibrahim, N. [Presentation]. "Biofuel experience in Malaysia: Oil palm", *Wetlands International Malaysia*, p. 12. [Online]. Available at: http://www.cbd.int/doc/meetings/agr/rwspubio-ap-01/other/rwspubio-ap-01-p-malaysia-09-11-27-en.pdf. [Accessed 1 October 2012].

UNEP. (2009). *Towards Sustainable Production and Use of Resources: Assessing Biofuels Report*, United Nations Environment Programme, Nairobi, p.54. [Online]. Available at: http://www.unep.org/resourcepanel/Publications/AssessingBiofuels/tabid/56055/Default.aspx. [Accessed 10 March 2012].

United Nations. (1992). *United Nations Framework Convention on Climate Change (A/AC.237/18 (Part II)/Add.1 and Corr.1).* [Online] Available at: https://stats.oecd.org/glossary/detail.asp?ID=6323 [Accessed 12 April 2015].

United Nations News Centre. (2013). *Malaysia must protect environment, indigenous rights as it reduces poverty — UN expert.* [Online]. Available at: http://www.un.org/apps/news/story.asp?NewsID=46774#.VSitcTuUeQx. [Accessed 11 April 2015].

USDA Foreign Agricultural Service. (2012). *USDA GAIN: Thailand Biofuels Annual 2012*, United States Department of Agriculture, Washington D.C.

Viet Nam News. (2012). "Ministry promotes cleaner fuel", *Viet Nam News*, 3 October. [Online]. Available at: http://vietnamnews.vn/Economy/230940/ministry-promotes-cleaner-fuel.html. [Accessed 20 November 2012].

Vissoor, O. and Spoor, M. (2011). "Land grabbing in post-Soviet Eurasia: the world's largest agricultural land reserves at stake", *Journal of Peasant Studies*, **38**(2), 299–323.

Wakker, E. (2013). "Leveraging product and capital flows to promote sustainability in the palm oil industry" in RSIS, *The Palm Oil Controversy in Southeast Asia: A Transnational Perspective*, ISEAS, Singapore.

Worldwatch Institute. (2007). "A Burning Issue", *World Watch Magazine*, **20**(4). [Online]. Available at: http://www.worldwatch.org/node/5136. [Accessed 26 September 2012].

Zhou, A. and Thomson, E. (2009). "The development of biofuels in Asia", *Applied Energy*, **86**(2009), S11–S20.

# Part 4

# Urban Studies

# 12

# ENERGY EFFICIENCY IN AFFORDABLE HOUSING: CAN IT WORK FOR INDEPENDENT RENTAL OWNERS

JULIA EMERSON and GEORGE OFORI

The increasing demand for rental housing in the United States is causing rental rates to rise, straining the already limited supply of ageing affordable housing. Currently, half of the multi-family stock is 30 years old or older, and properties built between the 1940s and 1960s cost 50 per cent more to operate than buildings built after 2000. With less efficient heating, cooling, plumbing and lighting systems, these older buildings will require investment over time in order to extend their life so that they can remain in the affordable housing stock. This would have an impact on overall energy consumption, and help reduce the energy–cost burden for both building owners and their residents. However, the multi-family housing industry, and in particular the Independent Rental Owners (IROs), do not actively pursue energy efficiency programs due to multiple barriers.

The purpose of this chapter is to discuss (1) the history of affordable housing in the United States and its importance today; (2) the necessity to preserve existing affordable housing stock; (3) how environmental changes can have positive economic impacts on older buildings; (4) barriers that IROs face in order to participate in energy–efficiency programmes; and (5) recommendations and actions that can be taken by IROs to address these issues.

Recommendations include the necessity for an integrated efficiency programme and support for IROs who lack time and resources to investigate and implement comprehensive programmes. There must be further support with sufficient public and private funding sources, including creative funding ideas, to bring in more participants. Another element that is not discussed in detail is the importance of human behaviour, attitudes and motivation. This warrants further investigation, as motivation of IROs and residents to participate in energy efficiency will be critical for a successful programme.

## 1. Introduction

### 1.1 *Energy and Sustainable Buildings — Energy Consumption in Cities*

Today, the urban population consumes the majority of the world's resources and has a significant impact on the environment. More than half of the world's people live in cities and urban areas, with the numbers being as high as 80 per cent in some developed countries. These populations account for 60–80 per cent of energy consumption and up to 75 per cent of carbon emissions (UNEP, 2012). Therefore, cities are an important element when addressing local, regional and even global environmental threats, such as climate change, energy and resource depletion (UNEP, 2012).

According to the United Nations Sustainable Buildings and Climate Initiative, if targets for greenhouse-gas emissions are to be met, carbon emission from the building-sector must be addressed (UNEP–SBCI, 2009). As increasing numbers of people throughout the world move into cities, there will continue to be pressure on energy resources.

Global energy consumption in the building sector has surpassed even that of the transportation and industrial sectors. In 2010, the United States consumed the second largest share of energy worldwide with 19 per cent of global consumption. Only China consumed more than the United States (U.S. Department of Energy, 2012).

Within the United States, the building sector accounted for 41 per cent of primary energy consumption. The majority of the usage came from Heating Ventilation Air Conditioning (HVAC) systems and lighting, which account for over 50 per cent of building-sector energy consumption (U.S. Department of Energy, 2011; EIA, 2012). Within the US building sector, 54 per cent of energy consumption was residential (U.S. Department of Energy, 2011).

In 2013, renters made up 35 per cent of all households, and there was continuing growth in the rental market. Although renters are more likely to be single, young, minority and low-income, with a median income of around half of owners' households, there is an increasing number of families and elderly also moving from owner households into renter households (Joint Center for Housing Studies, 2013). The Housing Vacancy Survey reported an increase of 1.1 million renters in 2011–2012 (U.S. Department of Commerce, 2013). Thus, the rental market has tightened and rates have increased, making housing less affordable for lower-income households (Joint Center for Housing Studies, 2013). Rent and utilities are two of the largest expenditures in affordable multifamily housing and as these numbers increase more renter households will bear the cost burden of energy inefficient buildings. (Benningfield Group, 2010).

## 1.2 *Multifamily Industry*

Multifamily housing in the United States is generally considered to be renter-occupied, and the standard industry definition of multifamily is a structure with five or more units (Follain, 1994). Many multifamily properties are managed by Independent Rental Owners (IROs) who typically have invested their own money, and are actively involved in the management of their properties (National Apartment Association, 2014). For IROs in the low-income multifamily industry, economic health is essential, especially if they own older buildings. The task of 'greening these smaller properties can be difficult as they are often a small operation and lack the funds required.

Many IROs do not actively pursue energy–efficiency programmes, and the impacts of poor participation in energy–conservation programmes can have a negative effect on the affordable multifamily housing stock. This is because these buildings are generally older and far less energy efficient, leading to higher utility bills for the people who can least afford it. The need for this type of housing is critical, but to keep rental prices affordable and competitive many building owners need to upgrade and retrofit these older buildings to become more energy efficient. Otherwise, these units will lose their economic value and will be taken out of service, further reducing an already insufficient affordable housing stock.

## 2. Affordable Housing in the United States

### 2.1 *History of Affordable Housing in the United States*

The federal standards for affordable housing in the United States indicate that a household should not spend more than 30 per cent of its income on rent (Economic Policy Program, Housing Commission, 2013). In the United States, nearly half of all renters are moderately rent burdened, that is, they pay more than 30 per cent of their income on rent. Of the 41 million renters, 25 per cent are considered severely rent burdened; they spend over 50 per cent of their income on rent (PD&R, 2011, United States Census Bureau, 2011).

The number of low-income families that have severe cost burdens hit a record high in 2011 with 68.7 per cent of households earning less than $15,000 per year and 30.9 per cent of households earning between $15,000 and $29,000 per year (Joint Center for Housing Studies, 2013). The increase in low income families, immigrants seeking a better life, gentrification and abandonment of housing stock have all added to the increasing shortage of affordable housing (Joint Center for Housing Studies, 2013; Serlin, 2013).

### 2.1.1 *The United States Housing Act of 1937*

Public and affordable housing in the United States dates back to 1937, when the federal government established the US Housing Act to provide rental assistance to the nation's low-income households. The program was intended for young, single people or married couples, who only needed temporary support until they could purchase their own houses. The original intent was not to create a long-term infrastructure, and therefore many social concerns which would later prove to be difficult to manage, were not given much consideration (Mueller, 2010).

### 2.1.2 *The Housing Act of 1949 (1949 Act)*

In the late 1940s, there was insufficient housing for returning World War II veterans. Hence, the Housing Act of 1949 was established for returning veterans and people who needed temporary assistance to get back on their feet (Katz, 2009). The 1949 Act allowed the wholesale demolition of slums and paved the way for new development. However, since there were no incentives to build affordable housing, economics drove commercial projects and housing for the middle class instead (Katz, 2009).

### 2.1.3 *The Housing and Community Development Act of 1974*

From the 1950s to the 1970s, the public housing stock had grown from 150,000 to over 1 million units (Economic Policy Program, Housing Commission, 2013) and the profile of the public-housing resident transitioned from young married couples with short-term plans to non-nuclear, single-head, low income households (Mueller, 2010). Older properties and escalating operating costs led to higher rents, which ultimately led to many families spending as much as 75 per cent of their income on rent and utilities. As a result, they had insufficient funds available for other basic necessities such as food, health care, education, and transportation (Economic Policy Program, Housing Commission, 2013).

Higher costs, corruption and poor management of public housing led to the Section 8 Program via the Housing and Community Development Act of 1974. This was a more flexible programme and funds were able to support new production, rehabilitation and tenant based rental assistance with existing properties (Economic Policy Program, Housing Commission, 2013). This helped retain stock but created few new developments.

By the 1980's, the cycle continued, with increasing operating expenses leading to higher rents, and more single head households who could not afford to pay their monthly rent while providing basic necessities for their families (Mueller, 2010).

The poverty rate in older, urban neighbourhoods rose by 54 per cent in the 1980s and 1990s, and approximately 71 per cent of the residents were African–American or Hispanic. Homelessness began to appear on a large scale (Katz, 2009). The economic hardship, combined with a lack of affordable housing, was responsible for 80 per cent of the instances of homelessness (Economic Policy Program, Housing Commission, 2013).

## 2.2 *Affordable Housing Today*

### 2.2.1 *Inadequate supply*

Today's affordable housing environment is far less than adequate, with the lack of supply being one of the primary issues. The National Low Income Housing Coalition (NLIHC) has reported that in 2013 there are currently only 30 affordable rental units available for every 100 low-income (LI) residents and 57 affordable rental units available for every 100 very-low-income (VLI) residents, whose annual income falls below 50 per cent of the HUD area median family income.

In 2011, there were 8.5 million families with what are defined as worst–case housing needs. These numbers have increased as a result of the nation's economic and housing market problems, where unemployment and foreclosures forced homeowners to become renters. Moreover, despite an increase in affordable housing, industry reports continue to highlight the mismatch between housing availability and need with inadequate supply in both regular and affordable rental markets (Serlin, 2013).

### 2.2.2 *"Competition" with open market units*

Competition with open markets is also having a negative impact on the overall situation. As the number of renter households continues to increase every year, with 7 million renters anticipated by 2020, supply and demand in the higher level rental markets will have a trickle down effect, tightening all rental classes, and eventually having an effect on affordable housing. Rental rates will increase as vacancy rates drop.

#### 2.2.3 *Low Income Housing Tax Credit (LIHTC) Program under threat*

The LIHTC programme was created with bipartisan support in 1986 as part of the Tax Reform Act. It replaced other tax subsidies and has encouraged developers and investors to restore or create new affordable housing units by providing a subsidy that makes up the difference in developmental costs and income generated due to restricted rents. In short, investors receive tax credits and reduce their tax liabilities, dollar for dollar (Economic Policy Program, Housing Commission, 2013). Unfortunately, this program and others like it are being cut due to Federal spending cuts, which will lead to even more stress on the already undersupplied affordable housing stock.

## 3. Preserving Existing Affordable Housing

The acute shortage of housing will continue to create a multitude of other social problems, such as concerns for quality of life and health, making clear the compelling need for additional affordable housing. While new affordable housing construction will help, the building rate is not high enough to keep up with demand. Therefore, it is necessary to not only develop additional housing, but also to maintain and upgrade the existing affordable housing supply (Norris, 2013).

### 3.1 *Existing Affordable Housing*

#### 3.1.1 *Public housing and HUD–assisted housing*

There are currently 1.2 million households living in 13,000 public housing developments owned by the federal government (Bamberger, 2010). HUD provides federal aid to the housing authorities, as well as technical and professional assistance in planning and maintaining the developments (HUD, undated). An additional 2.1 million households (approximately) live in privately owned housing that are subsidized by HUD with programmes such as Section 8 (Rice, 2011).

#### 3.1.2 *LIHTC housing*

The LIHTC programme is the largest federal programme that provides funds for the production and preservation of affordable housing. There are 2.2 million households living in LIHTC properties, and to qualify, income-level requirements have to be met. Residents with Section 8 voucher and other HUD subsidies are also able to live in these properties (Moelis Institute for Affordable Housing, 2012).

### 3.2 *Owners and Investors*

There is a diverse group of investors in today's market, from small business owners and non-profit organizations to large equity firms (Mallach, 2010). Larger developers and management companies currently dominate the affordable multifamily industry. In 2011, four of the five largest multifamily owners were affordable housing companies (Shaver, 2011).

There is also a significant percentage of IROs in the industry. The National Apartment Association states that up to 40 per cent of national membership is by IROs who own 500 units or less. Therefore, IROs are an important element in retaining affordable housing stock.

Because of the nature of the affordable housing industry, landlords are often restricted with the rents they can charge, yet must maintain their properties and manage the gap between income and cost. For smaller owners and IROs, the cost and rate of return are often matters of survival (Mallach, 2010). Therefore, in order to preserve this housing stock, both national and local governments must provide smaller owners with incentives such as grants, low-interest loans, tax credits and Section 8 vouchers. This will help close the gap between what tenants can afford and what landlords need to charge in order to provide quality, safe and sustainable housing.

## 4. Sustainability Principles and Affordable Housing

### 4.1 *The Importance of Preservation and Investment in Energy Efficiency Programmes*

Currently the burden of high utility costs falls on the renter, who in affordable housing situations is the least capable of affording those costs (HUD, 2011). Renters can pay up to four times more than the average of 4 per cent that is usually paid on energy bills. For example, in the 2007/08 fiscal year, 26 per cent of renter evictions in St Paul, Minnesota were due to utility cut offs. As these fees continue to rise, it will become increasingly difficult for vulnerable families to keep up with the costs (HUD, 2013).

From a landlord's perspective, as utility costs rise, there is greater pressure to increase rents, which in turn, reduces the availability of funds to upgrade to a more energy–efficient property. For example, between 2001 and 2009, rents increased by 7.6 percent and utilities rose by 22.7 percent (HUD, 2011).

Therefore, investment in energy efficient programs should include the preservation of older public and private affordable housing units (Bamberger, 2010; McKibbin *et al.*, 2012). Residents, owners, the government and utility companies will all benefit from investment in energy efficient retrofits.

#### 4.1.1 *Residents' and owners' savings*

A reduction in utility costs would benefit both residents who pay their own utility bills, and owners with operating and maintenance expenses. Reduction in cost can become even more beneficial to owners if they pay all the utilities on the property.

#### 4.1.2 *Federal savings*

The Department of Housing and Urban Development's "energy bill", which is the amount that HUD spends on utilities either through direct payment or allowances or subsidies in its portfolio of public and HUD-assisted (private owner) housing, reached the $6.6 billion mark in 2010. By improving energy efficiency through retrofit, HUD believes that it could save $1–2 billion, with an estimated usage reduction of 10–20 per cent (Bamberger, 2010; Rocky Mountain Institute, 2013).

#### 4.1.3 *Large-scale recruitment by utility companies*

Multifamily housing is a large homogeneous market that has potential for easy recruitment into energy–efficiency programmes. The industry has strong networks with national, state and city apartment associations. Energy companies can present to a large group of building owners more efficiently and in less time through the apartment associations (McKibbin, 2013).

The multifamily housing market can also be easier to upgrade than commercial businesses. In any given geographical area, many multifamily buildings will be similar in age, size and upgrade needs whereas each commercial business will have its own needs; for example, a restaurant and beauty salon will have different energy uses and needs (McKibbin, 2013).

#### 4.1.4 *Benefit to utility companies and reduction in carbon emissions*

Directing resources to multifamily energy–efficiency programmes allows utility companies to delay investment in power plants as well as transmission and distribution infrastructure, due to a reduction in demand on the entire system. Multifamily programmes also provide utility companies with economies of scale and an efficient way to achieve their state-required energy–efficiency goals (McKibbin, 2013).

## 4.2 *Barriers to Investing in Energy Efficiency*

Energy retrofits can save money, create energy–efficient buildings and extend the life of much-needed affordable housing units. It makes economic sense for owners to try to find ways to improve the efficiency of their buildings, yet there are large barriers that hinder progress in energy–efficiency in affordable housing, such as programme awareness, financial complexities, split incentives and fragmented leadership.

Building owners, particularly those in affordable housing, usually need financial incentives in order to proceed with energy–efficiency projects. Furthermore, studies report that multifamily affordable housing receives a disproportionately small share of available energy–efficiency funding (McKibbin *et al.*, 2012).

### 4.2.1 *Information and awareness*

Owners often lack clear information on energy–efficiency programmes in the affordable housing industry. This creates barriers on multiple levels (Bamberger, 2010; HUD, 2011; Living Cities and Institute for Sustainable Communities, 2011; McKibbin *et al.*, 2012).

Although programmes that encourage increased energy efficiency exist, there is much misinformation and lack of awareness about these programmes. Information is often presented in a way that is hard for building owners to understand, and many IROs do not have the resources or time to engage in these programmes (President of ABC, in person interviews, July 2013).

Also, programme eligibility is frequently unclear to potential participants. There are also differences in definitions and classifications from city to city. For example, some cities categorize 'multifamily' under commercial accounts while other cities categorize 'multifamily' separately. In addition, in some situations one type of property can be categorized under a 'commercial' account, while certain portions of the same property can be considered 'residential' (McKibbin, 2013).

### 4.2.2 *Financial barriers*

Capital for the upfront cost of a building retrofit is considered to be one of the largest barriers for multifamily-unit owners. High transaction costs and lack of direct grant opportunities can often exclude many properties from retrofit opportunities.

#### 4.2.2.1 Value of efficiency savings

The current housing finance system is reluctant to consider savings related to efficiency retrofit loans. Energy savings are not valued or appraised, even when utility costs are included in the operating cost projections when underwriting the risk of a rehabilitation project or new construction loan (Bamberger, 2010). However, the concept of valuing savings is changing. Government Sponsored Enterprises, such as Fannie Mae and Freddie Mac, (the nation's largest multifamily mortgage investor) now take into account utility cost savings and efficiencies when making lending decisions.

#### 4.2.2.2 Mixed funding

Pre-existing finance conditions can also cause difficulty if there is existing debt. Depending on the previous loan criteria, there may be senior debt holders who may have to approve a new loan or refinance. Others might impose large transaction costs to refinance. Smaller IROs do not usually have the resources to organise many of these requirements.

#### 4.2.2.3 Limited planning

Another setback for building owners is the short-term planning mentality. Many building owners know that if they retrofit their property they will see savings. Whether it is an efficiency investment with a return over 8–10 years or 1–2 years, they still struggle to plan for the long term when funds are generally tight.

#### 4.2.2.4 Green design

Both private and government lenders are also incorporating "green" design into their criteria. Developers seeking funding must deal with multiple lenders and can sometimes face multiple green certification requirements (Miller, 2013; Rocky Mountain Institute, 2013). For owners of affordable housing properties, many of the existing certification systems are beyond their financial means.

### 4.2.3 *Split incentives*

Split incentives occur when the individual responsible for investing in efficiency technology is different from the one responsible for the monthly bills. Some multifamily properties are individually metered and others have master meters.

If a resident pays their utilities individually or is charged for their electricity, there is little incentive for building owners to make energy–efficiency investments as they do not benefit directly from the savings. On the other hand, residents can show little incentive if they receive subsidies or live on a property where all bills are included in the monthly rent. This can hinder real potential savings. Energy initiatives in these scenarios are dependent on collaborative efforts between landlord and tenants.

### 4.2.4 *Fragmented leadership*

Leadership among federal, state and city authorities and utility companies is fragmented (McKibbin *et al.*, 2012). For example, HUD and the US Department of Energy share the responsibility for residential retrofit programmes, while the Treasury is responsible for tax credits associated with low-income housing, energy efficiency and renewables, and public benefit funds (Bamberger, 2010; McKibbin, 2013).

As utilities in the United States operate in a complex, heavily regulated and decentralised industry, which makes a national alignment between utilities and building owners improbable, greater collaboration between building owners and local utility companies is necessary (McKibbin *et al.*, 2012). The utility companies are, however, important and are considered to be key players in energy–efficiency efforts. In 2011, utility spending on oil and gas efficiency programmes rose to almost $7 billion. However, in 2008, almost 80 per cent of these contributions occurred in just 10 states (McKibbin *et al.*, 2012).

Programmes offered by states and cities can also vary significantly from simple incentives such as replacing light bulbs and adding controllable thermostats to more comprehensive incentives which include energy audits, financing and post-retrofit reviews (McKibbin *et al.*, 2012). The variations depend on the type of utility servicing the area, and incentives and constraints that they face with regards to regulatory circumstances within each state.

It is important for building owners to understand that there is no "one" single policy or federal programme that all building owners can join (Living

Cities and Institute for Sustainable Communities, 2011; McKibbin *et al.*, 2012). For that reason, owners need to engage in their local market and see what programmes are offered in their states.

## 5. Results and Discussion

A study was carried out through a series of interviews and surveys with an experienced IRO (ABC) in San Antonio, Texas with over 2000 affordable housing units. Through interviews and surveys with corporate staff, maintenance and operations, property managers and residents, it is clear that initiatives are available and that there is an increasing momentum in energy–efficiency programmes. However, there is concern as to whether they will be able to take advantage of these initiatives or whether future energy laws will become an additional cost in the much–needed affordable–housing industry.

### 5.1 *Interviews*

#### 5.1.1 *Management*

The company and staff believes that everyone deserves an opportunity to live in an affordable, healthy and safe environment. They understand that many residents struggle to pay rent even when they are subsidised with housing assistance, so they make attempts to keep rents competitive by keeping costs low. Preventive maintenance programmes focus on reducing water and utility costs. For example, air-conditioning and heating units are regularly cleaned for maximum efficient use, windows and doors are checked for air leakage, and appliances such as refrigerators are also checked for leakage.

By reducing costs in one area, funds can be used to improve other areas or keep rents affordable. The properties are old but still functional and are not due to expire any time soon, therefore preventative maintenance, constant improvement and diligent staff can keep 2104 affordable housing units in operation for families that need them.

#### 5.1.2 *Energy-efficiency programmes*

ABC staff expressed that there is often a "disconnect" between what is advertised by utility companies or the city and what is actually applicable to multifamily housing. They acknowledged that there are programmes with utility providers, but they are not usually given a point of contact. Also, they are often told that programmes advertised do not apply to multifamily housing, or, that they may qualify for the programme, but the time, requirements and costs of participation are higher than the benefits.

For example, there was a Weatherization Assistance Program (WAP) that included a variety of energy–efficiency measures, including work on the building envelope, heating and cooling systems, and appliances. The programme in their area, however, only applied to residential accounts.

There are also programmes that multifamily housing do qualify for, but these do not consider older master–metered properties. For example, the city has a refrigerator recycling programme. However, only two refrigerators can be recycled per address. This means that a property of 200 units on a master meter can only recycle two refrigerators per year.

### 5.1.3 *Awareness*

The preceding section does not mean that there are no programmes that can cover additional costs to make a retrofit financially feasible. The point is that IROs are not aware of these programmes and there does not seem to be a point of contact or dedicated agency, group or individual that building owners can communicate with.

Furthermore, there is little collaboration between utility companies and associations. ABC was not aware of any utility coordinator who worked with their local or state association.

When asked what actions need to be taken to increase participation they felt that a good example for building owners is for other participants to share their positive experiences through the associations. The associations should work with the utility companies to establish a point of contact and collaborate on reaching out to more building owners. Leading by example and having true benefits to share would encourage wider participation.

### 5.1.4 *Keep it simple*

Finally, "*keeping it simple*" was the wish that was repeated by all employees. If energy were considered a more critical resource there would be greater effort to conserve it. If utility companies made the process easier, more people would consider applying for incentives.

## 5.2 *Questionnaire-based surveys*

Managers, maintenance and local police officers all agreed that in order for any programme to work on any property there must be a successful relationship between all levels of staff and residents. The staff interacts with residents on a daily basis and can often give experienced insight on the problems they deal with and how they might implement a solution.

### 5.2.1 *Police*

Local police officers believed that collaboration would be difficult and only incentive programmes would work with residents. None of them were aware of any groups that work on environmental awareness except for an urban gardening group that would help low-income areas plant gardens (members of this group would often get robbed when coming to the low-income neighbourhoods to work on the gardens).

### 5.2.2 *Managers*

Managers had a guarded but more positive outlook about environmental changes on their properties. They felt that the changes would reduce utility costs which would help keep rents at an affordable rate. They also believed that it could influence residents in a positive way to encourage conservation and become a marketing tool to differentiate themselves from other properties.

This however would require time, awareness and education for both management and residents. Rather than the "top-down" approach, a "bottom-up" approach is necessary, as the change in behaviour must come from the resident.

### 5.2.3 *Maintenance*

Collaboration between managers and maintenance is equally important. The maintenance team continuously posts notices on upcoming improvements as well as classes for managers on electricity and plumbing issues, or maintenance problems that other staff should be aware of.

### 5.2.4 *Residents*

Residents were also interviewed on environmental topics, energy, water and recycling. Their responses, included in Table 12.1, show that 84 per cent of the group felt that the environment was important. Furthermore,

Table 12.1. Survey results.

| Questions | very | Some what | Not very | Not at all |
|---|---|---|---|---|
| 1. How important is the environment to you? | 84% | 12% | 3% | 1% |
| 2. How important is it for the property to conserve water and electricity? | 85% | 13% | 1% | 0% |
| 3. What types of things do you do to help conserve water and | | | | |
| 3.a. Limit shower time | 67% | | | |
| 3.b. Use clothing rack to dry clothes | 29% | | | |
| 3.c. Turn water off while brushing teeth | 78% | | | |
| 3.d. In winter, when leaving the room do you lower the thermostat? | 49% | | | |
| 3.e. In summer, when leaving the room do you raise the thermostat? | 37% | | | |
| 4. Have environental concerns change your behavor in any of the (*Please circle all that apply*) | | | | |
| 4.a Transportation | 56% | | | |
| 4.b Your home/city/location | 35% | | | |
| 4.c Recycling efforts on your property | 39% | | | |
| 5. How important is it for the properties to recycle? | 70% | 22% | 2% | 5% |
| 6. Are you currently recycling? | 41% | 21% | 4% | 34% |
| 7. If yes, what are you currently recycling? (please circle all that apply) | | | | |
| 7.a. Newspapers | 24% | | | |
| 7.b. Bottles | 32% | | | |
| 7.c. Cans | 48% | | | |
| 7.d. Mixed Paper | 12% | | | |
| 7.e. Cardboard | 13% | | | |
| 8. Have your received any information on recycling on your property? | 27% | 8% | 6% | 59% |
| 9. If yes, is the information easy to understand? | 49% | 16% | 8% | 26% |
| 10. Are the locations of the recycling bins convenient? | 33% | 13% | 4% | 50% |
| 11. What are the environmental issues most known to you? | | | | |
| 11.a. Trash/Litter | 73% | 18% | 6% | 3% |
| 11.b. Recycling | 74% | 20% | 6% | 0% |
| 11.c. Water | 76% | 11% | 7% | 5% |
| 11.d. Air Pollution | 77% | 12% | 11% | 0% |
| 11.e. Energy Issues | 69% | 15% | 10% | 5% |
| 11.f. Climate Change | 59% | 27% | 14% | 0% |
| 11.g. Transportation | 61% | 20% | 11% | 8% |
| 11.h. Green Technology | 53% | 29% | 18% | 0% |
| 11.i. Global Warming | 49% | 22% | 15% | 14% |
| 11.j. Polar Melt | 48% | 31% | 21% | 0% |
| 11.k. Motor Vehicle Pollution | 62% | 17% | 9% | 12% |
| 11.l. Industrial Pollution | 62% | 23% | 15% | 0% |
| 12. Are you interested in environmental education? | 44% | 21% | 18% | 17% |
| 13. Would you like more information on what you can do on your property? | 58% | 19% | 10% | 13% |

86 per cent felt that it is important to conserve water and electricity. Yet when asked what actions are taken to conserve energy, the numbers are lower; 66 per cent limited shower time, 29 per cent use clothing racks to dry clothing, 48 and 37 per cent alter the thermostat in winter and summer, respectively, when leaving the house.

It can also be seen that greater efforts were made to conserve water than energy. This may be attributed to the severe water problems that San Antonio faces. It may also be because the water company is proactive in promoting water conservation. These efforts are visible to residents, and so they are more conscious in conserving water (Figs. 12.1 and 12.2).

When asked for comments on what would help residents do more to conserve water and energy, 85 residents responded. The main categories from the responses showed residents were:

- Aware of conservation and gave general comments on the benefits of conservation but did not comment as to whether they are conserving (31 per cent);
- Aware and already conserving (19 per cent);
- Need greater awareness and more information on how to conserve (19 per cent);
- Felt that higher bills would prompt 11 per cent of residents to conserve;
- Believed keeping rent low would prompt 12 per cent of residents to conserve;

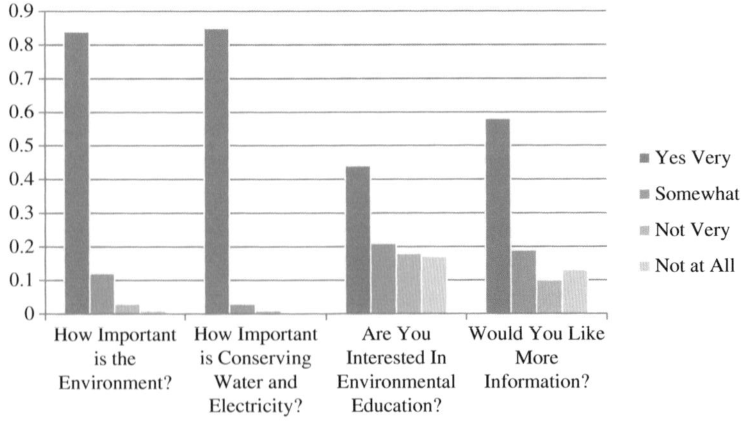


Fig. 12.1. Conservation questions.

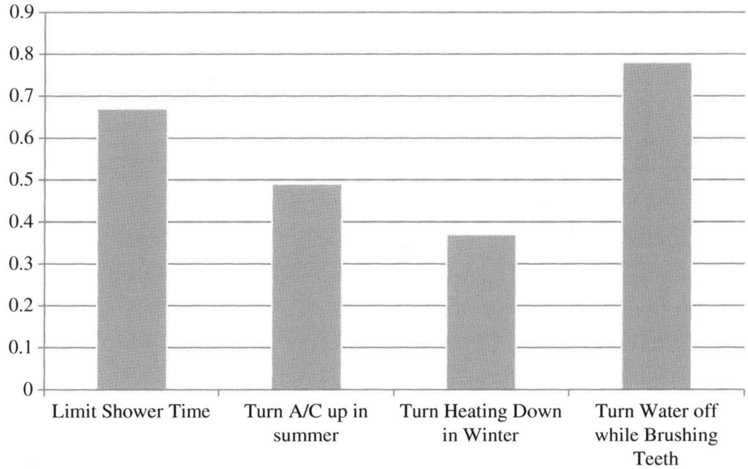


Fig. 12.2. What types of things do you do to conserve electricity?

## 5.3 *Recommendations*

The multifamily industry is complicated and has far-reaching impacts, one of which is the impact on national energy–consumption patterns. There are companies and smaller-scale building owners that would like to work towards more economic and energy–efficient buildings. But for IROs, the hurdles are often difficult to overcome. There is no "silver bullet" or "one size fits all" solution; rather, efforts must be made to find solutions to contribute towards energy conservation, while maintaining existing buildings and providing affordable housing for those in need. The social, economic and environmental costs of not working on a solution today will cost society far more in future.

### 5.3.1 *Lead by example*

The first recommendation will be to follow examples of leading cities and states in energy efficiency. Their experience shows programmes that succeed and others that have failed. Due to varying regulations, programme models can be adapted, drafted and implemented to satisfy local requirements. Multifamily needs to be included in energy efficiency programmes.

### 5.3.2 *Partner local multifamily associations with utilities and vendors*

Local multifamily associations are well networked. Both building owners and vendors/contractors can join the local, state and national associations, which give them access to member information. The utility companies and contractors must have greater presence in the local associations where they can work directly with decision–makers. Examples of current partnerships are shown in Box 12.1.

**Box 12.1 Leading by example: Austin Energy**

Austin Energy and the Austin Apartment Association have worked together for over 15 years. Austin Energy has been able to meet directly with owners, including smaller IROs, and adapt their programmes to increase participation in their energy–efficiency programmes. Austin is also the only city in Texas that is currently participating in the energy–benchmarking programme (Johnson, 2013).

The Massachusetts Low-Income Energy Affordability Network (LEAN) has developed a network of representatives from utilities, housing finance, community development, resident and owner communities to create the Multifamily Advisory Committee, which extended the outreach of the programmes to both public and private owners (Johnson, 2013).

### 5.3.3 *Provide dedicated staff for green initiative programmes*

All interviewees mentioned that there is often a "disconnect" between cities' green missions and their implementation. Local government, utilities and the apartment associations need to assign dedicated staff within their organisations to work together on green initiatives and become a point of contact for building owners.

### 5.3.4 *Ensure targets are measureable*

There is a critical shortage of information on building energy performance in the multifamily industry, creating difficulty on many levels with policy-makers, utility companies, lenders, building owners and residents. Such information is needed by policy-makers to formulate policy or incentives and appropriate financial vehicles for greater participation.

The multifamily housing industry is beginning to see more rules and regulations concerning energy performance benchmarking. Although this may be a difficult and sometimes unwelcome topic for the industry, it might hold part of the solution in creating a wider range of programmes that could apply to more multifamily housing and allow for greater transparency in the industry.

Cities and states are enacting or proposing laws that will require owners to track and report their energy usage (Miller, 2013b). It is suggested that new laws would focus initially on larger multifamily housing communities, which may have more resources to handle the new requirements. The Environmental Protection Agency and Fannie Mae (the nation's largest multifamily mortgage investor) have been working on a multifamily 1–100 energy scorecard similar to that currently used in commercial buildings. Also, cities such as Austin and New York are already reporting multifamily energy performance, which can affect occupancy if residents are unhappy with the ratings.

A great deal of communication and support will be needed to help building owners through this transition if this trend continues to grow (Miller, 2013). There are still many challenges that will need to be addressed in the benchmarking methods. The current method assesses buildings without regard for the type of multifamily buildings, such as mixed-use properties versus single use. Building owners, staff and residents will need to be aware of these issues or it could adversely affect occupancy rates.

### 5.3.5 *Provide a "one stop shop"*

Managers face difficulties in finding extra time to deal with matters such as a retrofit (MH, November 2013). Having just one point of contact would be necessary for more IROs to consider energy–efficiency programmes.

Many papers on energy efficiency in the multifamily housing industry have recommended establishing a "one stop shop" to provide owners with one point of contact which would help them with the application process, enable them to navigate through utility programmes, and help them to deal with issues such as energy assessments, the installation process, post-installation monitoring, education and awareness literature for residents, and verification of savings (Living Cities and Institute for Sustainable Communities, 2011; Barnes *et al.*, 2013; Johnson, 2013; McKibbin, 2013).

Such one-stop shops in cities such as Chicago (IL), Berkeley (CA), Oakland (CA) and Emeryville (CA) work with building owners to reduce transaction costs, level of stress and staff resources that are scarce in smaller companies (Johnson, 2013). Another group, Enterprise Green Communities, has been successful in creating a "one–stop website" that offers information for the affordable housing industry and can be a good starting point for IROs.

Furthermore, they understand the importance of community involvement, which is often overlooked in a successful retrofit project. Their checklists and retrofit toolkits can be useful for smaller companies providing there is a way to implement and follow through (Enterprise, 2014). Gathering the necessary data and implementing the project from start to finish will be a challenge due to financial and time restrictions. The utility companies or the city will need to assign people to help in this matter. An example of one such list:

- *Review portfolio*: Determine which property would benefit the most from an energy retrofit.
- *Conduct energy assessment*: Provide the owner with all the documents that would allow the auditor to obtain all utility information directly from the utility companies. This step is critical and would reduce time and stress, while increasing the possibility that the owner would allow the audit to proceed.
- *Perform ongoing monitoring of utility use after retrofit*: Set up an energy assessment that monitors energy usage in a building, and provides the building owner with a benchmark/starting point to see current energy usage; and the data necessary to show to lenders when applying for financial assistance.
- *Application process*: Provide the building owner with all federal, state and local qualifying programmes.
- *Consider all programmes*: Discuss both direct "no cost" installations with rebate programmes and comprehensive installations such as building envelope, HVAC units, roofing and windows. Owners need to know all the retrofit options along with all the financial options before making a final decision. Examples of items on the checklist that could be used by an IRO are shown in Table 12.2.
- *Navigate utility programmes*: Assist building owners when dealing with utilities.
- *Secure funding*: Provide building owners with information on all federal, state and housing financial assistance opportunities. Also, determine which utility criteria the building belongs to. Find the best financing programme that works within the building owner's financial parameters.

Table 12.2. Guidelines for IROs for energy-efficiency improvements.

| | |
|---|---|
| **Envelope** | |
| | Air leakage, |
| | Visual inspection |
| | Smoke pencil test |
| | Blower door test-in and test out |
| | Duct leakage test |
| | Infrared camera imaging |
| | Weather-stripping |
| | Windows replacement |
| | Roofing & R-3 insulation |
| **Hot water heaters** | |
| | Record hot water temperature at sink, bathtub & shower in units |
| | Measure delivery temperature of water leaving storage tank to first apartment |
| | Measure return temperature of water returning from the last apartment to the boiler |
| | Measure temperature inside the storage tank |
| | Measure water temperature inside the boiler |
| | Insulate hot water piping |
| | Insulate hot water tanks |
| | Low-flow fixtures |
| | Replace hot water pumps |
| | Install more efficient hot water heater |
| | Install solar water heater |
| **HVAC** | |
| | Programmable thermostats |
| | Install outdoor air temperature controls for boiler to allow better control |
| | Install Variable Frequency Drive (VFD) on heating and cooling |
| | Replace exhaust fan motors |
| | Use Energy Star air-conditioning systems |
| **Lighting** | |
| | Energy-efficient outdoor lighting (hallways, common areas, parking lots) |
| **Energy-efficient Appliances** | |

Notes: *Key*: Light shading = easier to implement; mid-tone shading = harder to implement (i.e. more comprehensive and more expensive); darkest shade = technology type. (Source: Enterprise, 2014).

- *Finalise depth of work, select contractor, ensure quality.*
- *Continue to monitor utility.*
- *Develop a preventative and regular maintenance plan.*
- *Educate staff and residents on improvements*: Programmes must be developed to educate, build awareness of, and work with residents and staff to turn the property into a conserving community. Such programmes will have a large impact on the success of the retrofits.

For IROs that do not have the resources for a complete retrofit, there are simpler and cheaper tasks that they can do. Table 12.2 is an example of one or many items that can be applied in-house and over a longer period of time if time and labour are issues. Although not as thorough as a full assessment and retrofit, efficiency and cost can be improved.

### 5.3.6 *Coordinate utility programmes for easier participation*

Multifamily properties generally deal with several utility companies for water, gas and electricity. Owners who are interested in comprehensive retrofits will have difficulty working with three different organisations, each with different points of contact and application procedures. Reducing time spent trying to coordinate amongst utilities would be welcomed by IROs. Examples of current integrated services by utility companies in Chicago are shown in Box 12.2.

**Box 12.2 Leading by example: Chicago electric and gas coordination**

Chicago area has four electric and gas utility companies — ComEd, North Shore Gas, Nicor Gas and People Gas. All four companies coordinate to provide integrated services to gain efficiency savings. Building owners with multifamily housing in more than one service territory can deal with one programme for all regions. All planning, implementation and reporting is done within the four utility companies and presented as one to the building owners. They provide free energy assessments, free direct installations such as lighting and weather-stripping, and discounted installation services using utility-company verified contractors. Contractors working with the utility companies do the paperwork and building owners receive rebates and incentives faster (Johnson, 2013).

### 5.3.7 *Encourage on-bill repayment for multifamily housing*

On–bill repayment allows for monthly repayment through the utility bill for energy–efficiency projects, as seen in Box 12.3. This however, is rarely offered to multifamily.

**Box 12.3 Leading by example: Public Service Electric and Gas**

Public Service Electric and Gas (PSE&G) Residential Multifamily Program uses on-bill financing. PSE&G pays all project costs up front, and provides a permanent cash incentive. The capital cost is converted to an expense and it is paid back through the utility bills with no interest over five years (Johnson, 2013).

### 5.3.8 *Provide information on utility programmes and public funding programmes when refinancing affordable housing*

Many affordable-housing projects were funded 30 to 40 years ago. Units due for refinancing and Tax Credit projects are able to exit the affordable-housing programme when their affordability requirements expire. This is an attractive option for building owners who currently own properties in gentrifying neighbourhoods. They can sell or change the classification of their properties to a higher grade but it reduces the affordable housing units available. Therefore, attractive financing and detailed information on retrofits and energy efficiency need to be provided to building owners and make it economically worthwhile to remain in affordable housing.

### 5.3.9 *Encourage Buy-In and Commitment of key stakeholders*

In order to move from energy–conservation objectives to implementation, all stakeholders must be involved, including residents. The residents are the end users of energy consumption and play a large part in defining the success of a programme. Cole (2011) suggests that technological solutions, economic proof and governance solutions are not sufficient to bring change. There must be greater social and organisational interaction among the stakeholders and an understanding of each other's motivations and drivers. Resident participation is a challenge for building owners to move from objective to implementation, but it is essential.

Creating programmes that enable residents to receive feedback on energy consumption coupled with innovative ideas to motivate them to take action to save can produce a successful collaborative effort between management and residents. Yet, awareness and education, which influence knowledge, attitude and behaviour, do not always lead to change. People need motivation to permanently change their energy–consumption behaviour (Cole, 2011; Palm, 2013). Information and advice might result in a temporary drop in consumption, but it is often not sustained (Bahaj and James, 2007). The ultimate goal needs to focus on permanently changing resident behaviour.

## 6. Conclusion and Further Investigation

Since this study was started in 2011, there has been progress in energy efficiency at national, state and local levels in the United States. For example, in 2013 some states including Mississippi and Louisiana started energy–efficiency programmes, while others are expanding their energy initiative programmes (Nadel, 2014). Also in 2013, nine US cities including Austin, Texas, passed energy–reporting benchmarking laws for multifamily properties. Awareness among local governments is on the rise and they are adopting comprehensive approaches to energy efficiency with new policies and initiatives (Nadel, 2014).

Although initiatives have become more prominent today, and there is an increasing momentum in energy–efficiency programmes, there is concern as to whether IROs and smaller building owners will be able to take advantage of these initiatives or whether future energy laws will impose additional costs in the affordable-housing sector.

Even though the process may be slow and complex, stakeholders in government, utility companies and housing must act upon the significant environmental, economical and social impact that energy efficiency can make in affordable housing and society. The process is good in theory, the problem remains on how to include "hard to reach" IROs and widen generally energy–efficiency participation and implementation.

First, due to the fragmentation of the industry, it is best to "start local" with high degrees of collaboration between representatives from the state, city, utility companies, apartment associations, building owners, contractors and residents' groups. A knowledgeable permanent working group should be formed to make the participation process simpler and over the long term. Each participant must understand how the other groups function in order to be effective as representatives.

Second, the efforts must be integrated and at a strategic level (Bamberger, 2010; Living Cities and Institute for Sustainable Communities, 2011; McKibbin *et al.*, 2012; Barnes *et al.*, 2013). All city building stock must be included. Creativity or a door-to-door approach will be necessary to reach IROs, as not all of them are registered with local apartment associations. Therefore, enlisting the help of local residents, local police, neighbourhood non-profit organizations, churches and other organised groups are another effective method to help include building owners that would usually be inaccessible or unaware of energy efficient programmes.

Third, there must be sufficient public and private funding sources, including creative funding ideas, to bring in more participants. Finally, there must also be a "green" workforce for these large-scale projects. Otherwise the simplicity of participating in a programme will be compromised.

### 6.1 *Avenues for Further Investigation*

Further investigation is required to understand how to motivate building owners and residents to participate in energy efficiency. Human behaviour, attitudes and motivation will be critical for a successful program. While inclusion in the process is one of the more successful approaches for motivating residents, motivating building owners in the long term may actually be the more difficult challenge. For many IROs, and for larger corporate owners, the decision is frequently based on the short–term bottom line, which does not lend itself to changes in human behaviours and attitudes.

Furthermore, how IROs implement these changes is vital. Many residents feel that they are powerless to effect changes in their lives and if IROs force additional rules on a property, it will be seen as a push and will be viewed negatively by residents.

It is important to empower residents and allow them be part of the conservation solution. Education and awareness will not be sufficient. Residents will need to make lifestyle changes, have the confidence that they are able to make a contribution to the environment and that their efforts count. The biggest obstacle with residents will be dealing with their own perceptions of themselves and their neighbourhoods.

Hastings (2009) discusses the "neighbourhood effect", and highlights the fact that these communities typically have lower expectations of themselves, as do service providers — such as those providing waste management, policing, and utility services. In time, this attitude can be changed. Low-income residents may often have minimal financial means, but the one thing that is important to them is respect. Therefore, the first step to participation is to help residents understand that their actions can make a difference and that their property owner and the community care.

The intention would be to "nudge" rather than push a new belief or philosophy onto residents. Incentives are also useful motivators but must progressively taper off, as the ultimate goal is to encourage a change in lifestyle towards conservation. These efforts will take time to take effect as it requires change in behaviour and perception, but they are possible to achieve (Cooper Marcus and Sarkissian, 1986; Sarkissian *et al.*, 2009).

In order for energy–conservation efforts to succeed, the impact of "neighbourhood effects" and community participation must be investigated further. In addition, long–term incentives for property owners must be developed in order to encourage them to make the long–term investment of capital and human resources in the efforts to drive greater participation in energy efficiency in older affordable housing units.

## References

Bahaj, A.S. and P.A.B. James. (2007). "Urban Energy Generation: The added value of photovoltaics in social housing", *Renewable and Sustainable Energy Reviews*, **11**(9), 2121–2136.

Bamberger, L. (2010). *Scaling the Nationwide Energy Retrofit of Affordable Multifamily Housing: Innovations and Policy Recommendations*, What Works Collaborative. [Online] Available at: http://www.brookings.edu/~/media/research/files/papers/2010/12/energy%20bamberger/12_energy_bamberger.pdf. [Accessed].

Barnes, C., S. Hodgin, J. Brew, and A. Karolides. (2013). *Superefficient Affordable Housing: Solutions for hurdles*, Rocky Mountain Institute, Snowmass, CO. [Online] Available at: http://www.rmi.org/PDF_Superefficient_Affordable_Housing-Solutions_to_Hurdles_RMI [Accessed 20 January 2014].

Benningfield Group Inc. [2010]. *Addendum Report: U.S. Multifamily Housing Stock Energy Efficiency Potential. HUD-Assisted, Low Income Housing Tax Credit, and Large Real Estate Investment Trust Properties.* [Online] Available at: http://www.benningfieldgroup.com/uploader/download.aspx?uuid=6e2b4aca-e907-4f8d-8fed-4bb06299ad9e. [Accessed].

Cadik, E. (2013). "NAHMA Presents — Inspiring new directions in providing niche affordable housing", *National Apartment Association Education Conference and Exposition*, San Diego CA, 19–22 June. [Online] Available at: http://naa.directionsav.com/presentation.asp?ID=118. [Accessed].

Cisneros, H. (2009). "A new moment for people and cities" in Cisneros, H. and Engdahl, L. (eds) *From Despair to Hope: HOPE VI and the New Promise of Public Housing in America's Cities.*, Brookings Institute Press, Washington, DC., pp. 3–13.

Cole, R.J. (2011). "Motivating stakeholders to deliver environmental change", *Building Research & Information*, **39**(5), 431–435.

Cooper Marcus, C. and W. Sarkissian. (1986). *Housing as if People Mattered: Site Design Guidelines for the Planning of Medium-density Family Housing*, University of California Press, Berkeley and Los Angeles, CA.

Economic Policy Program, Housing Commission. (2013). "Affordable rental housing", *Housing America's Future: New Directions for National Policy*, Bipartisan Policy Center, Washington, DC, pp. 81–108.
Enterprise. (). *Enterprise Green Multifamily Retrofit Program — Enterprise Community Partners Multifamily Retrofit Toolkit.* [Online] Available at: http://www.enterprisecommunity.com/solutions-and-innovation/enterprise-green-communities/resources/multifamily-retrofit#retro. [Accessed 14 January 2014].
Follain, J.R. (1994). "Some possible directions for research on multifamily housing", *Housing Policy Debate*, **5**(4), 533–568. DOI: 10.1080/10511482.1994.9521177.
Fuller, S.S. (2013). "The Trillion Dollar Apartment Industry.", *National Multi Housing Council, Washington, DC and National Apartment Association, Arlington, VA.* [Online] Available at: http://weareapartments.org/sites/default/files/FullerReportFinal.pdf [Accessed 23 October 2013].
Hastings, A. (2009). "Neighbourhood environmental services and neighbourhood 'effects': exploring the role of urban services in intensifying neighbourhood problems", *Housing Studies*, **24**(4), 503–524.
HUD (U.S. Department of Housing and Urban Development). (1996a). "Public/private partnerships for the mixed-finance development of public housing units", *Federal Register*, **61**(86), p. 19.
HUD. (1996b). "Notice of funding availability for public housing demolition, site revitalization, and replacement housing grants (HOPE VI), fiscal year 1996.", *Federal Register*, **61**(141), p. 38. [Online] Available at: http://www.hud.gov/offices/pih/programs/ph/hope6/grants/fy96/fy96nofa.pdf. [Accessed].
HUD. (1997). "Notice of funding availability for the revitalization of severely distressed public housing (HOPE VI), fiscal year 1997", *Federal Register*, **61**(), p. 18.
HUD. (1998). "Funding availability for revitalization of severely distressed public housing (HOPE VI), fiscal year 1998", *Federal Register*, 31 March. [Also released as HUD FY1998 Super NOFA: HOPE VI Revitalization NOFA].
HUD. (2007). *HUD historical background.* [Online] Available at: http://www.hud.gov/offices/adm/about/admguide/history.cfm [Accessed 7 January 2014].
HUD. (2008). *Trends in Housing Costs 1985–2005: and the 30-Percent-of-Income Standard*, Office of Policy Development and Research, U.S. Department of Housing and Urban Development, Washington, DC.
HUD. (2011). *Evidence Matters: Transforming Knowledge Into Housing and Community Development Policy*, Office of Policy Development and Research, U.S. Department of Housing and Urban Development, Washington, DC.
HUD. (2013). *Energy: Secretary Donovan's statement on clean energy.* [Online] Available at: http://portal.hud.gov:80/hudportal/HUD?src=/topics/energy [Accessed November 2013].
HUD. (). HUD Historical Background http://www.hud.gov/offices/adm/about/admguide/history.cfm (Last accessed 19 January 2014).

HUD. (2014). *Housing Choice Vouchers list.* [Online] Available at: http://portal.hud.gov/hudportal/HUD?src=/program_offices/public_indian_housing/programs/hcv/about/list [Accessed 14 January 2014].

HUD. (). *HUD's Public Housing Program.* [Online] Available at: http://portal.hud.gov:80/hudportal/HUD?src=/topics/rental_assistance/phprog [Accessed 14 January 2014].

Johnson, K. (2013). *Apartment Hunters: Programs Searching for Energy Savings in Multifamily Buildings*, American Council for an Energy Efficient Economy, Washington, DC. [Online] Available at: http://www.aceee.org/sites/default/files/publications/researchreports/e13n.pdf. [Accessed].

Joint Center for Housing Studies of Harvard University. (2008). *The State of the Nation's Housing 2008*, Joint Center for Housing Studies of Harvard University, Cambridge, MA. [Online] Available at: http://jchs.harvard.edu/sites/jchs.harvard.edu/files/son2008.pdf [Accessed 10 January 2014].

Joint Center for Housing Studies of Harvard University. (2013). *The State of the Nation's Housing 2013*, Joint Center for Housing Studies of Harvard University Cambridge, MA. [Online] Available at: http://www.jchs.harvard.edu/sites/jchs.harvard.edu/files/son2013_bw.pdf [Accessed 18 January 2014].

Katz, B. (2009). "The origins of HOPE VI" in Cisneros, H. and Engdahl, L. (eds) *From Despair to Hope: HOPE VI and the New Promise of Public Housing in America's Cities*, Brookings Institute Press, Washington, DC, pp. 15–30.

Living Cities and Institute for Sustainable Communities. (2011). *Scaling Up Building Energy Retrofitting in U.S. Cities: A Resource Guide for Local Leaders. Version 1.0*, Living Cities and Institute for Sustainable Communities, Montpelier, VT. [Online] Available at: http://www.iscvt.org/who_we_are/publications/Green_Boot_Camp_Resource_Guide.pdf [Accessed 14 January 2014].

Mallach, A. (2010). *Meeting the Challenge of Distressed Property Investors in America's Neighborhoods*, Local Initiative Support Coalition (LISC), New York.

McKibbin, A. (2013). *Engaging as Partners in Energy Efficiency: A Primer for Utilities on the Energy Efficiency Needs of Multifamily Buildings and their Owners*, CNT Energy and American Council for an Energy-Efficient Economy. [Online] Available at: http://www.cntenergy.org/media/CNT_ACEEE-Report-2013-Final.pdf [Accessed 10 January 2014].

McKibbin, A., A. Evens, S. Nadel and E. Mackres. (2012). *Engaging as Partners in Energy Efficiency: Multifamily Housing and Utilities*, ACEEE Report No. A122. CNT Energy and American Council for an Energy-Efficient Economy. [Online] Available at: http://www.cntenergy.org/media/Engaging-as-Partners-in-Energy-Efficiency-MF-Housing-and-Utilities-Final-012512.pdf [Accessed 10 January 2014].

Miller, L. (2013a). "What is the sequester?", *The Huffington Post*, 3 January. [Online] Available at: http://www.huffingtonpost.com/2013/03/01/what-is-the-sequester_n_2783917.html#slide=more282000. [Accessed 2013].

Miller, M. (2013b). "NAHMA presents — Innovative technology and green solutions in affordable housing", *Presentation at National Apartment Association Education Conference and Exposition*, San Diego, CA, 19–22 June. [Online] Available at: http://naa.directionsav.com/presentation.asp?ID=109. [Accessed].

Moelis Institute for Affordable Housing. (2012). "What Can We Learn About the Low-Income Housing Tax Credit Program by Looking at the Tenants?", *Moelis Institute for Affordable Housing Policy Brief*, Furman Center for Real Estate and Urban Policy and Moelis Institute for Affordable Housing Policy, New York. [Online] Available at: http://furmancenter.org/files/publications/LIHTC_Final_Policy_Brief_v2.pdf. [Accessed 2013].

Mueller, E.J. (2010). "Old apartments and new plans: reconciling planning and housing goals in two Texas cities.", *Community Development*, **41**(1), 121–140.

Nadel, S. (2014). *Significant Progress on Energy Efficiency in 2013 — And 2014 Could Be Even Better*, American Council for an Energy Efficiency Economy. [Online] Available at: http://www.aceee.org/blog/2014/01/significant-progress-energy-efficienc. [Accessed].

Norris, M. (2013). "Preserving affordable housing", *Affordable Housing Finance*, 21(3), p. 52.

Pagitsas, C. (2012). *Using sustainability to improve rental housing and preserve affordability. FM Commentary*, FannieMae. [Online] Available at: http://www.fanniemae.com/portal/about-us/media/commentary/011712-pagitsas.html. [Accessed December 2013].

Palm, J. (2013). "Energy efficiency in tenant-owners residences: The process of going from objective to implementation", *Housing Studies*, **28**(1), 57–73.

PD&R. (2011). *Worst Case Housing Needs 2011: Summary. Report to Congress*, Office of Policy Development and Research, US Department of Housing and Urban Development, Washington, DC.

Perez-Lombard, L., Ortiz, J. and Pout, C. (2008). "A review on buildings energy consumption information.", *Energy and Buildings*, **40**, 394–398.

Rice, D. (2011). *Section 8 Rental Assistance Programs are Not Growing as Share of HUD Budget: Proposed reforms would make Section 8 even more efficient*, Center on Budget and Policy Priorities, Washington, DC. [Online] Available at: http://www.cbpp.org/cms/?fa=view&id=3544 [Accessed 2013].

Rocky Mountain Institute. (2013). *Superefficient Affordable Housing: Solutions to Hurdles*, Rocky Mountain Institute, Snowmass, CO. [Online] Available at: http://www.rmi.org/PDF_Superefficient_Affordable_Housing-Solutions_to_Hurdles_RMI. [Accessed November 2013].

Sarkissian, W. Bunjamin-Mau, W., Cook, A., Walsh, K. and Vajda, S. (2009). *SpeakOut: A step-by-step guide to SpeakOuts and community workshops*, Earthscan. USA and UK.

Serlin, C. (2013). "Data reveals housing needs", *Affordable Housing Finance*, **21**(3),. [Online] Available at: http://www.housingfinance.com/housing-data/data-reveals-housing-needs.aspx?dfpzone=general. [Accessed January 2014].

Shaver, L. (2011). *Steady at the top: Affordable housing owners and REITs give the Top 50 Owners list a consistent, though slightly different, lineup*, Multifamily Executive. [Online] Available at: http://www.multifamilyexecutive.com/multifamily-trends/steady-at-the-top.aspx [Accessed 14 January 2014].

UNEP. (2012). *Sustainable Resource Efficient Cities — Making it Happen!*, United Nations Environment Programme Division of Technology, Industry and Economics, Paris.

UNEP SBCI (United Nations Environment Programme Sustainable Buildings & Climate Initiative). (2009). *Buildings and Climate Change: Summary for Decision-Makers*, UNEP DTIE Sustainable Consumption & Production Branch, Paris. [Online] Available at: http://www.unep.org/sbci/pdfs/SBCI-BC (2CSummary.pdf [Accessed 17 January 2014].

U.S. Bureau of the Census. (2007). *American Housing Survey 2007*, National data file. [Online] Available at: http://www.huduser.org/portal/datasets/ahs/ahsdata07.html [Accessed 15 January 2014].

U.S. Census Bureau. (2011). "Population and Housing Narrative Profile: 2009–2011", *American Community Survey 3-Year Estimates.*, U.S. Department of Commerce Economics and Statistics Administration. [Online] Available at: http://factfinder2.census.gov/faces/tableservices/jsf/pages/productview.xhtml?pid=ACS_11_3YR_NP01&prodType=narrative_profile [Accessed 16 January 2014].

U.S. Department of Commerce. (2013). "Residential Vacancies and Homeownership in the Third Quarter 2013.", *U.S. Census Bureau News*, 5 November. [Online] Available at: http://www.census.gov/housing/hvs/files/qtr313/q313press.pdf [Accessed 20 January 2014].

U.S. Department of Energy. (2012). *2011 Buildings Energy Data Book*, D&R International, . [Online] Available at: http://buildingsdatabook.eren.doe.gov/docs/DataBooks/2011_BEDB.pdf. [Accessed 16 January 2014].

U.S. Department of the Treasury and U.S. Department of Housing and Urban Development. (2011). *Reforming America's Housing Finance Market: A Report to Congress.* [Online] Available at: http://www.treasury.gov/initiatives/Documents/Reforming%20America's%Housing%Finance%20Market.pdf. [Accessed November 2013].

U.S. Environmental Protection Agency. (2013). *Inventory of U.S. Greenhouse Gas Emissions and Sinks: 1990–2011*, EPA 430-R-13-001. U.S. Environmental Protection Agency, Washington, DC. [Online] Available at: http://www.epa.gov/climatechange/Downloads/ghgemissions/US-GHG-Inventory-2013-Main-Text.pdf. [Accessed]

Young, R.A. (2012). *Stewardship of the Built Environment: Sustainability, Preservation, and Reuse*, Island Press, Washington, DC.

# 13

# THE POTENTIAL AND BENEFITS OF AN INCENTIVE SYSTEM TO PROMOTE ENVIRONMENTALLY FRIENDLY BEHAVIOUR IN RESIDENTS OF HDB FLATS

KOH KAI-JIE and GRACE WONG KHEI-MIE

Existing reward programmes (such as Grow your Recycling Incentives Now (GRIN)) that utilize non-monetary incentives to encourage recycling in Singapore have had limited success. To enhance its effectiveness, it could adopt a more structured and multi-disciplinary approach to achieve significant results through a change in behaviour. A general household survey was therefore conducted to find out the potential of another incentivising approach beyond the economic model perspective. The survey results identified that residents were motivated by monetary incentives at the individual level. Other significant factors identified included collective effort within the community and enhanced convenience in terms of information sharing and infrastructure. The benefits of a multi-disciplinary incentive system were discussed and it was ascertained that there was potential to develop this model further. It would involve five areas - legislation, incentives, information, involvement, and programmes that could be adopted for implementation by all Town Councils and other stakeholders. This could help achieve the environmental goals as set out in the Sustainable Blueprint 2030. Without a multi-disciplinary incentivising approach, these goals could be compromised if households continue to resist a change in their behaviour, as the tried and tested method of relying on information programmes have been slow in effecting a significant change so far.

## 1. Introduction

A sustainable planet is not possible without behaviours that conserve the planet's resources. The resource-costly life-styles that are characteristic of the current scene present a historic challenge. Never before have so many behaviours needed to change in so short a time. It remains a challenge to sustain any positive change in behaviours (De Young, 1993). When it comes to advice on the best way to promote and maintain environmentally friendly behaviours — such as responsible consumerism, recycling more and energy conservation — psychologists and behavioural economists offer differing but often complementary approaches. Some believe that appealing to people's self-interest is the best method. One study found that motivators of environmentally responsible behaviour are more effective when they have a direct impact on people's needs or concerns (De Young, 2003). Others say the field's emphasis on individual behaviours is misguided and instead advocate organisational, political and other large-scale changes.

### *Current Level of Environmentally Friendly Behaviour in Singapore*

In Singapore, a survey of 820 households conducted in January 2013 found that 69% of households wanted rewards for saving energy, 56% wanted time-of-use pricing and 51% asked for rebates for low-income and the elderly (*The Straits Times*, 20 Feb 2013). Therefore, it seems that incentives that appeal to their self-interest may encourage more people to change their behaviour to an environmentally friendly one. As a low-lying, densely populated island of about 5 million people in the tropics, Singapore is vulnerable to the impact of climate change (i.e., it is a key environmental issue). The people must be prepared for the possible effects with mitigation and adaptation efforts, such as living an environmentally friendly lifestyle. However, a public survey of 402 people aged between 21 to 34 years old in April 2011 found that less than 1% of them indicated the environmental issue as an issue that concerns them the most (*The Straits Times*, 16 April 2011). Since 1990, the Singapore Government has organized the 'Clean & Green Singapore (CGS)' campaign, with the aim of engaging the community and caring for the environment by inspiring the people to adopt an environmentally friendly lifestyle. This is essentially a public education exercise to motivate a change in behaviour by highlighting the benefits of collective gain. However, people are understood to be somewhat self-centered by nature and hence are more likely to be motivated by benefits that accrue directly to themselves than those that benefit everyone. Therefore, a carefully calibrated set of positive and negative incentives to align behaviour with consequences would need to be further developed in order to realise the vision of a "Sustainable Singapore". Sustainability encompasses not only the common notion of environmental sustainability, but needs to embrace social and cultural sustainability as well. A sustainable Singapore should be one that is socially inclusive, with strong bonds in the community, healthy social interaction, protection of the vulnerable, and respect for social diversity. While social interaction evolves organically, it can also benefit from institutions and programmes that support the development of healthy communal bonds and the building of trust relations.

## 2. Environmental Behaviour and its Influences

Governments and organisations have generally relied on public campaigns and programmes to promote environmentally friendly behaviour among the people. However, this approach has come under scrutiny for its assumption that inaction is determined by people's ignorance of the facts (Macnagthen and Urry, 1998). It was also argued that the scientisation and politicisation of environmental issues meant that individuals have become disillusioned by the nature of scientific information regarding environmental problems, such that individuals may rely on their own senses (or anecdotal experiences) to construct opinions on environmental change, rather than science. Consequently, individuals may seek other forms of knowledge based on local contextual understandings (Eden, 1998). Such knowledge from non-scientific and non-governmental sources can be interpreted by individuals in terms of their previous experiences — depending on how people link them to their everyday lives, and their relation to organisations and others (Myers and MacNagthen, 1998). It would then seem unlikely that information campaigns aimed at highlighting global, national concerns will have a significant impact on behaviour, given the embedded nature of local environment discourse, the alternative rhetorical situations that individuals encounter, along with a rejection of objective scientific knowledge from government sources. Also, by focusing on the rational framework of information dissemination, a one-way flow of information from science to policy and society has overlooked a greater range of influences (Steward, 2003). These suggest factors such as social and environmental values, situational factors and psychological variables can influence environmental action in people.

### *Social and Environmental Values*

Empirical research has demonstrated that there are essentially three continua of values that are of relevance to environmental behaviour. First, Stern *et al.* (1995) and Corraliza and Berenguer (2000) have demonstrated the utility of the two value dimensions identified by Schwartz and Blisky (1987) and Schwartz (1992), that could be placed on a value continua ranging from 'egoistic' to 'altruistic' and from 'conservative' to 'open to change'. The evidence provided by Stern *et al.* (1995) and Corraliza and

Berenguer (2000) showed that across a range of environmental actions, those individuals who were altruistic and who were open to change were more likely to take part in environmentally friendly behaviour. Second, there is a relational value continuum that places biocentrism at one end and anthropocentrism at the other. This incorporates the works of Schwartz (1992) and Dunlap and Van Liere (1978). Briefly, biocentrism conceptualises the human relationship with nature as egalitarian, and promotes a belief that nature has intrinsic value, whereas anthropocentrism can be characterized as the dominance of humans over nature, with nature being explicitly for human use and thus containing no intrinsic value. Third, there are the belief-driven values. O'Riordan (1985) conceptualised humans as ecocentrists who believe the solution to environmental problems lies in a working relationship with nature to resolve conflicts between society and nature, and technocentrists who believe that environmental problems are a result of a lack of technological development and argue that modernisation is the solution.

### *Situational Factors*

These refer to an individual's access to given services, their socio-demographic make-up and their knowledge and experience of the relevant behaviours. In general, research has predictably shown that those who have greater access to services, such as recycling schemes or opportunities to purchase environmentally friendly products, are more likely to be pro-environmental (Guagnano *et al.* 1995; Derksen and Gartell 1993). From a social-psychological perspective, Schahn and Holzer (1990) have distinguished between two types of knowledge when examining environmental action: abstract and concrete. The former relates to knowledge concerning environmental issues: problems, causes, solutions and so on. The latter relates to behavioural knowledge that can be utilised and acted upon. The source of knowledge, quite apart from its content, may also be significant as it relates to its frame of reference and how such information is conveyed (Stern, 1992).

### *Psychological Variables*

Research has provided a valuable insight into the personality and perceptual bases of environmental behaviour. Hopper and Nielsen (1991) and Van Liere and Dunlap (1978) have argued that environmental action can be conceptualized partly as "altruistic" behaviour (Schwartz 1977). In their study of recycling behaviour, Hopper and Nielsen demonstrated that those individuals who perceived what they termed an "environmental need" (or problem) and who felt morally obliged to act without any incentive or penalty were far more likely to recycle waste regularly than those who assigned the problem to other individuals. Acting for the benefit of others with no definitive personal benefit appears to be significant in influencing environmental action. On the other hand, De Young (1985–1986) has argued that intrinsic motivation plays a crucial role in motivating reduced consumption. Those individuals who gain satisfaction from environmental action, such as being frugal or recycling bottles, were more likely to maintain their behaviour after an initial start. Later work by De Young (1996) has provided a fascinating insight into the structure of intrinsic motivation, which focuses on the satisfaction and individual gain from environmental action, such as a sense of well-being and self-worth. These satisfactions appear to heavily influence the continued behavioural commitment of the individual. The influence of social pressure and others' behaviour can effect a change in environmental behaviour as well. Such a normative process is contingent upon an awareness of a social pressure to adopt positive environmental behaviour and may not be as effective in regard to more habitual action. It is also important that an individual perceives that the desired behavioural change will yield a positive outcome, or avoid a negative outcome. Otherwise, they will not see the benefits to utilising the extra time and effort to act in an environmentally friendly manner. Vining and Ebreo (1990) found that recycling behaviour was significantly affected by perceptions of the time allocated to recycling and the convenience of recycling facilities. Finally, the concept of environmental citizenship can exert a powerful influence on environmental behaviour (Selman, 1996).

## 3. Behavioural Economics: Incentives, Social Norms and Choices

In recent years, behavioural economists have begun to apply Plato's idea that human beings have two distinct decision-making systems. The Automatic System is intuitive, unconscious, uncontrolled, effortless, habitual and based on practice skills. This is the system of thinking we use in our day-to-day activities and when we have only a short time to make decisions. The other system is the Reflective System; it is controlled, slow, deliberate, conscious and effortful. Our reasoning, analytical skills and scientific methods are based on this system. All this while, conventional economic theory had assumed the individual to be making decisions based on the Reflective System (Low, 2012). However, reality has shown that the decision making process in an individual is far more complicated than a cost-benefit calculation (Low, 2012). Behavioural economists have begun to challenge the long-held assumption that individuals always respond in a rational, interest-maximising way to incentives (Tan and Low, 2012). Increasingly, it is important to consider the role of social norms when trying to understand the choices made by an individual. Sunstein (1996) offers a useful 3-part framework in this aspect: (1) intrinsic value, (2) reputational value, and (3) effect on self-image. These might not operate independently of one another. This framework helps us understand how social norms can be used to shape behaviour by operating as "psychological" taxes (in the form of social approval). In the area of environmental sustainability, countries like Denmark and Sweden have demonstrated that incentives (in the form of taxes and subsidies) are far more effective if supported by the right social norms and attitudes towards energy conservation. The people of these countries forged a national consensus on energy conservation, recycling and reducing waste during the post-war years when resources were scarce. It is recognised that incentives introduced will inevitably interact with existing social norms, and these social norms can work with or against incentives in achieving the goals (Tan and Low, 2012). Other than the influence of social norms, there are several other concepts from behavioural economics that can help to explain the psychology of environmentally friendly behaviours. They are image motivation, loss aversion, saliency or availability bias, mental accounting and discounting (Ong, 2012). The introduction of behavioural economics offers an insight into the conceptual framework of an incentive system that would enhance its effectiveness and become widely accepted. By taking into account people's image motivations and loss aversion, the structure of the incentive system will be better formulated. We can also aim to make intelligent use of the availability bias and to recognize how mental accounting might affect their receptivity to it, as well as to adjust for the tendency to discount the future excessively.

## 4. Case Study of GRIN (Grow your Recycling Incentives Now) Programme

This programme is the first on-line recycling rewards programme in Singapore to encourage recycling participation amongst residents from the

Pasir Ris–Tampines Sector, living in the domestic premises such as HDB estates and landed properties. It was started in July 2011 and managed by Veolia Environmental Services, a waste collection company. It encourages residents to register online and earn GRIN points, using them to redeem for privileges and gifts. With a minimum of 3 kg, the residents will be awarded one GRIN point for every kg of recyclables collected in their GRIN recycling bin. These points can also be donated to the GROW GREEN campaign to help raise environmental awareness. The participating companies have also declared their commitment to environmental protection, as well as help to raise environmental awareness among their staff. For domestic premises such as the HDB estates, the grassroots representatives such as the chairman of the residents committee or a representative appointed by chairman, from the residents committee, may sign up for the programme directly. They help to represent the residents in the same designated zone, as the programme is currently not applicable on a household basis. The residents are then encouraged to participate in the programme by depositing recyclables in the recycling bins within their HDB block so that more GRIN points can be accumulated for their respective zone. The representative will then redeem the points earned within their zone of HDB estates for gifts and vouchers from the company, and will be responsible for distributing the redeemed gifts and vouchers to the rest of the residents. The HDB residents can also find out how many recyclables were collected from their zone for each month through the website. The estimated 90,351 households involved in this program are grouped into 8 different constituencies. The breakdown of the flat types and household sizes by town are shown in Tables 13.1 and 13.2. In each constituency, there are individual zones that cover between 9 to 29 blocks of HDB flats.

A graphical analysis of the GRIN program was carried out based on the amount of recyclables collected from each zone from Aug 2011 to Aug 2012. The weightings were compared across the 12 months that data collection

Table 13.1. Breakdown of flat-types by town.

| HDB town | 1-room | 2-room | 3-room | 4-room | 5-room | Executive | Total households |
|---|---|---|---|---|---|---|---|
| Pasir-Ris | 0 | 27 | 108 | 10539 | 9033 | 7437 | 27144 |
| Tampines | 183 | 244 | 12084 | 26882 | 15780 | 5858 | 61031 |

*Source*: HDB Household Survey 2008.

Table 13.2. Household sizes.

| HDB Town | Average | Median |
|---|---|---|
| Pasir Ris | 3.9 | 4 |
| Tampines | 3.7 | 4 |

*Source*: HDB Household Survey 2008.

was carried out. From the preliminary analysis, there were some fluctuations in the amount of recyclables collected between the months for all the zones under study.

In order to mitigate to some extent the effects of these fluctuations, the 6-months average collection result for the first 6 months was compared against the average collection result for the next 6 months. Assuming that each household recycles about the same amount consistently over time, an increase in amount of recyclables measured over time would confirm that more households are being influenced towards adopting recycling as one of their regular behaviours. However, the exact increase in numbers of households would be difficult to determine, as there were no surveys conducted before and after the implementation of the program. Moreover, each HDB household was not required to register in order to participate in the program. The data collected would only reflect the collective effort of the community within the designated zones. The series shown in Graph 13.1 demonstrated that there was a general increase in the collection of recyclables in 43 zones out of the 59 zones studied. The remaining 16 zones show a trend of stagnant or decreasing collection over the 12 months period. The highest increase is about 16.5%. This suggests that the reward program has produced only a modest short-term behavioural change. The differences arose because some of the zones comprise a greater number of blocks than the rest of the zones. Overall, there seems to be minimal changes in the average amount of recyclables collected between the 1st half and 2nd half of the study period for all zones. The results show that the impact of the reward program would be more visible when implemented in larger zones where there is a greater potential of new households adopting the behaviour of recycling. For smaller zones, the lesser number of households would mean that the maximum amount of recyclables that could be collected would be reached much more quickly. The number of new households that could be encouraged to start recycling would also be low. It would also not be surprising to note that collection would drop due to a change in their consumption behaviour. As households consume

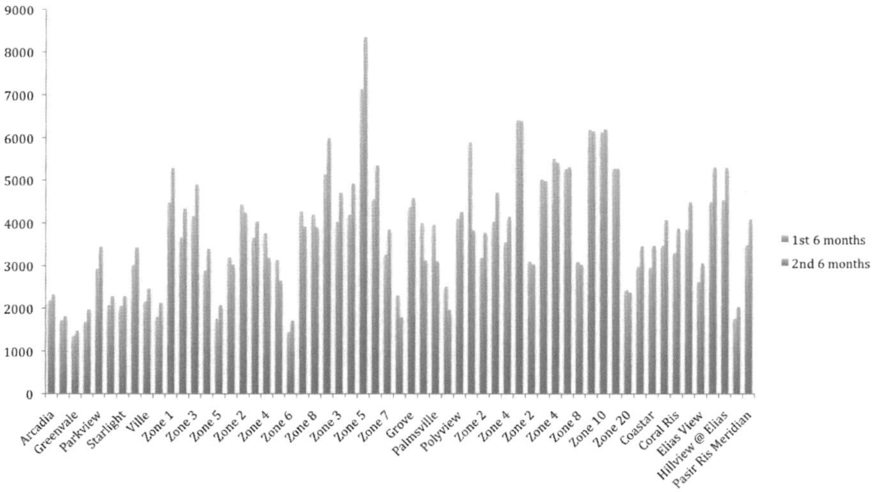


Graph 13.1. 6-months average collections of recyclables from Aug 11 to Aug 12 for all zones.
*Source*: GRIN Website.

less, they would recycle less as well. This change would be clearly visible in smaller zones, resulting in stagnant or declining trends in these zones. The fluctuating trends observed would also suggest that the reward program did not manage to consistently promote the behaviour of recycling. The inconsistencies show that the reward system had been unequal because households responded differently to the information presented to them. This case study would not be able to understand the individual responses of each household without a panel data. There was no prior collection data available before the program to be tracked over a period of time. From the analysis, all show a similar phenomenon of peak collection in January 2012. This could be attributed to the Chinese New Year celebrations in January 2012. The majority of Chinese households would be carrying out their house cleaning activities. Therefore, more household items and bulkier furniture tend to be contributed for recycling during this time. This phenomenon reflected the contextual or structural influences on environmentally significant behaviour as discussed by Stern (1999) in his paper. In this case, it was the cultural background of the households that drove them to engage in the environmentally friendly behaviour of recycling in January 2012. From the results of the analysis carried out for all the zones, the effectiveness of the reward program as an extrinsic factor to motivate more households to recycle is weak. Several factors could have contributed to this weakness. Firstly, it is a community-based program and the rewards obtained are allocated to all the households at a block of HDB flats. This limits the decision-making of each household to choose the reward it wants and forces it to accept the choice of the majority. Secondly, the administration of the reward program is managed by the representative of the grassroots organization and not by the individual household. As the household is not directly accountable for its actions, there is an absence of responsibility. Thirdly, the diversity of the households in terms of the manner of flat ownership, affluence and acquired capabilities made it difficult for a single reward program to attract the attention of every household. Generally, these factors would have affected the likelihood that any particular household will engage in the environmentally friendly behaviour of recycling. None of the zones show a significant jump in the amount of recyclables collected over the study period to demonstrate a significant change in behaviour of the households. This was a reward program that was supposed to be an incentive system that intervened in the contextual domain. However, the results have shown that the contextual forces are weak, and personal-domain variables seem to matter more in shaping pro-environmental behaviour in households. Despite the weakness of this reward program as an incentive system, it did manage to intervene quite modestly in the personal-domain as an information system. The online reward program provided a platform to present information on how and where to recycle. Households also can validate their actions by receiving monthly feedback on the amount of recyclables contributed towards this effort. It may have reinforced the behaviour of households to continue or increase their recycling efforts that is evident in the increase of recyclables collected in some zones.

## 5. Household Survey Results

A survey questionnaire was then designed to investigate the current level of environmentally friendly behaviour of Singapore's residents. The questionnaire was targeted at HDB residents, as a large majority of Singaporean residents live in HDB apartments. Out of the 83.5% of Singaporean households residing in HDB flats, 94.7% live in 3-room or larger flats (Singapore Department of Statistics, 2010a). Hence, the respondents were chosen from HDB flats of at least 3-room size. Cluster sampling was adopted as the sampling technique and the sampling parameters are shown in Table 13.3. The survey was conducted from July 2012 to Dec 2012.

Table 13.3. Actual parameters versus sample parameters.

| | Actual parameters (Statistics Singapore — Census of population 2010, statistical release 2) | | Sampling parameters (Bukit Merah, Jurong West, Punggol, Serangoon, Tampines & Woodlands towns) | |
|---|---|---|---|---|
| | No. of households as at 2010 | Percentage | No. of sampled households | Percentage |
| 3-room flats | 229,718 | 25.9 | 73 | 23 |
| 4-room flats | 365,423 | 41.1 | 177 | 56 |
| 5-room/Executive flats | 293,336 | 33.0 | 65 | 21 |
| Total | 888,477 | 100.0 | 315 | 100 |

*Source*: http://www.singstat.gov.sg/pubn/popn/c2010sr2/t1-7.pdf (retrieved on 2 Oct 12).

Environmental awareness has wide implications and encompasses many issues, including global warming and climate change, energy conservation and recycling. For the purpose of this research, respondents were asked 4 questions on affairs related to recycling, energy and water conservation so as to determine their level of awareness on local environmental initiatives. Awareness levels were high. 75% of the respondents could answer 3 or more questions correctly. Half of the households surveyed regularly practiced good energy saving habits pertaining to unplugging electronic devices, or switching them off when they are not in use. However, only 36% of households regularly switched off the lights, fans and air-conditioners when they leave the room. Higher awareness on energy efficient appliances helps explain why 63% of the households buy electrical appliances with the energy efficiency level of the device indicated. This could also be attributed to the Mandatory Energy Labeling Scheme (MELS) that was implemented for air-conditioners and refrigerators in Jan 2008 and clothes dryers in 2009. With more retailers offering these energy efficient appliances at a lower cost, households are more willing to purchase them. 46% of households never sort and recycle their garbage. This habit is attributed to the provision of a common refuse chute for households to dispose of their garbage. There is no infrastructure (for e.g., separate chutes for different types of recyclables) available to allow households to adopt the habit of sorting and recycling their common household waste. Most of them informed that they are not willing to reduce the amount of their household waste by sorting them and recycling more because they felt it was inconvenient and messy. Moreover, there is no penalty imposed should they fail to dispose less waste and recycle less. Instead, they believe that the waste management issue has minimal impact to the environment. 90% of household had donated their unwanted and reusable items to a charity or a thrift shop at least once. People are generally familiar with the habit of recycling, though only 29% of the households surveyed carried out this activity on a regular basis. As observed from the GRIN program, households only recycle more when they are cleaning out their homes before the New Year. Only 38% of the households surveyed regularly practiced good habits pertaining to printing on both sides of the paper, or on used paper that has been printed on one side, and making use of recycled paper. Printers that can conveniently print on both sides of a paper are usually more expensive, and people generally do not print their documents at home frequently. 91% of the households surveyed are generally familiar with the habit of water recycling and reducing their water consumption. It is encouraging to note that 62% of households expressed a high level of willingness towards environmental stewardship. These results were consistent with the observations that respondents wished to establish a positive self-image and communicate this image to others. The level of adoption of pro-environmental behaviour in areas of water, waste and energy conservation was also affected by the saliency bias. The simplest behaviour of switching off electronic appliances when they were not in use had the highest adoption rate of 50%. This could also be attributed to the mental accounting that cost savings could be achieved by using less electricity, which would result in a lower utilities bill. As for sorting and recycling their household waste, only 8% of respondents actively adopted this behaviour. This was consistent with the belief that no cost savings could be achieved, as they were not charged disposal fees accordingly. Although the households are aware of environmental issues and are willing to embrace an environmentally friendly lifestyle, it seems that as soon as a campaign or programme is over, complacency sets in and the issue is forgotten. Only 13% of households are confident of their own knowledge about the issues and efforts required of them. As 88% of the households spent more than $100 on their monthly utilities bill, it was observed that households would adopt energy and water conservation efforts so as to achieve significant cost savings. Based on the median household income of $3,480 (Ministry of Manpower Singapore, 2012), the utilities bill would constitute about 3 to 5 % of the household income. It was found that the same key barriers to overcome households' lack of motivation to adopt an environmentally friendly lifestyle were inconvenience, lack of collective action and monetary incentives. The results seem to suggest that *situational* factors seem to

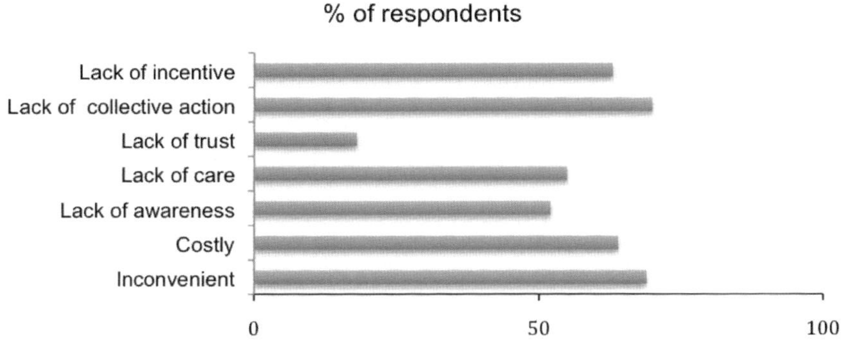


| | Inconvenient | Costly | Lack of awareness | Lack of care | Lack of trust | Lack of collective action | Lack of incentive |
|---|---|---|---|---|---|---|---|
| % of respondents | 69 | 64 | 52 | 55 | 18 | 70 | 63 |

Graph 13.2. Percentage of respondents and their answers on the likely reasons that people don't behave in an environmentally friendly manner (eg. Recycle more often, etc).

play a major role in changing their behaviour. Only 18% and 25% of households indicated a lack of trust and that it feels good to do the right things as shown in Graph 13.2 and Graph 13.3. This showed that they had not understood social and environmental values. 87% of the households indicated that they would adopt the particular environmentally friendly behaviour when there is an incentive system, and 89% would prefer a monetary reward system that can help to reduce their monthly utilities and conservancy bills as shown in Graph 13.4. This would suggest that households could be motivated by a positive incentive system to change and subsequently sustain their behaviour through economic instruments as shown in Graph 13.5.

It was observed that the motivations to change their behaviour were varied, with no strong bias towards any one or two factor. Interestingly, only 25% of surveyed households believe that others will change their behaviour because they feel good doing the right thing. This seems to suggest that altruism may not be perceived as a strong motivating factor, and that household seems to be driven more by extrinsic motivation than by intrinsic motivation. They could also be influenced by social and mass media as shown in Graph 13.6. With regard to their attitudes towards the family, close to 73% of them would inform and involve their family and friends in adopting environmentally friendly lifestyle choices as shown in Graph 13.7.

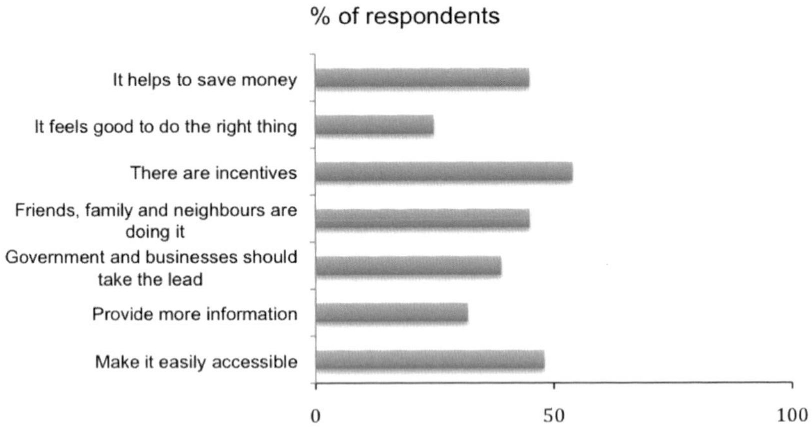


| | Make it easily accessible | Provide more information | Government and businesses should take the lead | Friends, family and neighbours are doing it | There are incentives | It feels good to do the right thing | It helps to save money |
|---|---|---|---|---|---|---|---|
| % of respondents | 48 | 32 | 39 | 45 | 54 | 25 | 45 |

Graph 13.3. Percentage of respondents and their answers on the likely reasons that people can be motivated to behave in an environmentally friendly manner (e.g. Recycle more often, etc).

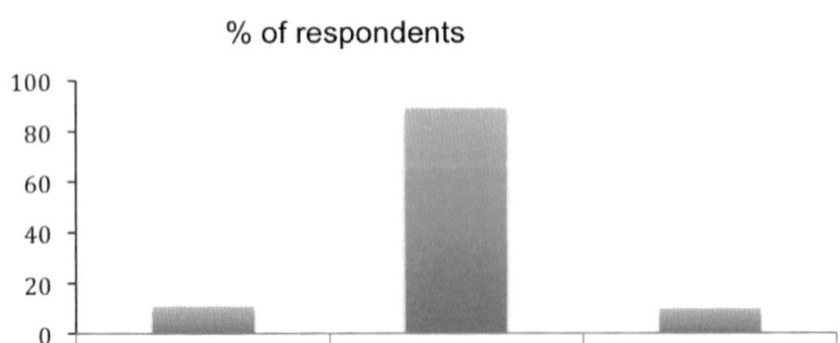


| | Points-based system involving non-monetary benefits | Monetary rewards to be used to reduce your monthly utilities and conservancy bills | Awards and membership in community and interest groups |
|---|---|---|---|
| % of respondents | 11 | 89 | 10 |

Graph 13.4. Percentage of respondents and their choice of an incentive system.

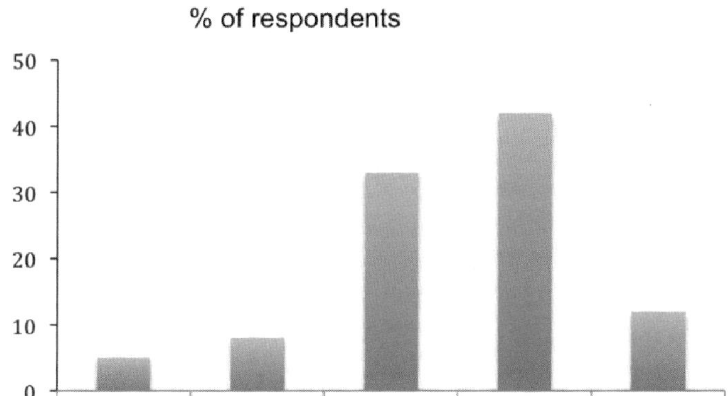


| | Very Low | Low | Average | High | Very High |
|---|---|---|---|---|---|
| % of respondents | 5 | 8 | 33 | 42 | 12 |

Graph 13.5. Percentage of respondents that would participate actively in environmentally friendly activities if there were an incentive system.

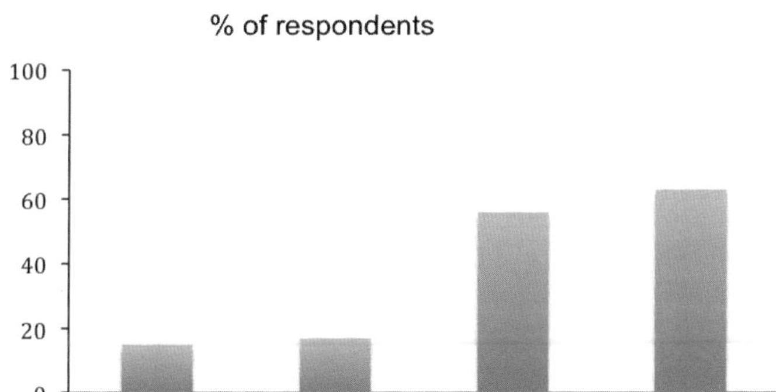


| | Door to door visit | Exhibitions | Social Media | Mass Media |
|---|---|---|---|---|
| % of respondents | 15 | 17 | 56 | 63 |

Graph 13.6. Percentage of respondents and their choice of outreach programs.

## 6. Recommendations

When we apply these results to Singapore, our approach can be summarized as below:

1. A short-term incentive and informative program to introduce a pro-environmental behaviour (for e.g., recycling at home).
2. To sustain the momentum, a 2-pronged approach is adopted separately for the homemakers, elderly and youths. For the homemakers and elderly,

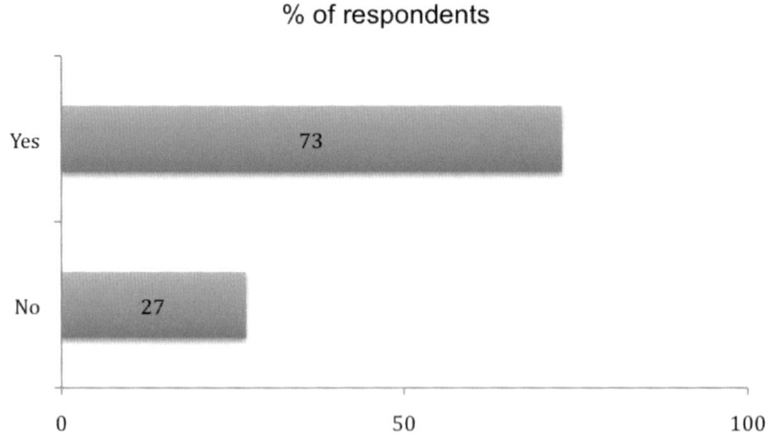


Graph 13.7. Percentage of respondents that would you help to spread the word or involve their family and friends.

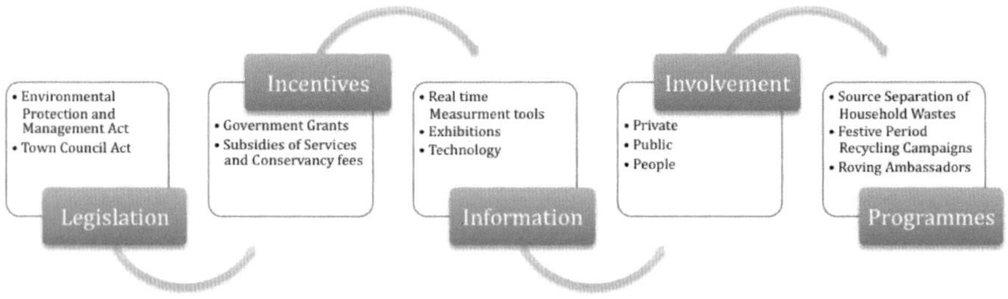


Fig. 13.1. Process flow of proposed model.

a local community-based movement or group is supported for face-to-face activities. For the youths, web technology-based platforms such as social media are used to form online communities to facilitate co-creation and sharing of knowledge.

3. Finally, the homemakers, elderly and youths will become key stakeholders who will work jointly with other private and public institutions to enhance the infrastructure network and organise programmes that continue to promote an environmentally friendly lifestyle.

This model will specify the requirements for legislations, the types of incentives, the modes of information transmissions, the roles of various stakeholders and programmes that could be promoted in the community. Figure 13.1 depicts the flow of the proposed model.

With this model, it is proposed that individual Town Councils establish their incentive systems with their first specific goal of achieving an increased rate of recycling within their estates. This would be aligned with the national target of 70% recycling rate for all households. With most of the boundaries defined, the proposed methodology will focus on providing them a working framework to start accounting for their incentive system immediately without worrying too much about complying with standards such as definition of boundaries and identifying the steps from scratch. To help the Town Councils, an Environment and Conservation Fund (ECF) Committee with an approved allocation of $500 million could be set up to support housing estates/residential buildings to acquire suitable infrastructure such as waste separation facilities. The fund also encourages the provision of waste separation and recovery facilities in public buildings to facilitate the occupants and visitors. The scope of this fund carries a broader area than just HDB flats, and should also cover other non-residential buildings belonging to public institutions and non-profit-making organisations (for e.g., tertiary educational, medical, religious and community facilities and premises, etc.) that perform various social functions. At the beginning, some form of environmental guidelines or code of practice that would educate Town Councils and residents on environmentalism should be implemented. The proposed environmental legislation amendment to the Town Council Act (Chapter 329A) need not necessarily penalise those who do not perform their environmental duties. Instead, it can come in the form of policy or management instruments that are made mandatory for Town Councils and residents to follow. Firstly, the amount of household waste generated should be measured and an annual reduction rate mandated, with penalties imposed for non-compliance. Conversely, the amount of recyclables collected is also measured and an annual rate of increase mandated. In addition, incentives such as government grants can be made available to the Town Councils for them to kick start initiatives that promote the rate of recycling.

As described, ECF will be granted as partial support to the set-up cost. For example, a limit of 50% of the total actual expenditure spent for the approved items, not exceeding a stipulated limit per block of flat, will be reimbursed. The funding support shall cover expenditure on procurement, delivery and installation of waste separation facilities, and equipment or tools to be used in the provision of infrastructure support only. A statutory body, which comprises mainly representatives from Town Council, waste collection companies and residents could be set up under the Environmental

Protection and Management Act to advise on the use of these funds. Under the Act, the National Environment Agency is the trustee responsible for the administration of the ECF.

The GRIN programme could be reorganized to be jointly organized with Town Councils, waste management companies, non-profit organizations such as Singapore Environment Council and public institutions such as the People's Association. For example, it would be modified to include a monetary incentive system that gives rebates on households' service and conservancy fees (S&C) or the universal waste collection fees (UF) so as to entice them to sign up. The Town Council would be tasked to administer the programme while the waste management company carries out the operations, such as maintaining the waste separation facilities in a clean and working condition. Concurrently, the People's Association, through its network of grassroots organisations, could promote the programme as well as recruit volunteers among the homemakers, elderly and youths. This may require about 1 year of support from the stakeholders before a group of volunteers could be formed to become co-stakeholders in this programme.

It is essential that residents are informed by way of notices, posters, circulars or online media of the recycling arrangements, including the location of recycling facilities near their flats, types of recyclables to be recovered and any specified time slots or days for collection. Meanwhile, the Town Councils and waste management companies could enhance the environmental infrastructure around the estates. These could be through providing a greater and more conspicuous environmental infrastructure around the estates, so as to encourage residents to be environmentally friendly. Examples of such infrastructure can be the provision of simple items like compartmentalised recycling bins for each household. From the second year onward, there should be a core group of volunteers who are ready to take on more responsibilities for this programme. The Town Council would play a supporting role of providing some financial support for new events and programmes to be organised by the volunteers. The volunteers could organise educational and publicity activities to increase residents' awareness and participation in recycling. For example, they could set up display panels and game booths during ECO-fairs, organize family workshops, and mobilise resident ambassadors to promote waste separation at source, within the community. As observed in our earlier results that peak collection of recyclables occur between the end of the previous year and the beginning of a new year, an annual collection and recycling campaign could be organised at this time. These recyclable items, such as electrical appliances, clothes, shoes, bags, books and toys could be donated to charities or sold at affordable prices at social enterprises (for e.g., second hand shops) that cater to low income families. Such a campaign could also help raise the intrinsic motivational value of residents who may not recycle otherwise.

The reward redemption system under the current GRIN programme has been found to be difficult to maintain due to varying expectations and desires of the residents. Therefore, a separate monetary and loyalty program that is tied up with an existing programme could be introduced instead. This could be administered through the People Association's Passion card programme that was introduced in April 2005. It is a membership card programme that residents can sign up to individually, which helps to connect them to their community by engaging them in activities by Community Clubs (CCs) and other business outlets. With an established membership base and partnerships with a wide variety of businesses, it would offer a sustainable choice to the residents to remain engaged in their effort in recycling. The Town Council and waste collection companies would have to arrange for the collection of waste paper, plastics and metals on a regular basis and keep the recyclables collection point in a clean and hygienic condition. The amount of recyclables collected should be recorded and reported by the waste collection companies through the use of an electronic tagging system, and the results audited by the statutory board. The results obtained would be used to track the effectiveness of the programme as well as to determine the amount of incentives in the form of grants and subsidies to be given to each Town Council. At the same time, regular newsletters would be issued to inform residents of their recycling achievements and serve as reminders for sustained participation. In conclusion, despite the limitations in the case study and surveys conducted, and some inherent shortcomings with an incentive system, they do not compromise the premise that the incentivising approach could be explored on a larger scale. It can be ascertained that there is significant unrealised potential in an incentive and information system now.

## References

Corraliza, J.A. and Berenguer, J. (2000). "Environmental values beliefs and actions: a situational approach", *Environment and Behavior*, **32**(6), 832–48.

De Young, R. (1985–1986). "Encouraging environmentally appropriate behavior: The role of intrinsic motivation", *Journal of Environmental Systems*, **15**(4), 281–292.

De Young, R. (1996). "Some Psychological Aspects of Reduced Consumption Behaviour: The Role of Intrinsic Satisfaction and Competence Motivation", *Environment and Behavior*, **28**(3), 358–409.

De Young, R. (2003). "Expanding and Evaluating Motives for Environmentally Responsible Behavior", *Journal of Social Issues*, **56**(3), 509–526.

Derksen, L. and Gartrell, J. (1993). "The Social Context of Recycling", *American Sociological Review*, **58**(3), 434–442.

Dunlap, R.E. and Van Liere, K.D. (1978). "The new environmental paradigm: a proposed measuring instrument and preliminary result", *Journal of environmental education*, **9**(2), 10–19.

Eden, S. (1998). "Environmental issues: knowledge, uncertainty and the environment", *Progress in Human Geography*, **22**(3), 425–432.

Guagnano, G.A., Stern, P.C., and Dietz, T. (1995). "Influences on Attitude–Behavior Relationships. A Natural Experiment with Curbside Recycling", *Environment and Behavior*, **27**(4), 699–718.

Hopper, J. and Nielsen, R.J.M. (1991). "Recycling as altruistic behaviour: Normative and behavioral strategies to expand participation in community recycling program", *Journal of Environment and Behaviour*, **23**(2), 195.

Macnagthen, P. and Urry, J. (1998). *Contested Natures.* London: Sage Publications Ltd.

Myers, G. and MacNaghten, P. (1998). "Rhetorics of environmental sustainability: commonplaces and places", *Environment and Planning A*, **30**(2), 333–353.

O'Riordan, T. (1985). "Research policy and review 6. Future directions for environmental policy", *Environment and Planning A*, **17**(11), 1431–1446.

Ong, P. (2012). "Can psychology save the planet and improve our environment?", *Behavioral Economics and Policy Design,* pp. 69–85.

Schahn, J. and Holzer, E. (1990). "Studies of individual environmental concern: The role of knowledge, gender, and background variables", *Environmental and Behaviour*, **22**(6).

Schwartz, S.H. (1977). "Normative influences on altruism", *Advances in Experimental Social Psychology*, **10**(1), 221–279.

Schwartz, S.H. (1992). "Universals in the content and structure of values: theoretical advances and empirical test in 20 countries", *Advances in Experimental Social Psychology*, **10**, 221–279.

Schwartz, S.H. and Blisky W. (1987). "Toward a psychological structure of human values", *Journal of Personality and Social Psychology*, **53**(1), 550–562.

Selman, P. (1996). *Local sustainability: Planning and managing ecologically sound places.* London: Paul Chapman.

Stern, P. (1992). "What psychology knows about energy conservation", *American Psychologist*, **47**(10), 1224–1232.

Stern, P. (1999). "Information, Incentives and Proenvironmental Consumer Behaviour", *Journal of Consumer Policy*, **22**(4), 461–478.

Stern, P.C. Dietz, T. and Guagnano, G.A. (1995). "The new ecological paradigm in social-psychological context", *Environment and Behavior*, **27**(6), 723–743.

Sunstein, C.R. (1996). "Social norms and social roles", *Columbia law review,* **96**(4), 903–968.

Tan, Charmaine and Donald, Low. (2012). "Incentives, Norms and Public Policy", *Behavioral Economics and Policy Design*, World Scientific Publishing (Singapore), pp. 35–49.

Vining, J. and Ebreo, A. (1990). "What makes a recycler? A comparison of recyclers and nonrecyclers", *Environment and Behaviour*, **22**(1), 55–73.

# 14

# A REVIEW OF ENVIRONMENTAL EDUCATION IN SINGAPORE — TOWARDS A CHANGE IN APPROACH

ANG GUORONG ALBERT and AUDREY CHIA

Environmental education programmes in Singapore over the past four decades are reviewed using environmental psychology models and perception studies. While Singapore's environmental efforts are well-covered at the broad policy level, more needs to be done at the individual level to encourage personal pro-environmental behaviour. For this to happen, the authors argue that a fundamental shift in mentality among policymakers in the approach to environmental education is necessary for pro-environmental behaviour to be sustained into the future.

## 1. Introduction

Environmental education is, indisputably, the crux to all good environmental policies. An ideal outcome of environmental education (EE) will see its recipients making pro-environmental choices as part of their lifestyle habits. This paper assesses the wholeness of Singapore's EE by reviewing it using psychological models posited through the decades, from the time of the formation of the United Nations Environment Programme (UNEP). Perception surveys in the literature that probe Singaporeans' attitudes towards the environment will also be briefly discussed to understand the outcome of EE efforts. The paper ends with a discussion and gives recommendations on environmental education in Singapore.

### *Environmental Education*

Environmental education has many definitions. When world leaders met in Tbilisi in 1977, the Declaration of the Tbilisi Intergovernmental Conference on Environmental Education read,

"Environmental education [...] should involve the individual in an active problem-solving process within the context of specific realities, and it should encourage initiative, a sense of responsibility and commitment to build a better tomorrow. By its very nature, environmental education can make a powerful contribution to the renovation of the educational process.

[...] The categories of environmental education objectives:

*Awareness*: to help social groups and individuals acquire an awareness of and sensitivity to the total environment and its allied problems.

*Knowledge*: to help social groups and individuals gain a variety of experience in, and acquire a basic understanding of, the environment and its associate problems.

*Attitudes*: to help social groups and individuals acquire a set of values and feelings of concern for the environment, and the motivation for actively participating in environmental improvement and protection.

*Skills*: to help social groups and individuals acquire the skills for identifying and solving environmental problems.

*Participation*: to provide social groups and individuals with an opportunity to be actively involved at all levels in working toward resolution of environmental problems".

(UNESCO, Tbilisi, 1977)

This paper takes a leaf from the above definitions to regard any form of strategies or programmes aimed towards the attainment or enabling of the above objectives — ***awareness***, ***knowledge***, ***attitudes***, ***skills***, and ***participation*** — in environmental betterment as constituting the process of environmental education.

## 2. Environmental Education in Singapore

### *First Twenty Years — Public Education*

Environmental Education in Singapore can be divided into two twenty-year periods. Starting from the formation of the Ministry of Environment in 1972 to coordinate environmental efforts, to the inflexion point in 1992 when the Singapore Green Plan was formulated for the UNFCCC Rio Conference, EE took different starting points. In the first twenty years, EE can be seen as 'coincidental' to economic initiatives while the post Singapore Green Plan decades saw environmental education being recognised and actively pursued in its own right.

Singapore recognised the importance of its environment early in nation-building. In order to tackle its environmental issues, the 'Keep Singapore

Clean' campaign was the first public environmental education scheme introduced in 1968. Various branches and units under Ministry of Health and Ministry of National Development ran campaigns related to the environment, for e.g., 'Tree Planting and Gardening' in 1970 and 'Save Water' in 1971. By 1972, the Ministry of Environment (ENV), a ministry dedicated to environmental issues, was set up to oversee issues of pollution control, sewerage, drainage and public health. The driving force for the formation of ENV was in essence, economic, as argued by Singapore's first Prime Minister, Lee Kuan Yew, "...If you put people in a good environment, you get people that are healthy and industrious, and they can work and think..." (Savage and Lau, 1993). Hence, environmental efforts were put alongside economic progress as a necessary factor of economic growth, and less for the appreciation of ecological conservation or sustainability.

Neo (2010) identified six areas commonly covered by ENV's environmental education efforts: "anti-littering, clean public toilets, dengue fever prevention, waste minimization and recycling, clean river, and food hygiene". These were direct responses to Singapore's needs and resource constraints, such as creating suitable conditions to attract foreign direct investments and limited space for landfills. Quick solutions were hence sought after while education of reasons and attitudes were secondary.

The modus operandi of these campaigns was: (1) creating awareness through circulation in the media, placing posters at public spaces and circulations to the young in schools; (2) monitoring public response; and (3) strict enforcement of relevant laws through fines. Where littering is concerned, positive disincentives went one step up by introducing public shaming. Since 1992, repeated offenders of littering were served with Corrective Work Order, which requires them to clean public places wearing a cleaner's attire. From 2014, video cameras have also been installed at 'hotspots' to apprehend persons responsible for littering.

### Second Twenty Years — Singapore Green Plan 1992

Two decades after the inception of ENV, the Singapore Green Plan (NEA, 1992a) was intentionally formulated and submitted to the 1992 Earth Summit in Rio de Janeiro. Titled 'The Singapore Green Plan — Towards a Model Green City', it laid bold ambitions of becoming a city that can be held up as a model for emulation by other cities. The Plan signalled a departure from the old ways where environmental directives were top-down and 'stick'-driven. A consultative approach was taken, and the outcomes identified environmental education as one of the key Action Programmes. Among other objectives, environmental education will be:

- Consciously integrated into the formal education system up to the tertiary levels;
- Targeted at the general public through campaigns in various media;
- Conducted through the involvement of institutions with large clout and audience, e.g., Armed Forces, Community Centres, consumer groups and businesses. (NEA, 1992b)

In a later paper submitted to the United Nations Framework Conventions on Climate Change (UNFCCC) specifically on its environmental education policy, NEA outlined its key strategies of building awareness, capacity and empowering actions through 3P partnerships (Public, People, and Private) (NEA, UNFCCC, 2009). The policy also called for platforms to showcase and recognise environmental players, as well as mechanisms to monitor and gather feedback. It is noteworthy that environmental movements are still government-led, as was the case for the formation of the Singapore Environment Council (SEC), one of the biggest (ironically) environmental NGO to date by membership and funding (Geh, 2008). SEC was set up by NEA to be its arm in engaging the People in 1992. This suggests either strong environmental advocacy by the government, or a lack of interest by Singaporeans on the state of the natural and built environment. Government-led initiatives continue to be strong, with two lead agencies identified for pushing environmental education — the Ministry of Education for the integration of environmental education into the formal curriculum, and NEA for spearheading all other forms of environmental education (private partners and NGO engagements).

*Blueprint on sustainable development*

An inter-ministerial committee on sustainable development was formed in January 2008 to develop a national framework and key strategies for Singapore's sustainable development. The details were documented in the 'Blueprint on Sustainable Development' (Blueprint). In the section on environmental education, programmes submitted earlier to the UNFCC were largely replicated. However, it spelled out the programme owners and their objectives more clearly. Of note were the inclusion of support for ground-up initiatives at the community grassroots levels and the promotion of networking among NGOs. The Blueprint is a strong indication of each ministry's commitment towards the environment and will lend legitimacy to environmental policies in years to come.

*Environmental education programme*

After careful review of programmes listed on websites of various statutory boards under the Ministry of Environment and Water Resources, Ministry of National Development, and the Blueprint on Sustainable Development (2009), components of EE in environmental programmes up to 2013 were listed below. Policies or programmes that qualify as environmental education will have to specifically enable environmental ***awareness***, ***knowledge***, ***attitude***, ***skill*** and ***participation*** of the populace towards pro-environmental behaviour. It is noteworthy that policies listed below result in pro-environment participation although no active decisions have to be made by the individual, as choice is not option. Fines and tariffs, however are still used to shape attitudes and behaviours at the individual level.

## 3. Environmental Psychology Models and Singapore's EE

Environmental psychology is a field of science that examines the transactions and interrelations between people and their socio-physical surroundings (Gifford, 2007; Bechtel and Churchman, 2002; Stokols, 1987; Canter and Craik, 1981), which include the built and natural environment, the use, and abuse of nature and natural resources, and sustainability-related behaviour. As the scope and application of this branch is broad, only models designed for pro-environmental behaviour will be discussed. By basing environmental programmes on some of these models, planners can reasonably design and evaluate outcomes of human-environment interactions and make suitable adjustments to promote pro-environment behaviour. As it is a relatively new branch of psychology, environmental psychology models are evolving with the development of new environmental issues and technology; hence, six of these models shall be used to review Singapore's EE. Where appropriate, Singapore's EE listed in Table 14.1 will be mapped against the models to provide a visual reference. While each model has its limitations and their validation was conducted under different contexts, their emphases on different behavioural pre-conditions and evolution do give us good standards to review Singapore's EE.

### *Linear Causality Models*

An early model advanced in the 1970s presented a rationalist linear framework. The model proposes that an individual who has knowledge of an

Table 14.1. EE policies or programmes initiated by the Singapore government.

| | |
|---|---|
| Awareness | • Environment awareness campaign<br>• Setting up interactive websites<br>• Coverage in school curriculum<br>• Clean and Green Singapore campaign<br>• Keep Singapore Clean |
| Knowledge | • Promoting community environment effort through CDCs<br>• Environment enrichment programmes<br>• Providing water consumption history to make informed future consumption<br>• Energy/Water-efficiency labels on appliances |
| Attitude | • Attitudes and values worded into school curriculum<br>• Added environmental indicators to School Excellence Model<br>(a system of appraisal of school's overall performance, to encourage emphasis on environment education in schools<br>• Holding competitions to encourage students to find solutions to environmental problems<br>• NEA outreach Programmes<br>• Fines and other penalties<br>• Extraneous expenditure, e.g. tariffs and fees |
| Skill | • '5-step Mozzie Wipe-out'<br>• NEA Learning Journeys |
| Participation | • Providing recycling bins<br>• Government promote grassroot engagement of environmental programme<br>• Promote use of recycled products, e.g. recycled compost and construction/demolition waste<br>• Efficient public transport<br>• Improved transport links<br>• Environmental Education Framework sets a roadmap for job scope<br>• Government cap annual increase of vehicle population from 3% to 0.5%<br>• Government active land use planning<br>• Having pedestrian-friendly environment to promote walking/cycling<br>• Making cost of private transport high through Vehicle Quota System<br>• Green Vehicle Rebate<br>• Providing funds and incentives for businesses to go green<br>• Promoting resource-efficient buildings<br>• Providing favourable tax-deductions for environmental capital expenditures<br>• Government invest in R&D on enviro-tech industry<br>• Stepping up efforts on enforcement against littering<br>• Imposition of fines and penalties for irresponsible environmental behaviour |

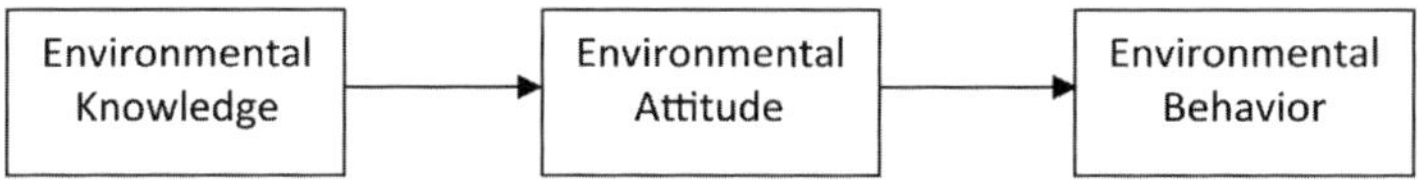


Fig. 14.1. Model of linear casuality.

issue can be expected to possess the common conscience to form an attitude towards the desirable outcome, or avoidance of the undesirable outcome, and hence behave in a way towards the eventual realisation of the outcome. Kollmuss and Agyeman (2002) presented this graphically in Fig. 14.1.

While this model seems logical, researchers have found it to be insufficient and too simplistic, as knowledge and awareness cannot be the sole conditions of pro-environmental behaviour. Diekmann and Preisendoerfer (1998) noted considerable discrepancies between environmental attitudes and behaviour in their citizen survey of Switzerland and Germany. Connell *et al.* (1999) and Kuhlemeier *et al.* (1999) in their study of Dutch youth noted the same phenomena. To some extent, Lee *et al.* (1998) also arrived at the same conclusion in Singapore. Though incomplete, it does bring knowledge and attitude as important factors to the fore in its time.

### *Model of Ecological Behaviour*

In the early 1980s, Fietkau and Kessel (1981) developed an elegant approach. In their model (Fig. 14.2), psychological and social factors were considered, yielding five factors that are instrumental to pro-environmental behaviours: (i) possibilities to act pro-environmentally, (ii) environmental attitudes and values, (iii) environmental knowledge, (iv) perceived consequences of behaviour, and (v) incentives for pro-environmental behaviour.

### *Model of Responsible Environmental Behaviour*

Four layers of conditions are posited to precede responsible environmental behaviours. As far as internal psychological factors go, the last stage is intention. However, intention is influenced by six others factors. (i) attitudinal variables deal with individuals' feelings — pro or against, in favour or disfavour — of a particular aspect of environment, (ii) *locus* of control, which refers to the amount of control an individual perceives his/herself to have over the outcome of an environmental issue, (iii) personal responsibility, which refers to the sense of duty or obligation towards the presented environmental issue. These three factors constitute personality factors, which only has partial influence towards intention.

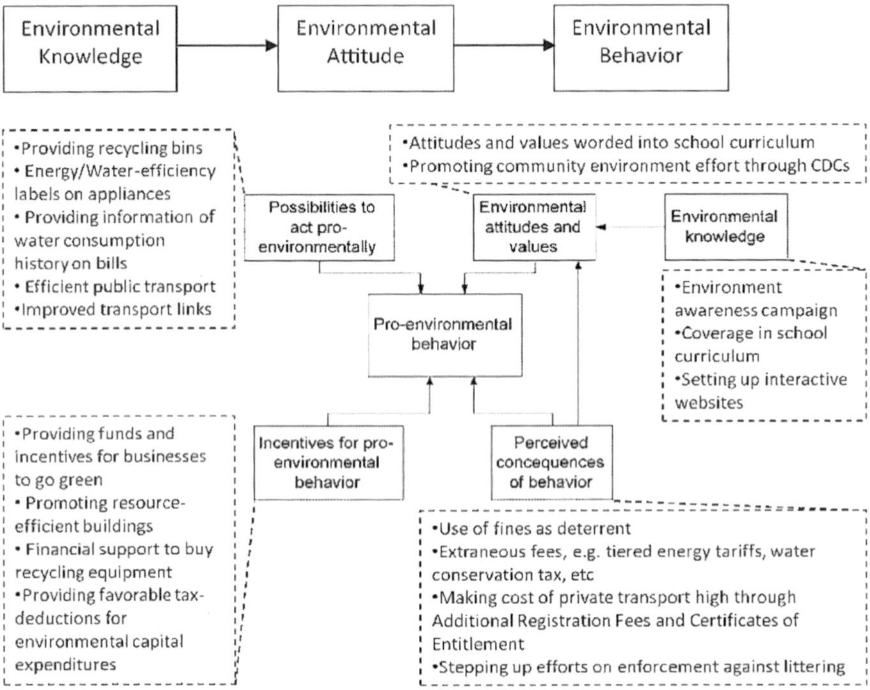


Fig. 14.2. EE mapped against Fietkau and Kessel's Model of Ecological Behavior (1981).

These personality factors, together with the plain cognition of issues, knowledge on 'what can be done', and 'how to do it' will trigger the intention to act. Conversely, lacking one or more factors will diminish the intention or kill it altogether. With internal psychology factors fulfilled, other socio-economic-geo-chronological factors have to be fulfilled. These are termed as "situation factors" by Hines *et al.* (1987). Situational factors such as economic constraints, social pressures and opportunities of competing agendas may sabotage the mind eventually take the irresponsible course of pollution the environment (Fig. 14.3).

## *Model of Pro-Environmental Behaviour*

A more complete, though by no means perfect, model emerged with internal and external factors as two large influence groups. The arrows indicated the inter-play between different psychological and sociological factors. Although no new factors were added, factors were characterised,

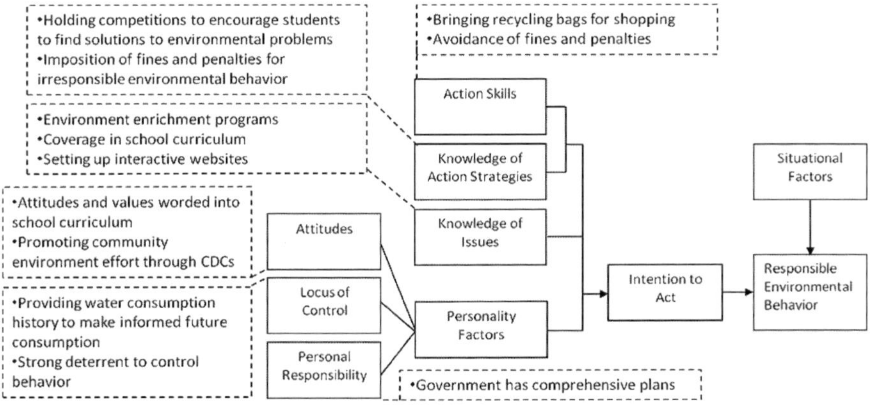


Fig. 14.3. EE mapped against Model of Responsible Environmental Behaviour, Hines, Hungerford and Tomera (1987).

and interactions between the factors were explained. An assumption that was previously ignored is also addressed — old behaviour pattern. In this model, Kollmuss and Agyeman (2002) elaborated on a number of complex feedback loops — one being between knowledge, emotions and attitudes/values. They argue that a lack of right knowledge causes emotional blocks towards the formation of positive attitudes/values, hence, the existing (presumably negative, or ignorant) attitudes/values will, in turn, play on negative emotions and block acceptance of new knowledge. In simple terms, this feedback loop explains the probable psychology of a stubborn and ignorant person. Such an individual trapped in this vicious feedback loop is described as lacking environmental consciousness versus awareness of environmental concerns. The root of this ignorance lies in misinformation that can only be corrected through patient re-education.

Next, at the macro level, collective environmental consciousness can effect indirect environmental actions, as their research found that individuals might not be moved into taking direct environmental actions but will vote for political figures running campaigns with an environmental thrust. Political parties with environmental issues high on their agenda will implement pro-environment policies which will, in turn, foster stronger environmental consciousness in their electorate.

The model warned that without the accomplishment of internal and external factors, individuals risk falling back to old behaviour patterns. As the age-old adage goes, "old habits die hard", it is reasonable not to assume individuals to be a piece of white canvas who value convenience

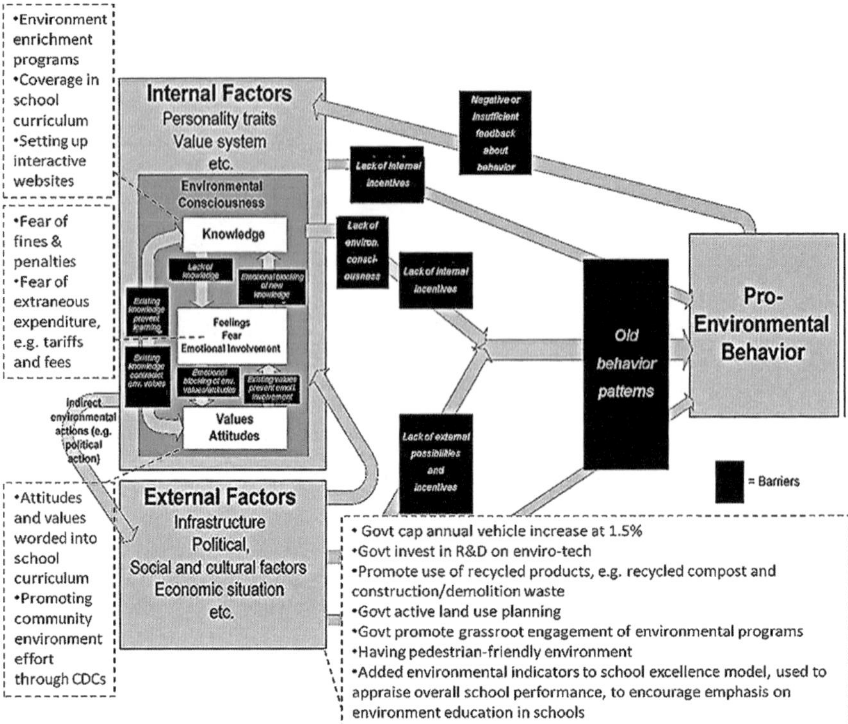


Fig. 14.4. EE mapped against Model of Pro-Environmental Behaviour, Kollmuss and Agyeman (2002).

and low-cost options more highly, relegating global issues as being 'too far away to concern me', and just fall back to the old ways which are more familiar and easier to perform. Similar to Blake (1999), Kollmuss and Agyeman (2002) used a 'barriers' approach to explain the non-fulfilment of an expected factor versus a direct causal approach. From a practitioner's point of view, the 'barriers' approach might be more useful in helping environmental programmers frame their scheme of work.

Lastly, this model suggests that it is natural human instinct to review their deeds as another feedback loop between behaviour and internal factors. If pro-environmental factors received negative or insufficient feedback, then internal psychological factors will be affected in terms of having a negative attitude henceforth, or alteration of taught knowledge. Therefore, there has to be positive feedback in terms of tangible or non-tangible factors, or a simple confirmation of the taught knowledge. For instance, after collecting aluminium cans for recycling and faithfully depositing them into the recycling bin, and one sees the trash collector mixing aluminium, paper, plastic wastes together with general organic waste repeatedly, one cannot be expected to continue the collection of used aluminium cans for long.

### *The Reasonable Person Model*

Kaplan and Kaplan conceived the Reasonable Person Model (Kaplan, 2000) to relate tendencies for altruistic pro-environment behaviour amidst self-interest at the individual level. It suggests that individuals with basic informational needs satisfied can be expected to make reasonable pro-environment choices that are not purely interest-motivated. Kaplan suggests three inter-supplementary informational domains for individual reasonableness: "model-building", "becoming effective" and having "meaningful actions" (Fig. 14.5).

At the personal level, individuals explore and understand environmental issues. However, when bombarded with excessive information, one does not only gravitate towards greater concern for their environment but also to a greater sense of helplessness (Corbett, 2003). To counter this feeling of helplessness or futility, multiple choices should be made

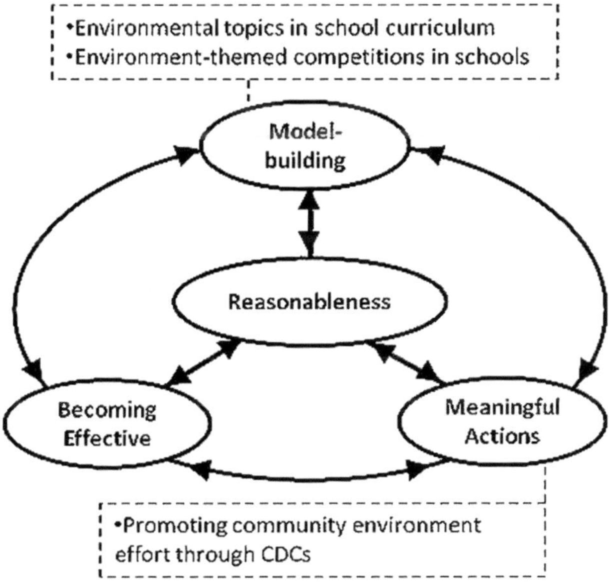


Fig. 14.5. EE Mapped against Reasonable Person Model, Kaplan and Kaplan (2002).

available so that individuals can explore those they have an interest in, and form cognitive mental models (action roadmaps). Thus inspired, participation in environmentally responsible actions allows one to be heard and feel needed, fulfilling an innate desire for reasonable pro-environmental choices and acceptance. Such kinds of participation are meaningful because they are motivated by self-interest, and enhance attitudes towards environment issues. To be effective sustainably — staying on the course of environmentally-responsible behaviours — requires sustained mental vitality, which can then summon directed attention on meaningful yet mostly uninteresting tasks; whilst distractions due to other more fascinating objects lead to mental fatigue and draw individuals away from continually making pro-environment choices. Kaplan and other studies suggests natural settings and meaningful activities as "significant sources of satisfaction" (De Young, 2013) for restoration from mental fatigue, thereby continually encouraging the behaviour of participating in environmentally-responsible work.

The above analysis shows that Singapore has a set of fairly thorough programmes to encourage its people to make pro-environmental choices, especially when compared with the Fietkau and Kessel (1981) model. When compared with the Hines *et al.* (1987) model, not enough has been done to foster personal responsibility and disable negative situational factors. Even so, it is fair that situational factors are arbitrary and are not possible to mitigate under diverse economic and social conditions. Based on the Kollmuss and Agyeman's model, one starts to see Singapore's programmes as having a much heavier focus on supporting external factors rather than internal factors, and using emotional involvement based on fears of financial penalty. Additionally, many programmes are government-led, and in most cases involve government agencies providing environment-friendly boosts, but little is aimed at giving ownership to the individual. This is reflected in the difficulty of mapping the EE programme onto Kaplan's Reasonable Person Model.

## 4. Surveys on EE Programmes

A review of EE in Singapore will not be complete without reference to survey studies of EE outcomes.

Tan *et al.* (1998) conducted a baseline study on the result of environmental education in Singapore through a knowledge, attitude and behaviour (KAB) survey on 1256 15- and 17-year-old students. In the survey, the researchers noted that Singapore's students were knowledgeable about general concepts regarding the environment (mean 78%), e.g., on global-warming, but were not proficient in explaining the facts (mean 68%), e.g., causes and factors. On attitudes, the mean attitude score was 66% with a standard deviation of 5.2. It is worthy to note that 74.4% of respondents strongly disagree(d) that it is "only the government's responsibility in conservation of natural resources", yet 57.5% of the respondents strongly agree(d) to "strong control by the government to reduce pollution problems". While one may interpret the responses as seeing the role of individuals and government as separate and distinct, there was also a hint of reliance on the government. On environmental behaviours, respondents scored fairly high on the overall (mean 70.5%) but on closer observation, the paper reports that students scored lowly in terms of being confident in committing to green consumerism (e.g. always reusing items like glass bottles, always choosing products with green labels, never buying products with lots of packaging) and environmental behaviours in general (e.g. never litter when no one is watching, always take short showers to save water, often attend environmental exhibitions). The authors further noted that survey scores were made up mostly from the response of "sometimes" environmental behaviours, which does not convincingly reflect that being environmental has become their way of life.

In their conclusion, the authors pointed out that "... Students' attitudes tended to be strongly positive with respect to items which are more general in nature and have less direct bearing on their lives. Such positive attitudes may be 'learned responses'. Therefore, there is a need to stress that environmental activities in schools should be designed to incorporate the affective domain". This conclusion seems to suggest that students tend to 'give the right answers', exhibiting social desirability bias when asked (suggesting that the empathetic reaction was a learned response their teachers have either modelled, or have taught them to have towards the environment, but that are not necessarily their own beliefs); but when it comes to taking actions, students are less inclined to commit since knowing does not equate to doing.

Taking account of the past forty years, much can be inferred from the approach taken by the Singapore government on environmental issues. Starting from a very strict monitoring and law enforcement approach towards a more persuasive (via various monetary incentive schemes) and educational stance, it probably realised that a high level of civic consciousness and a personal sense of ownership towards the environment is the sustainable way to go. In Ong *et al.* (2012), the author noted that Singapore spends more than SG$100 million on cleaning and litter removal, compared to SG$9.8 million by the municipal government of Yokohama, Japan, in 2008. Both port cities have comparable economic, demographic and geographical structures and sizes. Much of the difference in expenses is attributed to the attitudes and habits formed by the different approaches to environmental education in the two cities:

> "[In Japan,] all the schools incorporate daily cleaning as part of the school daily activities, plus proper separation of waste. [...] Through these cleaning activities, the basic principles that schools want to impart to the students are: 'Clean up the place you use', 'Cleaning is an important part of the day', and 'Respect the place you use'. In Singapore, such learning on keeping the environment clean is much less active and relegated to the reading of booklets and informational materials. Actual cleaning is left to the professional cleaners." (Ong *et al.*, 2012)

In 2011, the National Climate Change Secretariat of Singapore at the Prime Minister's Office released a public perception survey (NCCS, 2011) conducted on 1010 Singaporeans and Permanent Residents aged 15 years and above. Although the age and background of respondents are different from the study of Tan *et al.* (1998), and questions were focused on climate change, some notions of outcomes on environmental education in the public sphere and formal curriculum can be glimpsed, and comparisons made. Some key findings include: (i) 86% of respondents "felt a sense of responsibility in playing their part to address climate change", while 73.8% are "concerned about climate change"; (ii) 56.3% of the respondents feel that individuals are responsible for taking actions on climate change, 26.3% feel that the government should be responsible, 0.6% think businesses should be responsible to take actions on climate change, 1.5% cited NGOs and 5.3% said "others"; (iii) on trade-offs to address climate change, 58.1% responded that they are willing to "take steps to reduce climate change even if it involves significant cost", compared to 31% who are only willing to spend on changes that involve low costs.

The three perception studies, Tan *et al.* (1998), NCSS (2011) and Ong *et al.* (2012) did not ask the same questions to allow this study to make direct comparisons. However, from their results, at least two points can be understood: firstly, public and formal environmental education and law enforcement have achieved significant results in lowering the most overt pro-environmental indicator — littering, though less may have been achieved in the hearts of the people. Due to the good work of governmental agencies that is frequently reported in the media, there may be now a higher tendency for individuals to shed personal responsibility and rely on the government (4.2% by students in 1998 to 26.3% by Singaporeans 15 years and above in 2011) for environmental issues, e.g., in the provision of cleaners. It is not far-fetched to extrapolate this attitude to general pro-environmental behaviour. Secondly, as the generation which started to gain more environmental knowledge from their formal education starts to gain earning and consumption power, more is expected from businesses to be pro-environmental (10.6% of respondents, in NCSS (2011)), and be willing to spend on eco-friendly products goods and services "even if it involves significant cost". This is a strong indication of pro-environment behavioural choice, though possibly motivated by cost-savings from high electricity and water tariffs, and not environmental values.

On waste recycling in Singapore from 2001 to 2007, Neo (2010) noted that the government's institutional support has worked well for the implementation of the National Recycling Programme (NRP) but it criticised "how the environmentalist rationale for recycling is elided". Neo further noted that the recycling rate has stagnated at 51% from 2004 to 2007, while rates for glass and plastics (which are much more non-biodegradable) declined from 14% in 1997 to 12% in 2007; the increase in paper recycling made up the overall increase in recycling rate. Interestingly, he also made a contrast that while "the NRP reflects an erstwhile technocratic and top-down treatment, which through it addresses issues of convenience of [communitarian] incentives [such as recycle for charity], [it] ultimately neglects the role environmental values play in encouraging environmental behaviours like recycling".

## 5. Discussion and Recommendation

Through the review of perceptions surveys and the comparison with environmental psychology models used in this paper, it is suggested that environmental education in Singapore needs to give additional focus on the following pro-environmental factors:

- Provide meaningful direct experiences (as opposed to vicarious experiences) along with knowledge transmission in schools and public education campaigns
- Provide greater guidance on pedagogy for instilling environmental attitudes and values amongst students

- Encourage culture of positive/negative social feedback on pro/anti-environmental behaviour as perceived consequences of behaviour
- Continual provision of low-cost environmental infrastructure, e.g., recycling bins, incentives for clean-tech businesses, etc.
- Encourage personal responsibility toward environmental outcomes through transfer of environmental ownership and increased personal *locus* of control
- Have more programmes targeted at the individual, particularly on the affective domain

Although personal variables differ from individual to individual, and at first glance do seem amenable to have a directed campaign for influencing citizens at mass, Hines *et al.* (1987) singled out three personal variables that should be harnessed for each individual: attitude, *locus* of control, and personal responsibility. If a person commits an anti-environmental behaviour, he/she will be frowned upon by the various social systems; these social systems then become the eyes of law and the environment's conscience.

*Locus* of control is the perception an individual has over the amount of control s/he has towards the outcome of an issue, as mentioned previously. It is only natural that an individual will invest time and effort towards activities that are perceived to have a satisfactory outcome; otherwise it will be a waste of time and effort. This perception is contingent on two factors in the instance of paper recycling: (1) the fact that a piece of paper placed into the recycling bin will indeed make its way to a paper mill to be given another productive life, is especially reinforced when 'Made from recycled paper' is shown on products; and (2) the belief that his/her individual action of recycling one small piece of paper does help avert global warming. These require an awareness of the product life cycle, best achieved through direct experience, which schools can look into providing. Hence, environmental programmes have to be active, not passive, just as students in Japan have to be responsible for keeping their school environment clean, and not depend on cleaners. Through this, students will extend their sense of responsibility and ownership beyond their immediate surroundings to the larger environment.

## 6. Conclusion

In Singapore, the government takes the lead on environmental efforts, compared to many other countries where governments are compelled into action by environmental disasters and social pressure. As a result, this country stands out in many environmental indicators as far as other national priorities allow, for e.g., the lack of space for competing needs versus nature conservation. Ironically, while citizens of other countries lament about their government's lack of commitment to the environment, Singapore citizens enjoy strong government policies but tend to feel quite removed from personal actions, leaving environmental efforts to the government.

It is clear from this review that the next twenty years of environmental education in Singapore cannot be focused on ever more efficient enforcement measures and top-down initiatives to achieve narrow environmental performance indicators. It has to focus on sensitising its people to take up responsibility for the environment by imbuing attitudes of appreciation and habitual necessity with proper understanding of sustainable living. These require policymakers to make a mentality shift in their engagement from a technocratic approach, to an ecological and sustainability startpoint focused on the individual, so that unprompted and sustainable pro-environmental behaviour can result.

In Singapore and the world, everyone needs to exercise pro-environmental behaviour, and it takes a pro-active and holistic environmental education to achieve this. With hope and determination, Singapore can surely negotiate this discourse and emerge as a model sustainable city.

## 7. Appendices

1. Block-cleaning exercise led by Ang Mo Kio Member of Parliament in 1967.

Photo retrieved from Singapore National Archive, published by *The Straits Times* on 3 December 1967. http://a2o.nas.sg/picas/public/internetSearch/catalogueForm.jsp?id=763358&thesaurusFlag=Y&simpleSearch=clean&photographOption=1&Submit=Submit&pageNumber=40&total=1801

2. Various poster warning public against anti-health and environmental behaviours with threat of hefty fines or corrective work order.

(a) (b) (c)

Photo retrieved from personal blogsites:

(a) http://www.holland-village-singapore.com/singapore-laws/;
(b) http://travellersdiari.blogspot.sg/2013_02_01_archive.html;
(c) http://ramblerwithoutborders.blogspot.sg/2009/07/the-dos-and-donts-in-bangkok-public.html

## References

Adams, W.M. (2006). "The Future of Sustainability: Re-thinking Environment and Development in the Twenty-first Century", *Report of the IUCN Renowned Thinkers Meeting, 29–31 January, International Union for Conservation of Nature.*

ADB and CAI-Asia Center. (2006). *Country Synthesis Report on Urban Air Quality Management — Singapore, Discussion Draft.*, Asian Dvelopment Bank, Phillippines.

Agyeman, J. and Angus, B. (2003). "The Role of Civic Environmentalism in the Pursuit of Sustainable Communities", *Journal of Environmental Planning and Management*, **46**(3), 345–363.

Bechtel, R. and Churchman, A. [Eds.] (2002). *Handbook of Environmental Psychology,* Hoboken, New York, NJ.

Benson, A. (2012). "Great Concert, Trashy Ending", *The New Paper*, 11 December. [Online] Available at: http://news.asiaone.com/News/Latest%2BNews/Showbiz/Story/A1Story20121210-388794.html [Accessed 25 July 2013].

Bhattacharya, C.B., Sen, S. and Korschun, D. (2008). "Using Corporate Social Responsibility to Win the War for Talent", *MIT SLOAN MANAGEMENT REVIEW*, Winter, 37–44., Massachusetts Institute of Technology, USA.

Bickman, L. (1972). "Environmental Attitudes and Actions", *The Journal of Social Psychology*, **87**(2), 323–324.

Bradley, J.C., Waliczek, T.M., and Zajicek, J.M. (1999). "Relationship Between Environmental Knowledge and Environmental Attitude of High School Students", *The Journal of Environmental Education*, **30**(3), 17–21.

Canter, D.V., and Craik, K.H. (1981). "Environmental psychology". *Journal of Environmental Psychology*, **11**, 1–12.

Chang, C.H. (2013). "Advancing a framework for climate change education in Singapore through Teacher Professional Development", *HSSE Online,* **2**(1), 28–35.

Connell, S., and Fien, J. (1999). "If It Doesn't Directly Affect You, You Don't Think About It: a qualitative study of young people's environmental attitudes in two Australian cities", *Environmental Education Research*, **5**(1) 95–113,

Corbett, J.B. (2005). "Altruism, Self-Interest, and the Reasonable Person Model of Environmentally Responsible Behaviour", *Science Communication*, **26**(4), 368–389.

Curriculum Planning & Development Division (2001), *Primary Science Syllabus,, Ministry of Education, Singapore.*

De Young, R. (2013). "Environmental psychology overview" in Klein, S.R., and Huffman, A.H. (Eds.) *Green Organizations: Driving Change with 10 Psychology,* Routledge Academic, New York, pp. 17–45.

Department of Statistics (2013). *Singapore in figures 2013,* Singapore Department of Statistic, Singapore.

Diekmann, A. and Preisendoerfer, P. (1992). "Persoenliches Umweltverhalten: Die Diskrepanz zwischen Anspruch und Wirklichkeit, Koelner Zeitschrift fuer Soziologie und Sozialpsychologie (trans, *Journal of Sociology and Social Psychology* in German)", **44**, 226–251.

ENV (Ministry of the Environment). (1992a). *Singapore's National Report for the 1992 UN Conference on Environment and Development Preparatory Committee,* Ministry of the Environment, Singapore.

ENV (Ministry of the Environment). (1992b). *The Singapore Green Plan: Towards a Model Green City,* Ministry of the Environment, Singapore.

Fein, J. (2007). "Young people and the environment: Implications of a study of youth environmental attitudes and education in the Asia-Pacific region for curriculum reform during the UN decade of education for sustainable development", *Internationale Schulbuchforschung*(trans. *International Textbook Research* in German), **29**(2), 205–227.

Fietkau, H.J., and Kessel, H. (1981). *Umweltlernen* (trans. *Environmental Learning* in German), Hain, Königstein/Taunus.

Geh, M. (2008). "Growing Green Space, Civil Society and Government", *Social Space*, Singapore Management University, Singapore, **1**(1) 84–89.

Gifford, R. (2007). "Environmental Psychology and Sustainable Development: Expansion, Maturation, and Challenges", *Journal of Social Issues*, **63**1, 199–212

Heath, C., and Heath, D. (2010). *Switch: How to Change Things When Change Is Hard*, New York: Random House.

Hines, J.M., Hungerford, H.R., and Tomera, A.N. (1986–87). "Analysis and synthesis of research on responsible pro-environmental behavior: a meta-analysis", *The Journal of Environmental Education*, **18**(2), 1–8.

Hobson, K. (2005). "Considering "Green" Practices: NGOs and Singapore's Emergent Environmental-Political Landscape", *Journal of Social Issues in Southeast Asia*, **20**(2), 155–176.

Inter-Ministerial Committee on Sustainable Development. (2009). *Blueprint on Sustainable Development Singapore — A lively and liveable Singapore: Strategies for Sustainable Growth*, Ministry of Environment and Water Resources and Ministry of National Devrlopment, Singapore.

Kaplan, S. (2000). "Human Nature and Environmentally Responsible Behaviour", *Journal of Social Issues*, **56**(3), 491–580.

Kollmuss, A., and Agyeman, J. (2002). "Mind the Gap: Why do people act environmentally and what are the barriers to pro-environmental behavior?", *Environmental Education Research*, **8**(3), 239–260.

Kuhlemeier, H., Bergh, H., and Lagerweij, N. (1999). "Environmental knowledge, attitudes, and behavior in Dutch secondary education", *The Journal of Environmental Education*, **30**(2), 4–14.

Lee, C. (1994). Developing Environmental Consciousness in Singaporein Filho, W.L. (Eds), *Environmental Education in Small Island Developing States*, pp. 133–145. Vancouver: Commonwealth of Learning.

Leitmann, J. (1999). "Integrating the environment in urban development: Singapore as a model of good practice", *The World Bank, Urban Waste Management, Working paper series 7.*, Washington D.C, USA.

MEWR (2006). *The Singapore Green Plan 2012, 2006 Edition*, Ministry of Environment and Water Resources, Singapore.

MEWR (2012). *Key Environmental Statistics 2012*, Ministry of Environment and Water Resources, Singapore.

National Environment Agency, Singapore. (2009). *Environment Education in Singapore*, United Nations Framework Convention on Climte Change. Bonn, Germany.

NCCS (2011). *Public Perception Survey 2011, Annex B*, National Climate Change Secretariat, Singapore.

NEA (2013). *Factsheet on Littering.* [Online] Available at: http://app2.nea.gov.sg/docs/default-source/corporate/cos-2013/factsheet-on-littering.pdf?sfvrsn=0 [Accessed 31 July 2013].

NEA. (2013). *News Release — NEA Launches First Ever Youth for the Environment Day Forum to Crowdsource Solutions for a Greener Singapore.* [Online] Available at: http://app2.nea.gov.sg/corporate-functions/newsroom/news-releases/year/2013/month/4/nea-launches-first-ever-youth-for-the-environment-day-forum-to-crowdsource-solutions-for-a-greener-singapore- [Accessed 31 July 2013].

Neo, H. (2007). "Challenging the developmental state: Nature conservation in Singapore", *Asia Pacific Viewpoint*, **48**(2), 186–199.

Neo, H. (2010). "The Potential of Large-Scale Urban Waste Recycling: A Case Study of the National Recycling Programme in Singapore", *Society & Natural Resources: An International Journal*, **23**(9), 872–887.

Neo, H., and Savage, V.R. (2002). "Shades of Green, Fields of Gold: Representations, discourse and the politics of golf in Singapore", *Landscape Research*, **27**(4), 397–411.

Ong, I.B.L., and Sovacool, B.K. (2012). "A comparative study of littering and waste in Singapore and Japan" Resources *Conservation and Recycling*, **61**1, 35–42.

Perry, M., and Teng, T.S. (1999). "An overview of trends related to environmental reporting in Singapore", *Environmental Management and Health*, **10**(5), 310–320.

Perry, M., and Singh, S. (2001). "*Corporate Environmental Responsibility in Singapore and Malaysia. The Potential and Limits of Voluntary Initiatives" Technology, Business and Society Programme Paper Number 3. United Nations Research Institute for Social Development*, Geneva, Switzerland.

Proshansky, H.M. (1987). "The field of environmental psychology: securing its future", in Stokols, D. and Altman, I. (Eds) *Handbook of environmental psychology*, pp. 1467–1488. New York: John Wiley & Sons.

Savage, V.R. and Lau, S. (1993). "Green issues: Official policies and student awareness" in Briffett, C. and Sim, L.L. (Eds.) *Environmental Issues in Development and Conservation*. pp. 13–28, School of Building and Estate Management, National University of Singapore, Singapore.

Schultz, P.W. (2001). "The Structure Of Environmental Concern: Concern For Self, Other People, And The Biosphere", *Journal of Environmental Psychology*, **214**, 327–339.

Stokols, D., and Altman, I. (Eds.) (1987). *Handbook of Environmental Psychology*, Vol 1 & 2. New York John Wiley.

Steg, L., and Vlek, C. (2009). "Encouraging pro-environmental behavior: An integrative review and research agenda", *Journal of Environmental Psychology*, **291**, 309–317.

Tan, C.H. (2012) "Safeguard, nurture, cherish our environment, Singapore" National Environment Agency, Singapore.

Tan, G.C.I, Lee, K.E.C., and Goh, K.C. (1998). "A survey of environmental knowledge, attitudes and behaviour of students in Singapore.", *International Research in Geographical and Environmental Education*, **7**(3), 181–202.

Tan, L.H.H., and Neo, H. (2009). ""Community in Bloom": local participation of community gardens in urban Singapore, Local Environment", *The International Journal of Justice and Sustainability*, **14**(6), 529–539.

*The Straits Times*(1967). "MPs lead 1,000 in 'keep clean' campaign", *The Straits Times*, 30 July, p. 14. [Online] Available at: http://newspapers.nl.sg/Digitised/Article/straitstimes19670730-1.2.103.aspx [Accessed 4 July 2013].

UNESCAP. (2000). *Education, Information and Awareness*, Chapter 15, pp. 318–338, United Nations Economic and Social Commission for Asia and the Pacific, Bangkok, Thailand.

UNESCO. (1977). *Final Report, Intergovernmental Conference on Environmental Education*, United Nations Educational, Scientific and Cultural Organisation, Tbilisi (USSR).

URA. (2010). *URA Lifestyle Survey 2009*, Urban and Redevelopment Authority, Singapore.

Wong, F.B.K., and Stimpson, P. (2003). "Environmental Education in Singapore: A Curriculum for the Environment or in the National Interest?", *International Research in Geographical and Environmental Education*, **12**(2), 123–138.

World Wildlife Fund (2013). *International standards for Singapore schools with new environmental education programme.* [Online] Available at: http://www.wwf.sg/news_stories/?uNewsID=208422 [Accessed 20 July 2013].

Yang, G., Lam, C.C., and Wong, N.Y. (2010). "Developing an Instrument for Identifying Secondary Teachers' Beliefs About Education for Sustainable Development in China", *The Journal of Environmental Education*, **41**(4), 195–207.

Yeo, B.L. (1994). "The Singapore green plan" in: *AMICECOS Seminar on Urbanization, Sustainable Development and the Media: Singapore*, Apr 8–9, Asian Media Information and Communication Centre, Singapore.

# 15

# ENVIRONMENTAL LITERACY AMONG SECONDARY THREE STUDENTS IN A SINGAPORE SECONDARY SCHOOL

LOO HUI MIN and HARVEY NEO

This study was designed to investigate the level of environmental literacy in a Singapore secondary school. The aspects of knowledge, attitudes and behaviours among learners in environmental education in schools today provide important information for advancing environmental education tomorrow. This study seeks to establish baseline information that will shape recommendations on how environmental education can play a more focal role within a school curriculum, especially within Singapore's education system, where there is a renewed emphasis on 21st Century learners' competency skills and Character and Citizenship Education (CCE) in schools. A sample size of seventy four Secondary three students participated in this survey discussing the three aspects mentioned earlier. The results were analysed using SPSS software. The results showed that students who scored high in the environmental knowledge test tend to have informed attitudes towards the environment, and are more likely to engage in certain environmental-friendly behaviours.

## 1. Introduction

Environmental education aims to foster public awareness and concern about environmental issues and solutions by providing people with opportunities to acquire the knowledge, values, attitudes, commitment and skills needed to investigate issues, solve problems, and to protect and improve the environment (UNESCO, 1978). Over the years, the objectives and principles of environmental education have been translated into policies, programmes and resources at local and national levels.

The creation of an environmentally literate citizenry is an important aim of environmental education (Disinger and Roth, 1992). Environmental literacy is a prerequisite to the maintenance of, and improvement in, the quality of the environment, and it seeks to develop an active and informed citizenry committed to the values and practices of ecological and social sustainability (Fien, 1993). Volk, Hungerford and Tomera underscored the importance of environmental literacy as a need for a "citizenry that is competent to take action on critical environmental issues and with a will to take action" (Loubser *et al.*, 2001). In a study done by Hines, Hungerford, and Tomera (Minnesota Report Card on Environmental Literacy, 2008), they confirmed that besides attitudes and behavioural control, personal moral norm is a third predictor of pro-environmental behaviour intention where "problem awareness is an important but indirect determinant of pro-environmental intention". A study by Ballantyne *et al.* (1998) showed that students play an important role as "catalysts as they are tomorrow's leaders and stewards of the Earth".

Singapore is located 137 kilometres north of the Equator, at the tip of Peninsular Malaysia. With a land area of $714.3\,km^2$, and a population of 5.31 million (estimated figures for mid-2012), this makes Singapore one of the most densely populated nations in the world (Department of Singapore Statistics, 2012). Despite a lack of natural resources, Singapore used its strategic location, natural deep harbour, and its people to develop a strong economy based on trade and services. In recent years, Singapore has emphasised research and development in biomedical sciences, water and environmental technology, health care services, educational services, info-communications and the media, logistics and transport engineering.

Singapore's air and water quality are well within World Health Organisation's (WHO) standards (National Environment Agency, 2012). As a member of the global community, Singapore plays its part in translating her environmental conservation commitments into plans and actions.

As a party to the Kyoto Protocol on 11 July 2006, Singapore developed its National Climate Change Strategy in 2008 (National Climate Change Secretariat, 2012). Plans in the strategy include mitigating greenhouse gases emissions and assessing adaptation measures to address the impacts of climate change. As part of the commitment under Article 6 of the Convention on Biological Diversity (CBD), Singapore developed the National Biodiversity Strategy and Action Plan (National Parks Board, 2012).

The Sustainable Singapore Blueprint 2009[1] sets out the targets and four strategies for sustainable development by 2030. This is done through a 3P (People, Private and Public) partnership. The Singapore Green Plan addresses Singapore's commitment towards environmental sustainability. The Singapore Green Plan 2012 was crafted in 2002 and then revised in 2006. For this report, the Singapore Green Plan 2012 (2002

[1]A revised Sustainable Blueprint is expected to be released at the end of 2014.

and 2006 editions) will be referred to synonymously. Under SGP 2012, environmental education has been emphasised as a direction for the building of an environmentally proactive society (Ministry of Environment, 1993). Focusing on pollution prevention and the sustainable use of resources, the Ministry of the Environment and Water Resources (MEWR) seeks to raise public awareness on various environmental issues through public education, campaigns and environmental education in schools.

The Ministry of Education (MOE) writes the programme for environmental education in schools. As in many other educational systems, this is primarily through social studies, geography and science in secondary schools and through social studies in primary schools (Curriculum Planning and Development Division, CPDD, 2005; 2006; 2007). Singapore Environment Council has been organising the annual Schools Green Audit Awards since 2000, to encourage schools to cut down on energy and water wastage (Singapore Environment Council, 2009).

Since 1992, when Singapore presented the Green Plan at the Earth Summit, several studies were done to look at how aware the younger generation was of environmental issues. In 1993, the findings from a National University of Singapore (NUS) study showed that students' commitments to environmental-friendly behaviours were related to the level of environmental knowledge (Savage and Lau, 1993). A group of Science researchers from the National Institute of Education revealed that while teachers and students showed ready acceptance of environmental education topics in the curriculum, researchers felt that environmental education was unlikely to be seen as an independent subject at the time (Toh *et al.*, 1997). Similarly, a survey conducted by researchers in 1998 found that students gained most of their environmental knowledge from out-of-school sources rather than from general education at school (Tan *et al.*, 1998). From the interviews conducted with key agency officers and teachers involved in the development of Singapore's environmental education, Wong and Stimpson (2003) found that education is based on conventional disciplinary knowledge, with a practical approach for environmental concern. These factors are influenced by government consideration for its society.

Since the time of the surveys, significant curriculum innovations have taken place. More recently, there has been an emphasis on CCE that is rolled out in national schools. In lieu of the recent curriculum changes, it is timely to provide baseline information on the levels of environmental literacy in a Singapore secondary school.

## 2. Environmental Education: Origins, Definitions and Goals

Humans depend a lot on environmental capital by using it as both a source and a sink. Natural ecosystems provide essential life support services of climate regulation, air purification, waste decomposition, freshwater, food production, recreational enjoyment and aesthetic fulfilment (Millennium Ecosystem Assessment, 2005; Withgott and Laposata, 2012). In addition, the environment is a receptacle for wastes generated by societal consumption of environmental resources. For instance, natural vegetation act as sinks for carbon dioxide as they grow. Anthropogenic activities are causing increased rates and types of environmental changes, from local to global scales. These have resulted in environmental, economic and social challenges which have become part of everyone's daily lives (Francis, 2012; Kreike, 2010).

The past few decades have seen greater international emphasis on the educational dimensions of issues relating to environment and sustainable development. Since the 1960s, statements from international organisations such as UNESCO[2] and IUCN[3] have called for environmental problems to be included in environmental education. At the IUCN/UNESCO International Working Meeting on Environmental Education in School Curriculum held in 1970, one of the most widely used definitions of environmental education was formulated:

> Environmental education is the process of recognizing values and clarifying concepts in order to develop skills and attitudes necessary to understand and appreciate the interrelatedness among man, his culture and his biophysical surroundings. Environmental education also entails practice in decision-making and self-formulation of a code of behaviour about issues concerning environmental quality (IUCN, 1970).

Principle 19 of the 1972 United Nations (UN) Conference on Human Environment in Stockholm (UN, 1972) declared that "education in environmental matters, for the younger generation as well as adults, giving due consideration to the underprivileged, is essential in order to broaden the basis for an enlightened opinion and responsible conduct

[2]The United Nations Educational Scientific and Cultural Organisation (UNESCO) is an agency within the United Nations. Through international cooperation on educational, science and cultural programmes, it aims to promote social justice, peace, human rights and international security.
[3]International Union for the Conservation of Nature (IUCN) is a global conservation group that researches threatened species and coordinates practical conservation plans.

by individuals, enterprises and communities in protecting and improving the environment in its full human dimension". This conference reflected increasing global interest for the environment and led to the establishment of the UNEP.[4] Subsequent events in Belgrade and Tbilisi shaped the goals and nature of environmental education. The Belgrade Charter (UNESCO, 1975) produced the first inter-governmental statement on environmental education, listing down the aims, objectives, key concepts and guiding principles.

The plan to have different governmental representations in environmental policy and debate was materialised in Tbilisi, Georgia, USSR in 1977. At this event, many recommendations were made to apply the applications of environmental education in the formal and non-formal education (Palmer, 1998). The Tbilisi Report, based on the framework of the Belgrade Charter, has been the seminal influence on the development of environmental education and polices around the world. The five objectives under the Tbilisi Declaration are shown in Table 15.1.

Along with the upsurge of interest in environmental education, related fields of 'educations' have come to the fore: development education, citizenship education and education for sustainable development.

Table 15.1. Objectives of Environmental Education (UNESCO, 1978).

| Objectives | Actions |
|---|---|
| Awareness | To help social groups and individuals acquire an awareness of and sensitivity to the total environment and its allied problems. |
| Knowledge | To help social groups and individuals gain a variety of experience in, and acquire a basic understanding of, the environment and its associated problems. |
| Attitudes | To help social groups and individuals acquire a set of values and feelings of concern for the environment, and the motivation for actively participating in environmental improvement and protection. |
| Skills | To help social groups and individuals acquire the skills for identifying and solving environmental problems. |
| Participation | To provide social groups and individuals with an opportunity to be actively involved at all levels in working towards the resolution of environmental problems. |

[4]The United Nations Environment Programme (UNEP) is an international organisation established in 1972 to coordinate and spur activities to increase scientific understanding of environmental change and develop environmental management tools.

The different threads of environmental education have been generally seen in terms of a 'three-fold' structure:

- Education *about* the environment: basic knowledge and understanding of the environment and our complex relationships with it.
- Education *in* or *from* the environment: using the environment as a resource with emphasis both on planned inquiry and investigations, and also on providing opportunities for students to engage in first-hand experiences within it.
- Education *for* the environment: concerned with the values, attitudes and positive actions.

Palmer has attempted to present this in a model as shown in Figure 15.1 below.

The strategic objectives of ASEAN in addressing environmental education is found under Part II, section D3, paragraph 33 of the 2009 Blueprint on the ASEAN Socio-Cultural Community (2009). Regional cooperation on the environment in ASEAN dates back to 1977 with the drafting of the first ASEAN Sub-regional Environment Programme (ASEP I). This was followed by ASEP II (1982; 1987) and ASEP III (1988; 1992). ASEAN, in response to the international commitment, also

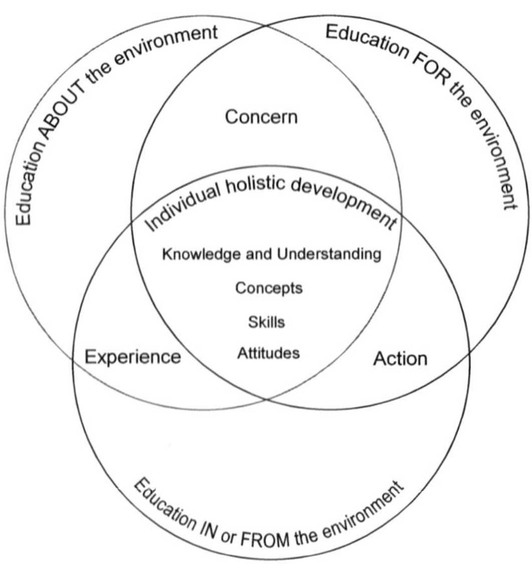


Fig. 15.1. Model for teaching and learning in environmental education (Palmer, 1998: 142).

formulated an Environmental Education Action Plan, i.e., AEEAP (2000; 2005) and AEEAP (2008; 2012). which serves as framework for the development and implementation of environmental education activities in ASEAN. Among the concerns of ASEAN member countries is how to promote greater awareness of local/national environmental problems among the population and the need for more baseline information on the extent to which national curricula in the basic education system include environmental education as a key integrating component for achieving sustainable development in the region.

### *Environmental Literacy and Environmental Education*

The terms environment and literacy were first coupled by Charles E. Roth in 1968 (Roth, 1992). Differing definitions of environmental literacy has emerged since the 1960s. This report adapts the definition by Reynolds, where environmental literacy is referred to as "an understanding of the environmental, social and economic dimensions of human-environment interactions, and the skills and ethics to translate this underlying understanding into life choices that promote the sustainable flourishing of diverse human communities and the ecological systems within which they are embedded" (Reynolds *et al.*, 2010). In other words, if citizens were to make sound decisions to improve the quality of our environmental capital, then environmentally literate citizenry is important.

It can be said that these definitions of environmental literacy fall within the five categories of objectives of environmental education (Table 15.1) and the guiding principles of environmental education. A study by Hines *et al.* (1986), which was based on Ajzen and Fishbein's theory of planned behaviour, found that knowledge of issues, knowledge of action strategies, locus of control, attitudes, verbal commitment, and an individual's sense of responsibility were associated with pro-environmental behaviours. Bamberg and Moser (2007) found similar relations between psycho-social determinants of pro-environmental behaviour.

## 3. School Environment: Singapore Education System, School Syllabus and Activities

Through the years, education has been instrumental in building capabilities for Singapore's economic growth, and also as a means of building social cohesion. Singapore's education system aims to nurture every child and help students to discover their talents, realise their fullest potential and develop a passion for life-long learning. Among the key strengths of the Singapore education system is holistic learning, focused on academic and non-academic areas (MOE, 2012). A primary school student spends, on average, about 30 hours in school per week, while a secondary school student may spend on average of 47. Therefore, the influence that the school has on the students is significant in affecting the students' attitudes and behaviours.

In recognising the new global challenges in the knowledge economy, MOE has introduced Curriculum 2015 (C2015) that develops student attributes, embedded in the "Confident Person", "Self-directed Learner", "Active Contributor", and "Concerned Citizen". A new curricular initiative, CCE, emphasizes values, social emotional competencies and civic literacy, global awareness and cross-cultural skills that are critical for character and citizenship development. The relative importance of the triple bottom line is best inferred from the CCE learning outcomes (MOE, 2011).

Under SGP 2012, environmental education has been emphasised as a direction for the building of an environmentally proactive society. MOE develops environmental education in schools based on environmental issues identified by MEWR. Environmental topics are infused into different school subjects in primary, secondary and post-secondary education.

Environmental activities organised by teachers or organisations spur students' interests in environmental issues. Schools can adopt a park or part of a beach (Sustainable Singapore Blueprint, 2009). Students in these participating schools are responsible for the maintenance and cleaning of these stretches of environment as part of their Values in Action (VIA) Programme. Learning Journeys (LJs) are also conducted as a way to enrich the educational experience. Some of the participating institutions include the National Environment Agency (NEA), National Parks Board (NParks) and Civil Aviation and Authority of Singapore (MOE, 2012).

Co-curricular Activities (CCA) in school also incorporate environmental activities into the Student Development Plan. Working landscapes such as farms and landfills allow students to contextualise their learning experience in a real-world setting (MOE, 2012). Seminars and competitions also help to heighten students' interests in environmental issues.

## 4. Methodology

A group of secondary three students in a Singapore secondary school was the target audience for the voluntary and anonymous written survey. Stratified random sampling was used in order to represent the existing education streams, i.e., Express, Normal Academic and Normal Technical. Requests were sent to the MOE Data Collection Centre and NUS Institutional Review Board (NUSIRB) prior to the start of the survey. In line with the MOE guidelines in conducting research in schools, the sample size was kept below 10% of the target population in the school. The survey results from the sample size of 74 students were analysed using SPSS software.

The survey instrument included similar questions from various sources (The Third Minnesota Report Card, 2002; 2004; 2008; Tan *et al.*, 1998; Williams, 1996). Questions were also developed specifically for this survey. The survey questions were divided into three sections:

*Section A*: To collect data about environmental knowledge, students were asked to indicate how much ***they think they know*** and ***actually know*** about the environment.

*Section B*: In order to collect data about environmental attitudes, students were posed a series of questions concerning their ***attitudes towards*** the environment.

*Section C*: To collect data about environmental behaviours, students were asked a series of questions about their environmental actions.

## 5. Results and Discussion

### *Environmental Knowledge*

Based on the general environmental knowledge questions, approximately 93% of the secondary three students have at least an average or basic knowledge about the environment (Figure 15.2). A score of seven or higher was used as a measure of average or basic knowledge (at least a grade of C). As the same set of general knowledge questions were not used, direct comparisons cannot be made at a group level. Instead, comparisons were made on individual questions (Figure 15.3).

Less than one-tenth of the respondents indicated they were highly familiar with the laws and related to the various environmental issues (Table 15.2). It is not surprising, given that at the time of the survey, there was no definitive content coverage of laws and regulations within the curriculum. These environmental regulations were also not widely published in the mass media.

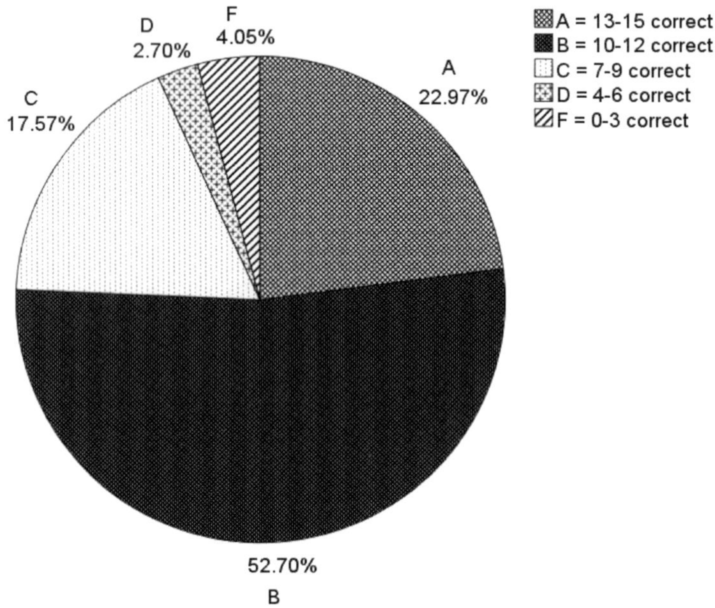


Fig. 15.2. The distribution of correct answers by secondary three students on fifteen general environmental knowledge questions.

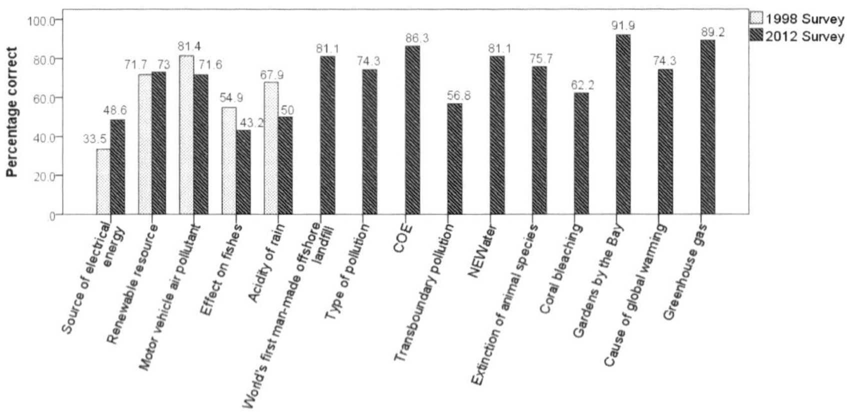


Fig. 15.3. Percentage of correct answers on fifteen general environmental knowledge questions used in the 1998 environmental awareness survey and 2012 survey.

In terms of the sources of environmental information (Tables 15.3 and 15.4), nearly half of the students reported that they always get information from the Internet (43.2%) and television (45.9%). The prevalence of technology and ease of getting information from these two sources could

Table 15.2. Number of secondary three students who report each level of familiarity with laws and regulations for specific environmental topics (Figures in brackets show the percentages).

| Environmental laws and regulations | High familiarity | Moderate familiarity | Low familiarity |
|---|---|---|---|
| Solid waste management | 7 (9.5%) | 26 (35.1%) | 41 (55.4%) |
| Air pollution | 4 (5.4%) | 33 (44.6%) | 37 (50.0%) |
| Water resource management | 4 (5.4%) | 36 (48.6%) | 34 (46.0%) |
| Nature conservation | 7 (9.5%) | 36 (48.6%) | 31 (41.9%) |
| Energy conservation | 4 (5.4%) | 30 (40.5%) | 40 (54.1%) |

Table 15.3. Frequency of using the various sources to get environmental information (Figures in brackets show the percentages).

| | Never | Seldom | Sometimes | Always |
|---|---|---|---|---|
| a) Newspapers (hard, soft copy) | 3 (4.1%) | 30 (40.5%) | 32 (43.2%) | 9 (12.2%) |
| b) Magazines (hard, soft copy) | 14 (18.9%) | 28 (37.8%) | 27 (36.5%) | 5 (6.8%) |
| c) Internet | 1 (1.4%) | 17 (23%) | 24 (32.4%) | 32 (43.2%) |
| d) Television | 4 (5.4%) | 4 (5.4%) | 32 (43.2%) | 34 (45.9%) |
| e) Radio | 19 (25.7%) | 32 (43.2%) | 19 (25.7%) | 4 (5.4%) |
| f) Conversation with friends or neighbours | 15 (20.3%) | 16 (21.6%) | 29 (39.2%) | 14 (18.9%) |
| g) Conversation with family members or guardians | 12 (16.2%) | 15 (20.3%) | 34 (45.9%) | 13 (17.6%) |
| h) Special environmental programmes in school | 7 (9.5%) | 25 (33.8%) | 37 (50%) | 5 (6.8%) |
| i) General education at school | 6 (8.1%) | 20 (27%) | 39 (52.7%) | 9 (12.2%) |
| j) Attending talks/exhibitions organised by other organisations | 24 (32.4%) | 18 (24.3%) | 25 (33.8%) | 7 (9.5%) |

be possible reasons why these were the most preferred choices. In terms of depth of environmental information gathered from the different sources, nearly 76% and 70% of the students reported that they got the information through programmes organised by the school or through formal curriculum in class. Lower levels of reporting are observed for conservation groups and the government agencies (i.e. 6.8% respectively). This could be due to lack of direct contact from these organisations.

### *Environmental Attitudes*

It is interesting to note that while nearly 42% to 55% (Table 15.2) of the students reported that they had low familiarity about the Singapore's

Table 15.4. Level of environmental information gathered from the sources (Figures in brackets show the percentages).

| | No information | Little information | Some information | Most of the information |
|---|---|---|---|---|
| a) Programmes organised by the school, e.g. LJs. | 1 (1.4%) | 17 (23%) | 41 (55.4%) | 15 (20.3%) |
| b) Subject(s) matter taught by teachers in class | 0 (0%) | 22 (29.7%) | 44 (59.5%) | 8 (10.8) |
| c) Conservation groups, e.g. Nature Society | 6 (8.1%) | 30 (40.5%) | 30 (40.5%) | 8 (6.8%) |
| d) Government agencies | 9 (12.2%) | 25 (33.8%) | 35 (47.3%) | 5 (6.8%) |

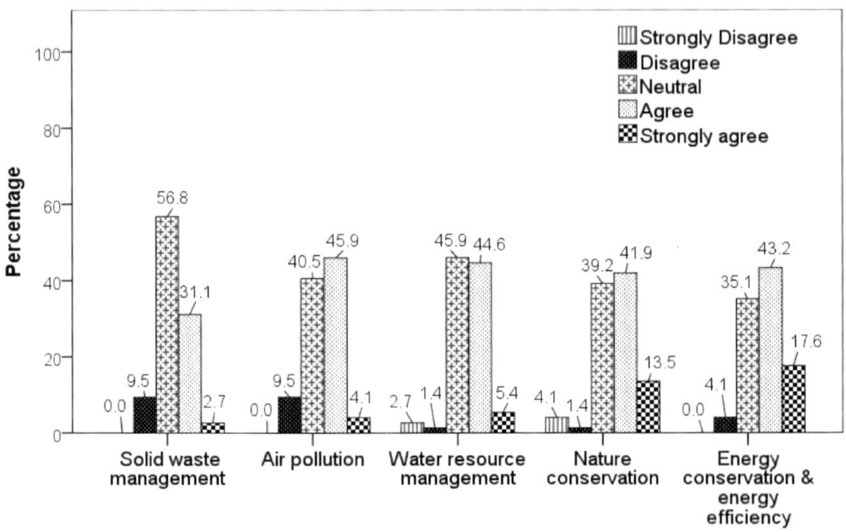


Fig. 15.4. Responses of students on the adequacy of the laws and regulations concerning solid waste management, air pollution, water resource management, nature conservation, energy conservation and energy efficiency in Singapore.

environmental laws and regulations, 31% to 46% of the students agreed that environmental legislation was adequate (Figure 15.4). Possible reasons for this observation might be that the students assumed the government took necessary steps in environmental protection legislation.

All students have indicated that they have taken deliberate action to improve the environment (Table 15.5a). 78.3% felt positive about the experiences and 49% of the students reported a 'very strong' or 'strong' desire to improve the environment (Tables 15.5b and c). The survey also

Table 15.5a. Responses of students when asked if they have taken actions to improve the environment (Figures in brackets show the percentages).

| | Yes | No |
|---|---|---|
| Have you taken any action to improve the environment? | 74 (100.0%) | 0 (0.0%) |

Table 15.5b. Feelings about their actions towards improving the environment (Figures in brackets show the percentages).

| | I regret being involved | I only feel okay about it | I feel happy the about experience | I feel really good and motivated to do more |
|---|---|---|---|---|
| How would you describe feelings about your actions improving the towards environment? | 1 (1.4%) | 15 (20.3%) | 40 (54.1%) | 18 (24.3%) |

Table 15.5c. Desire to improve the environment (Figures in brackets show the percentages).

| | Very weak | Weak | Medium | Strong | Very strong |
|---|---|---|---|---|---|
| Rate the desire to be involved in improving the environment | 0 (0.0%) | 1 (1.4%) | 37 (50.0%) | 32 (43.2%) | 4 (5.4%) |

Table 15.5d. Responses towards provision of environmental education in school (Figures in brackets show the percentages).

| | Yes | No |
|---|---|---|
| Do you think environmental education should be provided in your school? | 71 (95.9%) | 3 (4.1%) |

found that 95.9% of the students believe that schools should provide for environmental education (Table 15.5d). The majority of respondents indicated they thought financial support for environmental education should come from the government (Figure 15.5).

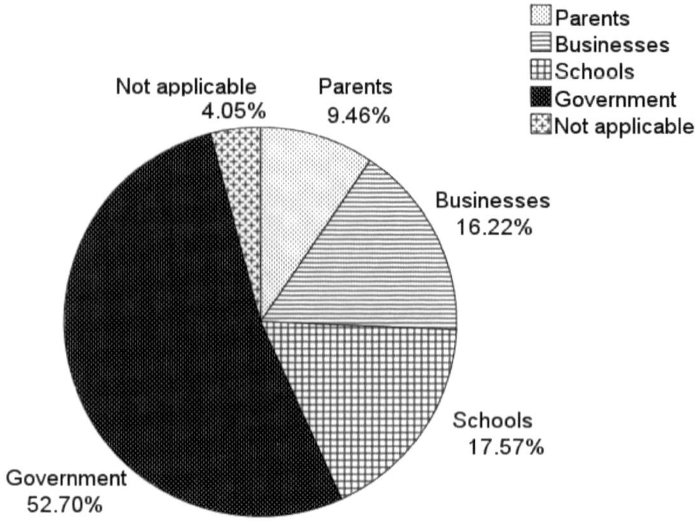


Fig. 15.5. Responses towards provision of funds for environmental education in school (Figures in brackets show the percentages).

### *Environmental Behaviours*

As with all self-reported behaviours, some caution is required in interpretation as there may not be a perfect correspondence between what people say they do and what they actually do.

Nearly 67.7% and 56.8% of those surveyed reported that they undertook most action in energy and water conservation (Table 15.6). In terms of reported buying behaviour, it is encouraging that 73% and 58% of the students indicated that they would consider buying products with green labels and products with lesser packaging respectively. In cases where a lower percentage of lack of action was reported, a greater percentage of students surveyed indicated that they would consider engaging in environmentally-friendly activities.

### *Environmental Literacy: Integrating Knowledge, Attitudes and Behaviours*

Respondents were asked how much they themselves feel they know about environmental issues and problems, where responses ranged from 1= "a lot," 2= "a fair amount," 3= "only a little," or 4= "not at all". Responses were then combined into three categories: "high" self- reported knowledge consisted of response 1, "moderate" self-reported knowledge was

Table 15.6. Self-reported environmental-related behaviours (Figures in brackets show the percentages).

| | Have done in the last 12 months | Would consider doing | Would not doing |
|---|---|---|---|
| a) Deciding for environmental reasons to re-use or recycle something instead of throwing it away | 35 (47.3%) | 36 (48.6%) | 3 (4.1%) |
| b) Making an effort to buy products with green labels | 18 (24.3%) | 54 (73.0%) | 2 (2.7%) |
| c) Taking part in a clean-up campaign or anti-litter scheme of some kind | 23 (31.1%) | 46 (62.2%) | 5 (6.8%) |
| d) Making a report or complaint you thought was bad for the environment | 11 (14.9%) | 47 (63.5%) | 16 (21.6%) |
| e) Making an effort to reduce the water usage in your home or school, e.g. turn off the tap when you are brushing your teeth | 42 (56.8%) | 28 (37.8%) | 4 (5.4%) |
| f) Making an effort to switch off electrical appliances when not in use or when you leave the room | 50 (67.7%) | 23 (31.1%) | 1 (1.4%) |
| g) Trying to encourage someone else to change an activity or practice you thought was harmful for the environment | 21 (28.4%) | 48 (64.9%) | 5 (6.8%) |
| h) Trying to get information for your own interest on a topic which you thought was relevant to protecting the environment | 9 (12.2%) | 61 (82.4%) | 4 (5.4%) |
| i) Making an effort to buy products that use less packaging | 27 (36.5%) | 43 (58.1%) | 4 (5.4%) |

response 2, and "low" self-reported knowledge consisted of responses 3 and 4.

Self-reported knowledge is not perfectly related to the score on the combined knowledge test (Table 15.7). 76.5% of the students who earned grade A on the knowledge test reported that they have moderate environmental knowledge. This percentage fluctuates as the grades earned on the knowledge test drops.

Table 15.7. Students' combined environmental knowledge grade and their self-reported knowledge of environmental issues and problems (Figures in brackets show the column percentages).

| Self-reported knowledge | Combined environmental knowledge grade/number correct* | | | | | Total |
|---|---|---|---|---|---|---|
| | A (13–15 correct) | B (10–13 correct) | C (7–9 correct) | D (4–6 correct) | F (0–3 correct) | |
| High | 0 (0%) | 8 (20.5%) | 1 (7.7%) | 0 (0%) | 0 (0%) | 9 (12.2%) |
| Moderate | 13 (76.5%) | 27 (69.2%) | 12 (92.3%) | 2 (100.0%) | 3 (100.0%) | 57 (77.0%) |
| Low | 4 (23.5%) | 4 (10.3%) | 0 (0%) | 0 (0%) | 0 (0%) | 8 (10.8%) |
| Total | 17 (100.0%) | 39 (100.0%) | 13 (100.0%) | 2 (100.0%) | 3 (100.0%) | 74 (100.0%) |

**Score out of fifteen combined environmental knowledge questions.*

Table 15.8. Students' environmental knowledge grades and environmental attitudes (Figures in brackets show the column percentages).

| Environmental attitudes | Combined environmental knowledge grade/number correct* | | | | | Total |
|---|---|---|---|---|---|---|
| | A (13–15 correct) | B (10–13 correct) | C (7–9 correct) | D (4–6 correct) | F (0–3 correct) | |
| High | 15 (88.2%) | 32 (82.1%) | 13 (100.0%) | 2 (100.0%) | 3 (100.0%) | 65 (87.8%) |
| Moderate | 2 (11.8%) | 7 (17.9%) | 0 (0%) | 0 (0%) | 0 (0%) | 9 (12.2%) |
| Low | 0 (0%) | 0 (0%) | 0 (0%) | 0 (0%) | 0 (0%) | 0 (0%) |
| Total | 17 (100.0%) | 39 (100.0%) | 13 (100.0%) | 2 (100.0%) | 3 (100.0%) | 74 (100%) |

**Score out of fifteen combined environmental knowledge questions.*

The 'environmental attitude scale' and 'environmental behaviour scale' were used to examine the possible relationships among general environmental knowledge on attitudes and behaviours.

*Environmental attitude scale*: This scale was constructed with questions in Section B. For each person, the mean of his/her response across the two items was calculated. "Low" attitudes include mean values of 1.00 to 1.49, "moderate" includes means of 1.50 to 2.90 and "high" includes mean of 3.00 to 4.50.

Students who correctly answered more questions on the knowledge test tended to have "high" environmental attitudes (Table 15.8). Out of those who earned an A on the knowledge questions, 88.2% of them indicated that they have "high" attitude. Five students (6.8% of the participants) who received a D or F grade indicated "high" environmental attitudes.

*Environmental behaviour scale*: This scale was constructed using questions in Table 15.6. In a manner similar to the attitude scale, the mean

Table 15.9. Students' environmental knowledge grades and environmental behaviour (Figures in brackets show the column percentages).

| Environmental behaviour | Combined environmental knowledge grade/number correct* | | | | | |
|---|---|---|---|---|---|---|
| | A (13–15 correct) | B (10–13 correct) | C (7–9 correct) | D (4–6 correct) | F (0–3 correct) | Total |
| High | 9 (52.9%) | 28 (71.8%) | 10 (76.9%) | 1 (50.0%) | 2 (66.7%) | 50 (67.6%) |
| Moderate | 7 (41.2%) | 11 (28.2%) | 3 (23.1%) | 1 (50.0%) | 1 (33.3%) | 23 (31.1%) |
| Low | 1 (5.9%) | 0 (0%) | 0 (0%) | 0 (0%) | 0 (0%) | 1 (1.4%) |
| Total | 17 (100.0%) | 39 (100.0%) | 13 (100.0%) | 2 (100.0%) | 3 (100.0%) | 74 (100%) |

**Score out of fifteen combined environmental knowledge questions.*

Table 15.10. Students' environmental behaviour and environmental attitudes (Figures in brackets show the column percentages).

| Environmental attitude | Environmental behaviour | | | |
|---|---|---|---|---|
| | High | Moderate | Low | Total |
| High | 48 (96.0%) | 17 (73.9%) | 0 (0%) | 65 (87.8%) |
| Moderate | 72 (4.0%) | 6 (26.1%) | 1 (100.0%) | 9 (12.2%) |
| Low | 0 (0%) | 0 (0%) | 0 (0%) | 0 (0%) |
| Total | 17 (100.0%) | 39 (100.0%) | 13 (100.0%) | 74 (100.0%) |

was calculated and categories were coded as follows: "high" (1.00 to 1.99), "medium" (2.00 to 2.49) and "low" (2.50 to 3.00).

67.7% of those who responded to these items were categorized with "high" behaviour (Table 15.9). The results that compared the environmental knowledge grade and environmental behaviour showed that almost half of the respondents who earned an A grade reported "high" behaviour (52.9%). About 72% of those who earned a B grade had "high" behaviour. It is interesting to note that 13 students who are categorised as "high" behaviour scored a grade of C or less. 96.0% of the students who were categorised as having "high" environmental behaviour tend to also report "high" environmental attitudes (Table 15.10).

The results show a connection exists between secondary three students' environmental knowledge and their self-reported attitudes and behaviours. Respondents who received a higher grade in environmental knowledge were more likely to have a positive environmental attitude. Even with a low environmental knowledge grade (i.e., Grade D or F), respondents seem to have a positive (high) attitude toward the environment. This may point to other factors, besides knowledge, that can help create positive attitudes. Students across all grades levels were more likely to perform high to medium pro-environmental behaviours. Those who have a high environmental attitude also are more likely to perform a high number of environmental behaviours. It does seem that the students seem to show their commitment to the environment by engaging in certain environmental behaviours. These students may do these activities because these behaviours are mandated, socially acceptable or because they can save money. It is difficult to tell what motivates them to engage in the environmental friendly behaviours from this survey.

### *Recommendations*

The results suggest that the secondary three students have been exposed to the environmental issues. The internet and television are the most preferred sources of getting environmental information. The majority of the students expressed positive experiences and reported "very strong or strong" desire to improve the environment. Students feel that government should be mainly responsible for giving financial support to environmental education. In comparing the categories of "have done in the last 12 months and "would consider doing", it was observed that in all cases (except water conservation and energy conservation), lower percentages of students reported that they have taken deliberate action to improve the environment in the last 12 months. Respondents who received a higher grade in environmental knowledge were more likely to have a positive environmental attitude. When examining environmental attitudes and behaviours, those who have a high environmental attitude also are more likely to perform a high number of environmental behaviours.

Engaging in environmental literacy on local and global activities is the next step in helping younger generations to develop a culture of care and concern for the environment. This can be done through creating sustainable learning communities within and across schools which focus on education *for*, *about* and *through* the environment. Educators can create opportunities to make connections with the natural world through hands-on learning. Continued pedagogical and curriculum innovations can help the younger generation to reconnect, redesign and realign our actions towards environmental stewardship.

The use of media spaces and technology can help in establishing and perpetuating official consciousness on new, under-appreciated or immediate environmental issues. Environmental activist, photographer and filmmaker Yann Arthus-Bertrand's aerial photographs of earth force humans to rethink our impacts on the environment. Documentaries such as *The Future of Food* can evoke responses and powerful debates on environmental issues. The internet and visual arts can be used as avenues of interactive genres to engage students in environmental issues.

More importantly, some degree of basic knowledge of environmental legislation should be introduced in schools to expose and illuminate the challenges of dealing with 21st century environmental issues. International and regional case studies can be infused in current topics in Humanities where students can discuss the challenges of applying environmental laws, for example, the Cartagena Protocol on Biosafety to the Convention on Biological Diversity[5] can be discussed under the topic of the Geography of Food.

A multidisciplinary programme that focuses on sustainability concerns may also be infused into existing school curriculum. For example, themes of ecosystem valuation and ecological footprint can be introduced in the topic of tourism. By building on an enquiry-based approach in teaching and learning, educators can help students to understand the aggregated impact of their individual choices and political positions. Alternatively, a stand-alone curriculum with modules such as environmental ethics can be structured, for students to participate in over a number of years. Students can be first exposed through situations that are artificially simple, culminating in real-life situations that are complex. For example, a topic on respect for wildlife can focus on the causes, effects of wildlife trade and a discussion on the role of Singapore as a member country of the CITES.[6] The explicit focus on environmental education as a stand-alone curriculum will enhance Singapore's commitment in pursuing sustainability.

[5]Cartagena Protocol on Biosafety to the Convention on Biological Diversity is an international agreement which aims to ensure the safe handling, transport and use of living modified organisms (LMOs) resulting from modern biotechnology, that may have adverse effects on biological diversity. It also takes into account risks to human health.
[6]Convention on International Trade in Endangered Species of Wild Fauna and Flora (CITES) is an international agreement signed by 180 nations designed to ensure that international trade in animals and plants does not threaten their survival in the wild.

More research is required on the combinations of knowledge, attitudes and behaviours in the creation of an environmentally literate school population. Information gained from focus group interviews helps to foreground the importance of understanding articulations of meanings by engaging participant voices, thus allowing spaces for the definition of problems and to create solutions that are meaningful to participants. In this way, it helps to present opportunities for social change by reflecting on social reality, behavioural and cultural norms.

## References

Association of Southeast Asian Nations (ASEAN). (2009). *Socio-Cultural Blueprint 2009–2015, at the 14th ASEAN Summit in Cha-Am, Thailand,* 1 March. [Online] Available at: http://www.aseansec.org/22336.pdf. Accessed on 29 July 2012.

Ballantyne, R., Connell, S., and Fien, J. (2006). "Students as catalysts of environmental change: A framework for researching intergenerational influence through environmental education", *Environmental Education Research*, **12**(3), 413–427.

Bamberg, S. and Moser, G. (2007). "Twenty years after Hines, Hungerford and Tomera: a new meta-analysis of psycho-social determinants if pro-environmental behaviour", *Journal of Environmental Psychology*, **27**(1), 14–25.

Curriculum Planning and Development Division (CPDD). (2005). *Social studies syllabus: Primary, from Curriculum Planning and Development Division*, Ministry of Education Singapore. [Online] Available at: http://www.moe.gov.sg. Accessed 26 July 2012.

Curriculum Planning and Development Division (CPDD). (2005). *Geography syllabus: Lower secondary: 2006, from Curriculum Planning and Development Division*, Ministry of Education Singapore. [Online] Available at: http://www.moe.gov.sg. Accessed 26 July 2012.

Curriculum Planning and Development Division (CPDD) (2007). *Science syllabus: Lower secondary: Express/Normal (Academic), from Curriculum Planning and Development Division*, Ministry of Education Singapore. [Online] Available at: http://www.moe.gov.sg. Accessed 26 July 2012.

Department of Statistics Singapore (2012). *Latest data.* [Online] Available at: http://www.singstat.gov.sg/stats/latestdata.html. Accessed 17 October 2012.

Disinger, J.F. and Roth, C.E. (1992). "Environmental Education research news", *The Environmentalist*, **12**(3), 165–168.

Fien, J. (1993). *Education for the Environment-critical curriculum theorising and environmental education.* Geelong: Deakin University Press.

Francis, R.A. (2012). *A handbook of global freshwater invasive species*, Abingdon: Earthscan.

Hines, J.M., Hungerford, H.R., and Tomera, A.N., (1986–1987). "Analysis and synthesis of research on responsible environmental behaviour: A meta-analysis", *Journal of Environmental Education*, **18**(2), 118.

International Union for the Conservation of Nature and Natural Resources (IUCN) (1970). *International Working Meeting on Environmental Education in the School Curriculum, Final report.* Gland: IUCN.

Kreike, E. (2010). *Deforestation and reforestation in Namibia: the global consequences of local contradictions.* Leiden: Koninklijke Brill NV.

Loubser, C.P., Swanepoel, C.H., and Chacko. C.P.C. (2001). "Concept formulation for environmental literacy", *South African Journal of Education*, **21**(4), 313–323.

Millennium Ecosystem Assessment (2005). *Ecosystems and Human Well-Being: Synthesis.* Washington, DC: Island Press.

Millennium Ecosystem Assessment (2005). *Ecosystems and Human Well-Being: Current State and Trends.* Washington, DC: Island Press.

Millennium Ecosystem Assessment (2005). *Ecosystems and Human Well Being: Scenarios.* Washington, DC: Island Press.

Millennium Ecosystem Assessment. (2005). *Ecosystems and Human Well-Being: Policy Responses.* Washington, DC: Island Press.

Millennium Ecosystem Assessment (2005). *Ecosystems and Human Well-Being: Multi-scale Assessments.* Washington, DC: Island Press.

Ministry of Education (MOE) (2011). "Opening Address by Mr HengSweeKeat, Minister for Education", *the 1st NIE-MOE Character and Citizenship Education Conference.* [Online] Available at: http://www.moe.gov.sg/media/speeches/2011/11/08/opening-address-by-mr-heng-swee-keat-at-1st-nie-moe-cce-conference.php. Accessed 12 October 2012.

Ministry of Education (MOE) (2011). *Subject Syllabus Character and Citizenship Education.* [Online] Available at: http://www.moe.gov.sg/education/syllabuses/character-citizenship-education/. Accessed on 6 December 2012.

Ministry of Education (MOE) (2012). *Our Education System.* [Online] Available at: http://www.moe.gov.sg/education/. Accessed 29 September 2012.

Ministry of Education (MOE) (2012). *National Education Learning Journey Programmes.* [Online] Available at: http://www.ne.edu.sg/. Accessed 10 October 2012.

Ministry of Environment and Water Resources (MEWR) and Ministry of National Development (2009). *Sustainable Singapore Blueprint.* [Online] Available at: http://app.nccs.gov.sg/data/resources/docs/Documents/Sustainable_Spore_Blueprint.pdf. Accessed 6 December 2012.

Ministry of Environment and Water Resources (MEWR) (2012). *Singapore Green Plan,* Ministry of Environment, Singapore.

Murphy, T.P. (2002). *The Minnesota Report Card on Environmental Literacy: A Benchmark Survey of Adult Environmental Knowledge, Attitudes and*

*Behaviour*. [Online] Available at http://www.seek.state.mn.us/eemn_b.cfm/. Accessed on 25 September 2012.

Murphy, T.P. (2004). *The second Minnesota report card on environmental literacy: A survey of adult environmental knowledge, attitudes and behaviour*, Minnesota Office of Environmental Assistance. [Online] Available at http://www.seek.state.mn.us/eemn_b.cfm/. Accessed on 25 September 2012.

Murphy, T.P. and Olson, A.M. (2008). *The third Minnesota report card on environmental literacy: A survey of adult environmental knowledge, attitudes and behaviour*, Minnesota Pollution Control Agency. [Online] Available at http://www.seek.state.mn.us/eemn_b.cfm/. Accessed on 25 September 2012.

National Climate Change Secretariat (2012). *National Climate Change Strategy 2012*. [Online] Available at: http://app.nccs.gov.sg/page.aspx?pageid=123. Accessed 27 July 2012.

National Environment Agency (NEA). (2012). *Future ready Transforming Singapore into a Clean, Green and Liveable City: NEA Annual Report 2011/12*. [Online] Available at: http://www.nea.gov.sg/cms/ar2012/pdf/nea_ar2012.pdf. Accessed 7 December 2012.

National Parks Board (NParks) (2009). *National Biodiversity Strategy and Action Plan,* National Parks Board, Singapore. [Online] Available at: http://www.nparks.gov.sg/cms/docs/nbc/NPark-booklet-final-4sep.pdf. Accessed 6 December 2012.

Palmer, J. (1998). *Environmental education in 21st century: Theory, practice, progress and promise*. New York: Routledge.

Reynolds, H.L., Brondizio, E.S., Robinson, J. M., Karpa, D., and Gross, B.L. (2010). (eds.) *Teaching Environmental Literacy: Across Campus and Across the Curriculum*. Indianapolis: Indiana University Press.

Roth, C.E. (1992). *Environmental Literacy: Its roots, evolution, and directions in the 1990s, Columbus,* OH: Eric Clearinghouse for Science, Mathematics Environmental Education, Ohio State University.

Savage, V.R. and Lau, S. (1993). "Green issues, official policies and student awareness", in Briffet, C., and Sim, L.L., (eds.) *Environmental Issues in Development and Conservation,* SNP Publisher, Singapore.

Singapore Environment Council (2012). *Schools Green Award*. [Online] Available at: http://www.sec.org.sg/sga/. Accessed 4 September 2012.

Tan, G., Lee, C., and Goh, K. (1998). "A survey of environmental knowledge, attitudes and behaviour of students in Singapore", *International Research in Geographical and Environmental Education*, **7**(3), 181–202.

Toh, K.A., Chin, L.F., and Boo, H.K. (1997). "Environmental education in a city state", *Journal of Science and Mathematics Education in Southeast Asia,* **20**(1), 58–69.

United Nations (UN) (1972). *Declaration of the United Nations conference on the human environment*. [Online] Available at: http://www.unep.org/Documents.Multilingual/Default.asp?DocumentID=97&ArticleID=1503. Accessed 28 September 2012.

United Nations Education Scientific Cultural Organisation (UNESCO) (1975). *UNESCO-UNEP International Environmental Workshop*, 1322 October,

Belgrade Charter. [Online] Available at: http://unesdoc.unesco.org/images/0001/000177/017772eb.pdf. Accessed 4 October 2012.

United Nations Education Scientific Cultural Organisation (UNESCO) (1978). *International Conference on Environmental Education: Tbilisi* (USSR), 14–16 October 1977 Final Report, UNESCO, Paris.

Williams, M. (ed.) (1996). *Understanding Geographical and Environmental Education: The role of research.* New York: Cassell Education.

Withgott, J. and Laposata, M. (2012). *Essential Environment.* New Jersey: Pearson Education Inc.

Wong, F.B.K. and Philip, S. (2003). "Environmental Education in Singapore; A curriculum for the environment or in the national interest?", *International Research in Geographical and Environmental Education*, **12**(3), 123–138.

# Part 5

# Climate Change

# 16

# POSSIBLE MARKET MECHANISMS FOR THE POST-KYOTO REGIME[1]

ELLEN MAY ZANORIA REYNES

Climate change and global warming has been a notable concern for the past few years. International governance has recognised this, with the UN General Assembly launching climate change negotiations in December 1990, which led to the UN Framework Convention on Climate Change (UNFCCC) and the Kyoto Protocol (KP). These established binding emission targets for industrialised countries, and a range of mechanisms to encourage cost-effective compliance. The climate mitigation market mechanisms available under the KP have been somewhat effective starting tools for bringing down emissions, as they have been the most rigorous and are the only comprehensive systems of compliance. However, with the first commitment period ending in 2012 for the EU Emissions Trading Scheme, it remains to be seen how implementation of climate change mitigation projects will move forward with the absence of a strong legally-binding international framework which creates the demand for these types of projects. This study looks into the existing market mechanisms available under the KP, followed by an analysis of how effective they have been. The study then touches on the different new market mechanisms (NMM) that are being proposed to scale-up climate change mitigation efforts. The conclusion covers important qualities that these NMMs need to have.

## Introduction

Climate change has been recognised as a notable concern, and from the 1992 Earth Summit in Rio de Janeiro, an important legally binding agreement on climate–related efforts was proposed.

The UN General Assembly initiated climate change negotiations in December 1990, giving birth to the UN Framework Convention on Climate Change (UNFCCC), which entered into force on 21 March 1994. By December 1997, the Kyoto Protocol was adopted, thus establishing binding

[1]This paper was supervised by Dr T S Gopi Rethinaraj, Lee Kuan Yew School of Public Policy, National University of Singapore.

emission targets for industrialised countries and a range of mechanisms to encourage cost-effective compliance. It wasn't until 16 February 2005, or the $10^{th}$ anniversary of the UNFCCC, that the Kyoto Protocol was brought into force after 55 Parties to the Convention, which accounted for at least 55% of the total global carbon dioxide ($CO_2$) emissions in 1990 of the Annex I[2] countries, ratified the protocol.

The Kyoto Protocol is generally seen as the first step towards a global emission reduction regime that will stabilise GHG emissions, and provides the essential architecture for any future international agreement on climate change. Its first commitment period ended in 2012, thus, a new international framework needs to be negotiated and ratified.

Since 2005, the Conference of Parties (COP) to the UNFCCC have been meeting, negotiating, and coming up with action plans and agreements every year with the goal of continuing the Kyoto Protocol. In 2007, the international community took a decisive step in Bali, Indonesia, by adopting the 'Bali Action Plan', which committed them to a follow-up agreement to the Kyoto Protocol by the time of their $15^{th}$ meeting (COP15) in Copenhagen, Denmark, in December 2009 (Kamiya, 2009). Unfortunately, the COP could not come to a consensus in Copenhagen.

In 2010, Canada, Japan and Russia stated that they would not commit to further Kyoto Protocol targets. COP16 in December 2010 resulted in the Cancun Agreements which consider the establishment of new market mechanisms "to enhance the cost-effectiveness of, and to promote, mitigation actions" (UNFCCC). COP17 in December 2011 in Durban resulted in the Durban Package which aims for the adoption of an agreement with 'legal force', latest, by 2015, for implementation in 2020, as well as the initiation of a second commitment period in the Kyoto Protocol which will enable the continuity of existing market mechanisms and instruments.

The absence of a strong legally-binding international framework for carbon emission reduction and climate change mitigation has led to the creation of several autonomous national initiatives and bilateral arrangements. It is important to look into the qualities of these initiatives that will allow for the trading of its products, the source of funds for the market to function, and the relevant market infrastructure that needs to be created.

[2]Industrialised Countries.

## Climate Change Mitigation Mechanisms

### *The UNFCCC*

Under the UN Framework Convention on Climate Change (UNFCCC), ratified by 186 countries, including the United States, nations committed to developing and implementing climate protection measures "appropriate for the specific conditions of each Party". In addition, industrialised ("Annex I") countries agreed to voluntarily reduce their GHG emissions to 1990 levels by 2000. The ultimate objective is to "stabilise greenhouse gases in the atmosphere at a level that would prevent dangerous anthropogenic interference with the climate systems. Such a level should be achieved within a time frame sufficient to allow ecosystems to adapt naturally to climate change, to ensure that food production is not threatened and to enable economic development to proceed in a sustainable manner." (UNFCCC).

The Kyoto Protocol[3] (KP) 'operationalises' the convention by setting mandatory emission limits of an average of 5% against 1990 levels over the commitment period of 2008–2012. There are two main substances of the KP: first, the binding emissions reduction commitments for developed country parties, and second, the flexible market mechanisms.

The emission targets, for the period 2008–2012, average 5.2 percent below 1990 levels. The binding emissions reduction targets meant that KP countries had to meet them largely through domestic actions such as national or sectoral policies. It also means that the allowance to pollute was limited, thus causing greenhouse gas emissions to become a new commodity commanding a price. This shows that the KP has now begun to internalise what was previously an un-priced environmental externality.

The second substance, the flexible market mechanisms of the KP, allows the KP bound countries to meet their emission reduction targets through three 'market-based mechanisms' that implement greenhouse gas abatement where it is most cost-effective, that is, in the developing world. These mechanisms are:

- Emissions Trading
- Clean Development Mechanism (CDM)
- Joint implementation (JI)

[3]Unlike the UNFCCC, Kyoto Protocol has a near-universal membership today comprising of 194 countries and 1 regional economic bloc (The EU), but is not ratified by the United States, which is a major emitter.

These mechanisms help achieve sustainable development through technology transfer and investment in developing countries, and help developed Parties meet their emission targets in a cost-effective way.

*The carbon market*

A new environmental commodity was created in the form of emission reduction units (ERUs), which are earned, used, and traded amongst the KP parties. These units refer to an amount of greenhouse gas that is reduced from the atmosphere. Since Carbon Dioxide is the most prevalent greenhouse gas, the traded units are simply referred to as carbon credits.

*Emissions trading*

A central authority, usually the Ministry of Environment of the Party, sets a limit on the amount of a pollutant that can be emitted. This limit is allocated or sold to firms in the form of emissions permits that allow them the right to emit or discharge a specific volume of pollutant. Firms need to have permits equivalent to their pollutant emissions, and their total emissions are limited to the level set by the government.

Emissions trading allows countries that have emission allowances to spare, to sell this excess capacity to countries that are over their targets. When firms exceed their allowed pollutant emissions level, they would have to purchase permits from those who have them in excess.

Emissions trading is under a cap-and trade system, which, when well-designed and implemented, will limit total emissions in a cost-effective manner. This has been tried in the Chicago Climate Exchange (CCX), the Regional Greenhouse Gas Initiative (RGGI), and in the European Union Emissions Trading Scheme (EU ETS).

*The European Union Emissions Trading Scheme — a closer look*

The EU ETS is the first international trading system for $CO_2$ emissions in the world, which started operations in 2005 and is the EU's main policy tool for cutting emissions. It covers about half of EU $CO_2$ emissions in thirty countries "in a region of the world that accounts for about 20 percent of global GDP and 17 percent of world energy-related $CO_2$ emissions" (Ellerman and Buchner, 2007). More than 11,000 factories, power stations, and other installations are required to monitor and report their $CO_2$ emissions and stay within its allowances (equivalent to one tonne of $CO_2$); if they exceed, they can purchase $CO_2$ credits from other installations or countries; if they have reduced more emissions than what is required, they can sell their excess credits. Penalties for violations increased from 40 Euros per ton of $CO_2$ in the first phase (January 2005 to December 2007) to 100 Euros in the second phase (January 2008–December 2012).

*Why did it fail?* In its first phase, the "setting [of] caps and allowances in member states" was decentralised (Kruger, Oates, and Pizer, 2007), with each member state responsible for proposing its own national carbon cap, subject to review by the European Commission. Expectedly, member states tried to be "generous with their allowances to protect their economic competitiveness" (Convery and Redmond, 2007), resulting in an aggregate cap that is higher than Business-As-Usual (BAU) emissions — meaning, more pollution permits were issued than actual emissions (Anderssen, 2007).

The absence of precise emissions data at the beginning of the programme may have contributed to the surplus of allowances. This effectively collapsed the prices of carbon, "creating no incentive to reduce emissions" (Anderssen, 2007).

The European Commission tried to avoid the problems of Phase I by enforcing more stringent caps through a centralised system and auctioning the permits (more than 60% shares), rather than allocating allowances freely. Figure 16.1 shows the tightened caps, though only by 2% when compared to 2005 verified emissions.

| Member State | 1st period cap | 2005 verified emissions | Proposed cap 2008-2012 | Cap allowed 2008-2012 (in relation to proposed) |
|---|---|---|---|---|
| Austria | 33.0 | 33.4 | 32.8 | 30.7 (93.6%) |
| Belgium | 62.1 | 55.58 | 63.3 | 58.5 (92.4%) |
| Bulgaria | 42.3 | 40.6 | 67.6 | 42.3 (62.6%) |
| Cyprus | 5.7 | 5.1 | 7.12 | 5.48 (77%) |
| Czech Rep. | 97.6 | 82.5 | 101.9 | 86.8 (85.2%) |
| Denmark | 33.5 | 26.5 | 24.5 | 24.5 (100%) |
| Estonia | 19 | 12.62 | 24.38 | 12.72 (52.2%) |
| Finland | 45.5 | 33.1 | 39.6 | 37.6 (94.8%) |
| France | 156.5 | 131.3 | 132.8 | 132.8 (100%) |
| Germany | 499 | 474 | 482 | 453.1 (94%) |

Fig. 16.1. EU ETS emissions and cap.

*Source*: Extract from the European Commission press release, 26/10/2007.

However, the global recession hit Europe particularly hard, resulting in lower economic activity, energy demand, actual $CO_2$ emissions, and therefore, demand for EU ETS allowances and thus, prices.

According to the European Commission, experience during the first two phases have made it clear that "besides underlining the need for verified data", "greater harmonisation within the EU ETS is imperative to ensure that the EU achieves its emissions reductions objectives at least cost, especially on "how the cap on overall emission allowances is set". "Furthermore, greater harmonisation, clarification and refinement are needed with respect to the scope of the system, the access to credits from emission-reduction projects outside the EU, the conditions for linking the EU ETS to emissions trading systems elsewhere and the monitoring, verification and reporting requirements."

The European Union plans to extend the EU ETS through Phase III, 2013–2020, with a centralised cap becoming increasingly stringent (20% below 1990 emissions), a larger share of the allowances subject to auctioning, tighter limits on the use of offsets, and unlimited banking of allowances between Phases II and III (Aldy and Stavins, 2011).

### *Carbon offsetting with international projects*

The EU ETS has allowed 'emission reductions' from international projects in the form of Clean Development Mechanism credits from developing countries, with the dual promises of emission reduction and sustainable development for the host country, and Joint Implementation credits from other developed countries. This was initially criticised as a way to avoid domestic action and the sourcing of cheaper credits outside with the difficulty "to verify which of these thousands of so-called off-set projects in the developing world are actually legitimate" (Henningsen, 2010).

### *Clean development mechanism*

The Clean Development Mechanism allows a country with an emission-reduction commitment to implement an emission-reduction project in developing countries. These projects — for example, energy efficiency measures in an industrial plant, or a wind farm as renewable energy source — can earn saleable carbon credits, which can be counted towards meeting the Kyoto targets of the developed Party who purchased them.

Operational since the beginning of 2006, the mechanism has already registered more than 4,500 projects,[4] and is anticipated to produce Certified Emission Reductions (CERs) amounting to more than 2.9 billion tonnes of $CO_2$ equivalent between 2008 and 2012. In total, 8871 CDM projects are now included in the pipeline, 4386 of the projects are now registered, while 1651 CDM projects have got CERs issued. 1,233 million CERs/year are expected, 19% of which come from wind projects, 26% from hydro projects, and 12% from projects related to industrial gases.

The mechanism stimulates investment in sustainable development and emission reductions in developing countries, while giving industrialised countries some flexibility in how they meet their emission reduction or limitation targets.

The United Nations Development Program acknowledges that CDM has "successfully delivered a significant amount of emissions reductions and has been a key financial instrument in stimulating mitigation activities in developing countries" (Alers, 2011). The World Bank recognises CDM as having exceeded expectations, not only in terms of the number of projects, but also in terms of awareness, as well as in identifying, assessing, implementing, monitoring, and verifying GHG mitigation/sequestration activities in different sectors, and capacity building in both the public and private sectors (Oppermann, January 2012).

CDM has also had its share of criticisms, and the take up of projects in developing countries have been somewhat slow, with the exception of China and India. Another criticism of CDM is that while 68% of the CDM projects are renewables, they account for 35% of the CERs only, as shown on Figure 16.2. Energy efficiency takes 9% of the CERs for Supply-side EE and only 1% for Demand-side EE. Industrial gases, notably hydrofluorocarbon-23 (HFC-23) and nitrous oxide ($N_2O$), as well as perfluorocarbon (PFC) projects only account for 2.0% of the projects but represent 28% of the CERs by 2012. HFC projects have since been reviewed and their registration postponed. They are also no longer allowed under the EU-ETS post 2012.

Another concern plaguing CDM is the risk of fraud. Several studies[5] have pointed out that some projects are financially viable by themselves and that the incentives received from CDM are not necessary for them to be implemented, as they would have happened anyway. This means there is a net increase in global emissions as the credits from these projects allow for

[4] As of July 2012.
[5] For example the study of Zhang and Wang in 2011.

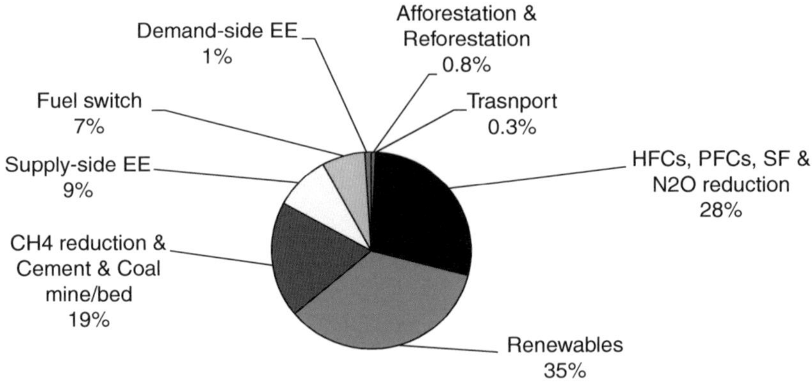


Fig. 16.2. CERs expected until 2012 from CDM projects in each sector.

*Source*: UNEP RISOE CDM/JI Project Pipeline Overview, August 2012.

higher levels of domestic emissions (when sold to the EU) without reducing emissions in the host developing countries.

However, CDM has had several successes on the ground. According to the PD Forum, "CDM projects help overcome financial and other barriers within their sectors, and help shift the baseline to higher efficiencies and lower emission alternatives, as is the case [for] the autonomous penetration of renewables within the electricity sector."

There are many examples where CDM has contributed towards technology transfer. In addition to engineering technology transfer, many CDM projects have also contributed to the development of financial capacity through demonstration of innovative financial models; training and education of staff and contractors, including in the implementation of monitoring systems and health and safety systems; the development of business models including improved Engineering, Procurement, and Construction and servicing contracts.[6]

CDM has also contributed to south-south technology transfer, where "project engineers work with local suppliers to improve the reliability of locally manufactured machinery". In the process, local manufacturers learn the benefits of using better quality equipment, for example, on the costs of spare part inventories and downtime.[7]

[6](PD Forum, 2012, p. 27).
[7](PD Forum, 2012, p. 27).

According to the UNEP-RISOE CDM Pipeline as of May 2012, 195 billion dollars' worth of investment in registered emission reduction projects has been leveraged, with a further 290 billion not yet registered. Therefore, CDM is by far the most successful international instrument for leveraging new and additional finance for mitigation.[8]

*On Sustainable Development.* CDM has resulted in a massive uptake of renewable energy and energy efficiency projects. This has "provided large amounts of additional jobs, reduced the use of fossil fuels", "extended grid power to more users and improved the reliability of power supplies" and, in some cases, "stimulated some sectors significantly". For example, the "Chinese wind sector gr[ew]... from relying almost completely on demonstration projects and ODA-funded projects, to being the biggest wind market in the world, including the biggest turbine manufacturers".[9]

According to the UNEP-RISOE (May 2012), the CDM Pipeline directly contributes more than 225 GW of renewable energy supply and nearly 100 GW of energy efficiency measures in host countries.

Schneider (2007, p. 73) also commented on the success of the CDM in reducing emissions from industrial plants and landfills. Many of the methane avoidance projects include the use of methane for energy production, helping an economy to reduce its reliance upon fossil fuels or unsustainable biomass.

### *Joint implementation*

Industrialised countries are allowed to carry out "joint implementation" projects with other developed countries to achieve emission reduction and earn carbon credits, which can be counted towards meeting its KP target. This enables a flexible and cost-efficient means to fulfill targets, while the host Party benefits from foreign investment and technology transfer.

Joint Implementation (JI) has seen a lower uptake of projects, with less than 400 projects registered and less than 250 issuing Emission Reductions Units (ERUs), as of August 2012. The total amount of ERUs expected annually is 195 million, with 870 million ERUs expected from projects in the pipeline until the end of 2012. While 22.4% of the 559 JI projects are renewables, they only take 5.9% of the ERUs. $CH_4$ reduction takes the majority (37%) of the ERUs, while energy efficiency comes a close second at 30% of the expected ERUs/year, having issued the most number of

[8](PD Forum, 2012, p. 23).
[9](PD Forum, 2012, p. 22).

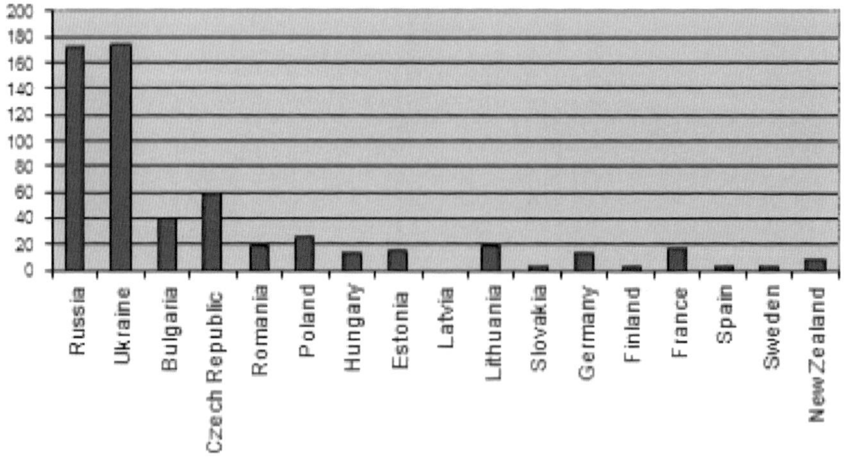


Fig. 16.3. Number of JI projects by host country.

*Source*: UNEP-RISOE. JI Projects. August 2012.

ERUs, at 39% of the 174 million ERUs, thus far. A total of 1,233 MW of hydro capacity has been installed by projects in Bulgaria, Russia, Czech Republic, Romania, Ukraine, Estonia, and Poland while 1787MW of wind capacity has been installed in Lithuania, Estonia, Poland, Bulgaria, New Zealand, Romania, and some parts of Eastern Europe. Figure 16.3 shows the distribution of JI projects among the host countries.

Through these market mechanisms (both CDM and JI), the private sector is included in the efforts against climate change by way of climate-friendly investments, technology transfer, staff skills development, labour, enhanced resource management, and the formation of a new market for the trade of environmental allowances/commodities. Bypassing older, dirtier technology for newer, cleaner infrastructure and systems for the low carbon development of developing Parties is also allowed at more cost-efficient levels, with obvious longer-term benefits.

### *What is Next in Line?*

As of November 2011, 141 Parties have expressed their intention to be listed as having agreed to the 2009 Copenhagen Accord with quantified economy-wide emissions targets for 2020. Table 16.1 shows the declared targets of major economies, though these commitments are largely conditional to a global and comprehensive agreement where all major economies

Table 16.1. Quantified economy-wide emissions targets for 2020.*

| Parties | Emissions reduction in 2020 | Base year |
|---|---|---|
| Australia | −5% up to −15% or −25%<br>Australia will reduce its greenhouse gas emissions by 25% on 2000 levels by 2020 if the world agrees to an ambitious global deal capable of stabilising levels of greenhouse gases in the atmosphere.<br>Australia will unconditionally reduce our emissions by 5% below 2000 levels by 2020, and by up to 15% by 2020, if there is a global agreement which falls short of securing atmospheric stabilisation at 450 ppm $CO_2$-eq and under which major developing economies commit to substantially restrain emissions, and advanced economies take on commitments comparable to Australia's. | 2000 |
| Canada | 17%, to be aligned with the final economy-wide emissions target of the United States in enacted legislation. | 2005 |
| EU and its Member States | 20%/30%. As part of a global and comprehensive agreement for the period beyond 2012, the EU reiterates its conditional offer to move to a 30% reduction by 2020 compared to 1990 levels, provided that other developed countries commit themselves to comparable emission reductions and that developing countries contribute adequately according to their responsibilities and respective capabilities. | 1990 |
| Japan | 25% reduction, which is premised on the establishment of a fair and effective international framework in which all major economies participate and with agreement by those economies on ambitious targets. | 1990 |
| Russian Federation | 15–25 % — depending on the following conditions:<br>— Appropriate accounting of the potential of Russia's forestry in the frame of contribution in meeting the obligations of the anthropogenic emissions reduction;?<br>— Undertaking by all major emitters the legally binding obligations to reduce anthropogenic GHG emissions. | 1990 |
| United States | In the range of 17%, in conformity with anticipated U.S. energy and climate legislation, recognising that the final target will be reported to the Secretariat in light of enacted legislation. The pathway set forth in pending legislation would entail a 30% reduction in 2025 and a 42% reduction in 2030, in line with the goal to reduce emissions 83% by 2050. | 2005 |

* *Note*: not a comprehensive list. Summarised from: UNFCCC, Appendix I — Quantified economy-wide emissions targets for 2020.

participate. This global agreement will require all major emitters, including fast emerging developing countries, to be legally bound to reduce emissions, according to their responsibilities and respective capabilities.

Increasingly, the BASIC (Brazil, South Africa, India, China) countries have become an emerging power in the climate negotiations, since they are the largest emerging economies, and thus, are the largest emitters from the group of developing countries who do not want to be imposed with caps on carbon. The European Union has announced that they will only allow carbon credits from projects in Least Developed Countries (LDCs) post-2012 into the EU Emissions Trading Scheme, leaving the developing countries without a clear market for their carbon credits.

A change of proposition was presented by the G77[10] and China at COP17, asserting that a second commitment period under the Kyoto Protocol must be undertaken, "fully legally binding and ratified. Non Kyoto Protocol Annex I Parties, have to also join these efforts, with comparable commitments." Unfortunately, Canada has formally withdrawn from the Kyoto Protocol (Vaughan, 2011) and Japan is continuing bilateral cooperation arrangements with some developing countries, independent of an international framework. Several countries also have their own initiatives to form new market mechanisms, while pursuing approaches through the existing allowable schemes under the Kyoto Protocol.

### *New Market Mechanisms*

A decision was made in Cancun in 2010 to consider the establishment of new market mechanisms "to enhance the cost-effectiveness of, and to promote, mitigation actions", complement nationally appropriate mitigation actions (NAMAs) by developing countries, and supplement the domestic mitigation efforts of developed countries. The discussions for market mechanisms revolve around general themes including shared vision, mitigation, adaptation, finance and technology, and builds on the existing mechanisms to scale up actions.

NAMA-based mechanisms can be in the form of sustainable development policies and measures, low carbon development plan, national

[10]The Group of 77 is the largest intergovernmental organization of developing countries in the United Nations, which provides the means for the countries of the South to articulate and promote their collective economic interests and enhance their joint negotiating capacity on all major international economic issues within the United Nations system, and promote South-South cooperation for development.

climate-related regulations and targets. For developed countries, a sectoral approach can be implemented whereby emission reductions are achieved in different industrial sectors and are credited or traded within that sector. Another option is to establish a domestic emissions trading, similar to the European Union Emissions Trading Scheme, and voluntary emission reduction (ER) crediting schemes which convert these ERs into useable and tradeable offsets. An internationally tradeable unit should represent 1 ton of $CO_2$-equivalent. Furthermore, carbon offset and carbon neutral schemes can also be established where domestic companies voluntarily offset their operational emissions through internal efforts or by purchasing offsets on the market. There are two types of market mechanisms, as shown on Figure 16.3:

- *Crediting* — Existing emissions of a broad segment of an economy will be checked against an agreed baseline for this segment (no-lose target).
- *Trading* — In accordance with a defined absolute target for a broad segment of an economy, emissions allowances will be issued.

The EU has suggested that for these new market mechanisms to succeed, a robust Monitoring, Reporting, and Verification (MRV) system has to be in place to avoid double counting with financial commitments and with existing mechanisms; these mechanisms must avoid leakage, and should be consistent with Sustainable Development policies. Furthermore, a Governance and Regulatory Structure is important where a Special Supervisory Board approves methodologies, oversees the accreditation of verifiers, approves Business As Usual emissions and crediting thresholds, issues credits, maintains the registry, etc. This new market-based mechanism allows for voluntary participation, complementing other means of support for NAMAs, stimulating mitigation across broad segments of the economy, safeguarding environmental integrity, ensuring a net decrease and/or avoidance of global GHG emissions, assisting developed countries to meet part of their mitigation targets, and ensuring good governance and robust market functioning. There should be clear rules regarding the complementary relationship between CDM and the new market-based mechanism.

### *Climate Deal at a Stalemate*

The BASIC countries and the world's largest emerging economies are aware of increasing international pressure to limit their emissions. China acknowledges that they are the largest emitters but insists on not taking

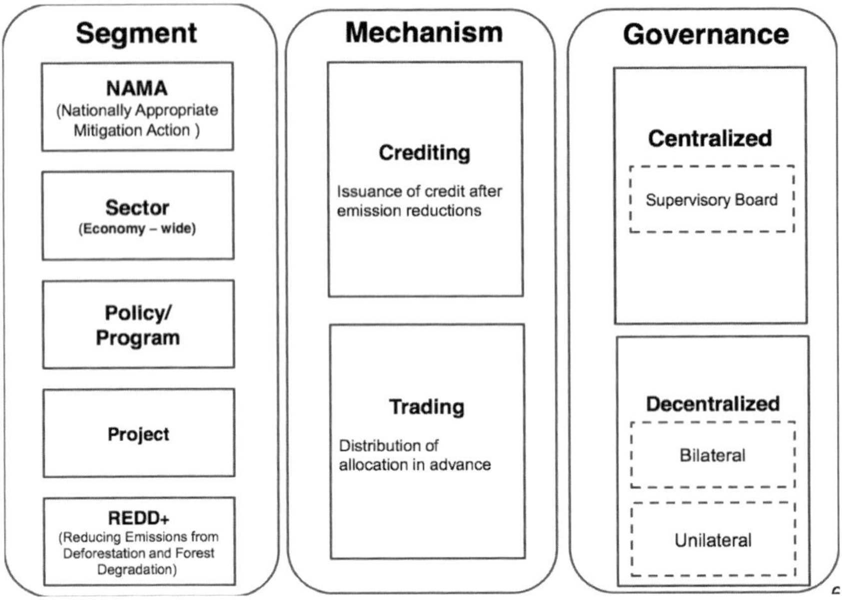


Fig. 16.4. Framework for new market mechanisms.

*Source*: IGES. New Market Mechanisms in Charts. (2011).

binding emissions targets and maintains the principle of "common but differentiated responsibilities".

The lack of binding constraints over giants such as the BASIC group and the refusal of the United States to join the protocol has caused other major economies such as Australia, Canada, Japan and Russia to refuse to sign on for a second commitment period of the KP. Instead, Japan wants "a single binding treaty" and encourages countries to implement the Copenhagen Accord, which is a "more effective framework" where 80% of global emissions are represented, compared to just 15% in the KP (Minamikawa, 2010). To keep to its 25% reduction target, Japan is forming its own domestic mechanisms (Japan-Voluntary Emission Reduction or J-VER) and international bilateral offset mechanisms (Office of Market Mechanisms, Climate Change Policy Division, Ministry of Environment, Japan).

China has its own commitment to reduce its $CO_2$ emissions intensity by 40–45% per unit of GDP by 2020 against 2005 levels, and plans to achieve this largely through low carbon development plans. India, on the other hand, has established new market mechanisms on energy efficiency and renewable energy.

Japan and Australia have shown that they are serious with their climate commitments with practical steps into its implementation. China and India are the largest carbon emitters with fast-emerging economies, and may contribute to significant emission reductions by incorporating these in their national development plans. Thus, it is worth looking at these countries' efforts, which can serve as models for other domestic or regional schemes.

### *COP17 in Durban — The "Extension of the Kyoto Protocol"*

In December 2011, more than 190 nations agreed on and adopted the "Durban Package" as an outcome of the UNFCCC's Seventeenth Conference of Parties (COP 17). The Package aims for "the widest possible cooperation by all countries and their participation in an effective and appropriate international response". This is the first time that both developed and developing countries have agreed to cooperate in an "inclusive collective action approach" (Morgan and Cameron, 2011). The goal is to adopt a "legal protocol, legal instrument or an agreed outcome with legal force", latest by 2015, for implementation in 2020.

In addition, countries have agreed to a second commitment period of the Kyoto Protocol that shall begin from January 1, 2013 and end either on December 31, 2017 or 31 December 2020, which will enable the continuity of existing market mechanisms and instruments. Parties who sign up to the Second Commitment Period will commit to reduce emissions by at least 25–40% below 1990 levels, by 2020. It is still unclear which additional countries will join the EU in this effort (*ibid*), major emitters such as the US, Japan, Russia and Canada will not sign on to it.

## Review, Analysis and Conclusions

An interesting concept brought forward by the UNFCCC is that efforts to combat climate change must be "on the basis of equity" and "common but differentiated responsibilities", and their respective capabilities must be considered while promoting sustainable development. This can be achieved by integrating these efforts with national development programmes, which take into account the fact that economic development is essential for adopting measures to address climate change. The UNFCCC recognised "that social and economic development and poverty eradication are the first and overriding priorities of developing countries and that a low-emission development strategy is indispensable to sustainable development."

The Kyoto Protocol compliance mechanism was designed to strengthen the Protocol's environmental integrity, support the carbon market's credibility and ensure transparency of accounting by Parties. It has been widely criticised as being insufficient to achieve the necessary global emission reduction efforts in order to stabilise the temperature to desired levels. However, it is among the most rigorous and perhaps the only comprehensive system of compliance for a multilateral environmental agreement.

The current dire economic situation of several developed Parties has forced governments to prioritise popular domestic issues, such as sovereign debt, the recession, unemployment, public health care, or even energy sourcing after the Fukushima incident (several governments have announced that they will discontinue or reconsider their nuclear programs). Further, the target-bound countries see that large emerging economies such as Brazil, India, and China are moving forward with carbon-intensive development and thus the need for these developing countries to also commit to their own emission reduction targets.

Now, the Conference of Parties to the UNFCCC is in the middle of a lengthy process of negotiation, where the international agreement to be instituted may still take a few more years, and yet more for its implementing mechanisms to be established. With the end of the first commitment period fast approaching and the super-economies[11] refusing to commit to a second commitment period, how the continuation of existing market mechanisms should be brought about are being deliberated, while new ones are being formed.

Japan is working towards a domestic emissions trading scheme and a voluntary emission reduction scheme, while arranging for bilateral agreements for offsets. Australia has just released its own carbon pricing legislation, which puts forward a carbon tax and a "cap and trade" scheme, forming a carbon market that they hope to unify with that of New Zealand and other international markets in the future. China has its own commitment of 40–45% reduction from 2005 and plans to achieve this through domestic initiatives such as a low carbon growth path, and a policy on increasing renewable energy power sources. India has the Renewable Energy Credit Scheme and the Perform Achieve Trade Scheme.

[11] Australia and Canada have become the latest Annex 1 countries to join Japan, Russia and the United States.

### *Analysis*

These new mechanisms are all in line with "nationally appropriate mitigation actions" in the form of national policies or targets, low carbon development plans, sectoral improvements and financial disincentives to pollute, by putting a price on carbon.

Sectoral approaches would, by definition, be better suited to achieving sector-wide transformations and might in particular give a boost to small-scale decentralised renewable energy, energy efficiency, and transport projects. This is especially useful for the building or transport sector, both of which have been historically difficult to involve in the existing carbon mechanisms. Sectoral approaches could potentially provide the necessary stimulus for governments of developing countries to introduce policies and measures such as building codes to redirect private sector investments in these sectors.

In order to create comprehensive mechanisms which protect environmental integrity, there is a need to consider the information requirements of each mechanism: the coverage or boundaries, potential for leakage, projected business as usual emissions, reference levels or sectoral targets and the estimated emissions reduction potential. A system for measurement, reporting and verification of emissions reductions or removals should also be established, as well as a registry that will issue and track them to avoid double counting. Finally, these mechanisms should be in line with the host country's sustainable development goals. This means that developing countries may have to pre-finance the upfront costs of establishing sectoral schemes and run the risk of not being able to recoup their costs through emissions trading. This limitation can probably be addressed if industrialised countries are willing to finance a significant part of the upfront costs.

All these different mechanisms need to simplify the process of achieving credits from emission reduction initiatives, ensure that they can be measured, monitored, and verified, and that they have enough parallelism amongst the different regional and country mechanisms to allow for trade between different markets, i.e., the equivalence of 1 credit with another. Renewable Energy Credits, Performance Achieve Trade Scheme, carbon offsetting, sectoral credits, should also be made clear as these goods being traded are intangible in nature and their markets are driven by market integrity. This is quite important to keep the private sector engaged and involved in climate action.

Some countries without commitments (Mexico, South Korea, Thailand) are actively considering the introduction of domestic multi-sector emissions trading schemes. These may gradually spread to other more economically advanced developing countries, and it may be possible to fully link such systems in developing countries with those of industrialised countries.

Furthermore, with the completion of several work programs and the emergence of new bodies at the UNFCCC expected in 2012, analysts expect the creation of a greater emphasis on implementation and transparency. Nations must now develop and deliver strategies to adopt an ambitious climate treaty, with a long-term horizon, in the next four years where citizens must support necessary actions in their economic and political systems.

## *Conclusion*

Market mechanisms should be a core part of a global agreement that assists both developed and developing countries in achieving their targets and commitments at the lowest overall cost. The pace of development of new market mechanisms (NMM) will depend on demand, generated by the various regional schemes being established now. A general framework of carbon should be incorporated in the domestic policies of countries such as Japan, Australia, China, and India.

NMM should be based on top-down rules as well as bottom-up implementation, as pilot projects and real experiments are keys to progress towards NMMs. Baselines for NMM should attempt to ensure the environmental integrity of credits and reflect the circumstances of developing countries, but in order to achieve this, the baseline determination should be less onerous. Incentives must be right for host countries, buyer countries, and the private sector, and risks must be acceptable in order to attract the required investments.

CDM still has an important role to play and provides important lessons for the design of new market mechanisms. There is no need to reinvent the wheel; but there is a need to build on and leverage existing capabilities and infrastructures, and make it more simple and operational.

The PD Forum proposes that there needs to be a "clear bridge" that connects the design of NMM to the CDM to facilitate a "seamless" scaling-up of carbon market instruments. (2012). Many elements and concepts developed under the CDM can play a key role in the design of these new mechanisms, i.e., MRV procedures, standardised baselines, definition of eligibility criteria, sophisticated tools and methodologies. The CDM is already a functioning mechanism and provides an operational governance framework in many developing countries, with 5,000–10,000 professionals and a massive body of methodological expertise behind them. Designated Operational Entities, tasked to audit a project, are accredited by an Accreditation Panel set up to ensure that they are competent in meeting the challenges of validating and verifying CDM projects; it is stringent and produces units which are considered fungible with the KP, the EU ETS and other compliance schemes (2012). New mechanisms should build on this CDM resource pool.

Furthermore, there is a need for urgent progress on the international agenda to avoid risks of fragmentation through bilateral and regional arrangements, and ultimately, the loss of legitimacy of the trading of environmental credits. New mechanisms must be simple and straightforward, consistent with environmental integrity and high quality, and different policy regimes: $CO_2$, Renewable Energy, and energy efficiency policies should be coherent.

Parallel mechanisms on Low Carbon Development of Cities, or Supply Chain Carbon Accounting in industries should be explored and tested. Currently, the EU Emissions Trading Scheme only covers power, some industries, and the aviation sector. There are still vast opportunities for putting a price on carbon in other industries, particularly on consumer products. This can hopefully change consumer behaviour and start conscious consumerism.

Finally, a global scaled-up private sector involvement, primarily thru market mechanisms, is essential to achieve long-term global emission reduction goals. As a reference, the Montreal Protocol was successful in reducing the use of ozone-depleting substances because the private sector was ready to achieve the targets through new technologies and market strategies.

According to UNFCCC Executive Secretary Christiana Figures at the Carbon Expo last May 2012, "Industry should step up its engagement in the international, negotiated response to climate change." Industry has a critical role to play, not just in helping to deliver the resources necessary to tackle climate change, but in spurring and shaping the international, negotiated response to climate change. "Demand more national policy action. Explore to the utmost those opportunities that allow you to increase the low carbon market share, thereby converting your activity in the market into a push for governments to take more ambitious political decisions. Most importantly, let governments know what you require so that you can engage more strongly in deploying low-carbon technology."

### *Recommendations*

The practicality of using internationally-employed market mechanisms in climate change mitigation activities has oft been discussed, not only for its cost-effectiveness but also for its intrinsic tendency to promote competitiveness amongst different market players and technologies. Internationally tradeable permits and a decentralised system of internationally-linked domestic cap-and-trade could "ensure internationally cost-effective emission mitigation" (Aldy and Stavins, 2011).

However, it is key to note that embedded in each of the host country's national plans for climate change are policies and targets that cover energy, technology and new project types, and may very well look into other financial instruments such as taxes. We can then explore alternative options for climate change mitigation policies and emission reduction initiatives, but only for the goal of giving an overview. Further studies can be conducted to explore these alternative options, specifically with the rise and scaling up of new technologies and financial schemes.

These alternative options for emission reductions can be broadly summarised below:

- Policy
  - Standards and Labels
  - Fiscal Stimulus
  - Carbon Taxes
- Technology Instruments
  - Renewable Energy portfolio
  - Unconventional technologies: Nuclear, Carbon Capture and Storage, BioFuels, Biomass

*Standards and Labels.* The ICTSD (Torney, 2009) has looked into different mitigation policies and determined that with the goal of setting a minimum benchmark with Standards and Labels, it may create barriers to market access for producers in developing countries with standards that are difficult to comply with. Developing countries should be involved in the process of setting these environmental standards, and should be provided with assistance by the more developed countries to build a capacity to conform to such standards. The advantage here is that "some niche markets, such as organic farming, can present new opportunities for [producers in] developing countr[ies]s".

*Fiscal stimulus packages:* these are likely to restrict market access for some producers in developing countries due to "buy local" provisions, but the goods in question are likely to be produced by only a few of the most economically advanced developing countries (Torney, 2009).

1. *Carbon Taxes.* Carbon taxes (*harmonised domestic taxes* or *adopt a uniform international tax*) force producers to internalise externalities, and tax revenues collected by the government could be utilised for a number of purposes. It is relatively easy to administer, and the tax rate can be modified as new scientific data becomes available. It would also promote technological innovation and improved production methods, to try and reduce costs of taxation (Joshi, 2010).

*Nuclear Energy.* The Fukushima incident created a sentiment in Europe and North America discouraging the expansion of their nuclear power sectors and encouraging eventual decommissioning. However, this will likely not have a significant impact on the expansion of the nuclear power sector in Asia. Several factors in Asia, including a fast-growing demand for energy, severe air pollution, global warming, and a heavy reliance on important fuels, has led to a need to diversify the region's energy mix.

With the current underdevelopment of clean renewables, renewable energy has limitations in providing reliable base-load power. Nuclear power can provide large scale power on a reliable basis. With the Fukushima incident, Japan and other Asian nations have committed to rechecking safety measures put in place to ensure the safe operation of nuclear power generators for years to come (Peimani, 2011).

*Renewables.* According to the International Energy Agency (IEA), renewable energy technologies, led by hydropower and wind, account for half of the new capacity installed to meet growing demand. Overall, modern renewables grow faster than any other energy form in relative terms, but in absolute terms total supply is still not close to the level of any single fossil fuel in 2035. Renewable energy subsidies can help cut GHG emissions by encouraging deployment of up to 3.4 Gtonnes by 2035.

*Policy Support for Renewables.* The most common policies for renewables are the feed-in tariffs and favourable business climate for green technologies, such as the wind sector in China. A benchmark price for electricity with compulsory purchase of electricity from renewable sources is also an effective policy, as is the case of Indonesia's geothermal power. Tax exemptions for certain renewable technologies, capital costs grants, and

R&D incentives are some of the other policies to provide economic support for renewables.

Carbon Capture and Storage (CCS) describes a technological process by which at least 90% of $CO_2$ emissions is captured from large stationary sources (that is, fossil fuel-fired power plants, heavy industry), transported to a suitable storage site, then stored in geological formations — safely and permanently — deep underground. IEA goes on to further say that CCS is a key abatement option, accounting for 18% of emissions, and estimates that the costs of achieving global climate objectives without CCS would be over 70% higher. According to the EU Zero Emissions Platform (2012), individual components of the CCS value chain are already proven — ready for scale-up and integration — but further R&D into next-generation technologies must also be initiated immediately to enable rapid and wide deployment post-2020.

*Biofuels and Biomass.* A wide range of biomass feedstock is available for biofuel and bioenergy production such as energy crops (e.g., miscanthus, jathropha, short-rotation coppice); wastes (e.g., waste oils, food processing wastes); agricultural residues (e.g., straw, corn stover); forestry residues; and novel feedstocks (e.g., algae). There are several routes of converting biomass into final energy products or chemicals: bio-chemical biofuels production, for e.g., fermentation to form biomethane or bioethanol; thermo-chemical production of biofuels, i.e., feedstocks gasified with oxygen or steam to form syngas; and biomass combustion for the production of electricity and heat (EU Zero Emissions Platform, 2012).

It is therefore important to set realistic CO2 reduction targets on the international stage, supported by implementable domestic regulations, allowing for several activities for each industry to make way for scaled-up private sector involvement.

## References

ABC News (2009). "Dalai Lama issues climate plea", *ABC Online,* 30 November. [Online] Available at: http://www.abc.net.au/news/2009-11-30/dalai-lama-issues-climate-plea/1162952. [Accessed 1 March 2012].

AFP (2011). "Emission cuts: You first, China tells West ahead of Cop17.", *Mail & Guardian Online.* 22 November. [Online] Available at: http://mg.co.za/article/2011-11-22-china-washes-hands-of-climate-change-responsibilities. [Accessed 28 February 2012].

Aldy, J.E. and Stavins, R.N. (2010). "The Promise and Problems of Pricing Carbon: Theory and Experience", *The Journal of Environment and Development, Special Issue,* **21**(2), 152–180.

Alers, M. (2012). *UNDP's input to the CDM Policy Dialogue.* [Online] Available at: http://www.cdmpolicydialogue.org/public_input. [Accessed 12 July 2012].

Allianz Knowledge. (2009). *Kyoto Protocol.* [Online] Available at: [Accessed 26 February 2012].

Andersson, M. (). *Europe's dirty secret: Why the EU Emissions Trading Scheme isn't working.* [Online] Available at: http://www.openeurope.org.uk/Content/Documents/PDFs/etsp2.pdf. [Accessed 3 August 2012].

Asian Development Bank (2009). *The Economics of Climate Change in Southeast Asia: A Regional Review.*

Beijing Government (2011). *Beijing 12th Five Year Plan.* [Online] Available at: http://www.bjpc.gov.cn/zt/125ny/nygh/erzhang_ny/201105/t804923.htm. [Accessed 8 August 2012].

California Air Resource Board (2011). *California ARB adopts key element of state climate plan.* [Online] Available at: http://www.arb.ca.gov/newsrel/newsrelease.php?id=245. [Accessed 28 July 2010].

Carbon Trust (2009). *Global Carbon Mechanisms: Emerging lessons and implications (CTC748).* [Online] Available at: http://www.carbontrust.com/resources/reports/advice/global-carbon-mechanisms [Accessed 8 August 2012].

Chen, K. (2011). "China to launch emissions markets in 6 regions by 2013", *ThomsonReuters.* [Online] Available at: http://www.pointcarbon.com/news/1.1527731 [Accessed 27 June 2011].

China–ASEAN Environmental Cooperation Centre (2011). "上海部署" '十二五' 节能减排工作 严格落实目标责任; *trans.* "Shanghai agency's '12-5' responsible for implementing strict steps towards energy conservation emissions reduction" (in Chinese). [Online] Available at: http://www.chinaaseanenv.org/zhxx/hjyw/271347.shtml. [Accessed 8 August 2012].

China Climate Exchanges [Online] Available at: www.cbeex.com.cn/; www.cneeex.com/; www.chinatcx.com.cn/. [Accessed 24 July 2012].

Chongqing Municipal People's Government [Online] Available at: http://www.cq.gov.cn/today/news/321517.htm. [Accessed 8 August 2012].

Climate Change and Carbon Finance Group (2011). "Asia Pacific Climate Change Policy Series", *Norton Rose.* [Online] Available at: http://www.nortonrose.com/au/knowledge/publications/30206/asia-pacific-climate-change-policy-series. [Accessed 1 March 2011].

Climate Change Secretariat (2012). 30 May. United Nations. [Online] Available at: https://cdm.unfccc.int/press/releases/2012_11.pdf. [Accessed 27 July 2012].

Climate Connect (2011). *Bilateral agreement between Australia and China on low-carbon economies.* [Online] Available at: http://www.climate-connect.co.uk/Home/?q=node/495 [Accessed 27 June 2011].

Connor, S. (2011). "Ex-climate sceptic now backs global warming", *The Independent,* 22 October. [Online] Available at: http://www.independent.co.uk/environment/climate-change/exclimate-sceptic-now-backs-global-warming-2374262.html. [Accessed 3 March 2012].

Convery, F., and Redmond, L. (2007). "Market and Price Developments in the European Union Emissions Trading Scheme", *Review of Environmental Economics and Policy,* **1**, 66–87.

Denmark (2009). *The Copenhagen Accord.* [Online] Available at: http://www.denmark.dk/NR/rdonlyres/C41B62AB-4688-4ACE-BB7B-F6D2C8AAEC20/0/copenhagen_accord.pdf. [Accessed 28 February 2012].

Department of Climate Change and Energy Efficiency (). *Actions under the International Forest Carbon Initiative.* [Online] Available at: http://www.climatechange.gov.au/government/initiatives/international-forest-carbon-initiative/action.aspx [Accessed 20 February 2012].

Ellerman, A.D., and Buchner, B.K. (2007). "The European Union Emissions Trading Scheme: Origins, Allocation, and Early Results", *Review of Environmental Economics and Policy,* **1**(1), 66–87.

European Commission (2008). *Questions and Answers on the revised EU Emissions Trading Sytem.* [Online] Available at: http://ec.europa.eu/clima/policies/ets/faq_en.htm. [Accessed 30 July 2012].

EU Zero Emissions Platform (2012). *Biomass with $CO_2$ Capture and Storage (Bio-CCS).* [Online] Available at: http://bellona.org/ccs/uploads/tx_weccontentelements/filedownload/EBTP__ZEP_Report_Bio-CCS_The_Way_Forward.pdf. [Accessed 10 August 2012].

Friends of the Earth (2010) *The EU Emissions Trading System: Failing to deliver.* [Online] Available at: http://ec.europa.eu/clima/consultations/docs/0005/registered/9825553393-31_friends_of_the_earth_europe_en.pdf. [Accessed 30 July 2012].

Government of China (2011). *China's 12$^{th}$ Five Year Plan.* [Online] Available at: http://www.gov.cn/2011lh/content_1825838.htm. [Accessed 3 August 2011].

Government of China (2011). *Energy Conservation and Emission Reduction 12$^{th}$ Five Year Work Plan, 2011.* [Online] Available at: http://www.gov.cn/zwgk/2011-09/07/content_1941731.htm [Accessed 3 August 2011].

Government of India. *India's National Action Plan for Climate Change.* [Online] Available at: http://pmindia.nic.in/climate_change.htm. [Accessed 8 August 2012].

Group of 77 and China (2011). *Statement of the Group of 77 and China by H.E. Ambassador Mr. Alberto Pedro D'Alotto, Vice Minister of Foreign Affairs of Argentina, at the Joint Hihg-Lenel Segment of the 17$^{th}$ Session of the COP of Climate Change Convention and the 17$^{th}$ Session of the COP serving as Meeting of the Parties to the Kyoto Protocol (COP 17/CMP7).* [Online] Available at: http://www.g77.org/statement/getstatement.php?id=111206a. [Accessed 26 February 2011].

Grubel, J. (2011). "Australia passes landmark price law.", *Reuters,* 8 November. [Online] Available at: http://www.reuters.com/article/2011/11/08/us-australia-carbon-idUSTRE7A60PO20111108. [Accessed 22 November 2011].

Guangdong Government (2011). *12$^{th}$ Five Year Plan.* [Online] Available at: http://www.rd.gd.cn/dhl/rdhyzy2/syjsic/dhwj/sewgh/201101/t20110126_114916.html. [Accessed 8 August 2012].

Henningsen, P. (2010). “The Great Collapse of the Chicago Climate Exchange.”, *21st Century Wire*, 18 November. [Online] Available at: http://21stcentury wire.com/2010/08/27/the-great-collapse-of-the-chicago-climate-exchange/. [Accessed 28 July 2012].

International Energy Agency (2011). *World Energy Outlook 2011 Factsheet.* [Online] Available at: http://www.worldenergyoutlook.org/media/weowebsite/2011/factsheets.pdf. [Accessed 12 August 2012].

Joshi, S. (2010). *Economic Instruments for Reducing Carbon Dioxide Emissions: An Exploratory Approach.* National University of Singapore, Singapore.

Jotzo, F. (2011). “Reaching for the sky?”, *Inside Indonesia.* [Online] Available at: http://www.insideindonesia.org/edition-105-jul-sep-2011/reaching-for-the-sky-18071461. [Accessed 20 February 2012].

Jowitt, J. (2010). “Bjørn Lomborg: the dissenting climate change voice who changed his tune“, *The Guardian*, 30 August [Online] Available at: http://www.guardian.co.uk/environment/2010/aug/30/bjorn-lomborg-climate-change-profile. [Accessed 3 March 2012].

Kamiya, S. (2009). “A brief history of climate talks: looking back, looking forward”, *The Japan Times Online*, 7 December. [Online] Available at: http://www.japantimes.co.jp/text/nn20091207f2.html. [Accessed 25 February 2012].

Kruger, J., Oates, W.E. and Pizer, W.A. (2007). “A Decentralization in the EU Emissions Trading Scheme and Lessons for Global Policy”, *Review of Environmental Economics and Policy*, **1**, 112–133.

Lang, C. (2010). “Norway-Indonesia forest deal: US$1 billion dollars worth of continued deforestation?”, *Redd-Monitor.org.* [Online] Available at: http://www.redd-monitor.org/2010/05/28/norway-indonesia-forest-deal-us1-billion-dollars-worth-of-continued-deforestation/. [Accessed 20 February 2012].

Liu, J. (2011). “Reining in China’s energy targets”, *China Dialogue.* [Online] Available at: http://www.chinadialogue.net/article/show/single/en/4138. [Accessed 22 November 2011].

Lütken, S., Fenhann, J., Hinostroza, M., Sharma, S. and Olsen, K.H. (2011). “Low Carbon Development Strategies”. *UNEP RISO Centre.* [Online] Available at: http://namapipeline.org/Publications/LowCarbonDevelopmentStrategies_NAMAprimer.pdf. [Accessed December 2011].

Lyster, R. (2011). “Explainer: Australia’s carbon price mechanism in six dot points”, *The Conservation*, 10 November. [Online] Available at: http://theconversation.edu.au/explainer-australias-carbon-price-mechanism-in-six-dot-points-4230. [Accessed 22 Nov 2011].

Minamikawa, H. (2010). “Statement by Japanese vice-minister for global environment, Hideki Minamikawa, UN Climate Change Conference in Cancún, December 2010”, *The Guardian UK*, 2 December [Online] Available at: http://www.guardian.co.uk/environment/2010/dec/02/japan-stance-kyoto-protocol. [Accessed 22 November 2011].

Ministry of Environment, Japan (2011). *Country Overview: Japan.* [Online] Available at: http://www.env.go.jp/en/earth/ets/mkt_mech/country-overview.pdf. [Accessed 22 November 2011].

Ministry of Environment, Japan (2011). *Bilateral Offset Credit Mechanism.* [Online] (Updated on 21 February 2012) Available at: http://www.env.go.jp/en/earth/ets/mkt_mech/bilateral_moej.pdf. [Accessed 27 July 2012].

Ministry of Finance and Australia Indonesia Partnership, Jakarta (2009). *Economic and fiscal policy strategies for climate change mitigation in Indonesia.* [Online] Available at: http://www.fiskal.depkeu.go.id/webbkf/siaranpers/siaranpdf%5CGreen%20Paper%20Final.pdf

Morgan, J. and Cameron, E. (2011). "Reflections on COP17 in Durban.", *WRI Insights.* [Online] Available at: http://insights.wri.org/news/2011/12/reflections-cop-17-durban. [Accessed 27 July 2012].

Müller, S., Marmion, A. and Beerepoot, M. (2011). "Renewable Energy Markets and Prospects by Region", *International Energy Agency*, November. [Online] Available at: http://www.iea.org/papers/2011/Renew_Regions.pdf. [Accessed 12 August 2012].

NDRC. [Online] Available at: http://www.gov.cn/zwgk/2011-09/07/content_1941731.htm. [Accessed 8 August 2012].

NDRC. [Online] Available at: http://www.ndrc.gov.cn/zcfb/zcfbtz/2011tz/t20120113_456506.htm. [Accessed 8 August 2012].

New Mechanisms Information Platform. (2011). [Online] (Updated 16 February 2015) Available at: http://mmechanisms.org/e/index.html. [Accessed 24 February 2012].

New Market Mechanisms in CHARTS. (2012). [Online] Available at: http://enviroscope.iges.or.jp/modules/envirolib/upload/3352/attach/new_mech_charts.pdf.

Norton Rose. (2011). *Asia Pacific climate change policy series: Thailand.* [Online] Available at: http://www.nortonrose.com/knowledge/publications/30206/asia-pacific-climate-change-policy-series. [Accessed January 2012].

OECD (2012). *South East Asian Economic Outlook 2011/2012.* OECD Publishing, p. 160. [Online] Available at: http://www.keepeek.com/Digital-Asset-Management/oecd/development/southeast-asian-economic-outlook-2011-12_9789264166882-en. [Accessed 27 February 2012].

Office of Market Mechanisms, Climate Change Policy Division, Ministry of the Environment, Japan. (2011). *Offset Credit (J-VER) Scheme.* [Online] Available at: http://www.env.go.jp/en/earth/ets/mkt_mech/j-ver_scheme.pdf. [Accessed 22 November 2011].

Office of Market Mechanisms, Climate Change Policy Division, Ministry of the Environment, Japan (2011). [Online] Available at: http://www.env.go.jp/en/earth/ets/mkt_mech/j-ver_scheme.pdf. [Accessed 28 February 2012].

Office of Natural Resources and Environmental Policy and Planning (2011). *Thailand's Second National Communication under UNFCCC,* p. 15. [Online] Available at: http://unfccc.int/files/national_reports/non-annex_i_natcom/submitted_natcom/application/pdf/snc_thailand.pdf. [Accessed 22 February 2012].

Oppermann, K. (2012). *World Bank's input to the CDM Policy Dialogue.* [Online] Available at: http://www.cdmpolicydialogue.org/public_input. [Accessed 12 July 2012].

Peimani, H. (2011). "Nuclear Energy in Asia: A Post-Fukushima Perspective", *IAGS Journal of Energy Security.* [Online] Available at: http://www.ensec.org/index.php?option=com_content&view=article&id= 309:nuclear-energy-in-asia-a-post-fukushima-perspective&catid=116:content0411& Itemid=375. [Accessed 10 August 2012].

Project Developer Forum (2012). *PD Forum Input to the CDM Policy Dialogue Stakeholder Questions: Questions to Project Participants.* [Online] Available at: http://www.pd-forum.net/files/1c01820934ee2392b64749f062380586.pdf. [Accessed 27 July 2012].

Reklev, S. (2011). "China considers absolute $CO_2$ caps on industries: report", *Thomson Reuters Online.* [Online] Available at: http://www.pointcarbon.com/news/1.1565075?date=20110804&sdtc=1. [Accessed 23 November 2011].

Reklev, S. (2011). "China's pilot carbon market faces fragmented start.", *Thomson Reuters Online.* [Online] Available at: http://www.pointcarbon.com/news/1.1528008. [Accessed 22 November 2011].

Schneider, L. (2007). *Is the CDM fulfilling its environmental and sustainable development objectives? An evaluation of the CDM and options for improvement* [Online] Available at: http://www.panda.org/what_we_do/footprint/climate_carbon_energy/energy_solutions/resources/?118000/An-evaluation-of-the-CDM-and-options-for-improvement [Accessed 20 April 2010].

Stavins, R. (2012). "Low prices a problem? Making sense of misleading talk about cap-and-trade in Europe and the USA", *Harvard Kennedy School Belfer Center for Science and International Affairs*, 25 April. [Online] Available at: http://www.robertstavinsblog.org/2012/04/25/low-prices-a-problem-making-sense-of-misleading-talk-about-cap-and-trade-in-europe-and-the-usa/. [Accessed 28 July 2012].

Stern, N. (2006). *Stern Review on the Economics of Climate Change.* [Online] Available at: http://webarchive.nationalarchives.gov.uk/+/http:/www.hm-treasury.gov.uk/sternreview_index.htm. [Accessed July 2012].

Thai TGO (2012). [Online] (Updated 26 April 2015) Available at: http://www.tgo.or.th/english/index.php?option=com_content&view=category&id =25&Itemid=28. [Accessed 28 February 2012].

Thailand Greenhouse Gas Management Office (2011). *Presentations.* [Online] Available at: http://www.tgo.or.th/english/index.php?option=com_content&view=category&id=51:publication&layout=blog&Itemid=68&layout=default. [Accessed 24 February 2012].

Thailand Greenhouse Gas Management Office (2011). *Technical Studies.* [Online] Available at: http://www.tgo.or.th/english/index.php?option=com_content&view=category&id=50:tgos-research-projects&Itemid=40&layout=default. [Accessed 24 February 2012].

Thailand Greenhouse Gas Management Office (2011). *TGO Carbon Monthly.* [Online] Available at: http://www.tgo.or.th/download/newsletter/2011/dec11/TGO_Carbon_Monthly_EN.pdf. [Accessed January 2012].

Thailand Greenhouse Gas Management Office (2011). *TGO Carbon Monthly.* [Online] Available at: http://www.tgo.or.th/english/index.php?option=com_content&view=category&id=36:newsletter&layout=blog&Itemid=50&layout=default. [Accessed 24 February 2012].

The Guardian (2011). "Canada pulls out of Kyoto protocol", *The Guardian*, 13 December. [Online] Available at: http://www.theguardian.com/environment/2011/dec/13/canada-pulls-out-kyoto-protocol. [Accessed 13 December 2011].

Tohoku Electric Power (2007). *Outline of Kyoto Mechanisms.* [Online] Updated 14 October 2014) Available at: http://www.tohoku-epco.co.jp/enviro/tea2006e/03/03f.html. [Accessed 26 February 2012].

Torney, D., and Gueye, M.K. (2009). "Climate Change Mitigation Policies in Selected OECD Countries", *ICTSD Global Platform on Climate Change, Trade and Sustainable Energy Issue,* Paper No. 8. [Online] Available at: http://ictsd.org/downloads/2010/01/climate-change-mitigation-policies-in-selected-oecd-countries.pdf. [Accessed 12 August 2012].

UNEP RISOE (2007) *CDM Pipeline.* [Online] (Updated 1 August 2012). Available at: http://cdmpipeline.org/.

UNEP RISOE (2011). *NAMA Pipeline.* [Online] (Updated 6 January 2012). Available at: http://namapipeline.org/.

UNEP RISOE (2012). *CDM Projects by host region.* [Online] Available at: http://www.cdmpipeline.org/cdm-projects-region.htm. [Accessed 12 August 2012].

UNFCCC (2012). "India's Development Needs and Mitigation Actions.", *UNFCCC Workshop on NAMAs.* [Online] Available at: http://unfccc.int/files/meetings/ad_hoc_working_groups/lca/application/pdf/india_ws.pdf. [Accessed August 8, 2012].

UNFCCC. Pdf. FCCC/CP16/2011 Paragraph 80.

UNFCCC. Pdf. FCCC/AWGLCA/2011/MISC.2/Add.4, page 2–6.

UNFCCC, The Clean Development Mechanism, http://cdm.unfccc.int/index.html

UNFCCC. (2005) [Online] Available at: http://unfccc.int/resource/docs/convkp/conveng.pdf. [Accessed 28 February 2012].

UNFCCC (2012). *CDM in numbers. Registered Projects by Host Country.* [Online] Available at: http://cdm.unfccc.int/Statistics/Registration/NumOfRegisteredProjByHostPartiesPieChart.html. [Accessed 3 August 2012].

UNFCCC (2012). *CDM in Numbers. CERs Issued by Host Country.* [Online] Available at: http://cdm.unfccc.int/Statistics/Issuance/CERsIssuedByHostPartyPieChart.html. [Accessed 8 August 2012].

UNFCCC (2011). *Appendix I — Quantified economy-wide emissions targets for 2020.* [Online] Available at: http://unfccc.int/meetings/copenhagen_dec_2009/items/5264.php. [Accessed 4 February 2012].

UNFCCC (1999). *Indonesia's National Action Plan for GHG Emission Reduction.* [Online] Available at: http://unfccc.int/resource/docs/natc/indonc1.pdf. (Accessed 24 February 2012).

UNFCCC (2001). *ISSUES IN THE NEGOTIATING PROCESS. A brief history of the climate change process.* [Online] (Updated 13 November 2013) Available at: http://unfccc.int/cop7/issues/briefhistory.html. [Accessed 1 March 2012].

UNFCCC (2007). *Glossary.* [Online] Available at: http://unfccc.int/essential_background/glossary/items/3666.php. [Accessed 7 August 2012].

Vaughan, A. (2011). "What does Canada's withdrawal from Kyoto protocol mean?", *The Guardian UK*, 13 December. [Online] Available at: http://www.guardian.co.uk/environment/2011/dec/13/canada-withdrawal-kyoto-protocol. [Accessed 24 February 2012].

World Monitor. (2006). *Asia and the Pacific.* [Online] Available at: http://www.worldmonitor.info/regions/asiapacific.html. [Accessed 28 February 2012].

Wuppertal Institute. (2011). *Domestic Emission Trading Systems in Developing Countries — State of Play and Future Prospects.* [Online] Available at: http://www.jiko-bmu.de/basisinformationen/publikationen/doc/1054.php. [Accessed December 2011].

Zhang, J., and Wang, C. (2011). "Co-Benefits and Additionality of the Clean Development Mechanism: An Empirical Analysis", *Journal of Environmental Economics and Management,* **62**, 140–154.

# 17

# PAST AND CONTEMPORARY PROPOSALS ON DIFFERENTIATION AND EQUITY: SHAPING THE 2015 CLIMATE AGREEMENT

MELISSA LOW and LIM LEI-THENG

Climate change is a global phenomenon and the greatest and most debated environmental threat facing humankind today. Scientific reports such as the Intergovernmental Panel on Climate Change (IPCC) Assessment Reports have called for the stabilisation of temperature increase to below 2°C relative to pre-industrial levels. Although this requires urgent and fundamental departure from business as usual, countries are still in disagreement on how to attribute the responsibility of dealing with climate change. Developed countries demand that all countries must do their part while developing countries insist that developed countries must take greater responsibility due to past actions. The last two decades have seen a shift in the thinking with regard to each party's responsibility towards climate change, stemming from the evolving interpretation of equity principles embedded in the United Nations Framework Convention on Climate Change (herein after referred to as 'UNFCCC' or 'the Convention'). A new form of differentiation has emerged that focuses more on countries' national circumstances and capabilities, departing from the traditional notions of historical responsibility, which accords greater burdens on developed countries. Recognising this new form of differentiation, in view of the future international legal regime on climate change, will be useful in identifying an equitable outcome for countries, so that the terms of a legal agreement come 2015 are more readily accepted.

## 1. Introduction

In December 2015, France will host the United Nations Framework Convention on Climate Change (UNFCCC) 21st Conference of the Parties (COP21), where it is expected that a new climate change deal will be struck. It will receive close to fifty thousand participants and international delegates. This will be an important milestone since it is hoped that for the first time in over 20 years of climate negotiations, all the nations of the world will be bound by a universal agreement on greenhouse gas (GHG) mitigation. The international community is on the cusp of defining the next stage of the collective effort to combat climate change under the Ad hoc Working Group on the Durban Platform for Enhanced Action (herein referred to as the "Durban Platform" or "ADP").[1] It is not unexpected that the equity principles such as Common but Differentiated Responsibilities and Respective Capabilities (CBDR & RC) have come to the fore once more. Although the world has changed a great deal since the adoption of key milestones in the area of climate change and sustainable development — the United Nations Conference on Sustainable Development (UNCSD) and adoption of the UNFCCC in 1992, and the Kyoto Protocol in 1997 — basic differences in wealth, perception and outlook persist among nations. In order to increase participation in the future international legal regime on climate change, soft law principles could be used to lead to a set of instruments with varying effects on the legal character of obligation on Parties in Paris. For example, countries which are particularly vulnerable to the effects of climate change could be exempt from strict mitigation commitments; be given financial or technical support for climate action; and be allowed longer timeframes for meeting actionable targets to deal with climate change. Though unclear if the 2015 outcome will contain legally binding obligations, Parties' positions in the ADP talks are already beginning to show where they stand on the equity debate.

In the 1980s, scientists warned that global warming beyond natural variability was occurring and that this was largely due to human activity, leading to the increase in anthropogenic emissions of greenhouse gases. International consensus soon developed that States should consider a legally binding agreement on climate change, which would address emissions of greenhouse gases not covered under the Ozone Layer protection regime. The events which followed, particularly the proceedings of the Second World Climate Conference in Geneva in 1990, highlighted the 'North-South' divide on how developed and developing countries viewed climate change. Disagreement led to the formation of an International Negotiating Committee (INC) for a Framework Convention on Climate Change established under the auspices of the United Nations General Assembly. The Convention was crafted, adopted in New York on 9 May 1992 and entered into force on 21 March 1994.

By the first Conference of the Parties (COP1) held in Berlin from 28 March to 7 April 1995, the UNFCCC had been ratified by 126 countries and the European Community, and 20 industrialized countries had already submitted their first national communications. In Berlin, a working group

[1] Decision 1/CP.17, 'Establishment of an Ad Hoc Working Group on the Durban Platform for Enhanced Action', UN Doc. FCCC/CP/2011/9/Add.1 (15 March 2012), paragraphs 2–6.

called the Ad Hoc Group on the Berlin Mandate (AGBM) was established, aimed at negotiating for stronger commitments on the part of industrialised countries in a protocol or any other legal instrument for adoption in Kyoto in 1997. At COP3 in 1997, the Kyoto Protocol (KP) was signed. The KP required 37 industrialised countries to reduce their aggregate emissions by 5.2% as compared with 1990 levels in its commitment period. Annex I of the Protocol contains the list of industrialised countries, and later these countries became collectively known as Annex I Parties. Notably, countries that were not included in this list ('Non-Annex I Parties') were not given any legally binding commitments to reduce their emissions on the basis of key equity principles embedded in the UNFCCC, including the principle of Common but Differentiated Responsibilities and Respective Capabilities (CBDR & RC).

## 2. Differentiation in the Climate Regime

Despite the weaknesses of the rest of the treaty, the principles found in Article 3 of the Convention are arguably the most important 'law' in the whole agreement because they prescribe how the regime for regulating climate change is to be developed by the parties. The main elements of this provision are worth quoting:

Article 3: Principles

In their actions to achieve the objective of the Convention and to implement its provisions, the parties shall be guided by, *inter alia*, the following:

1. The Parties should protect the climate system for the benefit of present and future generations of humankind, on the basis of equity and in accordance with their common but differentiated responsibilities ...
3. The Parties should take precautionary measures to anticipate, prevent, or minimise the causes of climate change and mitigate its adverse effects ...
4. The Parties have a right to, and should, promote sustainable development ...

The elements of Article 3 are drawn directly from the non-binding Rio Declaration on Environment and Development. They are not expressed in obligatory terms despite the obligatory wording of the *chapeau* sentence, though the use of 'should' qualifies their application by Parties. These principles are open-textured and leave considerable room for interpretation and elaboration. Given their explicit role as guidance and softer formulation, the principles in Article 3 are not necessarily binding rules which must be complied with, or which entail responsibility for breach if not complied with; yet, despite all these limitations, they are not legally irrelevant. It is significant that the relevance of Article 3 was reiterated in the mandate for the negotiation of the KP and is referred to in the preamble to the Protocol.

The force of differentiation brought about by these principles is seen in the Kyoto Protocol, which saw its genesis in the Berlin Mandate where developed countries eschewed targets for developing countries. Nothing less would have permitted the negotiations to move ahead. Since then, developing countries have been careful to refer to this principle in all contexts where obligations, real or potential, are discussed. This is true also for non-binding instruments adopted under the auspices of organs of the United Nations. Where a principle of environmental law is codified in treaty law, it assumes a different legal character as compared to ethical, moral or prudential principles. Yet principles have a slippery and contradictory quality — on the one hand presuming to guide action, yet on the other hand it is unclear in what circumstances they ought to be applied. Nevertheless, principles provide much guidance towards understanding Parties' interests when they invoke these principles in combination with one another.

Fairness is a crucial consideration in concluding multilateral agreements. The UNFCCC has been studied for its approach to fairness and the resultant differentiation of countries and their obligations. While some claim that differentiated responsibilities are necessary in the climate change regime, others are more critical and insist that the international community's reliance on differential obligations jeopardises the very purpose of international environmental law and that despite the increased number and reach of multilateral treaties, this still falls short of proof that a new normative 'principle' is in play, i.e., not a customary principle of international law and cannot be applied universally without question. The notion of differential treatment refers to instances where, because of pervasive differences or inequalities among states, formal legal equality and reciprocity are sidelined to accommodate extraneous factors. These include divergences in levels of economic development, different contributions to the creation of a problem or unequal capacities to tackle existing problems. The basis of differentiation is the recognition of the future development needs of developing countries, of other special needs and circumstances of certain countries, and of the different contributions of countries to the specific environmental problem at hand. Consequently, less stringent obligations, different timings in the application of provisions (grace periods or delayed implementation of obligations, or priority implementation in specially affected countries), and international assistance in terms of financing, capacity building or technology transfer are accorded to countries that are deemed to require such differential treatment.

Just as the Ad hoc Group on the Berlin Mandate (AGBM), operational from 1995-1997 paved the way for differentiation and equity to be discussed more broadly, the ADP is the current platform Parties use to shape the degree of differential treatment and legal form of the 2015 agreement.

## *Key Equity Principles*

Differentiation and equity are applied through Articles in the UNFCCC and Principles in the Rio Declaration from the UNCED in 1992. There are several key equity principles that Parties often refer to in submissions and interventions, but without contexts, it would be difficult to understand how Parties signal affiliation with certain principles in shaping their negotiation tactics and strategic interests. Nevertheless, these principles are important, and they provide much guidance towards understanding the interests of the Parties when they invoke these principles. Parties' positions in the ADP show where they stand on the equity debate.

Article 3.1 of the UNFCCC and the principle of **Common but Differentiated Responsibilities and Respective Capabilities (CBDR & RC)** emphasises that developed countries "should take the lead", given that it would not be morally, legally or economically feasible to expect developing countries to pay the supplemental costs associated with leapfrogging the carbon-intensive stage of industrialisation through which all developed countries have gone. Respective capabilities impose burdens upon developed States, while not necessarily bestowing upon them additional rights (French, 2000). It is a means of utilising the resources of the North for the benefit of the South. The principles of CBDR & RC are also expressed in Principle 7 of the Rio Declaration, calling for the delegation of heavier responsibilities to developed countries in response to "the pressure their societies place on the global environment", and effectively shift the focus from the needs of the poor to the wrongs of the rich. The principle of CBDR & RC is one of the most cited in negotiations. The very fact that the Kyoto Protocol offers differential legal arrangements for Annex I and non-Annex I Parties in terms of their mitigation commitment is illustrative of the recognition of this principle.

A country's contribution to past emissions is often highlighted in order to attribute greater responsibility to address climate change. Historical Responsibility finds expression in Article 4 of the UNFCCC and is operationalised by Article 3.14 of the Kyoto Protocol. According to the established concept of Historical Responsibility, the commitments of individual countries to take action on climate change are distributed based on the perceived effects of their past emissions as manifested in present climate change. This principle is intended to address relative inequalities, and has received substantial academic attention. A similar principle is the Polluter Pays Principle (PPP) found in Principle 16 of the Rio Declaration, which requires that the polluter bears the cost of pollution, with due regard to the public interest and without distorting international trade and investment. The PPP assumes that the party that has caused environmental damage should pay the cost of remediation. In the current climate regime, developed countries set Quantified Emissions Limitations and Reduction Objectives (QELROs) because of their Historical Responsibility, in accordance with the Polluter Pays Principle.

The Basic Needs Principle is encapsulated by Principle 3 of the Rio Declaration stating that "the right to development must be fulfilled so as to equitably meet developmental and environmental needs of present and future generations" and is concomitant with the Right to Sustainable Development and Intergenerational Equity, which has its roots in Article 3.4 of the Convention. The Basic Needs Principle is line with the notion that social and economic development is, for many developing countries, the overriding priority for developing countries. While not specifically mentioned in either the Convention or Kyoto Protocol, the principle of Poverty Eradication and loss of livelihoods featured strongly in several non-Annex I Parties' statements and interventions at the UNFCCC meetings, and is enshrined in Principle 5 of the Rio Declaration.

Another key equity principle is on national circumstances, found in Article 3.2 of the Convention. Furthermore, Article 4, paragraph 10 of the Convention notes the vulnerability of countries to the effects of climate change, emphasising their dependence on fossil fuels and lack of recourse to alternative forms of fuel for power generation. This last provision is an important one, which will be revisited in subsequent chapters in this thesis. Fossil fuel dependency is an important constraint on countries, both Annex I and non-Annex I alike. However, linked to the principles of CBDR and RC as well as national circumstances, Article 4.10 presents a unique consideration for countries that inherently lack the ability to switch to alternative energy sources and are therefore highly dependent on fossil fuels. Increasingly, in UNFCCC negotiations, such concerns have been raised by countries like Singapore, which claims to be "alternative energy disadvantaged". Countries to which Article 4.10 apply could, through their interpretations of the principles of differentiation and equity, show the way towards articulating a new legal agreement that would be able to effectively take into consideration this particular provision.

## 3. Contemporary Proposals on Differentiation and Equity

In the two years leading up to the Copenhagen Conference of 2009, the scheduled end of the Bali process, there were numerous attempts by developed countries to achieve parallelism between developed and some larger and more advanced developing country actions, such as China. Achieving this parallelism would effectively mean that there would be formal differentiation among developing countries, which currently does not exist under the Kyoto Protocol. The balance of power, it was argued, has changed dramatically since the UNFCCC and Kyoto Protocol were negotiated. Some developed countries noted that emerging powers like China, India, Brazil, and South Africa no longer merit the extent of differential treatment they were entitled to under the climate change regime. Future agreements, therefore, it was argued, should contain greater parity or 'parallelism' in commitments between developed and developing countries.

Equity principles are shaping the legal form of the 2015 Paris Outcome, most evidently in the formulation of the 'agreed outcome with legal force' in the Durban Mandate. While the terms 'a protocol' and 'another legal instrument' have clear precedents in international climate law as both terms were used separately in the 1995 Berlin Mandate and 1997 Kyoto Protocol, the novel phrase 'agreed outcome with legal force' has never been used in international climate law, and appears to be new to international law as well. Since the Bali meeting, Parties have been active in articulating their views, often hinged on principles and their interpretations, to the UNFCCC Secretariat and subsequently, the ADP Co-Chairs.

Notably, Decision 1/CP.17 critically left out CBDR, reflecting a departure from some seventeen years of UN-sponsored international negotiations on climate change, and serving to eliminate the Annex I/Non-Annex I distinction, which had been the centrepiece of international climate policy and negotiations since its adoption at COP1 in Berlin in 1995. This is no benign oversight. Through the two weeks of the conference, developed countries were unanimous in their insistence that any reference to CBDR must be qualified with a statement that this principle must be interpreted in the light of 'contemporary economic realities'. They were also insistent that the future regime must be 'applicable to all'. India, among other developing countries, argued in response that this would be tantamount to amending the UNFCCC.

The only way out of this impasse was to draft the text in such a way that it was rooted in the Convention — "under the Convention" — thereby implicitly engaging its principles, including the principle of CBDR. This, it was believed, would hold efforts to reinterpret and qualify this principle at bay. Nevertheless, the fact that the divisions over the application of this principle are such as to preclude even a rote invocation of it signals a likely recasting of differentiation in the future climate regime. Given the opposition of many developed countries to the perceived rigid Annex I-non-Annex I differentiation in the climate regime, the future regime is likely to countenance a more nuanced vision of differentiation between countries.

The use of the term 'applicable to all' further substantiates this point. Admittedly, merely because this instrument applies to all Parties does not necessarily imply that it applies symmetrically to all Parties. However, given the insistence by developed countries that the future regime contains mitigation obligations for all, and the conspicuous absence of the usual markers for differentiation — equity and CBDR — it appears that the goalposts on differentiation are likely to shift post-2020. This was a turning point in the UNFCCC negotiations, as Parties were given an opportunity to re-present their views on equity principles, further advancing understanding of the legal form of the 2015 agreement.

In the 2010 Cancun agreements, the notion of Equitable Access to Sustainable Development (EASD) was introduced in the context of a timeframe for global peaking of greenhouse gas emissions, where the COP recognised that the timeframe for peaking will be longer in developing countries, and that social and economic development and poverty eradication are the first and overriding priorities of developing countries. The following year in Durban, experts from China, India, Brazil and South Africa (BASIC Group) released a paper with suggestions on EASD, paving the way for a Workshop on EASD in 2012. The EASD Workshop held in Bonn, Germany on 12 June 2012 provided an opportunity for Parties to exchange views on EASD and equity in general, including on the following: the context for equity and EASD; the definition of equity; application of the equity principle; and possible next steps. The workshop allowed Parties to negotiate the right to development and their duties to control sources of environmental harm.

Once formal negotiations under the ADP got underway, this gave Parties and observer organisations the opportunity to submit information, views and proposals on the work of the ADP including, *inter alia*, mitigation, adaptation, finance, technology development and transfer, capacity-building, and transparency of action and support. Views on equity and differentiation were also expressed at meetings of the Parties, including additional sessions of the ADP. Several Parties and coalitions have made reference to CBDR in their most recent submissions relating to differentiation. The African Group suggested that the preamble to the 2015 agreement should make "specific reference to the guidance by science, equity, common but differentiated responsibilities and respective capabilities, and historical responsibility, whilst taking into account national circumstances" and the Independent Alliance of Latin America and the Caribbean (AILAC) group proposed that "all contributions are nationally determined; countries will determine what they can and want to put forward. The principles of sovereignty and CBDR & RC are inherent in this process". On the other hand, Japan and the European Union (EU) stressed that the principle of CBDR & RC has to be interpreted in a dynamic context and reflects evolving realities. The United States noted they would not support a post-2020 agreement based on a 1992-era bifurcated approach, given that it will be nearly 30 years old when the new agreement comes into effect. The Russian Federation similarly noted that current global socio-economic development realities should be accounted for, while one-sided interpretation of the principles should be avoided. In the interest of the objectives of the new instrument, the Russian Federation underscored that it is not acceptable to maintain the static division of countries as per the current UNFCCC categories. Categorisation should be based on objective, scientific information[2] shared by all countries. It should also be flexible, allowing for adjustments (where necessary) to be made to countries' commitments as their level of socio-economic development increases.

## 4. Recognising an Emerging Form of Differentiation

This chapter investigates an emerging form of differentiation by means of thorough examination of a series of publications, speeches and submissions

[2]This simply refers to science and technical information that clarify a country's GHG emissions and how much it intends to reduce GHG in a given time frame. Scientific information is usually measured using prescribed guidelines by UN bodies or the IPCC.

to the UNFCCC Secretariat and suggests that Singapore, an island city-state in Southeast Asia, has inadvertently managed to operationalise the principles of equity, CBDR and national circumstances. Withholding judgment, this chapter submits that countries that are dependent on fossil fuels for power generation and have lesser means to mitigate climate change by switching to alternative energy sources are more likely to come together if their needs and interests are met by being recognised and reflected in the new 2015 Agreement. A select group of countries, including Brunei Darussalam, Republic of Korea, Singapore, Trinidad and Tobago and the United Arab Emirates, have been identified as having similar interests at the UNFCCC in terms of fossil fuel dependency and lack of ability to switch to alternative energy sources for power generation. These countries account for 2.59% of the world's carbon dioxide emissions. Though this number appears small in the global context, they collectively emit 5.02% of the non-Annex I countries' total emissions. This chapter therefore aims to recognise that if such needs and circumstances are reflected in the new 2015 Agreement, it could be one way to get other countries with similar interests on board.

### *Country Grouping*

The above mentioned group of countries is disproportionately vulnerable to the effects of climate change, particularly in terms of their fossil fuel dependency and lack of ability to switch to alternative energy sources for power generation. They come from overlapping country groups that now exists in the negotiations. *Prima facie*, the sheer number of coalitions makes it more difficult for any one of them to be heard effectively. Similarly, the widening range of issues has implications for coordination among coalition members. In particular, divergent positions on various agenda items become more visible.

Brunei Darussalam, Republic of Korea, Singapore, Trinidad and Tobago and the United Arab Emirates Parties are members of the G77 & China (except Republic of Korea),[3] Alliance of Small Island States (AOSIS), and Small Island Developing States (SIDS), Environmental Integrity Group and Organisation of the Petroleum Exporting Countries

[3]The G77 is a UN intergovernmental caucus formed in the early 1960s in preparation for the 1964 UN Conference on Trade and Development. It derives its name from an original membership of 77 states, but the group now includes 132 states. China is not officially a member, but G77 texts are typically introduced jointly on behalf of the G77 and China.

(OPEC), though not all of them fall within the same groups. Brunei Darussalam, while an oil exporting country, is just part of the G77 & China but not OPEC. Singapore has 3 affiliations with the G77 & China, AOSIS and SIDS. Trinidad and Tobago is a member of the G77 & China and AOSIS groups and the United Arab Emirates is a member of both the G77 & China and OPEC. Upon examination of their submissions to the UNFCCC and oral interventions at meetings, these countries have referred to their vulnerabilities to the effects of climate change and the need for the new climate regime to take into account countries' national circumstances when deciding on how countries should share the responsibility to address climate change. Furthermore, several of these countries are advanced or emerging developing economies and have been singled out to perhaps contribute more to the global effort commensurate with their per capita income or other indicator of capability. Thus, it can be argued that reference to the principle of historical responsibility might not be sufficient for such a group of countries for an equitable distribution of mitigation responsibility and there seems to be a new form of differentiation towards an equitable outcome emerging in the current ADP track of the negotiations.

### *"Alternative Energy Disadvantaged" Finding Expression in Negotiating Texts*

An example of a country that has considered national circumstances in relation to renewable/alternative energy is Singapore, a city-state which aspires to be a climate-resilient global city, well-positioned for green growth. Having ratified both the Convention and Kyoto Protocol, Singapore set an unconditional emission reduction pledge of 7–11% below business-as-usual (BAU) levels by 2020 and is further prepared to reduce its emissions by 16% from the 2020 BAU level, should there be a global agreement on climate change where countries adhere to their targets in good faith.

For a small island state not endowed with any natural resources, this would require a major effort. The country's early mitigation actions and difficulties to switch to alternative forms of clean energy renders it 'alternative energy disadvantaged', defined as a state of lack of access, and the ability, to develop clean energy as a climate change mitigation action that does not compromise economic growth and national poverty reduction efforts (National Climate Change Strategy, 2012). Interestingly, Singapore's dependence on fossil fuels and lack of ability to switch to alternatives due to land size and population density is public knowledge. It is no secret that the creation of diplomatic space for negotiations is important for a small island with no natural resources. Multilateral disputes and various episodes in history highlight that in multilateral diplomacy, especially at conferences which are intended to codify legal norms and principles, Singapore's representatives must know from the start what its national interests on key issues are and not allow themselves to be intimidated by bigger states (Jayakumar, 2011). Countries like Singapore need to identify which other countries have similar interests and concerns in order to work with them to advance, or oppose, certain proposals. This can be a difficult and even frustrating process as countries are often locked in by regional group positions.

Mutual interests are clearly often the deciding factor for countries position to themselves with a group (such as an alliance), or as individuals when the time comes for them to raise a point. In Singapore's case, ensuring a fair and equitable process by which decisions are made is important to allow the country a chance to speak and to inform others — often bigger states — of its unique national circumstances. The UN can provide small countries like Singapore with opportunities to defend and advance its interests but when the opportunity does presents itself, diplomats must be swift to act and deft in multilateral diplomacy. This is probably what would happen in the time leading up to the 2015 Agreement. A country's position on differentiation and equity can therefore influence its negotiating tactic, compelling the Party to lobby countries with similar interests towards getting what they want out of the discussions. This is not unheard of in diplomacy.

Furthermore, the UNFCCC is not the only international organization that has adopted differentiated rules for groups of countries. In fact, several other multilateral environmental agreements, including the 1982 United Nations Convention on the Law of the Sea, the 1983 International Undertaking on Plant Genetic Resources, the 1987 Montreal Protocol on Substances that Deplete the Ozone Layer, the 1992 Convention on Biological Diversity and the 1994 United Nations Convention to Combat Desertification, have incorporated the notion of differentiated responsibility of states with respect to the protection of the environment. The basis of this differentiated responsibility is the recognition of the future development needs of poor countries, of other special needs and circumstances of certain countries, and of the different contribution of countries to the specific environmental problem at hand. The resulting differential treatment usually consists of less stringent obligations, different timing of the application of provisions (grace periods or delayed implementation of obligations, or priority implementation in specially affected countries), and international assistance in terms of financing, capacity building or technology transfer.

Drawing upon this notion, it could be argued that a new and emerging form of differentiation is taking shape towards the 2015 Agreement. As outlined in earlier chapters, this form of differentiation deviates from historical responsibility but hinges on interpretations of a country's specific needs and special circumstances — or simply, national circumstances. Singapore, in light of ADP negotiations, has been shown to elaborate on this principle, putting forth a strong case for its "alternative energy disadvantaged-ness", and most notably perhaps, is the fact that this has found expression in negotiating text at the UNFCCC, in the AWG-LCA.

At COP17 in Durban in December 2011, the Ad hoc Working Group on Long-term Cooperative Action under the Convention worked in informal groups toward the preparation of a comprehensive and balanced outcome to be presented to the Conference of the Parties for adoption at its seventeenth session (COP17). The document FCCC/AWGLCA/2011/CRP.39 reflects work undertaken at the fourth part of the fourteenth session of the AWG-LCA (AWG-LCA14), in order to carry forward ideas and proposals in areas where continued discussions were envisaged in 2012.

Section I, paragraphs 6 and 7 of document FCCC/AWGLCA/2011/CRP.39 recalls and recognizes that:

6. *Article 4, paragraph 10, of the Convention that recognises the serious difficulties faced by some Parties in switching to alternatives to the use of fossil fuels.*

7. *Parties that are alternative energy disadvantaged, as a result of a lack of access to renewable sources of energy, face significant constraints in undertaking mitigation actions ....*

Section I on Global Goals for substantially reducing global greenhouse gas emissions by 2050, paragraph 24 notes:

24. *Aspects that may be considered in this context include*:

(a) *Best available scientific knowledge*;
(b) *The objective, principles and provisions of the Convention, and the Bali Action Plan*;
(c) *Equity, common but differentiated responsibilities and respective capabilities, and sustainable development (equitable access to sustainable development which will ensure adequate time for social and economic*

*development for all developing countries; low-carbon development strategy is indispensable to sustainable development)*;

(d) *Socio-economic conditions and adequate time for economic development for all developing countries*;
(e) *The fact that all countries, especially developing countries, need access to the resources required to achieve sustainable social and economic development*;
(f) *The carbon budget, in the context of equitable access to global atmospheric space*;
(g) *The mitigation potential of "alternative energy disadvantaged" Parties which have serious difficulties in switching to alternatives to the use of fossil fuels (Article 4, paragraph 10, of the Convention)* (emphasis, mine);

Section I, paragraph 33 reiterates the concerns on elements of context:

33. *Aspects that [may] [should] be considered in this context include*:

(a) *Best available scientific knowledge*;
(b) *Time frame for peaking will be much longer in developing countries*;
(c) *Implementation of commitments under Article 4 paragraph 1, 2, 3, 5 and 7 of the Convention;*
(d) *Non-Annex I peaking is dependent on the level of support provided by Annex I Parties*;
(e) *Ensure sufficient time for and equitable access to sustainable development, and bearing in mind that social and economic development and poverty eradication are the first and overriding priorities of developing countries*;
(f) *Common but differentiated responsibilities and respective capabilities*;
(g) *Fair and equitable allocation of the atmospheric space, taking into account the criteria of historical climate debt and population*;
(h) *Relevant provisions and principles of the Convention, including the mitigation potential of alternative energy disadvantaged Parties which have serious difficulties in switching to alternatives to the use of fossil fuels* (emphasis, mine).

While these texts did not find their way into the final decisions adopted by the COP as part of the outcome of the work of the AWG-LCA (FCCC/CP/2011/9/Add.1), this gives a clear indication that the concept is emerging, and is starting to be taken seriously.

## 5. Shaping the 2015 Climate Change Agreement

Maintaining principled positions has its advantages, such as gaining respect. This helps in optimising opportunities to lobby other Parties and is part and parcel of diplomacy. Singapore's perseverance since the Bali Action Plan to make known its specific needs and special circumstances is admirable. However, in order to really have an impact in shaping the 2015 Agreement, countries need to come together to strengthen each others' positions. To illustrate this point, five countries were selected and grouped together on the basis of their similar national circumstances in their dependency on fossil fuel and lack of ability to switch to alternative sources of energy to mitigate climate change. These are: Brunei Darussalam, Republic of Korea, Singapore, Trinidad and Tobago and the United Arab Emirates. These countries are fossil fuel dependent and lack the ability to switch to alternative energy sources, as evidenced by Table 17.1.

These countries account for 2.59% of the world's carbon dioxide emissions. Though this number appears small in the global context, they collectively emit 5.02% of the non-Annex I countries' total emissions.

This brief look into potential country groupings and lobbying of interests by invoking principles of differentiation and equity show that substantive issues will now have to be considered in shaping the new 2015 Agreement. The UNFCCC will have to move beyond traditional notions of differentiation, such as historical responsibility, and addressing the concerns of developing countries that may now have to mitigate the effects of climate change by undertaking emission reduction pledges, targets or commitments. Evidently, countries like Brunei Darussalam, Republic of Korea, Singapore, Trinidad and Tobago and the United Arab Emirates are fossil fuel dependent and have serious problems switching to alternatives — though for varying reasons — and such specific needs and special circumstances will have to be taken into serious account.

One suggestion is that a new group of countries who are alternative energy disadvantaged could be given differentiated responsibilities. For instance, instead of a percentage reduction in greenhouse gas emissions like in the Kyoto Protocol's QELROs for Annex I Parties, the ADP could consider an outcome that reflects a different model of international law, which focuses on its catalytic and facilitative role by having a core agreement and a number of optional annexes among which countries could pick and choose. This model not only addresses the specific needs and special circumstances of Parties, but also allows flexibility, which would likely score well in terms of participation and compliance, but perhaps less

Table 17.1. Mix of energy sources per country.

| Party | Share in Total Primary Energy Supply (TPES) in 2009 (%) | | | | | | | | |
|---|---|---|---|---|---|---|---|---|---|
| | Coal | Oil | Gas | Nuclear | Hydropower | Geothermal | Solar/Wind/ Others | Biomass/ Waste | Electricity and heat |
| Brunei | 0 | 22 | 78 | 0 | 0 | 0 | 0 | 0 | 0 |
| Republic of Korea | 28 | 40 | 14 | 17 | 0 | 0 | 0 | 1 | 0 |
| Singapore | 0 | 61 | 38 | 0 | 0 | 0 | 0 | 0 | 0 |
| Trinidad & Tobago | 0 | 8 | 92 | 0 | 0 | 0 | 0 | 0 | 0 |
| United Arab Emirates | 0 | 18 | 82 | 0 | 0 | 0 | 0 | 0 | 0 |

(*Source*: IEA Energy Balance Tables).

Table 17.2. $CO_2$ emissions in 2009 for select countries.

| | $CO_2$ Emissions in 2009 (excludes land use change) |
|---|---|
| Party | (Mt $CO_2$) |
| Brunei | 8.12 |
| Republic of Korea | 515.46 |
| Singapore | 40.14 |
| Trinidad & Tobago | 40.17 |
| United Arab Emirates | 147.04 |
| World total | 28947.00 |
| Non-Annex I total | 14945.00 |

(*Source*: IEA Energy Balance Tables, 2012).

so in terms of stringency. The hope would be that, over time, participation and compliance would provide dynamic feedback, leading countries to ratchet up their level of ambition as they become more comfortable with the system. But whether this would happen at all, or would happen quickly enough to prevent dangerous climate change, is very uncertain. While such a facilitative approach taking into account this emerging form of differentiation is unlikely to lead to stabilisation of temperature increase to below 2°C relative to pre-industrial levels, such an outcome would be of great value and be equitable to most countries.

## 6. Conclusion

Leading up to the 2015 Agreement, Parties will have the tough job of building consensus on what they wish to see in the new legally binding outcome or protocol. Based on the analyses of various processes in the Convention dealing with differentiation and equity, one can understand how difficult this will be. Deep-rooted ideologies and references to principles and provisions of the Convention arguably ground Parties to their views. The ADP is an important milestone in the history of climate negotiations for three reasons:

(1) Parties are welcome to give their views on the basis for a fair outcome; the principles and provisions they draw on; lessons from other processes under the Convention and other multilateral processes; the scope, structure and design of the 2015 Agreement; and ways of defining and reflecting enhanced action.
(2) Parties are shaping the way that negotiations are being held, as evidenced by the calling for a workshop on EASD, the roundtable

setting of the ADP meetings and calls for contact groups in 2013 and 2014 for more focused discussions.

(3) The degree of transparency and openness of the process, such as having Co-Chairs to the ADP (one from a developed country and one from a developing country); invitation of experts to deliver views on the process and means of implementation; and the involvement of observers and civil society to partake in the ADP negotiations.

It is clear that the scale of the climate change challenge is enormous. Adding to the mix, uncertainties about the climatic system and the increased frequency of extreme weather events lend urgency to the issue. Perhaps the added complexity to climate change calls for a broader solution, and we might yet find this in the ADP process to build consensus toward the 2015 Agreement by getting Parties together to negotiate on what principles and provisions should aid in decision making, how such an agreement — applicable to all Parties — can be fair, equitable and ambitious enough for all (procedurally and decision-wise), and ultimately satisfy the needs and interests of all stakeholders.

The Principles of Equity, Common but Differentiated Responsibilities and Respective Capabilities (CBDR & RC), though important, have been left out of the text of the decision establishing the Durban Platform. Instead, the text establishes a goal of "ensuring the highest possible mitigation efforts by all Parties", reflecting a dramatic departure from some seventeen years of UN-sponsored international negotiations on climate change and which serves to eliminate the Annex I-Non-Annex I distinction, which had been the centerpiece of international climate policy and negotiations since its adoption at COP1 in Berlin in 1995.

This thesis has highlighted that past proposals have given way to reinvigorated arguments about differentiation and equity in the context of greenhouse gas emissions mitigation, in part due to the evolved negotiations process and overall greater accountability and transparency of the climate change regime. Parties are often compelled to act according to their domestic policy with many of their negotiation positions being an extension of the principles and preferences embedded in them, as illustrated in the chapters summarising the AGBM, EASD and ADP submissions. With this in mind, it is therefore unsurprising that many Parties wish to avoid being locked into a commitment that could be potentially burdensome and restrictive on growth. Collective effort on the part of individual countries and the international community over an extended period and across a range of activities is needed.

In conclusion, there is a need to recognise a new and emerging form of differentiation where countries that have explicitly noted their fossil fuel dependency and lack of ability to switch to alternative sources of energy to mitigate climate change could be grouped together and have their interests met through a facilitative outcome of the ADP which might involve a core agreement and a number of optional annexes, of which one might be tailored to address the specific needs and special circumstances of such "alternative energy disadvantaged" countries. This may assist in helping to lower the threshold of the entrance into force of the 2015 Agreement compared to that experienced with the Kyoto Protocol, where participation is optimised and yet not conceding on stringency and compliance over time. This is an important point as there needs to be a way to reconcile nationally proposed mitigation commitments with the required level of ambition as indicated by science to stabilise temperature increase to below 2°C relative to pre-industrial levels so as to avoid runaway climate change.

Proposed mitigation commitments need to be ambitious, quantifiable, transparent, robust and allow the comparison of Parties' efforts, including in terms of impacts on greenhouse gas emission reductions, before they are inscribed in the 2015 Agreement. The commitments should also be equitable, and fit with the low emissions development strategies of each Party. Step one must be about considering the options for achieving this. The key challenge will be how, in practical terms, this can be done in relation to mitigation commitments for all Parties in time for the 2015 Agreement. The key questions to consider are:

- Will nationally determined actions be ambitious enough to meet the 2°C threshold temperature of the Convention?
- What types of commitments could be considered?
- What would *ex ante* clarity elements look like in order to ensure transparency, quantification and comparability? Will developing countries accept an *ex ante* review process? Who would conduct it, and how?

The conclusion of a 2015 agreement requires that such difficult questions be addressed. Undoubtedly, since 2009 there have been some important re-alignments in historical allegiances. Many new negotiating groups and alliances have sprung up, some of them bridging the traditional developed and developing countries divide. The US has also recently entered into important bilateral agreements with China and India. Yet, there still exists deep tensions between those who advocate for maintaining the status quo, and those who want to move beyond the Kyoto Protocol's style of differentiation. Hard political bargaining will be necessary to broker a solution to the deadlock that affects international climate governance.

The results of the analyses from this paper provide an opportunity to discuss equity as the driver of the operationalisation of principles via commitments. My hope is that this conclusion would be useful to policy makers, diplomats, negotiators and scholars to assist in informing future decisions and positioning with regard to climate change policy in Singapore and beyond, as well as spark further research on finding the optimal climate policy architecture.

**List of Abbreviations**

| | |
|---|---|
| AGBM | Ad Hoc Group on the Berlin Mandate |
| AILAC | Independent Alliance of Latin America and the Caribbean (AILAC) |
| AOSIS | Alliance of Small Island States |
| AWG-ADP | Ad Hoc Working Group on the Durban Platform for Enhanced Action |
| AWG-LCA | Ad Hoc Working Group for Long-term Cooperative Action |
| AWG-KP | Ad Hoc Working Group Further Commitments for Annex I Parties under the Kyoto Protocol |
| BASIC | Brazil, South Africa, India and China |
| BAU | Business-as-usual |
| CBDR & RC | Common but Differentiated Responsibilities and Respective Capabilities |
| COP | Conference of Parties |
| EASD | Equitable Access to Sustainable Development |
| EU | European Union |
| G77 & China | Group of 77 and China |
| GDP | Gross Domestic Product |
| GHG | Greenhouse gases |
| INC | International Negotiating Committee (for a Framework Convention on Climate Change) |
| QELROs | Quantifiable Emissions Limitations and Reduction Objectives |
| UNCSD | United Nations Conference on Sustainable Development |
| UNFCCC | United Nations Framework Convention on Climate Change |

## References


Abbott, K.W., Keohane, R.O., Moravcsik, A., Slaughter, A. and Snidal, D. (2000). "The Concept of Legalization", *International Organization*, **54**(3), 401–419.

Ashton, J. and Wang, X. (2003). "Equity and Climate: In Principle and Practice", in Aldy, J.E., Ashton, J., Baron, R., Bodansky, D., Charnovitz, S., Diringer, E., Heller, T.C., Pershing, J., Shukla, P.R., Tubiana, L., Tudela, F. and Wang, X. (Eds.) (2003). *Beyond Kyoto: Advancing the International Effort against Climate Change*, Pew Center on Global Climate Change Working Paper Series, United States of America, pp. 1–22.

Bailer, S. (2012). "Strategy in the climate change negotiations: do democracies negotiate differently?", *Climate Policy*, **12**(5), 534–551.

Berk, M.M. and Den Elzen, M.G.J. (2001). "Options for differentiation of future commitments in climate policy, how to realise timely participation to meet stringent climate goals?", *Climate Policy*, **1**(4), 465–480.

Betzold, C., Castro, P. and Weiler, F. (2012). "AOSIS in the UNFCCC negotiations: from unity to fragmentation?", *Climate Policy*, **12**(5), 591–613.

Bhatti, Y., Lindskow, K. and Pedersen, L.H. (2010). "Burden-sharing and global climate negotiations: the case of the Kyoto Protocol", *Climate Policy*, **10** (2), 131–147.

Bodansky, D. (1993). "The United Nations Framework Convention on Climate Change: A Commentary", *Yale Journal of International Law*, **18**(2), 451–558.

Bodansky, D. (2001). "The History of the Global Climate Change Regime" in Luterbacher, U. and Sprinz, D.F. (Eds.) (2001). *International Relations and Global Climate Change*, Massachusetts Institute of Technology, United States of America.

Boisson de Chazournes, L. (2008) "United Nations Framework Convention on Climate Change" in *United Nations Audiovisual Library of International Law*, United Nations, pp. 1–7.

Bretschger, L. (2012). *Climate Policy and Equity Principles: Fair Burden Sharing in a Dynamic World*, Centre of Economic Research at ETH Zurich (CER-ETH) Policy Brief, Zurich, Switzerland, pp. 1–18.

Caron, D.D. (1991). "Protection of the Stratospheric Ozone Layer and the Structure of International Environmental Lawmaking", *Hastings International and Comparative Law Review*, **14**(4), 755–780.

Castro, P., Hörnlein, L. and Michaelowa, K. (2011). *Path dependence of negotiation structures in international organisations : the impact of Annex I membership on discussions within the UNFCCC*, Working Paper No. 67, Center for Comparative and International Studies, Zurich, Switzerland.

Cullet, P. (1999). "Toward a new paradigm of inter-state relations", *European Journal of International Law*, **10**(3), 549–582.

Cullet, P. (2010). "The Kyoto Protocol and Vulnerability: Human Rights and Equity Dimensions" in Stephen Humphreys (Ed.), *Human Rights and Climate Change*, Cambridge University Press, Cambridge, pp. 183–206.

Den Elzen, M.G.J. (2002). *Exploring post-Kyoto climate regimes for differentiation of commitments to stabilise greenhouse gas concentrations*, RIVM Report

728001020/2002, Climate Change Policy Support Project (M/728001 Ondersteuning Klimaabeleid), Dutch Ministry of Environment, The Netherlands.

Den Elzen, M.G.J., Schaeffer, M. and Lucas, P.L. (2005). "Differentiating future commitments on the basis of countries' relative historical responsibility for climate change: Uncertainties in the 'Brazilian proposal' in the context of a policy implementation", *Climate Change*, **71**(2), 277–301.

Doniger, D. (2009). *The Copenhagen Accord: A Big Step Forward.* [Online] Available at: http://switchboard.nrdc.org/blogs/ddoniger/the_copenhagen_accord_a_big_st.html [Accessed 10 March 2013]

Dupont, C. (1996). "Negotiation as coalition building", *International Negotiation*, **1**(1), 47–64.

Enting, I. and Law, R.M. (2002). *Characterizing Historical Responsibility for the Greenhouse Effect*, CSIRO Atmospheric Research Technical Paper No. 41, Commonwealth Scientific and Industrial Research Organisation (CSIRO), Australia.

Ewing, J.J. (2012). "From Kyoto to Durban: The Fits and Starts of Global Climate Change Negotiations", *NTS Insight February 2012.* [Online] Available at: http://www.rsis.edu.sg/nts/HTML-Newsletter/Insight/pdf/NTS_Insight_feb_1201.pdf [Accessed 6 February 2012]

Fischer, C. and Morgenstern, R. (2012). *Metrics for Evaluating Policy Commitments in a Fragmented World: The Challenges of Equity and Integrity*, Discussion Paper 08-17, Harvard Project on International Climate Agreements, Belfer Center for Science and International Affairs, Harvard Kennedy School.

Franck, T.M. (1999). *Fairness in International Law and Institutions*, Clarendon Press.

French, D. (2000). "Developing States and International Environmental Law: The Importance of Differentiated Responsibilities", *International and Comparative Law Quarterly*, **49**(1), 35–60.

Friman, M. (2007). *Historical responsibility in the UNFCCC*, Report 07:01, Centre for Climate Science and Policy Research (CSPR), Linköping University, Linköping, Sweden.

Friman, M. and Linnér, B. (2008). "Technology obscuring equity: historical responsibility in UNFCCC negotiations", *Climate Policy*, **8**(4), 339–354.

Garman, J. (2009). "Copenhagen — historic failure that will live in infamy: green activist's searing dispatch from Denmark", *The Independent*, [Online] Available at: http://www.independent.co.uk/voices/commentators/joss-garman-copenhagen–historic-failure-that-will-live-in-infamy-1845907.html [Accessed 6 February 2012]

Grubb, M., Hourcade, J-C. and Oberthur, S. (2011). *Keeping Kyoto: A study of approaches to maintaining the Kyoto Protocol on Climate Change*, Climate Strategies Report, London, United Kingdom.

Handl, G. (1990). "Environmental Security and Global Change: The Challenge to International Law", *Yearbook of International Environmental Law*, **1**(1), 3–33.

Hepburn, J. and Imran, A. (2005). *The Principle of Common but Differentiated Responsibilities, Legal Working Paper in the CISDL* — Recent Developments in International Law Related to Sustainable Development? Series, Centre for

International Sustainable Development Law, Montreal.

Heyward, M. (2007). "Equity and international climate change negotiations: a matter of perspective", *Climate Policy*, **7**(6), 518–534.

Höhne, N., Harnisch, J., Phylipsen, D., Blok, K. and Galleguillos, C. (2002). *Evolution of commitments under the UNFCCC: Involving newly industrialised economies and developing countries*, Ecofys for the German Federal Environmental Agency (Umweltbundesamt) FKZ 201 41 255, The Netherlands.

Höhne, N. and Blok, K. (2005). "Calculating historical contributions to climate change", *Climate Change*, **71**(2), 141–173.

Höhne, N., Den Elzen, M. and Weiss, M. (2006). "Common but differentiated convergence (CDC): a new conceptual approach to long-term climate policy", *Climate Policy*, **6**(2), 181–199.

Jayakumar, S. (2011). *Diplomacy: A Singapore Experience*, Straits Times Press, Singapore.

Matsui, Y. (2002). Some Aspects of the Principle of — Common but Differentiated Responsibilities, International, *Environmental Agreements: Politics, Law and Economics*, **2**(2), 151–171.

Michaelowa, K. and Michaelowa, A. (2012). "India as an emerging power in international climate negotiations", *Climate Policy*, **12**(5), 575–590.

Müller, B. (1998). *Justice in Global Warming Negotiations: How to Obtain a Procedurally Fair Compromise*, Oxford Institute for Energy Studies, United Kingdom.

Müller, B. (2001). "Fair Compromise in a Morally Complex World" in *The Pew Center 'Equity and Global Climate Change' Conference*, April 17–18 2011, Washington D.C., United States, pp. 1–14.

Müller, B., Höhne, N. and Ellermann, C. (2009). "Differentiating (Historic) Responsibilities for Climate Change", Summary Report, *Climate Policy*, **9**(6), 593–611.

Najam, A., Huq, S. and Sokona, Y. (2003). "Climate negotiations beyond Kyoto: developing countries concerns and interest", *Climate Policy*, **3**(3), 221–231.

National Climate Change Secretariat. (2008). *Singapore's National Climate Change Strategy,* Singapore, pp. 1–44.

National Climate Change Secretariat. (2010). *Singapore's Second National Communication Under the United Nations Framework Convention on Climate Change,* Singapore, pp. 1–76.

National Climate Change Secretariat. (2012). *National Climate Change Strategy 2012*, Singapore, pp. 1–140.

Olmstead, S.M. and Stavins, R.N. (2011). *Three Elements of a Post-2012 International Climate Policy Architecture*, Oxford University Press on behalf of the Association of Environmental and Resource Economists, United Kingdom.

Rajamani, L. (2000). "The principle of common but differentiated responsibilities and the balance of commitments under the climate regime", *Review of European Community and International Environmental Law*, **9**(2), 120–131.

Rajamani, L. (2006). *Differential Treatment in International Environmental Law*, Oxford University Press, United Kingdom.

Ringius, L., Torvanger, A. and Underdal, A. (2000). "Burden Differentiation: Fairness Principles and Proposals", Center for International Climate and Environmental Research and Energy Research Centre of the Netherlands (CICERO-ECN) Working Paper 1999:13, Oslo, Norway.

Ringius, L., Torvanger, A. and Underdal, A. (2002). "Burden Sharing and Fairness Principles in International Climate Policy", *International Environmental Agreements: Politics, Law and Economics*, **2**(1), 1–22.

Rive, N., Torvanger, A. and Fuglestvedt, J. S. (2006). "Climate agreements based on responsibility for global warming: Periodic updating, policy choices, and regional costs", *Global Environmental Change*, **16**(2), 182–194.

Shue, H. (1999). "Global Environment and International Inequality", *International Affairs*, **75**(3), 531–545.

Soltau, F. (2009). *Fairness in International Climate Change Law and Policy*, Cambridge University Press, United Kingdom.

Starkey, B., Boyer, M.A., Wilkenfeld, J. (2008). *Negotiating a Complex World: An Introduction to International Negotiation*, Rowman & Littlefield Publishers, Lanham, MD.

Stavins, R. (2012). *From Berlin to Doha*, The Environmental Forum, Environmental Law Institute, Washington, D.C.

Stone, C.D. (2004). "Common but Differentiated Responsibilities in International Law", *The American Journal of International Law*, **98**(2), 1704–1709.

Tolba, M. (1989). "A Step-by-Step Approach to Protecting the Atmosphere", *International Environmental Affairs*, **1** (4), 304–309.

Traber, D. (2010). *Preference attainment in Swiss legislative decision-making*, Paper prepared for the Annual Convention of the Swiss Political Science Association, University of Geneva, January 7–8, Geneva, Switzerland.

Victor, D.G. (1997). "The Regulation of Greenhouse Gases: Does Fairness Matter?" in: *Fair Weather? Equity Concerns in Climate Change*. Tóth, F. (ed.) Earthscan Publications, London, United Kingdom, pp. 193–205.

Weiler, F. (2012). "Determinants of bargaining success in the climate change negotiations", *Climate Policy*, **12**(5), 552–574.

Werksman, J. (2010). "Legal symmetry and legal differentiation under a future deal on climate", *Climate Policy*, **10**(6), 672–677.

Werksman, J. and Herbertson, K. (2009). *The Legal Character of National Actions and Commitments in a Copenhagen Agreement: Options and Implications*, WRI Working Paper, World Resources Institute, Washington DC.

Winkler, H., Brouns, B., Kartha, S. (2006). "Future mitigation commitments: Differentiating among non-Annex I countries", *Climate Policy*, **5**(5), 469–486.

Winkler, H. and Beaumont, J. (2010). "Fair and effective multilateralism in the post Copenhagen climate negotiations", *Climate Policy*, **10**(6), 638–654.

Winkler, H., Brouns, B. and Kartha, S. (2011). "Future mitigation commitments: differentiating among non-Annex I countries", *Climate Policy* **5**(5), 469–486.

United Nations Framework Convention on Climate Change. (1992). A/AC.237/18 (Part II)/Add.1 (1992) 31 I.L.M. 849.

UNFCCC. (2014). Decision 1/CP.19, 'Further advancing the Durban Platform' UN Doc. FCCC/CP/2013/10/Add.1.

UNFCCC. (2014), *Submission by China: China's Submission on the Work of the Ad Hoc Working Group on Durban Platform for Enhanced Action.* [Online] Available at: http://unfccc.int/files/bodies/application/pdf/20140305-submission_on_adp_by_china__without_cover_page.pdf [Accessed 14 October 2014].

UNFCCC. (2013) *Ad Hoc Working Group on the Durban Platform for Enhanced Action (ADP): Scope, design and structure of the future legally binding instrument; application of principles and ways of reflecting enhanced action of the Parties; and lessons learnt*, Environmental Integrity Group, comprising Liechtenstein, Mexico, Monaco, the Republic of Korea, and Switzerland. [Online] Available at: https://unfccc.int/files/documentation/submissions_ from_parties/adp/application/pdf/adp_eig_workstream_1_20130304.pdf [Accessed 27 October 2014].

# 18

# MITIGATING THE ENVIRONMENTAL IMPACT OF AIRCRAFT EMISSIONS THROUGH AN ECONOMIC THEORY — THE ENDOWMENT EFFECT

ZHANG QIANG[1]

Air emissions generated by aircraft result in health concerns and environmental impacts contributing to ozone depletion and climate change. As the aviation industry thrives, airplane induced emissions continue to contribute to anthropogenic impacts as the environment which pose an increasingly severe threat to sustainable development. To combat the environmental impacts from aircraft emissions, feasible and effective mitigation measures are required. The effective implementation of these measures plays a key role in reducing the adverse effects of air traffic on the environment. However, there are challenges in the road to effective implementation. This paper illustrates how the "endowment effect", an economic theory, can be applied to encourage airlines to opt for newly designed, more environmentally friendly aircraft and thus facilitate the reduction of airline-induced environmental impacts.

## 1. Introduction

Aircraft induced emissions are mainly caused by the combustion of jet fuel. Kerosene is presenting the most prevalent jet fuel for commercial aircraft. It comprises of different hydrocarbons and also includes additives such as antioxidants and corrosion inhibitors (ICAO, 2007b). Most air emissions produced by airplanes are released from the main engines, though auxiliary power units (APUs) also contribute to the emissions, especially at the ground level. When jet fuel is combusted, a range of exhaust components is generated and released into the ambient atmosphere, mainly in the forms of gases and particles. More specifically, the emissions include carbon dioxide ($CO_2$), water vapour ($H_2O$), carbon monoxide (CO), nitrogen oxides ($NO_x$), unburned or partially combusted hydrocarbons (UHCs), sulfur oxides ($SO_x$) and soot (Lee *et al.*, 2009). About 70% of the emissions

[1]The supervisor for this paper was Mr Rick Reidinger, Adjunct Senior Lecturer, School of Design & Environment, NUS.

are $CO_2$, while $H_2O$ constitutes a little less than 30%. All other gases, particulates and other trace components account for less than 1% (FAA, 2005).

Air emissions produced by aircraft result in a range of adverse impacts on both local and global communities and environments. Local impacts primarily consist of air pollution and the resultant health concerns. Large amounts of exhaust emissions, for example, nitrogen oxides ($NO_x$) and particulate matters (PM), are released from the aircraft's main engines and auxiliary power units (APUs) during the landing and take-off cycle, which takes place at up to 915 metres (3000 feet) above ground level (ICAO, 2010). These emissions pollute the air and pose threats to the health and welfare of local communities.

At the global level, air emissions from aircraft intensify the greenhouse effect, thereby contributing to global climate change and ozone depletion. Greenhouse gases emitted from aircraft, such as $CO_2$ and water vapour, directly worsen the greenhouse effect and global climate change. Other non-greenhouse gas exhaust components, such as nitrogen oxides ($NO_x$), sulfur oxides ($SO_x$), hydrocarbons (HC), aerosols and particulate matter (PM) emitted from aircraft also indirectly affect the Earth's surface temperature, and subsequently contribute to climate change through complex chemical reactions and microphysical processes.

Emissions from aircraft also affect the stratospheric ozone. Nitrogen oxides ($NO_x$) are the highest contributing substances to the perturbation of the stratospheric ozone concentrations. $NO_x$ in the stratosphere can catalytically split $O_3$ molecules (ozone) into oxygen molecules ($O_2$) in the presence of solar radiation. In addition to nitrogen oxides, sulfur species, water vapour and aerosols emitted by aircraft in the stratosphere may also contribute to ozone depletion.

## 2. Mitigation Measures

To mitigate the environmental impact resulting from airplane emissions, extensive efforts are required in the following aspects: aircraft and aviation related technologies, operational practices, infrastructure development, and positive economic measures.

### i. *Technological Improvement*

*Improving the design of engine aerodynamics*

The ability of the engines to convert the chemical energy present in sources of fuel into mechanical power is called thermal efficiency. This is measured by the ratio of power generated to the energy input, which is closely linked to the engine bypass ratio, and the pressure and temperature ratio (IPCC, 1999). It is reported that increasing the engine bypass ratio is regarded as the most practical method to raise both thermal and propulsive efficiencies (IPCC, 1999).

Engines with higher bypass ratios imply larger propulsor sizes (in terms of diameter), which may increase the weight of engines, and therefore the drag and installation penalties. This in turn may consume additional fuel and hence result in a trade-off — with the improvement in fuel efficiency generated by higher bypass ratios (Lee *et al.*, 2010). Therefore, an integrated development of the configuration of the engine and the nacelle, incorporating the air inlet, the engine core, and the exhaust nozzle, can be of great help in reducing air resistance.

Local drag effects can be decreased and the efficiency increased through "improvement[s] in the aerodynamics of engine-nacelle flows and changes to the shape and length of the inlet section".[2] Furthermore, lighter manufacturing materials are required to potentially decrease the weights of the engine and nacelle, and therefore reduce fuel consumption.

Several new engine concepts have been proposed; new engine concepts for example, rear mounted turbofans, open rotor engines and geared turbofans. Among these, the open rotor and the geared turbofan seem to be the most promising (Szodruch *et al.*, 2011). It is estimated that the open rotor engine can reduce fuel consumption by around 25% while the geared turbofan engine can decrease fuel burned by 10 to 15% (IATA, 2009a). Additionally, innovation in other components, such as engine turbine and nozzle design and configuration, will also improve fuel efficiency.

*Improving the combustor*

Besides the advances in the aforementioned aerodynamics of engines, combustor improvement can also lead to a significant increase in fuel efficiency, and reduce fuel consumption and air emissions.

Incoming air captured by the engine inlet will be pressurised by the compressor, and then enter the combustion chamber at a relatively high velocity. The combustor "is designed to allow time and space for the fuel and air to mix thoroughly and burn efficiently before entering the turbine stages".[3] As a result, the fuel efficiency is directly related to the

[2]IPCC, *Aviation and the Global Atmosphere*, 1999, p. 225.
[3]IPCC, *Aviation and the Global Atmosphere*, 1999, p. 234.

combustor's design. In general, increasing the pressure and temperature of the combustion chamber, amount of airflow, and the time for mixing the air and fuel, will affect combustion efficiency and reduce exhaust emissions.

Increasing the pressure, temperature, mass of airflow and combustion time, on the one hand, remarkably reduces fuel consumption; while on the other hand, considerably increasing the emissions of $NO_x$. This is because, when the temperature of the combustor is high enough, nitrogen and oxygen in the airflow will react with each other at a high rate to form thermal $NO_x$ (independent of the fuel type) posing a great challenge to combustor advancement. As a result, future engine combustor designs will focus on the improvement of $NO_x$ and $CO_2$ (fuel) tradeoffs. In order to further reduce $NO_x$ emissions, the International Civil Aviation Organization (ICAO) has set stringent medium and long term reduction goals, i.e., decreasing $NO_x$ emissions by 45% from the current standards by 2016, and improving the engine's emissions performance by 60% by 2026 (ICAO, 2007b). Achieving these goals will largely depend on developing low-emission combustor technologies. Since the increase in $NO_x$ emissions is mainly induced by high temperatures, this development will rely on the reduction of the temperature within the combustion chamber. In the report of ICAO (2007b), it is stated that the amount of $NO_x$ produced can be reduced by limiting the residence time (the time spent by the products of combustion at the highest temperatures). Currently, technology such as the Rich-Quench-Lean (RQL) system are used to reduce $NO_x$. In the RQL system, initial combustion takes place in the rich zone where the amount of air is less than that required, providing good running stability. Then, more air is added quickly to move combustion into the lean zone, where there is more air than that required, to quickly cool down the combustion to avoid substantial generation of $NO_x$ (ICAO, 2007b). Much of future combustor development will focus on ultra-low $NO_x$ technology to meet the requirements of ICAO emission standards.

*Weight reduction*

A vast majority of the fuel consumed by an aircraft is for the generation of thrust to overcome drag and hence, fly. As drag is proportional to the weight of the aircraft, decreasing the weight of aircraft will also reduce the drag, which further reduces the fuel burned to generate enough lift. Therefore, weight reduction is regarded as one of the most promising technological options to decrease the consumption of jet fuel (Lee *et al.*, 2010).

The total weight of an aircraft basically includes the weight of the aircraft itself (which is determined by the manufacturing materials used), its type, payload, and the weight of other elements that are required for the flight, such as fuel and food. Recent developments in aircraft manufacturing have seen the increasing application of advanced (lighter and stronger) structural materials. These have significantly contributed to the achievements of decreased fuel consumption and a further reduction in air emissions. For example, more and more aircraft designs have incorporated the use of composite materials, which are estimated to offer a 5% and 15% reduction in carbon dioxide emissions, currently and in the long-term, respectively (Szodruch *et al.*, 2011).

Future reduction of weight will still focus on the greater use of lighter and stronger materials, such as new aluminum alloys, hybrid alloys, titanium components, composites, and other innovative and smart materials (IPCC, 1999), in everything from the airframe to cabin components (OECD, 2012). Furthermore, a reduction in aircraft weight can be achieved through a range of innovative manufacturing techniques to decrease the number of rivets and local reinforcements necessary, like laser or friction welding (World Bank, 2012).

*Aerodynamic improvements*

Due to improvements in the aerodynamic performance of aircraft, less thrust is required to overcome the reduced drag, which will lower the fuel consumption and consequently the amount of air emissions. Currently, the greatest contributors to drag are lift-dependent and skin friction drag (ICAO, 2010).

The aerodynamic performance of aircraft has improved significantly to date. Its improvements include the remarkable reduction of lift-dependent drag by maximising effective wing span extension (ICAO, 2010). The advancement in wing tip shape, that evolved from dimple tips in the 1970s to small winglets in the 1980s, and then to large winglets (applied to Boeing aircraft) or sharklets (applied to Airbus aircraft) today, is regarded as the most important contributor to reducing lift-dependent drag. It is estimated that sharklets installed on the Airbus A320 family can reduce fuel consumption by as much as 3% (World Bank, 2012) by decreasing the turbulence and vortex generation by the wings (FAA, 2005).

The principles of approaches to reducing friction drag include, 1) decreasing turbulent skin friction by maintaining laminar flow,

2) minimising wetted areas while controlling flow separation, optimising surface intersections and fuselage aft-body shape, and 3) minimising manufacturing excrescences and advancing air flow devices (ICAO, 2010). Several technologies have been tested and new concepts, such as hybrid laminar flow technology and film surface grooves, are being assessed. Other innovative technologies and aircraft designs have been demonstrated to be able to reduce air resistance and improve the lift-to-drag ratio, and thus, lower fuel consumption. Examples include morphing technology that can freely change the airframe shape (Barbarino *et al.*, 2011), blended wing body, forward-swept wings and vertical tail planes.

*Aircraft systems improvements*

A large number of subsystems called aircraft systems, such as navigation, flight control, and in-flight entertainment systems, are employed in current aircraft to manage the flights. These systems contribute to a noticeable amount of fuel consumption and hence offer room for improvement to help reduce fuel consumption (IPCC, 1999). In the past, surface control systems were managed from the cockpit mechanically, through cables or heavy hydraulically-powered systems. These have been gradually substituted by electronically-controlled 'fly-by-wire' systems that operate using lighter and more powerful electrical systems. Further reduction in weight of surface control systems can be achieved through advancements in the design and weight of individual motors (World Bank, 2012). An extension in the capacity of fly-by-wire systems, which takes active pitch stability augmentation and wing load alleviation into account, can potentially enhance the overall aircraft efficiency by 1 to 3%. Greater application of the concept of an 'all electric' aircraft will further decrease the direct energy use for system operations, as well as the weight of these systems, which will in turn contribute to a reduction in fuel consumption.

*Aviation fuel improvement*

Kerosene, which is derived from crude oil, is currently used to power most modern airliners (ICAO, 2007b). The composition and proportion of jet fuel varies slightly in different countries. Hydrocarbons are the primary composition of jet fuel. Other products, such as olefins, sulfur, additives, and a trace amount of oxygenated organics and metals, can be found in jet fuel as well (ExxonMobil, 2005). As the emissions of carbon dioxide and water are directly related to the carbon to hydrogen ratio of the fuel (i.e., a fuel with a higher hydrogen percentage will generate less $CO_2$ and, correspondingly more $H_2O$), increasing the hydrogen percentage in jet fuel is one way to decrease carbon dioxide emission (IPCC, 1999). Several processes have been tested to produce low-sulfur jet fuel that may help reduce the emission of $SO_x$. For instance, a gas-to-liquid conversion process produces fuel that is virtually free of sulfur and aromatics, thereby reducing the emission of sulfur oxides and particulates (World Bank, 2012).

Apart from the improvement of kerosene, the development of alternative fuels is also being considered as a potential major contributor to the reduction of fossil fuel consumption and air emissions. Changing the fuel type or specifications in the future would require significant modification and at a high cost, as current aircraft and airport infrastructure are specially designed to optimise the use of kerosene (IPCC, 1999). Investigation and development in alternative fuels have also been encouraged due to concerns over issues ranging from environmental sustainability and supply stability, to the costs associated with conventional jet fuel (ICAO, 2010). However, the probability of the successful application of alternative fuels on a large scale, at least in the medium term, will depend heavily on their compatibility with the essential capabilities and requirements of existing aircraft and relevant infrastructure, such as energy density, safety and lubricity (IPCC, 1999).

### ii. *Advancing Operational Measures*

*Improving air traffic management*

The modernisation of air traffic management (ATM) offers remarkable potential in the reduction of fuel consumption and air emissions, because a large amount of fuel is consumed in 'taking off, landing, on landing approaches and flight paths including changes in direction and altitude'.[4] ATM can be improved through various processes, such as dynamic and flexible route management, flexible use of airspace, performance based navigation, and continuous descent arrival, optimising the use of jet fuel.

For example, fuel-optimal routings and altitude profiles can be deployed by advanced computer modelling, shortening the flight distance by reducing the changes in direction and altitudes, which in turn will reduce fuel consumption.

[4]OECD, 'Green Growth and the Future of Aviation', 2012, p. 15.

Continuous descent arrival (or approach) (CDA) allows for the optimisation of landing on a constant slope without requiring the use of flaps and landing gear, thereby reducing fuel burn and air emissions. Traditional stepped aircraft arrival processes require the consumption of more fuel to produce high thrust and power for flaps and landing gear (Frey and Kuo, 2007). It is reported that the "CDA Action Plan initiated by the International Air Transport Association (IATA) and other partners will save 500,000 tons of $CO_2$ through [the] implementation of CDA at 100 airports across Europe by the end of 2013".[5]

Moreover, the development and inter-operability of Global ATM Systems also play an important role in ATM improvement and reduction in fuel consumption. These systems will be interoperable for all users during all phases of flight to provide optimum operations (ICAO, 2010). Several examples can be found in different regions, such as the Single European Sky ATM Research (SESAR), the Next Generation Air Traffic Management system, and the Pearl River Delta ATM system (IATA, 2009a).

*Advancing airlines' operational measures*

Besides the improvement in air traffic management, which depends on air traffic control authorities, airlines can contribute to the reduction in fuel consumption and air emissions by developing their operational processes.

Replacing the use of auxiliary power units (APUs) of aircraft with ground-based power supplies during their parking at gates is considered to offer potential to save energy, and thus decrease air emissions on the ground level. APUs are employed to provide power to the aircraft when the main engines are shut down at airports. Even though the amount of fuel consumed by APUs differ from aircraft to aircraft, and it is amount relatively small compared to the total amount of fuel used during a full flight cycle, more and more airlines are increasing the use of ground-based equipment that supply lower-carbon electricity and preconditioned air to aircraft instead of APUs, thereby reducing noise and air emissions within the airport (Sustainable Aviation, 2012). For example, it is estimated that this substitution can save as much as 95% of the fuel consumed and emissions generated by APUs at Zurich Airport (IPCC, 1999), reduce about 0.6% of total $CO_2$ emissions from aviation relative to the baseline of 2010 in the UK (Sustainable Aviation, 2012), and decrease carbon emission at aircraft stands by as much as 85% at Hong Kong International Airport

[5]IATA, 'A Global Approach to Reducing Aviation Emissions', 2009a.

Table 18.1. Effect of load factor on fuel efficiency (MJ per passenger-km) of Tupolev Tu-154 (Source: IPCC, 1999).

| Load Factor (%) | Passengers | Fuel Burned per Passenger-km (kg) | MJ per Passenger-km |
|---|---|---|---|
| 19 | 32 | 0.21 | 8.9 |
| 51 | 84 | 0.08 | 3.6 |
| 70 | 115 | 0.065 | 2.9 |
| 82 | 135 | 0.053 | 2.3 |
| 100 | 164 | 0.051 | 2.0 |

(Hong Kong International Airport, 2011). In addition, switching off a certain number of engine(s) during taxiing is reported to further reduce fuel combustion.

Improving the load factor, i.e., the ratio of the payload actually carried to the maximum designed capacity of payload, can also decrease fuel consumption, which is measured by megajoules (MJ) per passenger per kilometre. The optimum fuel efficiency of an aircraft can be achieved when its load capacity is fully utilised. Table 18.1 shows that raising the load factor can reduce fuel consumption and, as a result, air emissions. Airlines can achieve a higher load factor through a variety of methods, such as increasingly sophisticated revenue management systems (Sustainable Aviation, 2012), which are estimated to reduce as much as 7.3 million $tCO_2eq$ in 2020 (OECD, 2012).

Several other operational measures have also been identified during the research for reducing fuel burn and air emissions. These approaches, for example, include "regular engine and airframe cleaning, the checking of door-seals for drag-inducing defects, and the removal of dents from the aircraft's external surfaces",[6] optimising the required fuel-load, and reducing the weight of in-flight entertainment equipment, seats, duty-free goods, and food and drinks. These methods have potential to save fuel consumption by about 1–2% (IPCC, 1999).

### iii. *Infrastructure Development*

Airport infrastructure characteristics also directly affect the fuel consumption and air emission from aircraft engines. As a result, the improvement in infrastructure at airports will potentially contribute to the optimisation of the environmental performance of aircraft.

[6]Sustainable Aviation, 'Sustainable Aviation CO2 Road-Map 2012', 2012, p.19.

The layout of airport infrastructure, such as runways, taxiways, and terminals, directly determines the distance and time of aircraft taxiing, which further affects the fuel consumption rate of the main engines and auxiliary power units (APUs). Studies have found that, a decrease in as much as 6% of emissions in a Landing/Take-off (LTO) cycle can be achieved by reducing 2 minutes of taxiing time (Kesgin, 2006). Furthermore, it is also estimated that an Airbus A320 can save about 12 kg and 15 kg of fuel per minute with CFM International's engines and International Aero Engines' V2500 engines, respectively, during its taxiing or idling period, while an Airbus A330 with Rolls-Royce Trent 772 engines can save approximately 34 kg of fuel per minute at idle (World Bank, 2012). Based on these findings, it is evident that shortening the distances of aircraft taxiing, by providing the most direct routes between runways and parking bays through sophisticated airport design and construction, offers potential to reduce fuel consumption.

As illustrated before, replacing the use of APUs with ground-based power and pre-conditioned air supply systems can further help lower the amount of fuel burned. However, currently, not all airports are equipped with these facilities, and even if there are such facilities, they may not cover all parking stands. Not all airports have implemented policies to restrict the use of APUs at gates. Furthermore, in some cases, some pilots may refuse to use a ground-based power supply owing to safety concerns. As a result, there is still noticeable potential to reduce aircraft fuel consumption by employing ground-based supply systems. In order to do so, relevant systems with reliable quality could be implemented at the global level, thereby encouraging and facilitating the use of alternative power and air supplies. Currently available systems include, for instance, ground power units (GPUs), fixed electrical ground power (FEGP), air climate units (ACUs), and mobile heating units (Zurich Airport, 2009).

Furthermore, should new technologies and innovative designs be commercially applied on a large scale in future, substantial or even complete enhancement or changes in the infrastructure can be expected. This is because existing aviation-related facilities are designed and developed to optimise the use of conventional kerosene fuel and aircraft. In order to optimise the application of innovations, such as alternative fuels and new aircraft models, relevant stakeholders, like aircraft manufacturers, airlines and airports, etc., should cooperate closely with each other, to efficiently and effectively research, design, and then upgrade existing infrastructure, or construct completely new facilities.

### iv. *Positive Economic Measures*

Along with the aforementioned measures, a variety of positive economic measures can also contribute to a further reduction in fuel consumption and air emissions both directly and indirectly. Unlike the conventional punitive regulations and policies, positive economic measures use market-based measures to facilitate stakeholders to internalise their externalities. Market-based measures (MBM) are "policy tools that are designed to achieve environmental goals at a lower cost and in a more flexible manner than traditional regulatory measures",[7] including, for example, emissions trading, emissions offsetting, and the clean development mechanism (CDM) (ICAO, 2007a). However, it is worth noting that emissions trading and emissions offsets need global coordination and agreement (IATA, 2009b) through ICAO (IATA, 2012).

*Emissions trading*

Emissions trading has been one of the most efficient measures to lower the emissions of greenhouse gases (GHG) through international co-operation, after the adoption of the Kyoto Protocol (OECD and IEA, 2001). Carbon emissions trading in particular has been successful. Conventional regulatory measures tend to set a target of emissions reduction to stakeholders, and if they fail to meet the target, they may get varied penalties — for instance, fines and/or business restrictions, etc., depending on specific contexts. Yet, participants may be reluctant to reduce their emissions. One reason is that with those who control their emissions well below the allowed levels, do not benefit directly from the reduction of emissions. Furthermore, these measures may not facilitate emissions reduction at all. For example, if most of the participants whose emissions contribute to a significant part of overall emissions fail to meet the targets, then the total level of emissions may not decrease significantly at all. However, unlike the punitive measures, in an emissions trading system, GHG emitting enterprises that produce GHG emissions higher than their designated levels can buy allowances from others, while those who generate fewer emissions than their allowed levels can either sell their surplus or save it for later. The idea here is that GHG emitting enterprises can meet their targets flexibly based on their own strategy, either by decreasing the amount of emissions or by purchasing allowances at a market price. They also can directly benefit from the

[7]ICAO, *Assembly Resolutions in Force* (as of 28 September 2007), 2007a, p. I-72.

emissions reduction through allowance selling, which will stimulate them to lower their emissions level so as to be able to sell the surplus. The total emissions reduction targets are achieved because they are determined by the overall ceiling on emissions. In other words, emissions trading systems integrate environmental aims and economic concerns, leading to a cost-effective approach to reducing emissions (EUROCONTROL, 2008).

At present, several emissions trading schemes have been established, including compulsory systems such as the European Union Emissions Trading System (ETS), the Regional Greenhouse Gas Initiative in USA (RGGI), New Zealand's Trading Scheme, the Tokyo Metropolitan Trading Scheme, and the New South Wales Greenhouse Gas Abatement Scheme in Australia, as well as voluntary trading systems. Such trading schemes have also been proposed in a range of other countries and regions. These emissions trading schemes tend to combat environmental impacts with cost-effective approaches through market-based methods (Perdan and Azapagic, 2011).

Among the six GHGs specified for reduction in the Protocol, namely carbon dioxide ($CO_2$), methane ($CH_4$), nitrous oxide ($N_2O$), sulfur hexafluoride ($SF_6$), hydrofluorocarbons (HFCs), and perfluorocarbons (PFCs), the limits on $CO_2$ have the most significant and direct influences on aviation (UNFCCC, 2008). This is because the emission of $CO_2$ accounts for some 70% of the total air emissions generated by aircraft. It is also reported that the average $CO_2$ emission from an aircraft is 190 g per km per passenger, as compared to 70, 45, and 30 g per km per passenger for a car (assuming there are 2 passengers), a train, and a bus, respectively (EUROCONTROL, 2008), indicating a substantially higher level of $CO_2$ emission from aviation when compared to other modes of transport. This also implies that there is scope for considerably lucrative markets for carbon emissions trading in the aviation industry, especially for airlines.

Despite the fact that international aviation emissions are excluded from the reduction targets of the Protocol (Preston *et al.*, 2012), the effectiveness of emissions trading in emissions reduction suggests that this market-based approach could be suitable for lowering aviation emissions. However, schemes to trade emissions from the aviation industry should be carried out very carefully, and with global coordination and agreement. The most important reason is that unilateral emissions trading systems or taxes established by governments may not allocate allowances equally. This is because civil aviation always involves international stakeholders, as even domestic flights may require international support, such as air navigation services. Therefore it is very challenging to determine how emissions should be allocated to Parties (Preston *et al.*, 2012), and how reduction measures are to be implemented at a local or regional level. As a result, including aviation emissions in these unilateral schemes, which are not effective incentives, will not help realise the goal of emission reduction. Furthermore, competition between participants will be distorted by the lack of harmonisation within these schemes (IATA, 2010). Apart from these, "a credible system of monitoring and verifying emissions reductions that allows for trading with minimal transaction costs would be needed to achieve the cost savings potential of these mechanisms".[8]

*Emissions offsetting*

Emissions offsetting could be employed to balance, cancel out or neutralise aviation emissions. In the context of greenhouse gas emissions reduction, offsetting is a compensation for GHGs, and "can be equivalent in part or in whole to the associated emissions, by financing a reduction in emissions elsewhere".[9] Emissions offsets can be achieved with various approaches, such as planting trees and investing in solar power, etc., through which participants can reduce their emissions to meet the emission cap, and further generate emission credits. Companies or individuals can then trade their emission credits in the market if certain criteria for the specific emissions trading scheme, such as monitoring, reporting, and verification, are met (IATA, 2008). Therefore, emissions offsetting is recognised as an immediate and pragmatic measure to combat environmental impacts of aircraft emissions in a cost-effective manner, at least in the short term (Cafferty, 2011).

As far as the aviation industry is concerned, emissions generated by aircraft can be compensated for by lowering a similar amount of emissions from a variety of other activities, including land use and forestry optimisation, industrial greenhouse gases offsets, methane capture and use in energy generation, energy efficiency enhancement, and renewable energy investment (IATA, 2008). These alternative projects offer flexibility to airlines in reducing emissions, based on their preferred business strategies, along with the advancement of aircraft technologies, more efficient operational management, and development of infrastructure. Airlines that cannot afford to introduce the most advanced technologies to limit emissions

[8]IPCC, *Aviation and the Global Atmosphere*, 1999, p. 347.
[9]IATA, 'Aviation Carbon Offset Programmes: IATA Guidelines and tookit', 2008, p.6.

will benefit the most from these offsetting schemes, because they can find a cost-effective project to offset their environmental footprint. The offset schemes could be mandatory or voluntary, and could be achieved either by airlines or by passengers. For example, IATA has established an industry-wide carbon offset scheme, called the IATA Carbon Offset Program, to facilitate member airlines in introducing an offset option to the customers. This had been employed by 19 airlines by 2012, thereby standardising the industry's carbon offsetting processes (IATA, 2012). Besides, there were some 30 IATA member airlines that had developed their carbon offset systems independently in 2011 (IATA, 2011).

### *Clean development mechanism (CDM)*

The clean development mechanism (CDM) was initially adopted in the Kyoto Protocol to encourage developed countries to invest in GHG emission reduction or sequestration projects in developing countries. This enables developed countries to gain emission credits that could be used to meet their own emission allowances (ICAO, 2007c).

In the context of the aviation industry, CDM implied that besides the aforementioned measures, airlines registered in developed countries can also indirectly decrease their own environmental footprint by financing or supporting technically GHG reduction activities in developing countries. These airlines could be awarded CDM credits which are derived from the implementation of emission mitigation programmess or from afforestation and reforestation projects in the recipient countries (ICAO, 2010). These credits could be sold to other participants looking to lower their environmental impact within either regulatory or voluntary carbon markets. A range of selected projects have potential for the application of CDM in the aviation industry, such as researching in alternative fuels, advancing aircraft technologies, and modernising airside vehicles (ICAO, 2010). The application of the mechanism in the aviation industry will not only benefit investor airlines in helping them to meet their emissions cap, but also enhance the capabilities of developing countries in combating environmental issues. However, like emissions trading systems and emissions offsetting schemes, employing CDM in the aviation industry should also not be treated in a discriminatory way. The primary focus of aviation CDM should be to serve the interests of the industry itself, so as to ensure equity and non-discrimination. Since ICAO can adjust general CDM to meet the specific targets of aviation while aiding developing countries 'not only financially, but also through technology transfer and capacity building', it is recognised as the appropriate institution to manage aviation CDM (ICAO, 2010).

## 3. Challenges in Implementing Mitigation Measures

Substantial investments will be required to implement these mitigation measures efficiently and effectively. More specifically, it is estimated that about $1.6 trillion will be needed to achieve a carbon neutral growth from 2020, out of which approximately $1.5 trillion will be spent by airlines on upgrading their fleet alone (IATA, 2009a). This results in remarkably high costs for airlines to introduce the newly built, more environmentally-friendly airplanes, which may result in reluctance from airlines in replacing their existing aircraft models.

Moreover, airlines have optimised the operation of their existing fleet to minimise their operating costs. Besides the substantial direct investment required in acquiring new aircraft models, noticeable costs will result from various operation related activities, such as pilot acquisition/conversion, cabin crew and ground staff training, route planning, airport infrastructure upgrading, etc. Therefore, airlines may prefer the acquisition of aircraft models that are in their existing fleet rather than the more environmentally friendly airplanes that are new to them.

Based on the above factors, airlines may still prefer existing aircraft models which are less environmentally friendly when purchasing or leasing airplanes to add to their fleet. If the newly designed aircraft models cannot replace the existing models on a large scale, it is very challenging to achieve carbon-neutral growth from 2020. In order to achieve this goal, approaches to changing the acquisition behaviours of airlines must be identified to facilitate the implementation of mitigation measures. This paper proposes one approach, that is the application of the endowment effect and indifference curves to shift the consumption preference of the airlines from preferring existing aircraft models to opting for lower emission models.

## 4. Endowment Effect and Indifference Curves

In behavioural economic theory, once an individual is endowed with a product or object, he/she will expect to be repaid a sum larger than the sacrifice required to obtain it. This implies that the value which a person is willing to accept (WTA) for a product ('good') is consistently higher than his/her willingness to pay (WTP) for it. This disparity between

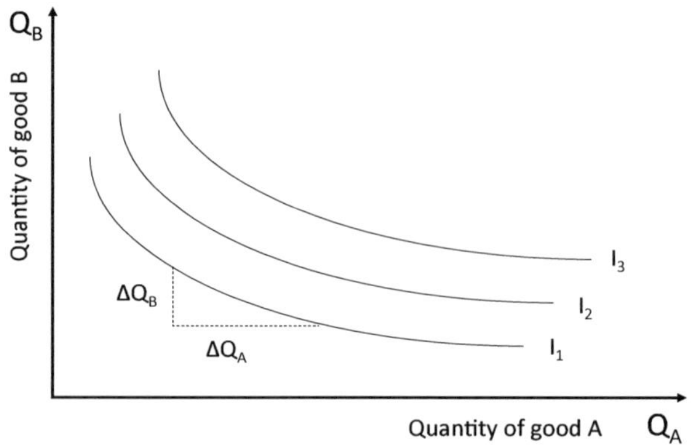


Fig. 18.1. Indifference curves.

WTA and WTP is called the endowment effect (or divestiture aversion) (Thampapillai, 2002). This further implies that, assuming that a person has just two types of goods to choose from, namely, A and B, and he/she prefers A to B, then more units of B will be required for him/her to give up one unit of A.

An indifference curve describes, at each point on the curve, that the utility of the bundle of goods to an individual is indifferent. That is, if we look at Figure 18.1 and take indifference curve 1 (labeled as $I_1$) as an example, reducing the amount of A by $\Delta Q_A$ while increasing the amount of B by $\Delta Q_B$ will result in the same utility of goods to the customer, and vice versa. This is because of the endowment effect. As stated above, to let a person who has his/her own consumption behaviour decrease or forgo the number of A by $\Delta Q_A$, a $\Delta Q_B$ amount of B will be required for compensation so as to maintain a constant utility of goods. Should a customer have different budgets, then the consumer will also have many more indifference curves with a similar shape, such as those in Figure 18.1, i.e., $I_1$, $I_2$, and $I_3$.

Again, assuming that a person only has two types of goods to choose from, if his/her budget should be fixed at a specific time, a budget line can be derived. For example, if he/she has a fixed budget of $100, and the prices of A and B are $5 and $10 respectively, then the budget line (shown as a dash line) can be determined, and is shown in Figure 18.2.

Normally, a customer tends to maximise his/her utility by striving to be on the highest possible indifference curve, and this maximisation is determined by his/her budget. As a result, when the prices of goods and the budget are fixed, the point of tangency of the budget line and the

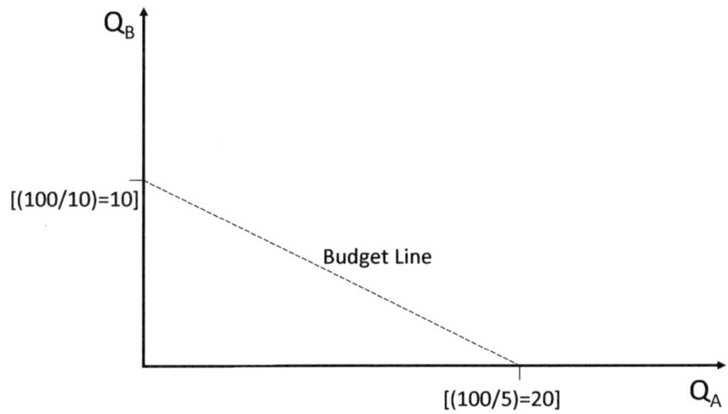


Fig. 18.2. Budget line.

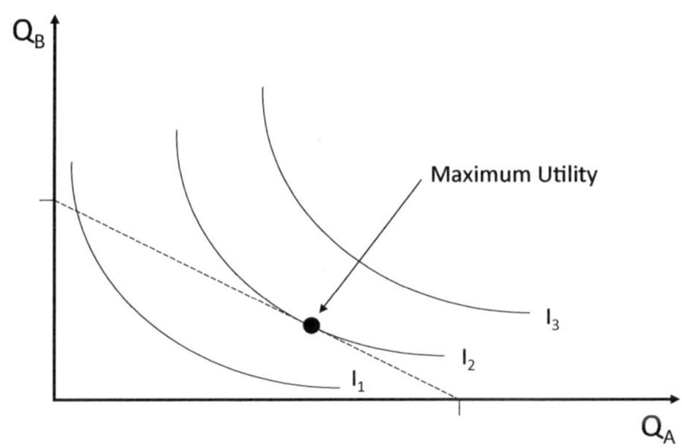


Fig. 18.3. Tangible maximum utility.

highest attainable indifference curve is the maximum utility the customer can achieve, as shown in Figure 18.3 (Thampapillai, 2002).

Now, if a consumer prefers A to B, because A is noticeably cheaper than B, and his/her budget is fixed, then his/her consumption behaviour can be derived, including his/her indifference curves (labeled as $I_{11}$, $I_{12}$ and $I_{13}$), budget line (a dash line) and maximum utility (dot a), which is shown in Figure 18.4.

If the budget and the price of A are held constant, and the price of B is reduced, a new budget line and tangency will emerge, which are represented by the solid line and dot b, respectively, in Figure 18.5.

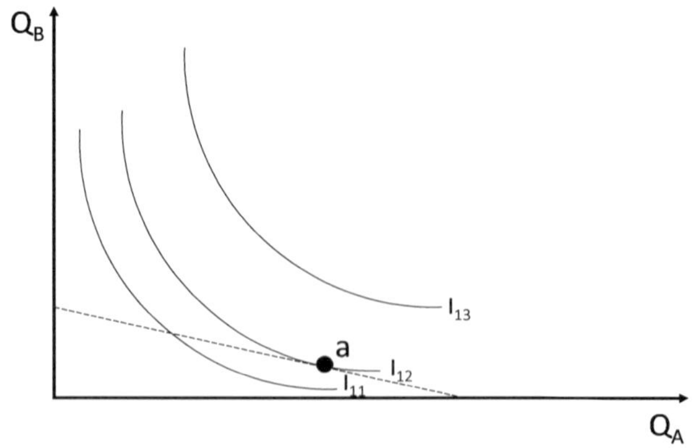


Fig. 18.4. Initial consumption behavior.

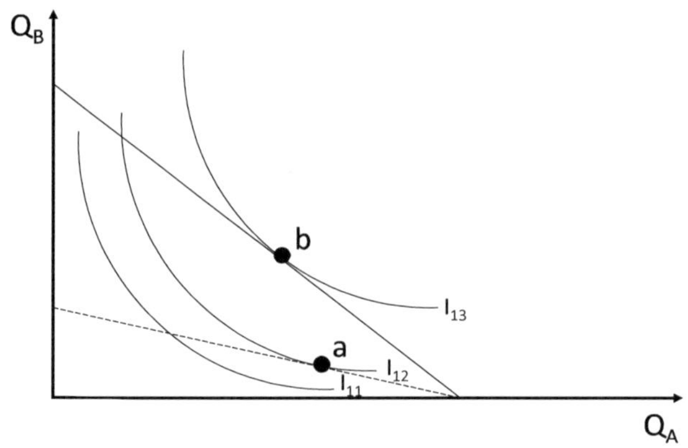


Fig. 18.5. New budget line (the solid line) and tangency (dot b).

This is where the endowment effect will influence the customer's consumption behaviour. At the initial tangency (dot a), the customer purchases much more A than B to get the maximum utility, where many more units of B are needed as compensation to make the consumer give up one unit of A, due to the endowment effect. Likewise, when the tangent point moves from 'a' to 'b', the customer consumes more B and less A to optimise the utility. This change in consumption behaviour will gradually make the consumer prefer B to A, thus changing his/her indifference curves. New indifference curves, such as the dot curves labelled $I_{21}$ to $I_{23}$ in Figure 18.6, can be

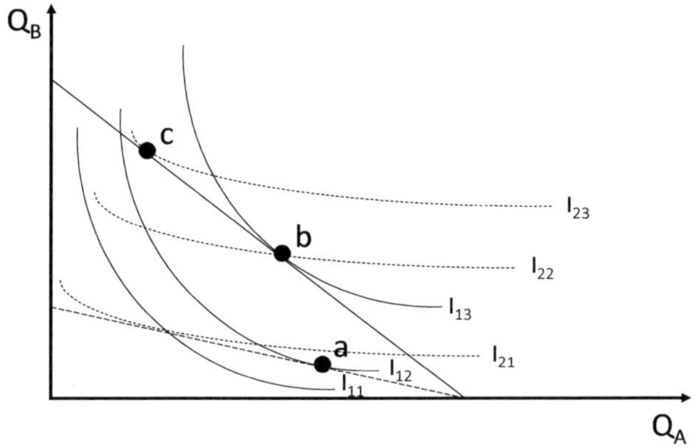


Fig. 18.6. New difference curves (dot curves) and new tangency (dot c).

depicted to reflect the changed consumption behaviour. With the new indifference curves, more units of A are required to compensate the giving up of one unit of B. As stated above, customers tend to get the maximum utility at a point where the indifference curve is tangential to the budget line. Therefore, dot c represents the new optimal utility where the budget line is tangential to one of the new indifference curves (see Figure 18.6).

As a result of reducing the price of B for a certain period of time, the consumer has gradually changed his/her consumption behaviour to favour B. Now, even if the price of B is increased to its initial price, the customer's consumption behaviour will not go back to its original pattern where he/she consumes much more A than B. That is because, when the price of B is back to its original price, the budget line will correspondingly return to the original budget line (the dash line). With the changed budget line (the dash line) and the new indifference curves (dot curves), the new maximum utility will be at point d rather than point a, which is the original maximum utility (see Figure 18.7). At point d (the new maximum utility), the customer will purchase more B and less A than he/she would do at point a (the original maximum utility).

The whole process essentially changes the maximum utility for the customer from the initial tangent point a to b, then to c, and lastly to d. The last tangency here, which is point d, indicates that the customer consumes more B and less A than he/she would initially do at point a. In other words, the customer's consumption behaviour has been changed from preferring A only to purchasing less A and more B, or even preferring B

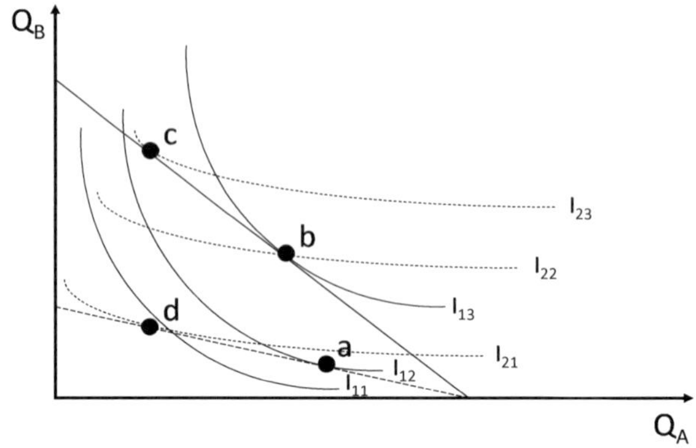


Fig. 18.7. New budget line (dash line, the same as the initial budget line) and new maximum utility (dot d).

only. That is achieved by employing the endowment effect and indifference curves, and through changing the price of B. A significant advantage here is that the theory can minimise the economic burdens on every party involved while changing the consumption behaviours of customers. That is because the budget of the customer and the price of A do not change at all, and the price of B only decreases initially, but eventually returns to the initial price.

## 5. Application to Aviation Industry to Reduce Emissions

### i. *Implication on Sustainability*

The processes illustrated above offer significant potential to improve environmental sustainability. The idea here is to replace A and B in Figure 18.7 with un-environmentally friendly goods (UEF) and environmentally friendly goods (EF) respectively, while keeping the other factors the same as those illustrated above (see Figure 18.8). Then, customers' purchasing preferences can be changed to consume more EF and less UEF by employing the aforementioned processes.

The initial maximum utility of the two types of goods, namely UEF and EF, to the consumers is achieved at point a (see Figure 18.8). Then, the tangency can be moved from a to b, where customers already purchase more EF and less UEF, due to a reduction in the price of EF, while the price of UEF and customers' budgets are held constant. After that, new indifference curves, the dot curves, will form due to the endowment

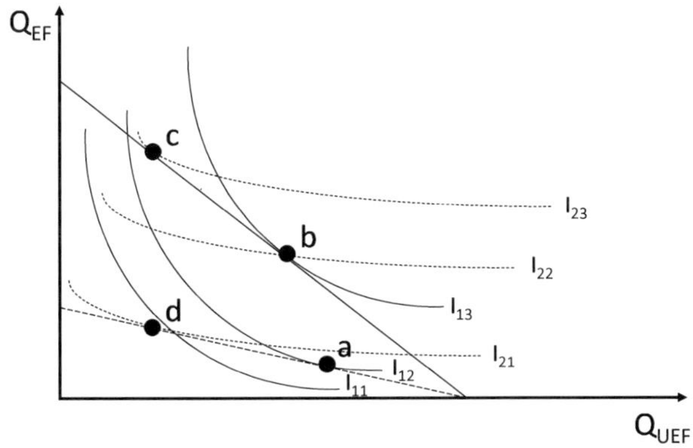


Fig. 18.8. Indifference curves, budget line and maximum utility.

effect, resulting in a new tangency that is c, where much more EF are consumed than UEF. At this point, even if the price of EF increases to its initial price, which also brings the budget line back to its initial place, the tangency will not move back to its initial point which is ‘a’, but rather, move to ‘d’, where customers still consume more EF and less UEF than they would initially do at point ‘a’. As a result, customers’ consumption behaviour is changed, leading to a more sustainable way of consumption, which will benefit the environment, the society as well as the economy.

### ii. *Further Implication on Aircraft Emission Reduction*

The approach stated above can be applied to the aviation industry to reduce aircraft emissions. As described above, future emissions reduction will still largely rely on improvements in aircraft technology. However, existing aircraft tend to be cheaper than their newly designed successors that are expected to increase fuel-efficiency and reduce air emissions. Table 18.2 below shows some examples to illustrate the differences in prices between existing airplanes and their respective successors.

As a result, some airlines may not actively upgrade their fleet. Instead, they may optimise the use of existing aircraft to maximise their profit, regardless of the higher emissions. This will considerably hamper the potential of advanced technologies in emission reduction, especially when we take the relatively long life cycle of aircraft into account, which can be up to 50 years. To encourage airlines to introduce low emission aircraft more actively, the endowment effect and indifference curves can be employed.

Table 18.2. Examples of differences in average prices between aircraft with higher emission (first column) and newly designed, less polluting aircraft (successors) in 2012 (Source: Airbus and Boeing).

| Existing aircraft | Price (Million USD) | Successor | Price (Million USD) | Increase in Price (%) |
|---|---|---|---|---|
| Airbus A319 | 80.7 | Airbus A319neo | 88.8 | 10.03 |
| Boeing 737–900 ER | 94.6 | Boeing 737 MAX 9 | 107.3 | 13.42 |
| Boeing 767–300 ER | 182.8 | Boeing 787–9 | 243.6 | 33.26 |
| Boeing 747–400 | 244.0 | Boeing 747–8I | 351.4 | 44.02 |

That is, to make airlines prefer low-emission aircraft, aircraft manufacturers could initially decrease the prices of newly designed aircraft. After the purchasing behaviours of the airlines have been changed to favour low-emission aircraft, the prices can be increased back to their initial values, while the airlines are likely to continue to prefer these aircraft to higher emission ones.

Noticeable advantages in this process can be identified. First, the budgets of aircraft operators will not be changed — in other words, by employing the endowment effect and indifference curves, changing their consumption behaviours to favour newly-designed aircraft will not result in additional financial burdens. Second, the benefits to aircraft manufacturers may increase in the long run. Although there is an initial decrease in prices, there is an increase in consumption, which may generate even more revenue. The prices will be increased to the initial value, while consumption will still be higher than its initial level. As a result, more revenue and net profit can be achieved. Third, based on the first two advantages, both aircraft operators and manufacturers can benefit from this process eventually. Therefore, their active participation can be expected in this approach. Fourth, with respect to the first three benefits, lower emissions and environmental impacts can also be achieved by increasing the number of low-emission aircraft employed. Nevertheless, certain uncertainties can also be expected due to the lack of data, with the major concerns being the quantification of price reduction, process time and actual benefits to stakeholders and the environment.

## 6. Conclusion

The aviation industry produces a range of air emissions. These emissions have been contributing to adverse effects on the welfare of all lives on the earth. The emissions also worsen the anthropogenic impacts on the environment, such as global warming, climate change and ozone depletion. To reduce the emissions and environmental impacts from aircraft, comprehensive measures have been identified. Mitigation measures include designing and manufacturing more environmentally-friendly aircraft, advancing operational systems, upgrading infrastructure, and optimising the use of market-based measures. However, the mitigation measures require substantial investments from the stakeholders in the aviation industry, and airlines in particular, resulting in challenges in implementing the mitigation measures. To facilitate the implementation, an economic theory called the endowment effect can be used, which can change the consumption behaviours of airlines to favour the newly designed, more environmentally-friendly aircraft. As existing airplanes that generate more air emissions are replaced by the lower emission aircraft on a large scale, environmental impacts from the aviation industry can be expected to be mitigated.

## References

Airbus (2012). *New Airbus aircraft list prices for 2012.* [Online] Available at: http://www.airbus.com/newsevents/news-events-single/detail/new-airbus-aircraft-list-prices-for-2012/, [Accessed 14 March 2014].

Anger, A. and Köhler, J. (2010). "Including Aviation Emissions in the EU ETS: Much ado about Nothing? A Review", *Transport Policy*, **17**(1), 38–46.

Barbarino, S., Bilgen, O., Ajaj, R.M., Friswell, M.I., Inman, D.J. (2011). "A Review of Morphing Aircraft", *Journal of Intelligent Material Systems and Structures*, **22**(9), 823–877.

Blakey, S., Rye, L. and Wilson, C.W. (2011). "Aviation Gas Turbine Alternative Fuels: A Review", *Proceedings of the Combustion Institute*, **33**(2), 2863–2885.

Boeing, retrieved from http://www.boeing.com/commercial/prices/, accessed on 14 March 2014. Now available at http://www.aircraftcompare.com/manufacture-aircraft/Boeing/01.

Cafferty, C. (2011). "Is the Sky the Limit for Carbon Offsetting? Corporate Responsibility, Consumer Sovereignty and Commitment in the Airline Industry", [Online] Available at: http://www.qantas.com.au/infodetail/about/environment/dissertation-study-on-the-airline-industry.pdf, [Accessed 23 January 2013].

Daley, B. (2010). *Air Transport and the Environment*, Ashgate Publishing LimitedSurrey.

EUROCONTROL (2008). "Comprehensive Impact Analysis of the Kyoto Protocol on Air Transport", [Online] Available at: http://www.EUROCONTROL.int/crds/gallery/content/public/thesis/8209-CRDSRNDUNV-LES.pdf, [Accessed 11 March 2013].

ExxonMobil (2005). "World Jet Fuel Specifications", [Online] Available at: http://www.exxonmobil.com/AviationGlobal/Files/WorldJetFuelSpecifications2005.pdf, [Accessed 11 March 2013].

FAA (2005). "Aviation & Emissions: A Primer", [Online] Available at: http://www.faa.gov/regulations_policies/policy_guidance/envir_policy/media/aeprimer.pdf, [Accessed 16 January 2013].

Fowlie, M.L. (2009). "Incomplete Environmental Regulation, Imperfect Competition, and Emissions Leakage", *American Economic Journal: Economic Policy*, **1**(2), 72–112.

Frey, H.C. and Kuo, P.Y. (2007). "Best Practices Guidebook for Greenhouse Gas Reductions in Freight Transportation-Final Report", [Online] Available at: http://www4.ncsu.edu/~frey/reports/Frey_Kuo_071004.pdf, [Accessed 17 January 2013].

Frömming, C., Ponater, M., Burkhardt, U., Stenke, A., Pechtl, S. and Sausen, R. (2011). "Sensitivity of Contrail Coverage and Contrail Radiative Forcing to Selected Key Parameters", *Atmospheric Environment*, **45**(7), 1483–1490.

Hong Kong International Airport. (2011). *Environmental Experts Share Views on Mitigating Aviation Emissions.* [Online] Available at: http://www.hongkongairport.com/eng/media/press-releases/pr_1038.html, [Accessed 10 March 2013].

IATA. (2008). "Aviation Carbon Offset Programmes: IATA Guidelines and Toolkit", [Online] Available at: http://www.iata.org/pressroom/Documents/Carbon_Offset_Guidelines_May2008.pdf, [Accessed 11 March 2013].

IATA. (2009a). "A Global Approach to Reducing Aviation Emissions", [Online] Available at: http://www.iata.org/SiteCollectionDocuments/Documents/Global_Approach_Reducing_Emissions_251109web.pdf, [Accessed 17 January 2013].

IATA. (2009b). "The IATA Technology Roadmap Report", [Online] Available at: http://www.iata.org/whatwedo/environment/Documents/technology-roadmap.pdf, [Accessed 11 March 2013].

IATA. (2009c). "The IATA Technology Roadmap Report — Technical Annex", [Online] Available at: http://www.iata.org/whatwedo/environment/Documents/technology-roadmap-annex.pdf, [Accessed 19 March 2013].

IATA. (2010). "Annual Report 2010", [Online] Available at: http://www.iata.org/pressroom/Documents/IATAAnnualReport2010.pdf, [Accessed 10 March 2013].

IATA. (2011). "Annual Report 2011", [Online] Available at: http://www.iata.org/pressroom/Documents/annual-report-2011.pdf, [Accessed 10 March 2013].

IATA. (2012). "Annual Review 2012", [Online] Available at: http://www.iata.org/about/Documents/annual-review-2012.pdf, [Accessed 3 March 2013].

ICAO (2007a). "Assembly Resolutions in Force (as of 28 September 2007)", [Online] Available at: http://legacy.icao.int/icao/en/env2010/A36_Res22.pdf, [Accessed 15 January 2013].

ICAO (2007b). "Environmental Report 2007", [Online] Available at: http://legacy.icao.int/icao/en/env2010/pubs/Env_Report_07.pdf, [Accessed 18 January 2013].

ICAO (2007c). "Report on Voluntary Emissions Trading for Aviation (VETS Report)", [Online] Available at: http://www.icao.int/environmental-protection/Documents/Measures/vets_report1.pdf, [Accessed 3 March 2013].

ICAO (2010). "Environmental Report 2010", [Online] Available at: http://www.icao.int/environmental-protection/Documents/Publications/ENV_Report_2010.pdf, [Accessed 18 January 2013].

ICAO (2012). "Emissions Databank", [Online] (Updated January 2012). Available at: http://easa.europa.eu/environment/edb/docs/edb-emissions-databank.xls, [Accessed 3 February 2013].

ICAO (2012). "Facts and Figures", [Online]. Available at: http://www.icao.int/sustainability/Pages/Facts-Figures_WorldEconomyData.aspx, [Accessed on 15 March 2013].

IPCC (1999). *Aviation and the Global Atmosphere*, Cambridge University Press, Cambridge.

IPCC (2007a). *Climate Change 2007: Mitigation of Climate Change*, Cambridge University Press, New York.

IPCC (2007b). *Climate Change 2007: The Physical Science Basis*, Cambridge University Press, New York.

Kesgin, U. (2006). "Aircraft emissions at Turkish airports", *Energy*, **31**(2–3), 372–384.

Knetsch, J.L. (1989). "The Endowment Effect and Evidence of Nonreversible Indifference Curves", *The American Economic Review*, **79**(5), 1277–1284.

Kundu, A.K. (2010). *Aircraft Design*, Cambridge University Press, New York.

Lee, D.S., Fahey, D.W., Forster, P.M., Newton, P.J., Wit, R.C.N., Lim, L.L., Owen, B., Sausen, R. (2009). "Aviation and Global Climate Change in the 21st Century", *Atmospheric Environment*, **43**(22–23) 3520–3537.

Lee, D.S., Pitari, G., Grewe, V., Gierens, K., Penner, J.E., Petzold, A., Prather, M.J., Schumann, U., Bais, A., Berntsen, T., lachetti, D., Lim, L.L., Sausen, R. (2010). "Transport Impacts on Atmosphere and Climate: Aviation", *Atmospheric Environment*, **44**(37), 4678–4734.

Lee, D.S. (2004). "The Impact of Aviation on Climate", in Hester, R.E. and Harrison, R.M. (eds.) *Transport and the Environment, Issues in Environmental Science and Technology*, Vol. No. 20, Royal Society of Chemistry, Cambridge, pp. 1–23.

Lee, J.J. and Mo, J. (2011). "Analysis of Technological Innovation and Environmental Performance Improvement in Aviation Sector", *International Journal of Environmental Research and Public Health*, **8**(9), 3777–3795.

Marais, K. and Waitz, I.A. (2009). "Air Transport and the Environment" in *The Global Airline Industry*, Wiley, 405–440.

NASA (2011). "Sulfur Dioxide", [Online] Available at: http://eospso.gsfc.nasa.gov/ftp_docs/SO2poster_508.pdf, [Accessed 4 February 2013].

OECD and IEA (2001). *International Emission Trading From Concept to Reality*, OECD Publishing, Paris.

OECD (2012). "Green Growth and the Future of Aviation", [Online] Available at: http://www.oecd.org/sd-roundtable/papersandpublications/49482790.pdf, [Accessed 3 March 2013].

Perdan, S. and Azapagic, A. (2011). "Carbon trading: Current Schemes and Future Developments", *Energy Policy*, **39**(10), 6040–6054.

Preston, H., Lee, D.S., Hooper, P.D. (2012). "The Inclusion of the Aviation Sector within the European Union's Emissions Trading Scheme: What are the Prospects for a More Sustainable Aviation Industry?", *Environmental Development*, **2**, 48–56.

Ryerson, M.S. and Hansen, M. (2010). "The Potential of Turboprops for Reducing Aviation Fuel Consumption", *Transportation Research Part D: Transport and Environment*, **15**(6), 305–314.

Sustainable Aviation (2012). "Sustainable Aviation $CO_2$ Road-Map 2012", [Online] Available at: http://www.sustainableaviation.co.uk/wp-content/uploads/SA-$CO_2$-Road-Map-full-report-280212.pdf, [Accessed 8 March 2013].

Szodruch, J., Grimme, W., Blumrich, F., Schmid, R. (2011). "Next Generation Single-Aisle Aircraft — Requirements and Technological Solutions", *Journal of Air Transport Management*, **17**(1), 33–39.

Thampapillai, D.J. (2002). *Environmental Economics: Concepts, Methods, and Policies*, Oxford University Press, Melbourne.

UNFCCC (2008). "Kyoto Protocol Reference Manual — on Accounting of Emissions and Assigned Amount", [Online] Available at: http://unfccc.int/resource/docs/publications/08_unfccc_kp_ref_manual.pdf, [Accessed 1 March 2013].

UNFCCC (2010). "The Right Flightpath to Reduce Aviation Emissions", [Online] Available at: http://www.enviro.aero/Content/Upload/File/AviationPositionPaper_COP16_normalprinter.pdf, [Accessed 3 March 2013].

University of Oxford (?). "The Aircraft at End of Life Sector: a Preliminary Study", [Online] Available at: http://users.ox.ac.uk/~pgrant/Airplane%20end%20of%20life.pdf, [Accessed 19 January 2013].

Whitelegg, J. (2000). "Aviation: the Social, Economic and Environmental Impact of Flying", [Online] Available at: http://waag.co.nz/pdf/res0006.pdf, [Accessed 16 January 2013].

WHO (2006). "Air Quality Guidelines — Global Update 2005: Particulate Matter, Ozone, Nitrogen Dioxide, and Sulfur Dioxide", [Online] Available at: http://www.euro.who.int/__data/assets/pdf_file/0005/78638/E90038.pdf, [Accessed 8 February 2013].

WHO (1999). *Environmental Health Criteria 213: Carbon Monoxide* (Second Edition), World Health Organization, Geneva.

Wilson C.W., Petzold A., Nyeki, S., Schumann, U., Zellner, R. (2004). "Measurement and Prediction of Emissions of Aerosols and Gaseous Precursors from Gas Turbine Engines (PartEmis): an Overview", *Aerospace Science and Technology*, **8**(2), 131–143.

World Bank (2012). "Air Transport and Energy Efficiency", [Online] Available at: http://siteresources.worldbank.org/INTAIRTRANSPORT/Resources/TP38.pdf, [Accessed 19 January 2013].

Zurich Airport (2009). "Fixed Energy Systems for Aircraft at Zurich Airport", [Online] Available at: http://www.zurich-airport.com/Portaldata/2/Resources/documents_unternehmen/umwelt_und_laerm/Technical_Report_Aircraft_Fixed_Ground_Power_Systems_2009.pdf, [Accessed 10 March 2013].